4

EDITION

AN INVITATION TO
ENVIRONMENTAL
SOCIOLOGY

For my father
Moses David Bell (1923–1996)
Writer, scholar, teacher
Washtub-bass player extraordinaire
A lover of people and the land
He fished without hooks

AN INVITATION TO
ENVIRONMENTAL
SOCIOLOGY

MICHAEL MAYERFELD BELL
University of Wisconsin—Madison

With Michael S. Carolan

With Illustrations by Matthew Raboin and Matthew Robinson

Los Angeles | London | New Delhi
Singapore | Washington DC

FOR INFORMATION:

SAGE Publications, Inc.
2455 Teller Road
Thousand Oaks, California 91320
E-mail: order@sagepub.com

SAGE Publications Ltd.
1 Oliver's Yard
55 City Road
London EC1Y 1SP
United Kingdom

SAGE Publications India Pvt. Ltd.
B 1/I 1 Mohan Cooperative Industrial Area
Mathura Road, New Delhi 110 044
India

SAGE Publications Asia-Pacific Pte. Ltd.
33 Pekin Street #02-01
Far East Square
Singapore 048763

Acquisitions Editor: David Repetto
Editorial Assistant: Maggie Stanley
Production Editor: Karen Wiley
Copy Editor: Teresa Herlinger
Typesetter: C&M Digitals (P) Ltd.
Proofreader: Jennifer Gritt
Indexer: Kathleen Paparchontis
Cover Designer: Gail Buschman
Marketing Manager: Erica DeLuca
Permissions Editor: Karen Ehrmann

Printed in the United States of America

Library of Congress Cataloging-in-Publication Data

Bell, Michael, 1957-

An invitation to environmental sociology / Michael Mayerfeld Bell. —4th ed.

p. cm.
Includes bibliographical references and index.

ISBN 978-1-4129-9053-0 (pbk.)

1. Environmental sociology. 2. Environmental responsibility. 3. Environmental ethics. I. Title.

GE195.B46 2012 363.7—dc23 2011020083

This book is printed on acid-free paper.

11 12 13 14 15 10 9 8 7 6 5 4 3 2 1

Brief Contents

Detailed Contents

Preface

I t is a delight to offer to readers this fourth edition of *An Invitation to Environmental Sociology*—indeed, it is many delights. I am gratified that the book remains of sufficient usefulness to warrant a new edition. But far more satisfying is that environmental scholarship has continued to deepen within sociology and related social sciences, and that interest in environmental sociology has continued to broaden within society at large. Each year, the flow of scholarly books and articles increases. So, too, does the volume of texts oriented to students and the number of courses in which students can encounter the field. Environmental sociology has more to say now, and more people who want to hear and contribute to what it has to say. Amid the continuing indications of alarming and accelerating stress on the sustainability, justice, and beauty of our environment, we should remember to appreciate good and hopeful signs like these. There is care and concern out there—lots of it.

But even if environmental sociology has more to say, it is important that we scholars don't drone on about it. With that caution in mind, in this edition, I have focused on integrating new scholarship without lengthening the book. I've achieved this goal for the most part. This edition is only very slightly longer than the previous, and does not contain any new chapters. The wonderful production people at Pine Forge Press have also worked to keep the same inviting feel of the third edition, retaining the evocative artwork of Matthew Raboin and Matthew Robinson, which readers admired. But there is new material in all the chapters, and quite a bit in a few of them, as well as some reorganization.

Here's a chapter by chapter overview of the main changes.

Chapter 1: Environmental Problems and Society. This is the "big numbers" chapter, synthesizing the numerous efforts of government and nongovernment agencies, national and international, to monitor the environmental condition of the planet. I have always sought to keep these figures as up to date as possible, and have worked hard to incorporate the latest information once again. Every section has changes as a result. The biggest addition is a new section drawing together the challenges to sustainability posed by energy production and use. Plus, I have pulled (and updated) some of the material on environmental justice from the third edition's "Body and Health" chapter and moved it here, so as to provide a more comprehensive look at environmental justice from the start. I have also once again renamed the third of the three central environmental challenges—what I now term the "beauty of ecology" but called the "rights and beauty of habitat" in the third edition and the "rights and beauty of nature" in the first two editions. The challenge of *rights* is redundant with the notion of "environmental justice," unless we take a narrowly anthropocentric view. And while I moved to the word *habitat* in the third edition to avoid the philosophical tangle of the word *nature,* I think it brought its own problems. *Beauty* and *ecology* seem to me wonderful words that we would do well to think of together. I think I've finally got the phrasing right.

Chapter 2: Consumption and Materialism. Michael Jackson and the excesses of Neverland Ranch lead off the revised version of this chapter, in place of Liberace and his 30 cars and 18 pianos. The section on "Green Advertising" got a remake too. And the discussion of international comparisons of the relationship between wealth and happiness got an update, keeping abreast of the considerable recent attention to this question by scholars and in the media. Otherwise, this chapter is much the same.

Chapter 3: Money and Machines. The main change here is a rewrite of the section on "The Needs of Money," highlighting what I like to call the "wage–price gap" and the "problem of the original capitalist," which should give students a more vivid understanding of the dynamics of capitalist economies. I've also foregrounded the work of John Bellamy Foster, Brett Clark, and Richard York on "metabolic rift." The section on "The Dialogue of State and Market" now introduces the important concept of avoiding "zero-sum" thinking. In the second half of the chapter, on technology, I briefly introduce a discussion of Bruno Latour and his colleagues on "actor network theory," which in previous editions only made an appearance later in the book. Their work on technology is vital to have here as well. Elsewhere in the chapter, I updated a lot of facts and figures, especially those about cars.

Chapter 4: Population and Development. For this edition, I have brought in the work of William Catton and his notion of "overshoot," as I ought to have done long ago, in the discussion of Malthusianism and anti-Malthusianism. Plus, I have extensively updated the figures on population growth, international debt, and food production.

Chapter 5: Body and Health. I extensively revised this chapter. As I noted above, I moved much of the review of environmental justice studies into the first chapter. This made room for a new section on the sociology of food and one on the sociology of "mobilities" and "environmental flows," both of which have been much discussed in the scholarship of the last few years. The third edition had only a brief mention of "mobilities" and "environmental flows," and only in the last chapter. And the third edition had a lot on agriculture, as does this edition, but very little directly on food. I have also dialed back my use of the term *invironment* and have abandoned completely the phrase "invironmental justice" as confusing and unnecessary. This chapter still includes a discussion of pesticides and health, but I have rewritten those passages almost entirely, in light of much recent (and shocking) research. In the final section on the sociology of environmental justice, I revised most of the discussion of utilitarianism, which was a bit simplistic and one-sided in earlier editions.

Chapter 6: The Ideology of Environmental Domination. The main changes to this chapter are an updated and improved discussion of the Weberian hypothesis about the relationship between Protestantism and capitalism and a discussion of the gendered metaphors used in media accounts of the Gulf Oil Spill.

Chapter 7: The Ideology of Environmental Concern. This edition updates the results of national and international polls of environmental concern and discusses their relationship to the Great Recession. I also include a new account of the widening partisan divide in environmental concern in the United States. Plus, I tweaked the discussion of ecological modernization, in light of new scholarship.

Chapter 8: The Human Nature of Nature. The main change here is I added a new section on the social construction of "environmental nonproblems," focusing on climate change skepticism. I also updated the discussion of the realist–constructionist debate and improved the discussion of actor network theory.

Chapter 9: The Rationality of Risk. This edition updates the discussion of Ulrich Beck's "risk society" thesis, based on his recent publications, but otherwise the chapter is little changed.

Chapter 10: Mobilizing the Ecological Society. I clarified the discussion in a few sections, but otherwise left this chapter unchanged.

Chapter 11: Governing the Ecological Society. This final chapter got a major makeover in the previous edition, and it still looks pretty good to me. I made a few changes nonetheless. My account of my own little environmental sins got an update. I added passages introducing the ideas of "plenitude" and "conscious consumers." I put in a passage discussing how at least 35 percent of the U.S. economy is not based on profit maximization. And I added a few other small points here and there.

With each edition, the community of people who contributed significantly to the making and remaking of this book happily grows. They include Marilyn Aronoff, Megan Berry, Mike Carolan, Richard Coon, Eileen Curry, Kate Entwhistle, Carla Freeman, Will Goudy, Wendy Griswold, Ram Guha, Valerie Gunter, Windy Just, Philip Lowe, Anne Martin, Margaret Munyae, Ted Napier, Pam Ozaroff, Eric Pallant, Bob Penner, Jake Peterson, Stephanie Prescott, Matt Raboin, Chris Ray, David Repetto, Jennifer Rezek, Matt Robinson, Sanford Robinson, Lucas Rockwell, Steve Rutter, Robert Schaeffer, Becky Smith, Kaelyn Stiles, Pam Suwinsky, Hilary Talbot, Astrid Virding, Christine Vatovec, Jerry Westby, Martin Whitby, and Rachel Woodward. I want to give special mention to Teresa Herlinger, the marvelous copy editor for this edition, and to Maggie Stanley for her efficiency, good editorial sense, and deft sense of how to prod an overcommitted author along. Over the years, it's been a lot of folks. It couldn't have happened— nor would I have wanted it to have been—any other way.

And let me close once again by thanking my children, Sam and Eleanor, and my wife, Diane Bell Mayerfeld, for the nonstop seminar of our lives. It will always be the ecological dialogue I cherish the most.

MMB
Madison, Wisconsin

CHAPTER 1

Environmental Problems and Society

Without self-understanding we cannot hope for enduring solutions to environmental problems, which are fundamentally human problems.

—Yi-Fu Tuan, 1974

"**P**ass the hominy, please."

It was a lovely brunch, with fruit salad, homemade coffee cake, a great pan of scrambled eggs, bread, butter, jam, coffee, tea—and hominy grits. Our friends Dan and Sarah had invited my wife and me and our son over that morning to meet some friends of theirs. The grown-ups sat around the dining room table, and the kids (four in all) careened from their own table in the kitchen to the pile of toys in the living room, and often into each other. Each family had contributed something to the feast before us. It was all good food, but for some reason the hominy grits (which I had never had before) was the most popular.

There was a pleasant mix of personalities, and the adults soon got into one of those excited chats that leads in an irreproducible way from one topic to another, as unfamiliar people seek to get to know each other a bit better. Eventually, the inevitable question came my way: "So, what do you do?"

"I'm an environmental sociologist."

"Environmental sociology. That's interesting. I've never heard of it. What does sociology have to do with the environment?"

I used to think, during earlier editions, that the point of this book was to answer that question—a question I often used to get, as in this breakfast conversation from many years ago. (My children, like this book, are much older now.) Today, I sense a change in general attitudes. Now I don't get so many blank looks when I say I'm an environmental sociologist. Most people I meet have still never heard of the field, but more and more of them immediately get the basic idea behind it: that society and environment are interrelated.

And more and more, the people I meet recognize that this interrelation has to confront some significant problems, perhaps the most fundamental problems facing the future of life, human and otherwise. They readily understand that environmental problems are not only problems of technology and industry, of ecology and biology, of pollution control and pollution prevention. Environmental problems are also social problems. Environmental problems are problems *for* society—problems that threaten our existing patterns of social organization and social thought. Environmental problems are as well problems *of* society—problems that challenge us to change those patterns of organization and thought. Increasingly, we appreciate that it is people who create environmental problems and it is people who must resolve them.

That recognition is good news. But we've sure got a lot to do, and in this work we'll need the insights of all the disciplines—the biophysical sciences, the social sciences, and the humanities. There is an environmental dimension to all knowledge. The way I now understand the point of this book is to bring the sociological imagination to this necessarily pan-disciplinary conversation.

A good place to begin, I think, is to offer a definition of *environmental sociology*. Here goes: *Environmental sociology is the study of community in the largest possible sense.* People, other animals, land, water, air—all of these are closely interconnected. Together they form a kind of solidarity, what we have come to call ecology. As in any community, there are also conflicts in the midst of the interconnections. Environmental sociology studies this largest of communities with an eye to understanding the origins of, and proposing solutions to, these all-too-real social and biophysical conflicts.

But who are environmental sociologists? My view of who is a large community in itself—a large community of scholars from many social science disciplines that share this passion for studying community in the largest possible sense. Some might call themselves "environmental geographers" or "environmental anthropologists" or "environmental economists" or "environmental psychologists." Or they might prefer to think of themselves as "political ecologists" or "social ecologists" or "human ecologists" or "ecological economists." What is important is the passion, not the disciplinary label. Increasingly, academic conferences focus on an issue like global warming or sustainable consumption or sustainable agriculture or environmental justice, not on a specific discipline's take on it. The research papers that come out of these conferences similarly cite scholars from across this wide spectrum. We all have our starting points, of course, our distinctive angles of vision to bring to the conversation, which is great. That is how, and why, one learns from others. But it is the goals that matter, not the starting points. In this book, I discuss contributions from scholars with all these many different departments on their business cards. It is all environmental sociology.

One of environmental sociology's most basic contributions to studying the conflicts behind environmental problems is to point out the pivotal role of social inequality. Not only are the effects of environmental problems distributed unequally across the human community, but social inequality is deeply involved in causing those problems. Social inequality is both a product and a producer of global warming, pollution, overconsumption, resource depletion, habitat loss, risky technology, and rapid population growth. As well, social inequality influences how we envision what our environmental problems are. And most fundamentally, it can influence how we envision nature itself, for inequality shapes our social experiences, and our social experiences shape all our knowledge.

Which returns us to the question of community. Social inequality cannot be understood apart from the communities in which it takes place. We need, then, to make the study of community the central task of environmental sociology. Ecology is often described as the study of natural communities. Sociology is often described as the study of human communities. Environmental sociology is the study of both together, the single commons of the Earth we humans share, sometimes grudgingly, with others—other people, other forms of life, and the rocks and water and soil and air that support all life. Environmental sociology is the study of this, the biggest community of all.

Joining the Dialogue

The biggest community of all: Then clearly, the topic of environmental sociology is vast. Not even a book the length of this one can cover all of it, at least not in any detail. In the pages to come, I will take up the main conversations about the state of relations within this vast community. I won't take up all the side conversations, but I will invite the reader into a good many of them, in order to trace how the larger debates play out in particular neighborhoods of discussion and investigation. Continually, though, the book will return to the front pages of debate, the better to bring the local and the global, the particular and the general, into better communication.

For the most part, this first chapter considers the front pages—of environmental sociology; of the environmental predicament; and, in this section, quite literally of the book itself. (These are the front pages of the book, after all.) After this introduction, the book falls into three parts:

The Material: How consumption, the economy, technology, development, population, and the health of our bodies shape our environmental conditions

The Ideal: How culture, ideology, moral values, risk, and social experience influence the way we think about and act toward the environment

The Practical: How we can bring about a more ecological society, taking the relations of the material and the ideal into account

Of course, it is not possible to fully separate these three topics. The deep union of the material, the ideal, and the practical is one of the most important truths that environmental sociology has to offer. The parts of the book represent only a sequence of emphases, not rigid conceptual boundaries. A number of themes running throughout the book help unite the parts:

- The dialogic, or interactive and unfinished, character of causality in environmental sociology
- The interplay of material and ideal factors with each other, constituting the practical conditions of lived experience
- The central role of social inequality in environmental conflict
- The connections between the local and the global
- The power of the metaphor of community for understanding these social and ecological dynamics
- The important influence of political institutions and commitments on our environmental practices

The Ecology of Dialogue

By approaching environmental sociology in this way, I hope to bridge a long-standing dispute among scholars about the relationship between environment and society. *Realists* argue that environmental problems cannot be understood apart from the threats posed by the way we have organized our societies, including the organization of ecologic relations. They believe that we can ill afford to ignore the material truth of organizational problems and their ecologic consequences. *Constructionists* do not necessarily disagree, but they emphasize the influence of social life on how we conceptualize those problems, or the lack of those problems. Constructionists focus on the ideological origins of environmental problems—including their very definition *as* problems (or as nonproblems). A realist might say, for example, that global warming is a dangerous consequence of how we currently organize the economic side of social life. A constructionist might say that in order to recognize the danger—or even the existence—of global warming, we must wear the appropriate conceptual and ideological eyeglasses. Although the debate sometimes gets quite abstract, it has important consequences. Realists argue that the practical thing to do is to solve the social organizational issues behind environmental problems, like the way land use laws and current technologies encourage the overuse of cars. Constructionists argue that the first step must be to understand our environmental ideologies, with all their insights and oversights, lest our solutions lead to still other conflicts.[1]

Fundamentally, the realist–constructionist debate is over materialist versus idealist explanations of social life. I mean "materialist" here in the philosophical sense of emphasizing the material conditions of life, not in the sense of material acquisitiveness. And I similarly mean "idealist" in the philosophical sense of emphasizing the role of ideas, not in the sense of what is the best or highest. The tension between materialist and idealist explanations is itself a centuries-old philosophical dispute, one that perhaps all cultural traditions have grappled with in one way or another. An ancient fable from India expresses the tension well. A group of blind people encounters an elephant for the first time. One grabs the elephant's tail and says, "An elephant is like a snake!" Another grabs a leg and says, "An elephant is like a tree!" A third grabs an ear and says, "An elephant is like a big leaf!" To the materialist, the fable shows how misinformed all three blind people are, for a sighted person can plainly see how the "snake," "tree," and "big leaf" connect together into what an elephant really is. To the idealist, the fable says that we all have our ideological blindnesses and there is no fully sighted person who can see the whole elephant—that we are all blind people wildly grasping at the elusive truth of the world.

The approach I take to this ancient debate is that the material and the ideal dimensions of the environment depend upon and interact with each other and together constitute the practical conditions of our lives. What we believe depends on what we see and feel, and what we see and feel depends on what we believe. It is not a matter of either/or; rather, it is a matter of both together. Each side helps constitute and reconstitute the other, in a process that will never, we must hope, finish. I term this mutual and unfinalizable interrelationship *ecological dialogue*.[2] Throughout the book, I consider the constant conversation between the material and ideal dimensions of this never-ending dialogue of life and how our environmental practices emerge from it.

Ecological dialogue is also a way to conceptualize power—to conceptualize the environmental relations that shape our scope for action: our ability to do, to think, to be. These relations of power include both the organizational factors of materiality and the knowledge factors of our ideas—which, in turn, shape each other. By using the word *dialogue,* I don't mean that everything in this interrelationship is happy and respectful, smooth and trouble-free, or even that it always should be. Dialogue is not a state we reach when we have overcome power; it only happens *because of* power. There is often conflict involved, which is one of the main ways that the material and the ideal continually reshape each other and express themselves in our practices of living. And conflict is not necessarily a bad thing. Sometimes it is exactly what is needed to get us to pay attention. But neither is power all kicking and yelling. There is much cooperative and complementary action in the dialogue of ecology, much conviviality that we relish and that constantly changes us. We experience power in cooperative and complementary action, too. Nor is power necessarily a bad thing. (Imagine for a moment having no power at all in your life and what an awful circumstance that would be.) It's a matter of what power does and how and why, and the legitimacy of its balances and imbalances. These are moral questions that we need to continually ask and re-ask.

Maybe a diagram will help. Have a look at Figure 1.1, a kind of environmental sociological updating of the *Taijitu,* the ancient Chinese yin–yang icon. The *Taijitu* suggests that the world is constituted through the interaction of yin and yang, which together create a unity of earth and heaven—or in more Western terms, of the material and the ideal. Often the *Taijitu* is interpreted to mean that yin and yang are opposites, but the black dot in the white side and the white dot in the black side are supposed to indicate that each is the seed of the other. Also, the *Taijitu* indicates the

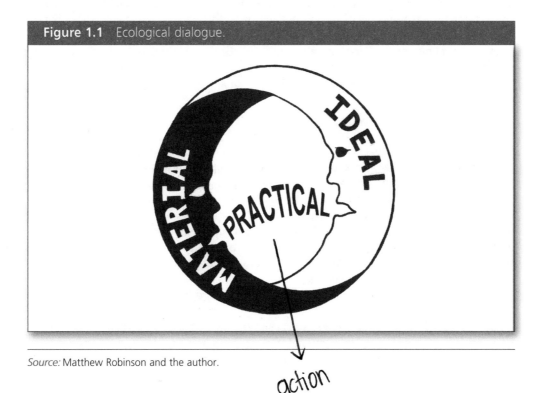

Figure 1.1 Ecological dialogue.

MATERIAL

IDEAL

PRACTICAL

action

Source: Matthew Robinson and the author.

interactiveness of yin and yang through curved inter-nesting of the two sides, instead of a straight line dividing yin and yang into oppositional hemispheres. It's one of history's great images.

But from the perspective of ecological dialogue, the *Taijitu* represents the world as overly unified, static, and finished. Figure 1.1 suggests the changing, unfinished, and sometimes conflictual character of the world through showing the material and the ideal as two partial moon faces in practical dialogue with each other. Together, the moons of the "material" and the "ideal," which tuck together in a basket weave at their edges, make a circle and a kind of ecological holism. That holism is always unfinished, though, and thus never fully whole, which the diagram represents through the open space between the partial moons. But the open space is not empty. Rather, it is an active space of interchange, interaction, and interrelation through the "practical"—the ideas and materialisms we put into joint practice. Some of that practice may be conflictual, and some may be cooperative and complementary. Through it, the ideal and material shape each other and change each other, shaping and changing the practical at the same time. To further represent this mutual constitution of the material and the ideal, through the relations of the practical, Figure 1.1 takes the seed imagery of the *Taijitu* and converts it to eyes, one of the central organs of communication, with a black eye on the white side and a white eye on the black side. Plus, the imagery of the moon faces is meant to suggest the motion of light and shadow across the ever-unfinished holism, like phases of a moon, as white becomes black and black becomes white over time.

I'm not completely satisfied with the diagram. But at least, I hope Figure 1.1 is thought-provoking. There are also many resonances between Figure 1.1 and the cover illustration of the book, Marc Chagall's great painting *I and the Village*. I'll leave it to each reader to puzzle those connections out in his or her own way.

But the most important connection to puzzle out is simply that we are connected, although by no means completely so. Like power and conflict, this incompleteness of connection can also be okay, for it provides the crucial open space for change. It also provides for difference, and that is okay, too. If everything were connected up into one great sameness, there would be no connections to make—no need, and no possibility, for dialogue and its endless creativity. I mean this, too, by the phrase "ecological dialogue": that we should see ourselves as part of a creative community of the Earth and all its inhabitants, ever working out our ever-changing samenesses and differences, connections and disconnections, in the practical art of living.[3] The biggest community of all is thus the biggest dialogue of all.

The Dialogue of Scholarship

Let me also make it clear that this book takes an activist position with regard to environmental problems and the way we think about them. We often look to scholars to provide an unbiased perspective on issues that concern us, and we sometimes regard an active commitment to a political position as cause for suspicion about just how scientific that perspective is. Yet, as many have argued, it is not possible to escape political implications.[4] Everyone has concerns for and interests in the condition of our world and our society. Such concerns and interests are what guide us all every day, and scholars are no different from anyone else in this regard. Nor should they be any different. Such concerns and interests are not necessarily a problem for scholarship. On the contrary, they are the whole reason for scholarship.

This does not mean that anything goes—that any perspective is just as academically valid as any other because all knowledge is only opinion and we are all entitled to our own opinions. Scholarship is opinion, of course, but it is a special kind of opinion. What scholarship means is being critical, careful, honest, open, straightforward, and responsible in one's opinions—in what one claims is valid knowledge. One needs to reason critically and carefully, to be honest about the reasons one suggests to others, to be open to the reasons others suggest, to be straight-forward about one's political reasons, and to be responsible in the kinds of reasons one promotes. Being honest, open, and straightforward with each other about our careful, critical reasons is the only academically responsible thing to do.[5]

Therefore, it is best for me to be straightforward about why I think environmental sociology is an important topic of study: I believe there are serious environmental problems that need concerted attention—now, not later. And I believe environmental issues are closely intertwined with a host of social issues, most of them at least in part manifestations of social inequality and the challenges inequality poses for community. Addressing these intertwinings, manifestations, and challenges is in everyone's interests. We will all benefit, I believe, by reconsidering the present state of ecological dialogue.

My perspective, particularly the focus on social inequality, coincides more closely with the current politics of the Left than the Right. Yet issues of the environment and inequality cut across traditional political boundaries, as later chapters discuss. The evidence and arguments that I offer in this book should be of interest to anyone committed to careful, critical reasoning. In any event, we should not let political differences stop us from engaging in dialogue about ecological dialogue.

Nevertheless, you, the reader, should be aware that I indeed have a moral and political perspective and that it unavoidably informs what I have written here. Keep that in mind as you carefully and critically evaluate what is in this book. But it is also your scholarly responsibility to be open to the reasoning I present and to have honest reasons for disagreeing.

Sustainability

Let us now turn to some of the reasons that lead many people to believe there is cause for considerable concern about the current condition of ecological dialogue: the challenges to *sustainability, justice,* and the *beauty of ecology.* These, the three central environmental issues, will already be well-known to some readers. Still, as these considerations underlie the rest of the book, it is appropriate to pause and review them here, beginning first with sustainability.

How long can we keep doing what we're doing? This is the essential question of sustainability. The length of the list of threats to environmental sustainability is, at the very least, unnerving. True, much is unknown, and some have exaggerated the dangers we face. Consequently, there is considerable controversy about the long-term consequences of humanity's continuing transformation of the Earth, as the headlines and blogs and podcasts every week demonstrate. But much relevant evidence has been gathered, and some have underestimated the dangers involved. It is therefore prudent that we all pay close attention to the potential challenges to sustainability.

Energy

How much longer can we keep doing what we're doing with regard to energy? Not any longer at all. In almost everyone's view these days, we want more energy than we have—or at least more than we can easily get. The issues of this mismatch confront the world already. Rising costs. Pollution of land, air, and water. Declining stocks of some sources. Competition for space to produce energy. Tense international politics and even, say some, war.

What to do? When you don't have enough of something, there are two basic ways to go: Get more or use less. Or maybe do both. There is a caveat, too, especially with regard to energy: Make sure that any way you go is clean, safe, and just. Given our record with energy recently, we'll have to inspect our options with care. Let's begin with the "get more" approach.

First, let's review where we get energy from now, as of 2008 (see Figure 1.2). About 37 percent of the world's energy supply comes from oil, the most of any source. Coal and peat are next at a combined 27 percent, followed by gas at 21 percent. Given that oil and gas generally come together or from closely related geologies, it is worth pointing out that together they amount to 54 percent of the world's energy supply. Then add in coal and peat, and we're up to 81 percent of our energy coming from fossil carbon. That's a lot of fossil carbon. And then add in what the International Energy Agency (the keeper of these statistics) calls "combustible renewables and waste"—firewood, ethanol, and other biofuels, plus whatever else people can get to burn, like municipal solid waste and animal dung—at 10 percent. That's a lot of total carbon. Combined, we're up to a 91 percent carbon energy economy.

The other 9 percent? Six percent is nuclear energy and 2 percent is hydropower. The rest is so quantitatively insignificant that the International Energy Agency lumps it all into a single "other" category of less than 1 percent: mostly wind, solar, and geothermal.

Can we get more? There are a lot of unknowns of geology and technology here. Plus, a lot of money and jobs hang on this question, so clear and straight answers are hard to come by.

Much attention has been given of late to the contention that we have now reached a "peak oil" state, fulfilling M. King Hubbert's prediction in the 1950s that we would soon see terminal decline in oil and gas production, albeit a few decades later than Hubbert thought. There are still

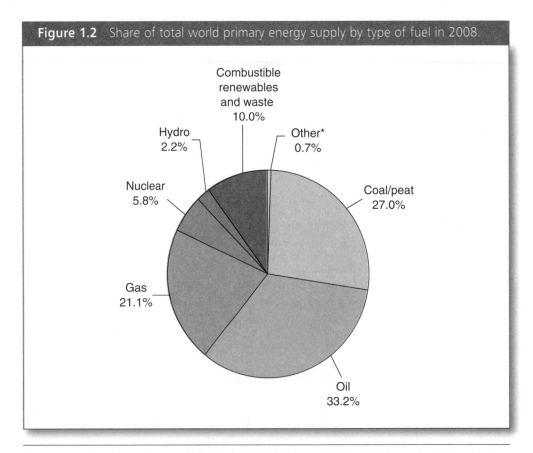

Figure 1.2 Share of total world primary energy supply by type of fuel in 2008.

Combustible
renewables
and waste
10.0%

Hydro
2.2%

Other*
0.7%

Nuclear
5.8%

Coal/peat
27.0%

Gas
21.1%

Oil
33.2%

Source: International Energy Agency. 2010. *2010 Key World Energy Statistics*. Paris: International Energy Agency.

substantial reserves of oil and gas in the world, and still some regions that have not been fully prospected. But the big and easy petroleum fields appear to have been pretty much all found. Although oil and gas interests often argue that there is no need for alarm, it seems likely that the finds of the future will be mostly smaller and harder to access, raising costs and lowering yields. The evidence? Crude oil production has been basically flat since 2004, and has even begun to decline. In the 30 years from 1979 to 2009, it grew considerably more slowly than the world economy, rising roughly 20 percent while the world economy grew two and a half times larger (see Figure 1.3). Consequently, the proportion of the world's energy supplied by oil and gas has declined from about 70 percent in 1979 to 54 percent in 2008.

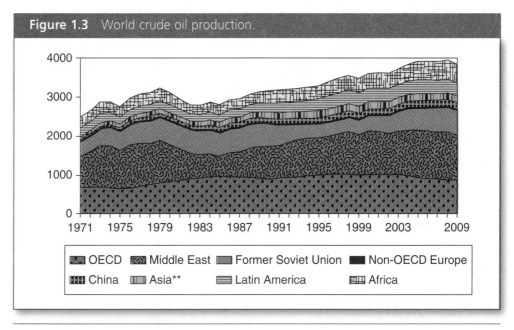

Figure 1.3 World crude oil production.

Source: International Energy Agency, 2010.

Moreover, it is becoming increasingly acknowledged that oil and gas, especially oil, is dirty and dangerous. The huge Gulf of Mexico spill in the spring and summer of 2010 from the explosion of the Deepwater Horizon drilling platform focused worldwide attention on the costs of oil and gas that we don't see directly in our bills. Many of the environmental troubles that the rest of this chapter describes can be traced back to fossil fuels. Global warming. Smog. Particulates. Acid rain. Social inequalities in who bears the brunt of pollution.

Plus, it seems that much of the supply of oil and gas that we will need to rely on in the future is from the dirtiest and most dangerous sources. Canada and Venezuela boast huge reserves of what used to be known as "tar sands," but recently has come to have the more polite names "oil sands" or "bituminous sands." Tar sands is more accurate to describe the form of these deposits in the ground: thick, rigid, and sticky, in need of vast investments in digging equipment for surface mining and heating equipment for pumping it out through steam injection, which makes the tar flow. The resulting landscape is not pretty. Conventional oil sources are also increasingly not so conventional, as they present new technological and environmental challenges. The Deepwater Horizon well was part of a push into deeper waters, further offshore, where water pressure is higher and infrastructure is chancier. Many potential reserves on land underlie areas where it is colder, hotter, or more remote—which is why they haven't been tapped already.

In a world that often seems to agree on little, politicians from across the political spectrum now speak of the need to transition to other energy sources. Presidents George W. Bush and Barack Obama in the United States, Prime Ministers David Cameron and Gordon Brown in Britain, and Chancellors Angela Merkel and Gerhard Schröder in Germany all have promoted reducing dependence on oil and finding alternative sources, despite their wide political differences on other issues.

How about coal, the next biggest of our current sources? There is still a lot of it in the ground, to be sure. But coal is infamously dirty. It is a carbon-based fuel, after all, which means it leads to global warming, smog, acid rain, particulates, and most of the rest of our carbon woes. Plus, coal has some special zingers of its own. Take the continued despoliation of land from coal mining. Take the billions of gallons of hot water discharged from coal-fired power plants into surrounding lakes and rivers. Take the hundreds of thousands of tons of highly toxic ash and sludge from smokestack scrubbers that a typical coal-fired power plant produces each year. Take the airborne mercury deposition from coal-fired power plants that has led to health guidelines on how many wild-caught fish from lakes in the U.S. Midwest one can safely eat. Or take the continued loss of miners' lives, like the 104 miners that died in a coal mine explosion on November 21, 2009, in China's Heilongjiang Province—and these were only a small proportion of the 2,600 mine fatalities in China in 2009 alone.[6] Some argue that new technologies will allow us to speak of "clean coal," which doesn't contribute to global warming and other carbon troubles. Perhaps so. And if so, this is very welcome news. But most of the special zingers of coal would still remain.

Maybe nuclear energy, then? Many say that, after the 2011 Fukushima Daiichi reactor leak in Japan, nuclear energy should clearly be off the table. It's just way too risky. But others say the new plant designs are much safer, and the operation of the older ones has improved considerably. Sure, the Fukushima meltdowns led to horrible and long-lasting problems, and the Chernobyl and Three Mile Island accidents were terrible too—especially Chernobyl, which killed several thousand as a direct result of the explosion, and is expected in time to cause at least another 4,000 deaths (some say tens of thousands more) due to radiation exposure.[7] Even so, say advocates, most of the 441 nuclear power plants around the world have good safety records.[8] The French get about three-quarters of their electricity from nuclear power, for example, and they haven't had serious problems. In the face of global warming and the other problems with a carbon economy, nuclear energy is worth the risks, the argument goes: the risks of radiation, of terrorism, of nuclear proliferation and warfare, of plant malfunctions, of earthquakes, of tsunamis, of tornados, of hurricanes, of storing the waste for 100,000 years, and of problems no one has thought of yet. In addition, right in the present, nuclear power also despoils land and depresses property values through mining, reclaiming, and plant siting. Still, maybe all that is better than floods, droughts, heat waves, strip-mining, air pollution, oil spills, coal mine accidents, and the rest of the carbon economy mess. Personally, I don't think so. But whether or not the risks are worth it, of this we can all be sure: The situation can't be good if the choices are so bad.

Or are they so bad? Are renewables a realistic option? Maybe we can power a twenty-first–century and even a twenty-second– and twenty-third–century economy with the sun, the wind, the water, the tide, the heat of the Earth, and the living power of biofuels. Environmentalists have been saying this for years, and not without reason. Most countries certainly do not make a lot of effort in these regards now, except with hydropower and increasingly with wind and biofuels. But some countries have made huge progress, especially Germany, which as of 2009 was getting 10 percent of its energy from renewables, and 16 percent of its electricity, thanks to policies like "feed-in tariffs" that require utilities to buy from renewable sources.[9] In Germany now, it is utterly routine to see a house with photovoltaic solar panels on the roof.

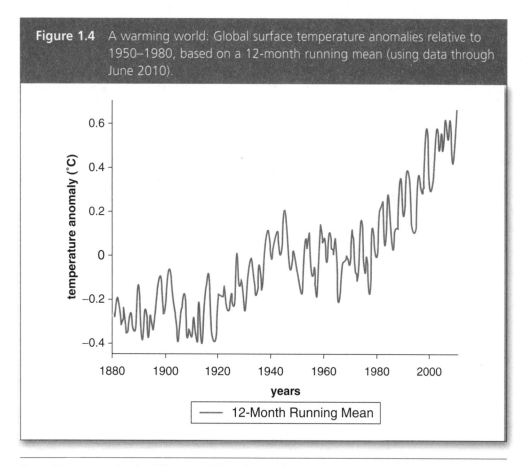

Figure 1.4 A warming world: Global surface temperature anomalies relative to 1950–1980, based on a 12-month running mean (using data through June 2010).

Source: Hansen, J. R., Ruedy, Mki. Sato, and K. Lo (In press.).

Why is it happening? You'd have to be living in a cave not to have heard by now that scientists place the blame most squarely on carbon dioxide emissions from fossil fuel use. The excess carbon dioxide in turn leads to an increased "greenhouse effect" through the ability of carbon dioxide to trap heat that would otherwise radiate out into space. The greenhouse effect is not a new discovery. Scientists figured out 150 years ago that the Earth would be a cold and barren place without it. But too much of a good thing is, well, too much of a good thing.

However, extra carbon dioxide accounts for only about half of human-induced climate "forcing," as climatologists say. Other greenhouse gasses like methane, nitrous oxide, chloro-fluorocarbons (CFCs), and ozone, as well as the soot or "black carbon" released by the myriad combustion processes of human activity, together account for roughly as much forcing as carbon dioxide does.[21] However, most of these forcings also come about through the burning of carbon-based fuels, directly or indirectly. Here's where a lot of the controversy comes, of course. The great engine of modern life is currently utterly dependent on carbon-based fuels, as the previous section discusses.

That is pretty scary, considering the effects that global warming is likely to bring about. Take what sea-level rise will mean for the settlement patterns of humanity. Sea level has already risen significantly as the climate has warmed. The Intergovernmental Panel on Climate Change (IPCC) projects that the average sea level will rise another 0.18 to 0.59 meters—half a foot to 2 feet—by the beginning of the twenty-second century, as glaciers and the ice caps melt and as ocean water heats up and expands.[22] That may not seem like all that much, unless you happen to live in a place like New Orleans, Amsterdam, or the low-lying Pacific Island nations of Tuvalu and Kiribati. Extensive regions of low-lying coastal land (where much of the world's human population lives) would be in grave danger of flooding during storm surges— or even being submerged underwater. It is not inconceivable that Tuvalu and Kiribati could be washed away. In view of the threat, the New Zealand government has already made plans for accepting immigrants displaced from Tuvalu.[23]

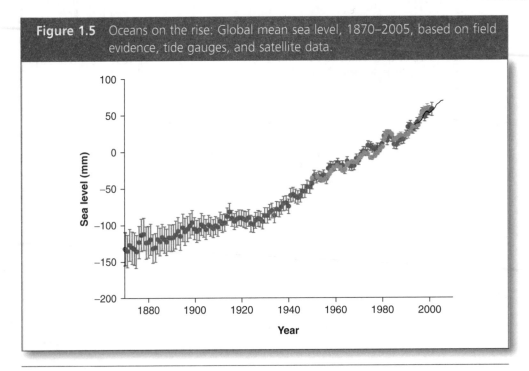

Figure 1.5   Oceans on the rise: Global mean sea level, 1870–2005, based on field evidence, tide gauges, and satellite data.

Source: IPCC (2007b).

Or consider the ecological disruptions global warming will bring. Coral reef dieback.[24] Increased wildfires.[25] Oceans so acidic that shellfish cannot make shells. Increased risk of extinction for up to 30 percent of species.[26] Gradual replacement of tropical forests with savanna in eastern Amazonia.[27] Increased drinking-water shortages and heat waves. Increased drought stress.[28] Increased incidence of disease, as the new conditions would likely be more hospitable to mosquitoes, ticks, rodents, bacteria, and viruses.[29] Increased incidence of damaging storms, with consequences for shoreline ecologies, including human ones like New Orleans.[30]

The consequences for agriculture would be complex. Some farming areas will likely be stricken with drier conditions. For example, the IPCC projects that rain-fed agriculture in Africa could be down as much as 50 percent in yields by 2020.[31] Farmers in Iowa, the leading corn-producing region in the United States, might have to switch over to wheat and drought-tolerant corn varieties, which would mean overall declines in food production per acre.[32] On the other hand, some regions will likely receive more rain. Yet many of these regions do not have the same quality of soil as, say, Iowa. To add to the complexity, carbon dioxide can stimulate growth in some crop plants; one study has found a 17-percent yield boost in soybeans.[33] However, this stimulation may not result in actual increased crop yields because of other limiting factors, such as low rainfall, poor soil conditions, and the existence of other pollutants in the air.[34]

Indeed, the predictions are coming true already. We are seeing an increase in heat waves, and resulting fatalities and property loss are on the rise. The spell of four days that peaked at around 100 degrees Fahrenheit between July 12 and 15, 1995, blamed for 739 deaths in Chicago.[35] The 2002 heat wave from April to May in India, which killed 1,000 people.[36] The even more horrific 2003 heat wave in Europe, estimated to have killed an astounding 35,000 to 50,000 people in France, Germany, and other European countries.[37] The July 2007 Western North American heat wave and drought that eventually led to the devastating wildfires of Southern California in October of 2007, which burned 1,500 houses and 400 million acres and forced 1 million people to evacuate their homes.[38] The summer of 2010 heat wave in Russia, the hottest on record there, which touched off devastating forest fires and is estimated to have caused $15 billion in damage and to have killed at least 15,000 people, and possibly as many as the European heat wave of 2003.[39]

The occurrence of devastating storms and floods is also way up. The Great Storm of 1987 in Britain. Hurricane Mitch in 1998. Katrina in 2005. The Mumbai floods of 2005. The British floods of 2007. The Tennessee, Iowa, Pakistan, China, India, Brazil, Nigeria, Central Europe and Northeast Australia floods of 2010. From the too wet to the too windy to the too hot to the too dry, the global rate of severe weather events now is twice what it was in the early 1980s, with 600 to 700 weather-related disasters per year.[40] The number of named Atlantic storms per year is now double what it was 100 years ago and 50 percent higher than it was 50 years ago.[41] The insurance industry is quite worried about the upsurge in claims that has resulted.[42] In most years now, worldwide economic losses from catastrophic weather events top $50 billion, which is 10 times the 1970s rate and 16 times the 1960s rate, measured in 2006 constant dollars.[43] In 2005, the worst year yet, worldwide losses were $220 billion.[44] In the United States alone, every year from 2004 to 2009 saw at least five weather-related disasters that topped $1 billion in damage (in 2007 dollars). In the 1980s, most years saw from none to two, with only one year (1983) as high as four (see Figure 1.6).

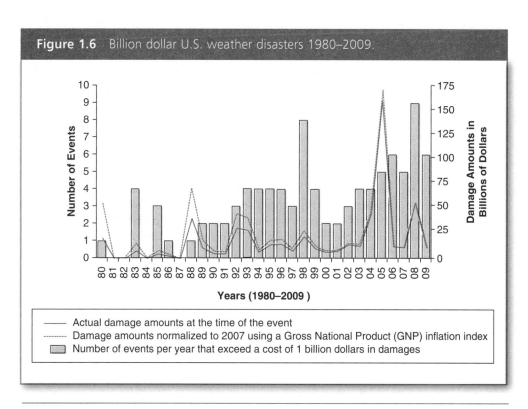

Figure 1.6 Billion dollar U.S. weather disasters 1980–2009.

—— Actual damage amounts at the time of the event
-------- Damage amounts normalized to 2007 using a Gross National Product (GNP) inflation index
▨ Number of events per year that exceed a cost of 1 billion dollars in damages

Source: NRDC (2010).

There's more. Meteorologists are becoming increasingly alarmed about the successive break-offs of ever-larger chunks of high-latitude ice shelves, such as the Rhode Island–sized patch of the Larsen B Ice Shelf that disintegrated during February 2002 and the 41-square-mile Ayles Ice Shelf that broke free of Canada's Ellesmere Island in the summer of 2005.[45] At the North Pole, there isn't as much ice to begin with, but it is showing worrisome signs of losing what there is. The years 2007, 2008, and 2009 recorded the three lowest extents of the Arctic ice cap on record.[46] This is especially worrying because less white surface cover on the Earth means less solar energy is reflected back out to space, heating the planet even more. By 2008, the mean thickness of Arctic sea ice had dropped 48 percent since 1958, from 11.9 feet to 6.2 feet.[47] There are now frequent sizable stretches of open water at the Arctic ice cap during the summer, including a 10-mile by 3-mile-wide stretch quite close to the North Pole itself in 2001, although scientists say this open water may have been there for some time.[48] But the northern ice cap is clearly retreating and thinning. Researchers from the National Aeronautics and Space Administration (NASA) predict that by 2040, we will have no Arctic ice cap at all at the height of the Northern Hemisphere summer; the British Antarctic Survey thinks it will happen even earlier than that.[49]

Then there are the implications for disease. Warmer world weather has been implicated in the resurgence of cholera in Latin America in 1991 and pneumonic plague in India in 1994, and in the outbreak of a hantavirus epidemic in the U.S. Southwest in 1994. Scientists are wondering if global warming is a factor in about 10 other diseases that resurged or reemerged in the 1990s.[50] Increased allergy complaints may be due to global warming, one study suggests.[51] A 2009 report by the World Health Organization concludes, based on a review of many studies, that climate change increases malaria, dengue, diarrhea, Lyme disease, tick-borne encephalitis, and food-borne pathogens such as salmonella.[52] In areas where a population's disease resistance is already weakened by the AIDS virus, these increases will be particularly problematic.[53]

Meanwhile, greenhouse gas emissions continue to rise. Annual mean carbon dioxide, as measured at Hawaii's Mauna Loa Observatory as of 2010, is up to 390 parts per million in the atmosphere.[54] In the mid–eighteenth century, the number was about 277 parts per million, according to data from ice cores drilled in Antarctica.[55] But growth still hasn't leveled out, despite the initial efforts of many nations around the world. Recently, the concentration has been going up about 2 parts per million per year, as we continue to force the climate, and push our luck.[56]

You could think of human-induced climate forcings as acting like extra blankets on a warm night, gradually stifling the planet. I say "on a warm night" because solar radiation is also on the rise, adding a climate forcing about a tenth as strong as human-induced forcings.[57] This is, at the very least, bad timing. Consequently, taking together all the forcings that are warming as well as cooling the planet—and there are indeed a few working in the direction of cooling, such as increased reflectivity back into outer space from increased cloudiness—the IPCC estimates that average temperatures will rise 1.1 to 6.4 degrees Celsius in the twenty-first century, depending on the scenario and model.[58] These are enormous increases when you consider that an average drop of 6 degrees Celsius caused the ice ages, covering much of the northern latitudes with a mile-thick sheet of ice.[59] Climate is a touchy thing. A few degrees average change one way or the other can make quite a difference. In this case, we could be on the verge of many centuries of generally lousy sleeping weather—and circumstances much more ominous than that.

Think about it the next hot summer evening as you ponder whether you should crank the air conditioner up another notch, causing your local utility to burn just that much more carbon-based fuel, and to release that much more smog and soot to generate the necessary electricity.[60] More cooling for you will mean more heating for all of us.

The Two Ozone Problems

There are several other threats to our atmosphere. While perhaps not quite as drastic in their potential consequences as global warming, they are plenty drastic enough for considerable concern. Two of these threats involve ozone, although in quite different ways.

Ozone forms when groups of three oxygen atoms bond together into single molecules, which chemists signify as O_3. Most atmospheric oxygen is in the form of two bonded oxygen atoms, or O_2, but a vital layer of O_3 in the upper atmosphere helps protect life on the Earth's surface from the effects of the sun's ultraviolet radiation. Ultraviolet light can cause skin cancer, promote cataracts, damage immune systems, and disrupt ecosystems. Were there no ozone layer in the upper atmosphere, life on Earth would have evolved in quite different ways—if indeed it had begun at all. In any event, current life forms are not equipped to tolerate much more ultraviolet radiation than the surface of the Earth currently receives. We badly need the upper-atmosphere ozone layer.

In 1974, two chemists, Mario Molina and Sherwood Rowland, proposed that CFCs—which, as we have seen, are also a potent global-warming forcing—could be reacting with the ozone layer and breaking it down. Molina and Rowland predicted that CFCs could ultimately make their way into the upper atmosphere and attack the integrity of the ozone layer. In 1985, scientists poring over satellite imagery of the atmosphere over Antarctica discovered (almost accidentally) that the ozone layer over the South Pole had, in fact, grown dangerously depleted.

Many studies later, we now know that this "ozone hole," as it has come to be called, is dramatically large. We also know that it changes in size with the seasons, has a much smaller mate over the North Pole, and stretches to some degree everywhere on the planet except the tropics. In fact, it's really not a hole. It is more accurate to say that, outside the tropics, the ozone layer is *depleted,* particularly over the South Pole. (See Chapter 8 for an extensive discussion about the metaphor of an ozone "hole.") At times, the layer depletes to as low as 25 percent of the levels observed in the 1970s.[61] Most worrisome is that the area of high depletion might spread to heavily populated areas. In 2000, the high-depletion area passed over the tip of South America for 9 days, including the Chilean city of Punta Arenas. The perimeter of the hole skirts Punta Arenas most years now.[62] Australians and New Zealanders have yet to experience this, but they're plenty worried. Levels of depletion there are already worse than in other populated regions, skin cancer rates are the highest in the world, and classes in "sun health" have become an essential feature of the school curriculum.[63] And in Punta Arenas, the world's most southerly city, skin cancer rates shot up 66 percent between 1994 and 2001.[64]

This is scary stuff. Skin cancer is no fun. But it has galvanized a truculent world into unusually cooperative action.[65] In 1987, the major industrial countries signed the first of a series of agreements, known as the Montreal Protocol, to reduce the production of CFCs. As a result of these agreements, CFC production for use in these countries ended on December 31, 1995, and ended throughout the world on December 31, 2010.

There is more good news to report: The ozone hole is no longer increasing. Since 2000, the amount of ozone at the poles has been essentially stable.[66] It will be many decades until the depletion is repaired, however. The ozone-damaging chlorine that CFCs contain remains resident in the atmosphere for some time, and the HCFCs (hydrochlorofluorocarbons) that industrial countries first turned to as a substitute also damage the ozone layer to some extent. Plus, like CFCs, HCFCs are a potent greenhouse gas. Chlorine-free "Greenfreeze" refrigerants do not damage the upper-atmosphere ozone layer and do not contribute to global warming. Greenfreeze technology now dominates the refrigerator market in Europe and is taking hold in South America, Japan, China, and elsewhere. But Greenfreeze is just now reaching the North American market and, at this writing, is still not approved for use in its refrigerators—for reasons of protectionism, argue some.[67] Meanwhile, an extensive black market in CFCs has arisen in North America.[68] Thus, the current expert view is that ozone depletion will be with us until the middle of the century at least, and likely longer than that.[69]

The banning of CFC production and resulting stabilization of the ozone hole is nevertheless one of the great success stories of the environmental movement, and perhaps the greatest. Despite our differences, sometimes we can achieve the international cooperation necessary to make major progress on big problems. We know we can because, in the case of CFCs, we have done it.

Much less progress, however, has been made on the other ozone problem: ozone at ground level. Hardly a city in the world is free of a frequent brown haze above which only the tallest buildings rise (see Figure 1.7). Ozone is the principal component of this brown smog that has become an unpleasantly familiar feature of modern urban life.

Figure 1.7   Mexico City disappears in the smog, trapped by the mountains that surround the city, December 23, 2009. Unacceptable levels of ground-level ozone, as defined by the World Health Organization, occur about 180 days a year. But 20 years ago, before a huge clean-up campaign, the figure used to be over 300 days a year.

Source: Author.

Ground-level ozone forms when sunlight glares down on a city's dirty air. As a result of fossil fuel combustion, cars and factories discharge large volumes of a whole array of nitrogen oxide compounds. NO_X (pronounced "knocks") is the usual term for this varied nitrous mixture. In sunlight, NO_X reacts with volatile organic compounds (or VOCs) to produce ozone. (The VOCs themselves are also produced during fossil fuel combustion, as well as by off-gassing from drying paint and by various industrial processes.) If the day is warm and still, this ozone will hug the ground. Because it needs sunlight to form, scientists often call the resulting haze "photochemical smog."

Although we need ozone up high to protect us from the sun, down low, in the inhabited part of the atmosphere, ozone burns the lung tissue of animals and the leaf tissue of plants. This can kill. A 2006 study estimated that over 100,000 Americans suffer premature death each year because of photochemical smog.[70] Stop for a moment: That's a huge number of premature deaths. A 2004 study found that even small differences in ozone concentration have measurable effects on mortality.[71] Smog alerts have become an everyday feature of big-city life in all industrial countries. Walking and bicycling are increasingly unhealthful and unpleasant—driving people even more into their cars and causing even more smog. In Mexico City, long a victim of some of the world's worst ozone pollution, a 2006 study showed that smog even compromised residents' sense of smell.[72] When it drifts out of the city and into the countryside, smog also reduces crop production and damages forests. For example, soybeans suffer a 20 percent loss in yield due to ozone—not an insignificant amount in a hungry world.[73]

To put the matter simply, there's too much ozone down low, not enough up high, and no way to pump ozone from down here to up there.

Particulates and Acid Rain

Big cities and their surrounding suburbs also face the hazard of fine particulates in the air. These particles are microscopic—the definition of "fine particulates" is particles 2.5 microns or smaller in size, much smaller than the diameter of a human hair—and they penetrate deeply into lung tissue. In contrast to the brownish color of photochemical smog, fine particulates envelop cities with a whitish smog. About half of these particulates are basically dust, mainly released because of poor fuel combustion in cars, trucks, power plants, wood stoves, and outdoor burning, or kicked up by traffic, construction, and wind erosion from farms. Most of the rest are tiny pieces and droplets of sulfates, nitrates, and VOCs formed in the atmosphere following the burning of fossil fuels, such as the coal used for electric generation; together, these are called "secondary" particulates.[74] Ammonium and ammonium compounds also contribute significantly to secondary fine-particulate pollution, mainly due to emissions from livestock and fertilizers.

According to a 2006 study, over 160,000 Americans die prematurely each year due to fine particulates.[75] Stop again: That's 160,000 premature deaths. Another study found that in American cities with the most fine particulates, residents are 15 to 17 percent more likely to die prematurely.[76] A study in Sydney, Australia, found premature death rates to be double even those of American studies.[77] Children in Los Angeles who live closer to roads have decreased lung capacity, largely because of fine particulates.[78] Fine particulates also increase heart attack rates, according to another 2006 study, which, along with studies of lung capacity and asthma effects, helps explain the higher death rates associated with areas that have higher levels of fine particulates.[79] This is serious stuff, even worse than the much better-known problem of photochemical smog.

And then there's acid rain. This is an issue that has largely dropped from sight, after a flurry of concern in the 1970s and early 1980s over sharp declines in the populations of some fish and frogs and extensive signs of plant stress and dieback in many forests. But acid rain is still falling from the sky, despite substantial efforts to reduce acidifying emissions of sulfur dioxide and NOx (which also have other dangerous impacts, as we have seen). These pollutants combine with water in the atmosphere to acidify rain, resulting in direct damage to plant tissues, as well the leaching of nutrients from soil and the acidification of lake waters, which, in turn, affect most wildlife—particularly in areas with normally acidic conditions, where ecosystems have less capacity to buffer the effects of acid fallout. When things get bad enough, lakes die and trees refuse to grow, like the miles of blasted heath in the acid deposition zone surrounding the old nickel smelters in Sudbury, Ontario. The situation is especially severe in northern Europe, where more than 90 percent of natural ecosystems have been damaged by acid rain; a year 2000 survey by the European Union found that 22 percent of all trees in Europe have lost 25 percent or more of their leaves.[80] Conditions are also quite worrisome in much of Canada and in China. One 2007 study found defoliation rates as high as 40 percent in some Chinese forests.[81]

Efforts to reduce acidifying emissions of sulfur dioxide and NOx have made a difference in some regions. There does seem to be some slight improvement in the condition of Canadian lakes and forests—but only slight.[82] Deposition rates for sulfate from rain are down considerably in much of the United States. But there are areas of the country, mostly in Ohio, Indiana, and Illinois, that as of 2008 were still receiving more than 25 kilograms per hectare of sulphate from the sky each year—10 times more than falls in the western United States.[83] A 2000 review of the scientific literature by the federal government's General Accounting Office found that the condition of lakes in New England and the Adirondack Mountains of New York was either stable or getting worse, but none seemed to be improving.[84] Some 43 percent of Adirondack lakes are expected to be acidic by 2040—up from the 19 percent observed to be acidic in 1984.[85] Between 1992 and 1999, the condition of trees in Europe did improve in 15 percent of the test sites, but deteriorated in a further 30 percent of sites.[86] In Taiwan, the Central Weather Bureau has registered increasingly severe acid rain events in recent years.[87]

In other words, the situation is at best a mixed bag. Why, after so many years of effort, does acid rain still threaten? Technological improvements, international treaties, and domestic legislation have all contributed to a sharp decline in sulfur emissions in most countries. But we have made little overall progress in reducing nitrogen emissions. Industry's nitrogen emissions have been reduced, but these advances have been overwhelmed by increased emissions from automobiles and trucks as the world comes to rely ever more on these highly polluting forms of transportation.[88] Plus, there is evidence that the ability of sensitive ecosystems to handle acid rain has been damaged such that slight improvement in the acidity of rain often does not result in any improvement in the condition of lakes and forests.[89]

Acid rain is still a big problem.

Threats to Land and Water

There's a well-known saying about land: They aren't making any more of it. The same is true of water. And in a way, there is less of both each year as the expansion of industry, agriculture, and development erodes and pollutes what we have, reducing the world's capacity to sustain life.

Consider soil erosion in the United States. Soil erodes from farmland at least 10 times faster than it can be replaced by ecological processes, according to a 2006 study.[90] Despite decades of work in reducing soil erosion, largely in response to the lessons of the Dust Bowl, it still takes a bushel of soil erosion to grow a bushel of corn.[91] The Conservation Reserve Program, implemented by the U.S. Congress in 1985, has resulted in significant improvements by offering farmers 10- to 15-year contracts to take the most erodible land out of production. Many farmers have also switched to much less erosive cropping practices. Consequently, soil erosion dropped 43 percent from 1982 to 2007, water erosion dropped from 4.0 to 2.7 tons per acre, and wind erosion dropped from 3.3 to 2.1 tons per acre.[92] But those numbers are still way too high, most observers inside and outside of agriculture agree.

Elsewhere, the situation is equally grim. Soil erosion exceeds replacement rates on a third of the world's agricultural land.[93] Worldwide, almost a quarter—23 percent—of cropland, pasture lands, forests, and woodlands have become degraded.[94] True, fertilizers can make up for some of the production losses that come from eroded soils, at least in the short term, but only at increased cost to farmers and with increased energy use from the production of fertilizer and the application of it to fields—and increased water pollution as the fertilizer washes off into streams, rivers, and groundwater.

Soil erosion is only one of many serious threats to farmland. Much of the twentieth century's gain in crop production was due to irrigation. But irrigation can also salinize soils. Because most irrigation occurs in parched regions, the abundant sunlight of dry climates evaporates much of the water away, leaving salts behind. In China, nearly half of the cropland is irrigated, and 15 percent of the irrigated land is affected by salinization. In the United States, only about 10 percent of cropland is irrigated, but almost a quarter of that 10 percent has experienced salinization. In Egypt, virtually 100 percent of cropland is irrigated, and almost a third of it is affected by salinization.[95]

Irrigation can also waterlog poorly drained soils. Clearing of land is doing the same thing in Australia. Once the land is cleared of its native woodland and bush, rates of transpiration—the pumping of water through the leaves of plants, enabling plants to "breathe"—slow down. Water tables in the dry wheat belt of Western Australia are rising by up to one meter a year, waterlogging these poorly drained soils. This, in turn, can lead to salinization as waterlogged soils bake in the sun. A 2006 estimate found that 16 percent of Australia's agricultural land has been degraded by salinization.[96] Thus, over-irrigation can turn soils both swampy and salty at the same time.

Irrigation of cropland, combined with the growing thirst of cities, is leading to an even more fundamental problem: a lack of water. The World Bank reports that 700 million people in 43 countries face "water stress"—that is, demand for water in their regions exceeds availability, at least temporarily.[97] Another depressingly vogue term is "peak water" to describe the moment when overall demand for water in a country or region exceeds the rate of replenishment, and not just temporarily. Some 15 nations, mainly in the Middle East, now use more water than their climates can replenish.[98] How do people in these regions survive? Mainly by importing food, a strategy that leaves them dependent on world markets for what policy analysts have begun calling "virtual water": the water it took to produce a crop or a product. Even in countries not classified as water-stressed, the situation is increasingly dire. Take the United States and Mexico. By the time it reaches the ocean in the Gulf of California, the Colorado is probably the world's most famous "non-river," for not a running drop remains after the farms and cities of the United States and Mexico have drunk their fill. Further development in the regions dependent on the Colorado River will require water from other sources—and it is not obvious where those generally dry territories can easily find them—or greatly improved efficiency in current water use.

In the Murray-Darling Basin of Australia, the country's richest agricultural region, the story is much the same. Today, only one fifth of the water that enters the basin's rivers is still there by the time the Murray reaches the sea, and the comparative trickle of water that remains is salty and prone to bacterial blooms and fish kills.[99] Perhaps the most dramatic example of overuse of water sources is the Aral Sea in central Asia—once the world's fourth-largest lake. Diversion for irrigation has reduced the Aral's surface area by 90 percent. Salinity has quadrupled; former fishing ports lie miles inland; thousands of square miles of lake bottom have turned to desert; the 24 species of native fish are all gone because of the salinity, as are half the bird and mammal species; and the region's economy has collapsed.[100] By 2000, the Aral had divided into two separate bodies of water: what are now called the North Aral Sea (also known as the Small Aral Sea) and the South Aral Sea.[101] But those names are already dated, for the Aral later divided into four even smaller bodies of water, and, as of 2010, one of those—the Eastern Lobe of the South Aral Sea—had almost completely dried up. The European Space Agency predicts the entire South Aral Sea will be gone by 2020.[102] Some good news is that a dike funded by the World Bank has begun restoring some of the lake area in the North Aral Sea.[103] But the overall situation remains very grim.

Not only surface water, but groundwater too is being rapidly depleted. Over-irrigation can lead to rising water tables and the waterlogging of soils in some regions, but the more general problem is falling water tables from the depletion of groundwater stocks. Around the world, extraction of groundwater for cities and farms is exceeding replenishment rates. Recent production gains in agriculture in India have relied heavily on irrigation from groundwater, but now, because of what one observer has called a "race to the bottom of the aquifer," water levels have dropped in some 90 percent of wells in the Indian state of Gujarat, and every year more wells run dry.[104] There are reports of villagers having to pump from as deep as 800 feet to get water.[105] In the dry Great Plains of the United States, farmers pump the famous Ogallala Aquifer 8 times faster than it recharges from precipitation, endangering 15 percent of U.S. corn and wheat production and 25 percent of U.S. cotton production. Nearly one-fifth of the Ogallala's water reserves have already been pumped out, and the taps have had to be turned off in many places.[106] In the North China Plain, a major grain-producing area, water tables have been dropping at the rate of 3 to 5 feet each year, due to overdraw for irrigation.[107] In some regions, the lowering of water tables is causing major land subsidence. Downtown Mexico City has dropped nearly 25 feet.[108] Some parts of the Central Valley of California and the Hebei Province surrounding Beijing have dropped similar amounts.[109] Venice has dropped 10 centimeters because of pumping the freshwater aquifer beneath it—which may not sound like a lot, but for a city at the water line, that is an alarming figure.[110]

Overextraction can degrade the quality of the groundwater that remains. The main threat here again is salinization, either through the overapplication of irrigated water to the land's surface or through the invasion of seawater into shrinking groundwater aquifers. Ten percent of wells in Israel have already been abandoned because of seawater invasion, and many more will soon have to be given up.[111] In the Indian state of Gujarat, half the hand-pumped wells are now salty.[112] When farmers overapply irrigated water, the salinization of the soil can be carried down into the aquifer, as the water percolates down past crop roots. In many areas, only some 30 to 40 percent of irrigated water actually reaches crops, with the rest being lost through evaporation and percolation, promoting salinization of groundwater. In the lower Indus River Valley of India and Pakistan, the situation is so bad that engineers have installed an expensive system of pumps and surface drains to carry some of the salinized groundwater away to the sea.[113]

Much of the freshwater that remains is badly polluted. Some years ago, in 1992, Donella Meadows, Denis Meadows, and Jorgen Randers calculated that "the amount of water made unusable by pollution is almost as great as the amount actually used by the human economy."[114] They also noted then that we are very close to using, or making unusable, all the easily accessible freshwater—freshwater that is close to where people live (as opposed to rivers in the Arctic, say) and that can be stored in rivers, lakes, and aquifers (as opposed to the huge amounts of freshwater lost to the sea during seasonal floods, which cannot be easily stored).[115] The situation around the world today remains dire. The remaining margin for growth in freshwater use is disturbingly narrow.

Cleaning up water pollution is one way to increase that vital margin, and industrial water pollution has diminished in many areas, particularly in the wealthier countries. We have also made progress in controlling agricultural water pollution. But we still have a long way to go. From 1950 to 2001, farmers across the world upped their use of commercial fertilizers 8-fold and their use of pesticides 32-fold.[116] In the United States, the development of stronger pesticides for a number of years led to substantial drops in the number of pounds of pesticides farmers applied. After the 1980s, pesticide use in the United States began to rise again, with about a 10 percent increase through 2001.[117] No agency has tabulated national or worldwide statistics since then, but local reports suggest that pesticide use is now rising rapidly. Many developing countries are continuing a "green revolution" approach to food production, using all available agricultural chemistry. And in wealthy nations, use has increased with the widespread planting of herbicide-tolerant GMO crops like "Round Up–ready" corn and soybeans—that is, crops with a gene spliced in that lets farmers increase their use of Round Up, a popular herbicide, without hurting the crop. The resulting runoff continues to threaten the safety of many drinking water supplies. As Chapter 5 discusses in detail, many pesticides are quite hazardous for human health. Excess nitrogen fertilizer in the water is, too. We all need something to eat and something to drink, but some of our efforts at maintaining food production put us in the untenable position of trading food to eat for water to drink.

Or are we trading them both away? In addition to the threats to agricultural production caused by soil erosion, salinization, waterlogging, and water shortages, we are losing considerable amounts of productive farmland to the expansion of roads and suburbs, particularly in the wealthiest nations. Cities need food; thus, the sensible place to build a city is in the midst of productive agricultural land. And that is just what people have done for centuries. But the advent of the automobile made possible (although not inevitable) the sprawling forms of low-density development so characteristic of the modern city. The result is that cities now gobble up not only food but also the best land for growing it. The problem is worst in the United States, which has both a large proportion of the world's best agricultural land and some of the world's most land-consuming patterns of development. Some 86 percent of fruit and vegetable production and 63 percent of dairy production come from urban counties or from those adjacent to urban counties. A 2002 study found that the United States loses about 2 acres of farmland every minute, or about 1.2 million acres per year, an area the size of Delaware.[118] Given that the United States has almost a billion acres of agricultural land, this may not seem worth worrying about. But in most cases, the nation is losing its best land, and in the places where it is most needed: close to where people live.

Then factor into the calculation the effects of global warming, photochemical smog, and acid rain on crop production. Now add some major issues I have not even mentioned: increased resistance of pests to pesticides, declining response of crops to fertilizer increases, the tremendous energy inputs of modern agriculture, loss of genetic diversity, desertification due to overgrazing, and pesticide residues in food. It is no wonder that increases in agricultural production have been falling behind increases in human population. The result is that after decades of steady increases, world grain production per person per year declined from the historical high of 343 kilograms in 1985 to 297 kilograms in 2002, the first year below 300 kilograms since 1970.[119] Production per person has picked up a bit since 2002, due to a number of world-record harvests in subsequent years, but the 1985 figure of 343 kilograms per person has not been exceeded. The needle has basically been stuck for 25 years. Plus, the growth of biofuels threatens to divert a sizable proportion of grain away from food, as many have tried to point out.[120]

Let's face it. We're eating up the world. An increasingly popular way to represent our overconsumption on an ecological scale is ecological footprint analysis, which converts all the various demands we make on the Earth's ecosystems to a measure of area (see Figure 1.8). Since about 1975, our collective footprint has been larger than the Earth itself. We are now demanding about 1.5 Earths. We are provided with only one. You can't eat your Earth and have it, too.

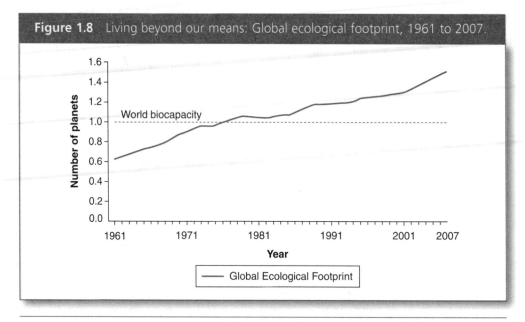

Figure 1.8 Living beyond our means: Global ecological footprint, 1961 to 2007.

Source: Global Footprint Network. 2010.

Environmental Justice

On the morning of January 4, 1993, an estimated 300,000 Ogoni rallied together. The protesters waved green twigs as they listened to speeches by Ken Saro-Wiwa, a famous Ogoni writer, and others. With such a huge turnout, the Ogoni—a small African ethnic group, numbering only a half million in all—hoped that finally someone would pay attention to the mess that Shell Oil Company had made of their section of Nigeria. Leaking pipelines. Oil blowouts that showered on nearby villages. Disrupted field drainage systems. (Much of Ogoniland has to be drained to be farmed.) Fish kills. Gas flares that fouled the air. Water so polluted that just wearing clothes washed in it caused rashes. Acid rain from the gas flares so bad that the zinc roofs people in the area favor for their houses corroded away after a year. Meanwhile, the profits flowed overseas to Shell and to the notoriously corrupt Nigerian government. The Ogoni got only the pollution.[121]

Such open protest by the relatively powerless is a courageous act. And for the Ogoni, the consequences were swift and severe. During the next 2 years, Nigerian soldiers oversaw the ransacking of Ogoni villages, the killing of about 2,000 Ogoni people, and the torture and displacement of thousands more.[122] Much of the terror was carried out by people from neighboring regions whom the soldiers forced or otherwise enticed into violence so that the government could portray the repression as ethnic rivalry.[123] The army also sealed the borders of Ogoniland, and no one was let in or out without government permission. Ken Saro-Wiwa and other Ogoni leaders were repeatedly arrested and interrogated. Finally, the government trumped up a murder charge against Saro-Wiwa and eight others and, despite a storm of objection from the rest of the world, executed them on November 10, 1995.[124] The torture and killing of protestors continued for some years. For example, on June 15, 2001, police shot Friday Nwiido, a few weeks after Nwiido had led a peaceful protest against the April 29, 2001, oil blowout that rained Shell's crude onto the surrounding countryside for 9 days.[125] Recently, there have been efforts to broker a peace process for the region among Shell, the Nigerian government, the United

Nations Environment Programme (UNEP), and the Ogoni people. But as of this writing, the situation remains chaotic with the recent killing of a 20-year-old Ogoni man by Shell's private security forces, a major oil fire on Ogoni farmland, and the displacement of about 1,700 Ogoni fishing families after a series of oil spills in adjacent creeks.[126]

The Ogoni experience is a vivid example of a common worldwide pattern: Those with the least power get the most pollution.

The Ogoni experience is also an outrage, as virtually the entire world agrees.[127] This outrage is a reminder of another of the three central issues of environmentalism: the frequent and tragic challenges to *environmental justice*. There is a striking unevenness in the distribution of environmental costs and environmental benefits—in the distribution of what might be termed *environmental bads* and *environmental goods*.[128] Global warming, sea level rises, ozone depletion, photochemical smog, fine-particulate smog, acid rain, soil erosion, salinization, waterlogging, desertification, loss of genetic diversity, loss of farmland to development, water shortages, and water pollution: These have a potential impact on everyone's lives. But the well-to-do and well-connected are generally in a better position to avoid the worst consequences of environmental problems, and often to avoid the consequences entirely.

Who Gets the Bads?

One prominent basis of being well-connected is a person's social heritage, as a large number of sociological studies have depressingly documented, and as everyday social experience routinely proves. Within issues of environmental justice, there are special challenges of *environmental racism*—that is, social heritage differences in the distribution of environmental bads, due to either intentional or institutional reasons.

For example, much research in environmental racism has shown that people of color are more likely than not to live in communities with hazardous waste problems. In 1987, the United Church of Christ's Commission for Racial Justice released the first of two controversial reports. Based on studies of zip codes, the reports concluded that African Americans and other people of color were 2 to 3 times as likely as other Americans to live in communities with commercial hazardous waste landfills.[129] A 1992 study found that 3 percent of all Whites and 11 percent of all minorities in the Detroit region live within a mile of hazardous waste facilities—a difference of a factor of nearly 4.[130]

Findings like these were central to the emergence in the early 1990s of the *environmental justice movement*. Originally a largely grassroots movement of local activists concerned about pollution in their neighborhoods, environmental justice now has a prominent place on the agenda of most national and international environmental organizations. Environmental justice has become one of the central civil rights issues in the United States and elsewhere, helping create a political climate for change.[131] The U.S. government, on occasion, has taken these issues seriously and has undertaken several self-studies. As a result, in 1992, the U.S. Environmental Protection Agency admitted that it may sometimes have been discriminatory in its siting and regulatory decisions, and in 1994, President Clinton signed Executive Order 12898, which requires all federal agencies to work toward environmental justice.[132]

Some studies of hazardous waste siting, however, have found that social class predicts who gets the bads better than race does. For example, a 2000 study by researchers from the University of Massachusetts found a strong association of hazardous waste facilities with blue-collar working-class communities, and only a slight association with African Americans (and only for African Americans living in rural areas).[133] Researchers are coming around to the view that a "is-it-race-or-income" debate is a bit beside the point, however.[134] For one thing, within the United States at least, race and income closely correspond and intertwine. To talk about one is largely to talk about the other. And when one is dealing with statistical categories that necessarily do a bit of jamming and cramming of the variety of the world to get it into the precise boxes needed for numerical calculation, some dimensions of things get compromised. Nevertheless, of 27 empirical studies of environmental justice in the United States published between 1998 and 2007, a total of 9 found race was significant,[135] 7 found social class was significant,[136] and 11 showed that both race and social class were significant factors.[137] Moreover, every single one found evidence of environmental inequality. More and more studies find the same pattern. In 2010 alone, 10 new studies of environmental justice in the United States all confirmed these results.[138]

They found that Los Angeles schools with high proportions of minority students tend to be located in areas with high levels of airborne toxics.[139] They found that in Florida, people of color face much higher odds that their homes are located near a toxic chemical plant—up to 5 times higher, in some cases.[140] They found that in Michigan, poor people and people of color are more likely to live in areas subjected to the toxic releases registered in the U.S. Environmental Protection Agency's Toxic Release Inventory.[141] They found that industrial-scale hog farms in Missouri are more likely to be located in counties with lower average income.[142] They found that in Massachusetts, low-income communities experience 8.5 times as many chemical releases from industry than high-income communities and that communities with a high proportion of minorities receive 10 times as many releases as communities with a low minority proportion.[143] They found that poor people across the United States experience higher levels of ambient and indoor air pollution, worse drinking water quality, and more ambient noise (from streets and highways, for example) where they live.[144] They found that people of color disproportionately hold the dirtiest and most dangerous jobs in the United States, and typically are poorly paid for their sacrifices.[145] And can we ever forget who was living in the Lower Ninth Ward of New Orleans on August 29, 2005, as Katrina's winds began to gust and the levees began to breach?

But whether along race or class lines or any other dimension of social difference, such biases are a challenge to the environmental and justice we all have a right to enjoy.

One of those other dimensions of social difference is whether one lives in a rich country or a poor one. Take the hazardous waste crisis, for instance. Wealthy countries are now finding that there is more to disposing of garbage than simply putting it in a can on the curb. One response has been to pay others to take it. We now have a lively international trade, much of it illegal, in waste too hazardous for rich countries to dispose of at home.

There has been considerable protest about this practice. In 1988, Nigeria even went so far as to commandeer an Italian freighter with the intent of loading it up with thousands of barrels of toxics that had arrived from Italy under suspicious circumstances and shipping it back to Europe. After a heated diplomatic dispute, the waste—which turned out to originate in 10 European countries and the United States—was loaded on board the *Karin B.*, a West German ship, and sent back to Italy. But harbor officials in Ravenna, Italy, where the waste was supposed to go, refused the load because of vigorous local opposition to it. The *Karin B.* was later refused entry in Cadiz, Spain, and banned from French and British ports, where it also tried to land. Months later it was finally accepted into Italy.[146]

In 1989, in response to diplomatic crises like these, 105 countries signed the Basel Convention, which is supposed to control international toxic shipments. Yet loopholes are large enough and enforcement lax enough that these shipments still go on. INTERPOL, with its 188 member countries, holds regular conferences to try to control this and other forms of international environmental crime, as does the 150-nation International Network for Environmental Compliance and Enforcement. They've had some success, but there is much that they don't catch, unfortunately.

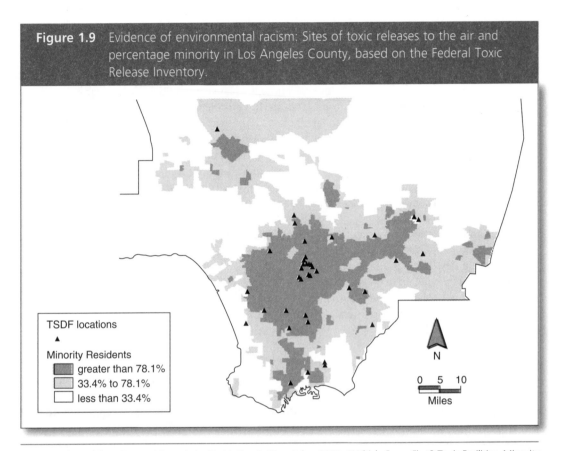

Figure 1.9 Evidence of environmental racism: Sites of toxic releases to the air and percentage minority in Los Angeles County, based on the Federal Toxic Release Inventory.

TSDF locations

▲

Minority Residents

greater than 78.1%

33.4% to 78.1%

less than 33.4%

N

0 5 10

Miles

Source: Adapted from Pastor, Manuel, Jr., Sadd, Jim, & Hipp, John. 2001. "Which Came First? Toxic Facilities, Minority Move-In, and Environmental Justice." *Journal of Urban Affairs* 23(1):1–21.

Plus, much that most people would regard as environmental crime is perfectly legal. Perhaps most glaringly, no international conventions currently stop companies from merely relocating their most hazardous production practices to poorer countries, or from purchasing from companies that use the laxer environmental and labor regulations and enforcement in most poorer countries to save on production costs. Like Union Carbide, which operated the infamous pesticide plant at Bhopal, India, that killed over 5,000 people in a single night, due to a chemical leak on December 2, 1984. (See Chapter 5 for the details on what happened.) Like the many companies that buy from the textile, toy, and electronics factories of China, which have so badly polluted the land, water, and people of the "factory to the world." Or like the companies that buy from the sweatshops of Southeast Asia, India, Africa, and Latin America. Many of our industrial practices expose workers—generally those on the production line, as opposed to those in the head office—to environmental hazards. Exporting hazardous jobs does not lessen the degree of environmental inequality involved, however.

All this seems to take place far away—until a toxic disaster happens in your own community. The growing placelessness of the marketplace makes it easy to overlook the devastating impact untempered industrialism can have on the daily lives of the farmworker applying alachlor in the field and the factory worker running a noisy machine on a dirty and dangerous assembly line. When we shop, we meet a product's retailers, usually not the people who made it, and the product itself tells no tales.

Who Gets the Goods?

Environmental justice also concerns patterns of inequality in the distribution of environmental goods. These patterns are usually closely associated with inequality in the distribution of wealth. Thus, those who are concerned about environmental justice often point to the huge inequalities in average income between countries. Here are the numbers, based on gross national income (GNI) per capita in 2009 in U.S. dollars.[147]

The average annual income in the world is $8,728. In contrast, the average annual income in the world's 20 wealthiest countries is $48,874. In the United States, it is $46,360. The richest country in the world is Norway, at $84,640 per capita. But once we take into account the cost of living, Luxembourg noses out Norway in per capita buying power at $59,550 to $54,880.[148] (The United States ranks as the fifth-richest country in the world in per capita buying power at $45,640 annual income.)

With all that income flowing to the top, hardly any is left for those on the bottom. The 2.4 billion people of the 50 poorest nations average just $794 per capita per year—hardly more than $2 a day. The 8 million people of Burundi have the lowest average: just $150 per capita per year. That's less than 50 cents a day for the average Burundian. The situation is hardly better for the people of the Democratic Republic of the Congo: just $160. In Liberia, it's also $160. True, the cost of living is unusually low in those countries. That $150 annual income in Burundi buys about what $390 buys in the United States. But $390 is still not very much. Imagine living on so little.

Moreover, despite the many advances in technology and the change to a more market-oriented world economy—and some say because of these advances and this change, as Chapter 3 discusses—income inequality has dramatically increased in recent decades. In 1960, the fifth of the world's people living in its richest countries commanded 30 times as much of the world's income as the fifth of people living in the poorest countries—a figure that, in most people's view, was bad enough.[149] Roughly 100 years earlier, in 1879, it was 7 to 1.[150] But today, that richest fifth commands at least 66 times as much of the world's income as the poorest fifth, and probably more.[151]

These figures are all based on averages for the populations of whole countries. But there are also substantial levels of inequality *within* countries. Typically, the income differential between the richest 20 percent and poorest 20 percent within a country is 7 to 1 or less.[152] In many poor and middle-income countries, however, the numbers are far higher. The situation is most extreme in Sierra Leone, where the richest fifth command 57.6 times the income of the poorest fifth. In another half-dozen countries, such as Brazil and South Africa, the ratio is 30 to 1 or higher. In 13 other countries, it is 15 to 1 or higher.[153]

Although there is usually less inequality in wealthy countries, some do exceed the world norm of 7 to 1. In Germany, the ratio of richest to poorest 20 percent is 8 to 1. In the United States, it is 9 to 1. In fact, the United States has the most unequal income distribution of all 26 OECD (Organisation for Economic Co-operation and Development) nations, once tax policies are taken into account.[154] Interestingly, the situation in the United States represents a historical reversal. In the 1920s (the first decade for which these figures are available), the United States was one of the most economically egalitarian countries, giving America the image of the land of opportunity. In comparison, most European countries, such as Britain, were more wealth stratified at the time.[155] Today, European countries are all less stratified, in most cases much less so—such as the 4-to-1 figures for the Scandinavian countries and the 5-to-1 and 6-to-1 figures for France, Belgium, Switzerland, Spain, and the Netherlands. The lowest figure in the world is for Japan, 3.4 to 1.[156] Most countries with a Muslim majority also have quite egalitarian income ratios.[157]

Inequality within countries means that the 66-to-1 ratio of income between the fifth of people living in the richest countries and the fifth living in the poorest understates the level of global inequality. If the richest fifth of the world population from all countries, rich and poor, were put together, their income would likely total 150 times that of the poorest fifth of the world's population (see Figure 1.10).[158]

Consequently, taking the world's population as a whole, the number of very poor people is staggering. The World Bank defines "extreme poverty" as living on $1.25 a day or less in terms of local purchasing power. As of 2005, some 1.4 billion people live in this deplorable condition, and the number has almost certainly increased since then as a result of the Great Recession.[159] The good news is that there have been substantial improvements. In 1990, a total of 1.8 billion were living on less than $1.25 a day.[160] Moreover, world population has risen

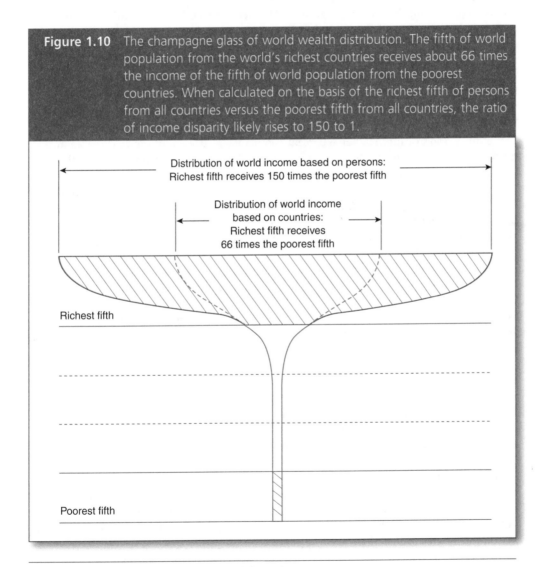

Figure 1.10 The champagne glass of world wealth distribution. The fifth of world population from the world's richest countries receives about 66 times the income of the fifth of world population from the poorest countries. When calculated on the basis of the richest fifth of persons from all countries versus the poorest fifth from all countries, the ratio of income disparity likely rises to 150 to 1.

Distribution of world income based on persons:
Richest fifth receives 150 times the poorest fifth

Distribution of world income based on countries:
Richest fifth receives 66 times the poorest fifth

Richest fifth

Poorest fifth

Source: Based on Korten (1995) and World Bank (2007a).

quite a bit since then, so the proportion of the poor has actually fallen even further. Thus, there have been some encouraging changes. But there are also significant regional differences. Due to rapid population growth, the numbers of the very poor have been on the rise in India and Africa over this time period even as the proportion of the very poor has declined. Almost all of the numerical gains in poverty alleviation have been in just one country, albeit a very large one: China.[161] And most of those who used to be on the very bottom haven't moved up very far. In fact, the overall number of those living on $2 a day or less has actually gone up a bit since 1981—by about 100 million.[162]

Income isn't the same as wealth, though. One's command of riches can come in many forms: savings accounts, land, buildings, possessions, investments, and more. The discrepancy in distribution of environmental goods gets even more extreme when we calculate it by wealth instead of income because the wealth of the poor is usually pretty much only in the form of income, as their assets are so minimal. For example, in the United States, the firebrand filmmaker Michael Moore made headlines by declaring that "just 400 Americans—400—have more wealth than half of all Americans combined." (He made this statement in a speech in my city, Madison, Wisconsin, on March 5, 2011.) Closer inspection using figures for 2010 shows this is actually an understatement. The 400 richest have a combined wealth of $1.37 trillion. The combined wealth of the poorest 60 percent of American households—more than half, and totaling roughly 100 million households in all—is $1.26 trillion, or 2.3 percent of the United States' total net worth.[163]

Global figures show much the same pattern. A 2008 study by United Nations University, based on 2000 data, found that the top 1 percent of the world's population commands 40 percent of the world's wealth. The top 5 percent commands 71 percent of the world's wealth, and the top 10 percent commands 85 percent. This study didn't provide a ratio of the wealth of the top and bottom 20 percent, as I calculated above for income. However, the ratios the study's authors did provide show how much more extreme the difference in wealth is versus the difference in income. The figures show that the top 10 percent is 400 times wealthier than the entire bottom 50 percent (not just the bottom 10 percent or 20 percent), and the top 1 percent is almost 2,000 times richer than the entire bottom 50 percent. The entire bottom 50 percent owns just 1.1 percent of the world's wealth.[164]

The wealth of the world's richest people is staggering. As of 2010, the world had 1,011 billionaires worth a combined $3.6 trillion dollars, or an average of about $3.5 billion each.[165] Now consider the wealth of the 2.4 billion people living in the world's 50 poorest countries. As their assets are so minimal, we can, roughly speaking, consider their annual income to be the same as their wealth. Pretty much all they've got is what they make. Their annual income together amounts to $1.9 trillion, barely half the wealth of the 1,011 richest people.[166] Now let's assume that the poor do usually have a few assets—some clothes, tools, housing of some sort, perhaps a radio and some other saleable possessions—and estimate the value of those as equal to their income. That brings their wealth up to $3.8 trillion, almost the same as the wealth of the 1,011 richest individuals. Think of it: 1,011 people as wealthy as 2.4 billion people put together.

That's a rough calculation, of course. So let's narrow it down to one person, Carlos Slim Helú, the richest person in the world as of 2010 with $53.5 billion, having slightly passed Bill Gates's $53.0 billion. And let's make the comparison straight on annual income this time. Slim's assets increased by $18.5 billion during 2009.[167] The individual national incomes of the 72 smallest economies were less than that in 2009.[168] In other words, Carlos Slim Helú is a mid-sized national economy all on his own. Taken together, his income that year was roughly the same as that of the 27 smallest national economies put together. Let me put it more plainly: He made more money in 2009 than 27 entire countries did.

The wealth of the average person in the rich countries leads to a substantial global consumption gap. The average person in the rich countries consumes 3 times as much grain, fish, and fresh water; 6 times as much meat; 10 times as much energy and timber; 13 times as much iron and steel; and 14 times as much paper as the average resident of a poor country. And that average person from a rich country uses 18 times as much in chemicals along the way.[169] These consumption figures are lower than the 66-to-1 income differential because the comparison here is between the roughly 20 percent of the world's people who live in industrial countries and the roughly 80 percent who don't—not the richest fifth and poorest fifth of countries. If the 60 percent in the middle were removed from the calculations, the consumption gap for many of these items would probably reach or exceed the 66-to-1 ratio of income. (For some items, however, it would not—even a very wealthy person can eat only so much grain, fish, and meat.)[170]

Along with the consumption gap comes an equally significant pollution gap. The wealthy of the world create far more pollution per capita than do the poor. For example, in the rich countries, per capita emissions of carbon dioxide are 12 times higher than in poor countries.[171] Moreover, the rich countries are also more able to arrange their circumstances such that effects of the pollution they cause are not as significantly felt locally, as with the export of toxic wastes and dirty forms of manufacturing noted earlier.

The consequences of these differentials are serious indeed. The Global Information and Early Warning System of the Food and Agriculture Organization regularly reports that 30 or more countries at any one time are in need of external food assistance.[172] As of 2010, some 925 million people in the world are undernourished, a figure that has generally remained flat in percentage terms since the mid-1990s at around 13 to 14 percent, but took a sharp spike upwards as a result of the Great Recession.[173] Twenty-five percent of the world's children under age 5 are malnourished.[174] In the 39 poorest countries, the figure is 28 percent. India has the highest figure: 44 percent.[175] Hunger and malnutrition annually cause the death of almost 6 million children before they reach the age of 5.[176] Because of rampant malnourishment, adults face a reduced capacity to work and children grow up smaller, have trouble learning, and experience lifelong damage to their mental capacities.[177]

Many of the world's poor find it difficult to protect themselves from environmental "bads." As of 2006, an estimated 825 million people live in slums, generally in shelter that does not adequately protect them from such environmental hazards as rain, snow, heat, cold, filth, and rats and other disease-carrying pests.[178] And the number is rising fast; by 2030, it could be 2 billion, the United Nations Human Settlements Agency projects.[179] Moreover, the world's poor are more likely to live on steep slopes prone to landslides and in low-lying areas prone to floods. Over 700 million lack access to safe drinking water.[180] Some 18 percent of the world's population, about 1.5 billion people, do not have any form of sanitation—no toilets or even latrines.[181] The poor also typically find themselves relegated to the least productive farmland, undermining their capacity to provide themselves with sufficient food (as well as income). Compounding the situation are the common associations between poor communities and increased levels of pollution and between poverty and environmentally hazardous working conditions.

It is also possible to have too much of the good things in life. In the United Kingdom, 66 percent of men and 57 percent of women are now either overweight or obese.[182] From 1993 to 2008, the prevalence of obesity in the United Kingdom shot up from 16 to 24 percent of women and 13 to 25 percent of men.[183] The situation in the United States is even worse, with some 73.7 percent of all adults being overweight, obese, or extremely obese in 2008.[184] Adult obesity in the United States has more than tripled since 1962, to 34.3 percent, and for children aged 6 to 11 it has gone up by almost a factor of 5.[185] Other wealthy countries have also experienced rapid rises as lifestyles have become more sedentary and calorie intake has increased. The diseases associated with too much food are increasing as well: diabetes (especially type II), hypertension, heart disease, stroke, and many forms of cancer.

But the problem of being overweight is not limited to the wealthy nations. Weight problems are rising dramatically in poorer nations, as people increasingly take up more sedentary lives there, too, and as food consumption shifts more into the marketplace and away from home production, making healthier foods less readily available for the poor. The World Health Organization (WHO) estimates that, worldwide, 1.6 billion adults were overweight and 400 million were obese, as of 2005.[186] More than 30 percent of adults in Egypt and Kuwait are obese.[187] The problem is particularly pronounced in urban areas. In China, the obesity rate as of 2002 was 7 percent, double what it was in 1992. But the rate was double in the cities in comparison with the countryside.[188] In some cities in China, 20 percent or more are obese.[189] In urban Samoa, as many as 75 percent of adults are obese—not just overweight, but obese.[190] With excess weight comes its many deleterious effects on health. Yet the world's wealthy are generally better able to protect themselves from the consequences of high weight. Medical treatments for diabetes, circulation problems, and cancer are far less accessible for the poor.

Considering these stark facts, it comes as no surprise that people in the wealthy countries live an average of over two decades longer than those in the poor countries—80.3 years versus 57.7—despite great advances in the availability of medical care.[191] In 14 very poor countries, the average person has no better than a 50 percent chance of reaching age 50.[192] In half a dozen countries, 20 percent or more won't even make it to age 5.[193] The good news is that, in recent years, the life expectancy gap between rich and poor has closed a good bit. But it remains wide and stark.

Within-country differences in income have a substantial impact on the quality of life for the poor, even in rich countries. In the United States, some 3.5 million Americans experience a period of homelessness during the year, about one-third of them children, according to a 2009 estimate based on several studies.[194] A 2007 study estimated that 7.7 percent of the British, 4 percent of Italians, and 3.4 of Belgians experience homelessness at some point in their lives.[195] Typically, some 5,000 in France, 20,000 in Germany, and 8,000 in Spain are "sleeping rough," with no roof at all.[196]

Hunger can also exist in conditions of prosperity. Take the United States, for example. Some 14.7 percent of American households experienced *food insecurity* in 2009, the highest figure the U.S. Department of Agriculture (USDA) had recorded since it began tracking this statistic in 1995. As a result, people are forced to reduce the "quality, variety, or desirability" of their diet, without necessarily experiencing hunger, according to the USDA definition of food insecurity. But some 5.7 percent of American households experienced "very low food insecurity" during the year, meaning they experienced hunger—what the USDA defines as "multiple indications of disrupted eating patterns and reduced food intake." Three-fourths of those 5.7 percent faced hunger in 3 or more months of the year. As of September 2010, a total of 43 million people were receiving food stamps, or 14 percent of the U.S. population. During the 2009–2010 school year, the USDA provided 2.9 billion free lunches to U.S. schoolchildren from poor families, and another half billion reduced-price lunches.[197]

Food, shelter, longevity—these are the most basic of benefits we can expect from our environment. Yet people's capabilities to attain them are highly unequal. As Tom Anthanasiou has observed, ours is a "divided planet."[198]

Environmental Justice for All

But you don't have to be poor to experience environmental injustice. (Some argue that you don't have to be human either—that all living things can experience it.) Many environmental hazards cross social boundaries as they cross bodily ones.

Take the nine people, including journalist Bill Moyers, who in 2003 volunteered to let Mount Sinai Hospital researchers search their bodies for traces of industrial chemicals and pollutants—chemicals that their own bodies did not make. None of the nine had jobs that exposed them to hazardous chemicals in their workplace, and none of them lived near industrial facilities; these were middle-class and upper-class folks. Yet when researchers took blood and urine samples, they found in the volunteers' bodies an average of 91 different chemical pollutants. Among these chemicals, the volunteers averaged 53 that cause cancer, 62 neurotoxins, 53 immune system disrupters, 55 that cause birth defects or disrupt the body's normal development, and 34 that damage hearing. (Many of these chemicals have more than one effect.) Of course, these chemicals were present in only trace amounts, and the researchers used sophisticated equipment to detect them. But they were there. And although this was a comprehensive assessment of individual *body burden,* as toxicologists call it, there were many kinds of common chemical pollutants that the researchers were not able to study. Indeed, some 80,000 chemicals circulate in products on the market in the United States today, and only a few hundred of them have been screened for their safety.[199] So it is likely that 91 was a low estimate of the number of trace pollutants.[200]

Can trace amounts sometimes amount to something? Many observers now think unfortunately yes. Increasingly, the leading medical journals are filling up with studies that link environmental chemicals with a host of diseases. Not all the studies show this link. But more and more do. For decades, cancer researchers had estimated that environmental factors account for 2 to 4 percent of all cancers, and have attributed most cancer to inheritance and pathogens, matters which are largely unavoidable and therefore apolitical.[201] Then in 2010, the President's Cancer Panel—appointed earlier by President George W. Bush—declared that those low estimates are "woefully out of date" and that "the true burden of environmentally induced cancer has been grossly underestimated."[202] These were controversial statements, and many voices rushed to rebut them, including the American Cancer Society. However, many voices, such as the Science and Environmental Health Network, also rushed to support them.[203]

No matter how wealthy you are, you can't run far enough, or build a gated community secure enough, to escape the body burden of industrialism. True, the wealthy are better able to avoid these effects through buying organic food and working cleaner jobs. And they are better able to deal with the consequences through better health care. There is definitely considerable inequality in the impact of industrialism's dirty side. But it wouldn't make it just if there were some way to divide the impact equally. Even if everyone suffers from something that is preventable, it is still preventable suffering. Environmental justice is an issue for us all.

The Beauty of Ecology

"A thing is right when it tends to preserve the integrity, stability, and beauty of the biotic community. It is wrong when it tends otherwise."[204] These are probably the most famous lines ever written by Aldo Leopold, one of the most important figures in the history of the environmental movement. Leopold's words direct our attention to a sense of community—to a sense of *ecos,* of home, of the habitat we share with so many others. Understood in this way, sustainability and environmental justice concern not only the conditions of human life but also the conditions of the lives of nonhumans. I think that's what Leopold was getting at with the words "integrity" and "stability." Integrity sounds to me like justice, and stability sounds like sustainability. And to talk about community is to talk about the interdependence of justice and sustainability for all.

Leopold also directs our attention to a word that is certainly one of the hardest of all to define but is no less significant for that difficulty: beauty. (Indeed, the difficulty of describing this area of environmental concern has led me to change the term I use for it from earlier editions of this book.[205]) For many, and myself included, an essential part of beauty is the sustainability and justice of what we behold. To speak about the *beauty of ecology*, then, is to speak about every living thing's right to a home, a habitat, that is sustainably beautiful and beautifully sustainable.

Threats to the integrity, stability, and beauty of ecology are manifold. Take the loss of species. For example, of the 10,027 known species of birds, some 12 percent were threatened with extinction as of 2010.[206] Many have already gone; the passenger pigeon, the dodo, the ivory-billed woodpecker, and the 11 species of moa are only some of the best known. Since 1800, a total of 103 have gone extinct.[207] New Zealand has perhaps been the hardest hit. Before people arrived around 1300 CE, New Zealand's birds had no mammalian predators, and thus no evolutionary pressure to adapt to them. Since then, half the bird species of the North and South Islands have disappeared, including all 11 species of moa—among them the wondrous *Dinornis robustus* and *Dinornis novaezelandiae* which grew to 500 pounds and 12 feet tall.[208]

Estimates of extinction rates for all species vary widely because we still do not have a good count of how many there are, or ever were. Many species are still unknown or survive in such low numbers that they are hard to study. But even the low estimates are staggering. Perhaps the most widely regarded account, based entirely on individual assessments for each species, is the "Red List" of the World Conservation Union, known by the acronym IUCN, a 140-nation organization (see Figure 1.11). As of 2010, the Red List registered 18,351 species as threatened with extinction. In addition to the 12 percent of bird species, extinction is now a real and present possibility for 20 percent of mammal species, 4 percent of fish, 5 percent of reptiles, and 29 percent of amphibians, according to the 2010 Red List.[209] But while the IUCN has reviewed the status of all bird species and about 53 percent of all vertebrate species, very little is yet known about the status of invertebrates and plants. All told, as of 2010, the IUCN had evaluated the status of only about 3 percent of known species.[210] Most have been barely studied.

The overall extinction rate is thus in the realm of educated guesswork, given the spotty data we have. Richard Leakey, the famous paleontologist, is one who has made a try. He suggests that we could lose as many as 50 percent of all species on Earth in the next 100 years, largely because of very high rates of extinction among invertebrates, the group we know the least about. (For example, nearly half of the insect species that the IUCN has assessed are threatened. However, evaluations often focus on species at risk.) If Leakey is anywhere near right, that would put the current period of extinction on the same scale as the one that did in most dinosaurs and much of everything else 65 million years ago, and four earlier periods that had a similar effect on the Earth's living things. That's why Leakey and Roger Lewin call the current period the "sixth extinction."[211] When we add in the extinction of subspecies and sub-varieties, the decreasing diversity of planetary life is even more dramatic.

Of course, species have always come and gone, as Charles Darwin famously observed in his theory of natural selection. But the rate of these losses has greatly increased since the beginning of the Industrial Revolution. Some have disappeared because of habitat loss, as forestlands have been cleared, grasslands plowed, and wetlands drained and filled. Some have suffered from pollution of their habitat. Some have found themselves with no defenses against animals, plants, and diseases that humans have brought, often unintentionally, from other regions of the world into their habitat. The Earth is a single, gigantic preserve for life, and we have not been honoring its boundaries and protecting its inhabitants.

The loss of species is an instrumental issue of sustainability. The leaking global gene pool means a declining genetic resource base for the development of new crops, drugs, and chemicals.

Figure 1.11 A leaking gene pool: The IUCN "Red List" of threatened species, an annually updated inventory.

	Number of described species	Number of species evaluated by 2010	Number of threatened species in 2010	Number threatened in 2010, as % of species described	Number threatened in 2010, as % of species evaluated
Vertebrates					
Mammals	5,491	4,863	1,094	20%	22%
Birds	10,027	9,956	1,217	12%	12%
Reptiles	9,205	1,385	422	5%	30%
Amphibians	6,638	5,915	1,808	29%	31%
Fishes	31,800	3,119	1,201	4%	39%
Subtotal	63,161	25,238	5,742	10%	23%
Invertebrates					
Insects	1,000,000	1,255	623	0.07%	50%
Molluscs	85,000	2,212	978	1.21%	44%
Crustaceans	47,000	553	460	1.15%	83%
Corals	2,175	13	5	0.23%	38%
Arachnids	102,248	33	19	0.02%	58%
Velvet Worms	165	11	9	5%	82%
Horseshoe Crabs	4	4	0	0.0%	0%
Others	130,200	52	24	0.03%	46%
Subtotal	1,305,250	9,526	2,904	0.0%	30%
Plants					
Mosses	16,236	101	80	0.0%	79%
Ferns and Allies	12,000	243	148	1%	61%
Gymnosperms	1,052	926	371	35%	40%
Flowering Plants	268,000	11,584	8,116	3%	70%
Green Algae	4,242	2	0	0.0%	0%
Red Algae	6,144	58	9	0.1%	16%
Subtotal	307,674	12,914	8,724	3%	68%
Fungi & Protists					
Lichens	17,000	2	2	0.01%	100%
Mushrooms	31,496	1	1	0.003%	100%
Brown Algae	3,127	15	6	0.2%	40%
Subtotal	51,623	18	9	0.0%	50%
TOTAL	1,727,708	55,926	18,351	1%	33%

Source: IUCN (2010).

In addition, most ecologists suspect that decreased diversity destabilizes ecosystems—ecosystems that we, too, need to survive. But the ethical and aesthetic impact of the loss of so many forms of life may be as great, if not greater.

The loss is not only one of forms of life but also of forms of landscape. Take deforestation. The world has lost about half of its original area of forestland.[212] Between 2000 and 2010, the loss continued as some 13 million hectares of forested landscapes were converted to other uses every year—an area about the size of Costa Rica.[213] Some areas have reforested, though, mainly through replanting, such that the net loss of forested land was about 5 million hectares a year during 2000 to 2010. These numbers represent considerable improvement over the previous decade, when forestland was being converted to other types of land use at a rate of 18 million hectares a year and reforestation was running at about 10 million hectares a year, for a net loss of 8 million hectares a year. So we've seen almost a 40 percent improvement, which is certainly encouraging. Nonetheless, Africa lost 4.9 percent of its forestland from 2000 to 2010, as did South America, including Brazil and the Amazon. Only about one third of the world's remaining forestlands are what ecologists call primary forests, little disturbed by human use. Alarmingly, 80 percent the continuing net loss—some 4 million hectares annually—is of these primary forests, with their richness of species and habitat.[214] Replanted forests are poor substitutes for the woods they replace, at least in terms of biodiversity—the ecological equivalent of exchanging the paintings in the Louvre for a permanent display of engineering blueprints.

There's another loss, too—the disappearance of a kind of quiet intimacy with the Earth, the sense of being connected to the land and to each other through land. It is a common complaint that modern technology removes us from contact with a greater, wilder, and somehow realer reality. This removal, it should be said, has been the whole point of modern technology, but some have come to wonder whether our lives are emptier because of it. A romantic concern, perhaps. But do we want a world without romance?

Moreover, the loss of quiet intimacy is not merely a philosophical matter. It is physical, too. We in the industrialized world are seldom away from the sound of machines, and we generally interact with the world by means of machines. Got something to do? Get a machine. Try to escape from the constant sound of machines? Good luck. Saturday morning in the suburbs, and the lawn mowers and leaf blowers are at it. Late into the summer night, the air-conditioners hum and the highways growl. Out in the countryside, the situation is often no better: tractors, snowmobiles, Jet Skis, motorcycles, passing airplanes, chain saws, all manner of power tools, and the nearly inescapable sound of the highway except in the remotest locations. Back at the office, the lights buzz, the computer whines, the air-handling equipment rushes with a constant Darth Vader exhale, and the traffic—always the traffic—invades the sanctum of the ear with an ever-present tinnitus of technology. And we hardly seem to notice. We have lost our hearing, our hearing for habitat.

Finally, there's the question of our right to make such great transformations in the world. Nothing lasts forever, of course. Over millions of years, even a mountain is worn away by erosion. Wind, rain, ice, and changes in temperature constantly sculpt the land, and the shape of Earth's surface constantly changes as a result. But geologists now recognize humans to be the most significant erosive force on the planet.[215] Agriculture, forest cutting, road building, mining, construction, landscaping, and the weathering effects of acid rain—all these have resulted in enormous increases in the amount of sediment that rivers carry into the oceans. We wield the biggest sculptor's chisel now. Perhaps it is our right. If so, then it is also our responsibility.

Although I have not covered the question of the beauty of ecology in much detail, let me conclude here. After all these pages on our environmental problems, I'm exhausted. Maybe that's the most difficult environmental problem of all: There are so many of them.

Figure 1.12 The beauty of ecology: Sunrise over Grenadier Island, St. Lawrence River, 2007.

Source: Author.

The Social Constitution of Environmental Problems

These matters seemed quite remote at that lovely brunch as we loaded up our plates with fruit salad, coffee cake, scrambled eggs, bread, butter, and those great hominy grits—remote but ironic. Here I was amid a group of three families whose incomes, although not unusually high by Western standards, were sufficient to command a brunch that two centuries ago would have been seen as lavish even by royalty. And what was I doing? Once I had finished explaining environmental sociology, I was reaching for seconds.

My point is not that there is something wrong with pigging out every once in a while. Nor is it that consumption is necessarily a bad thing. (To consume is to live.) Rather, I tell this story to highlight how social circumstances can lead to the sidelining of concern and action about environmental consequences.

Everything we do has environmental implications, as responsible citizens recognize today. Although at the time sociology's relevance for the environment was less widely recognized, the adults at that brunch a decade ago were genuinely interested in my explanation of environmental sociology. They listened avidly for the 2 minutes the dynamics of politeness grant in such a setting. But the currents of social life quickly washed over the momentary island of recognition and concern that my explanation had created. Soon we were all reaching for more of the eggs from chicken-factory hens, the fruit salad with its bananas raised on deforested land and picked by laborers poorly protected from pesticides, the coffee cake made with butter and milk from a dairy herd likely hundreds of energy-consuming miles away, and the grits made from corn grown at the price of one bushel of soil erosion for one bushel of grain. Meanwhile, the conversation moved on to other matters. And shortly, we guests got into our cars and drove home, spewing greenhouse gases, smog, and fine particulates all along the way.

A completely ordinary brunch. But do you refuse to invite friends over because you cannot easily get environmentally friendly ingredients for the dishes you know how to prepare? Do you refuse a host's food because it was produced in a damaging and unjust manner? And do you refuse an invitation to brunch because the buses don't run very regularly on a Sunday morning, because a 20-minute ride in the bike trailer in below-freezing weather seems too harsh and long for your 5-year-old, and because no one nearby enough to carpool with is coming to the party? Do you refuse, especially when you have your own car sitting in the driveway, as is almost certainly the case in the United States?

Likely not.

What leads to this sidelining of environmental concern and action is the same thing that manufactures environmental problems to begin with: the *social constitution of daily life*—how we as a human community institute the many structures and motivations that pattern our days, making some actions convenient and immediately sensible and other actions not.[216] Caught in the flow of society, we carry on and carry on and carry on, perhaps pausing when we can to get a view of where we're eventually headed, but in the main just trying to keep afloat, to be sociable, and to get to where we want to go on time. Our lives are guided by the possibilities our social situation presents to us and by our vision of what those possibilities are—that vision itself being guided in particular directions by our social situation. That is to say, it is a matter of the social organization of our material conditions, the ideas we bring to bear upon them, and the practices we therefore enact. Yet the environmental implications of those conditions, ideas, and practices are seldom a prominent part of how we socially constitute our situation. Instead, that constitution typically depends most on a more immediate presence in our lives: other people.

We need, I believe, to consider the environment as an equally immediate (and more social) presence in our lives. That does not mean we need to be always thinking about the environmental consequences of what we do. As an environmental sociologist, I cannot expect this— especially when I don't always think about environmental consequences myself. People have lots and lots of other concerns. Nor should it be necessary to think constantly about environmental consequences. Rather, what is necessary is to think carefully about how we as a community constitute the circumstances in which people make environmentally significant decisions. What is necessary is to create social situations in which people take the environmentally appropriate action, even when, as will often be the case, they are not at that moment consciously considering the environmental consequences of those actions. What is necessary is to reconstitute our situations so what we daily find ourselves doing compromises neither our social nor our environmental lives.

The challenge of environmental sociology is to illuminate the issues such a reconstitution must consider.

PART I

The Material

CHAPTER 2

Consumption and Materialism

I shop, therefore I am.

—Conceptual artist Barbara Kruger, 1987

Michael Joseph Jackson will be remembered for many things, but not for austerity. What does one do with the proceeds of 750 million record sales?[1] Here's what the King of Pop's view was: Go on $250,000 shopping sprees.[2] Spend about a million dollars a month on your living expenses.[3] Buy yourself a 2,676-acre ranch for $30 million. Outfit your ranch with a zoo complete with tigers, giraffes, and orangutans; an amusement park with 16 rides including a Ferris wheel, bumper cars, and an ornate carousel; and not one but two railroads plus a vast fantasy railway station.[4] And fit out your 13,000-square-foot, 25-room home with a dragon's lair of stuff—so much stuff that it takes a 1,058-page catalog in five volumes (available for $1,000 as a boxed set) to describe only a portion of it.[5] The astounding excess of Jackson's home life had long been rumored. But after years of financial mismanagement and an insatiable appetite for antiques, clothes, statues, paintings, toys, pinball and video games, and—apparently, because he had 10,000 of them—books, Jackson overspent even his lavish income. So he had to quasi-sell the ranch to a company he partly owned and put the 1,058 pages of items up for public auction (thus the catalog).[6] He was able to cancel the auction at the last minute, for he was shortly to go on a new 50-date concert tour, with a good chance of returning to solvency and his lavish ways. But on June 25, 2009, he died, to the lasting sorrow of generations of fans.

We are not all Michael Jacksons. Nor would we likely be, even if we all had the income from 750 million record sales. Few of us are so unabashedly materialistic. But it is certainly the case that, aside from the poor, nearly everyone today consumes more than is necessary to survive, and some of us a lot more. Yet we are not satisfied.

Why? What do we want it all for? This is an important sociological question, for it relates to all three of the central issues of environmentalism: sustainability, environmental justice, and the beauty of ecology. Even if few of us would want Michael Jackson's Neverland life, can society and the Earth sustain the appetites we *do* have without impoverishing others and without damaging our collective habitat?

One common answer to the question of why we consume so much more than we need is that people are greedy. In the words of Mahatma Gandhi, "The world has enough for everybody's need, but not enough for everybody's greed."[7] There is much wisdom in Gandhi's aphorism. Better sharing of the world's resources would go a long way toward resolving the three central issues of environmentalism, and this is a theme that subsequent chapters of this book continually return to.

But Gandhi was also implying something else, both in this statement and in the way he lived his remarkable life: It is possible to change our feelings of greed if we make an effort to do so. "Greed" is socially highly variable. There are those who make millions a year and give quite a bit of it away. There are also people who truly have no interest in wealth. The variability of greed indicates that it is not an immutable natural fact. How we arrange our social lives can markedly influence the character of the "greed" we feel within ourselves. The kinds of goods that our greed desires vary markedly across time and culture. It is also important to see that the motivations behind the pursuit of material pleasures are complex.

This chapter begins the exploration of the material dimensions of environmental sociology by considering the *social psychology of consumption* and the complex and variable pleasures that underlie it.

The Material Basis of the Human Condition

We have bodies. We need to eat, we need shelter, and we generally need some kind of clothing. Certain inputs and outputs are essential to all living bodies, which means that no body can exist without interacting with its environment. As Karl Marx, the still-controversial nineteenth-century philosopher, observed, "The worker can create nothing without *nature*, without the *sensuous external world*. It is the material on which his labor is manifested, in which it is active, from which and by means of which it produces."[8] But can there be any controversy here? Who could long deny—and yet still live—that we must produce our livelihood from what Marx called the "sensuous external world," or what is commonly termed today the *environment*? (By *sensuous*, Marx meant what we experience through our senses.)

Marx also observed that "the mode of production of material life conditions the social, political and intellectual life process in general."[9] In other words, there are many ways that societies can arrange material production from the environment, and these arrangements have great consequences for how we live, even how we think. Our ecology is our economy, and our economy is our society.

You don't have to be a Marxist to appreciate the centrality of material forces in social life. Marx's basic idea is probably as widely accepted as any idea could be. Bill Clinton—a liberal, perhaps, but certainly no Marxist—put it well with the famous motto of his 1992 U.S. presidential campaign: "It's the economy, stupid." Or, as we might put the point more generally, "It's the material, stupid."

Ecological Dialogue

But we must be wary of the simplistic clarity of a purely materialist perspective. *Material* factors always depend upon *ideal* factors. The converse is equally true. Our ideals are shaped by the material conditions of our lives, and our material conditions are shaped by our ideals. You can only do what you can do. But what you can do is as much a matter of what you know, believe, and value—all ideal factors—as it is a matter of what your material circumstances are. Moreover, what your material circumstances are depends in large measure on what you know, believe, and value. If you don't know about germs, you are far more likely to do things that leave you vulnerable to their effects and to have different cultural values of cleanliness. And what you know, believe, and value depends on your material circumstances. If you live in the Arctic, you are likely to know quite a bit about ice and snow and probably will value it as more than a source of occasional recreation. It's a dialogue—a constant interplay of factors that condition and influence each other, a never-ending conversation between the material and ideal dimensions of social life.[10]

The concept of dialogue provides an alternative to the mechanical, hammer-and-nails notion of causality that the social sciences for many years attempted to borrow from that ultimate materialist science, physics. In social life, causality is rarely, if ever, a one-way street, and the material is rarely, if ever, all that is involved. In fact, mechanical materialism probably isn't even good physics, as physicists themselves now argue, due to the complexity of the universe.[11] Contemporary ecologists make related arguments about the way mechanical thinking gives us a dangerously reductionist image of the world as a series of parts, ticking one into the other. Rather than a mechanical realm of linear causes and effects, the more ecological view is that all life—not just social life—is an interactive phenomenon in which causes cause effects and effects effect causes, blurring the boundary between them. When we call something in social life a "cause" or an "effect," we are intellectually, and artificially, arresting this constant interplay for a moment. A material cause or effect and an ideal cause or effect are mere intellectual "moments" in the endless ecological dialogue (see Figure 1.1 in Chapter 1).

An interactive understanding of causality is also a more open one. Materialist analyses like conventional science's mechanical reductionism, as well as some of the materialist social theories this section of the book discusses, tend to be deterministic. In this view, the world is what it is because of what it is. What happened is what could happen. It's all just the product of the ticking of the materialist clock. But a dialogic view imagines that, as Alfred Koestler put it, "neither parts nor wholes exist in an absolute sense."[12] There are no little gears that we can pop out of the clock to inspect, each on its own, nor a clock that we can stand back and admire as a whole. Parts are part of the other parts; wholes are part of the other wholes. Each is the other. And there is much jumble and tumble everywhere, confusion and conflict. The world isn't all worked out and neatly put together. Life isn't determined; rather it is constantly de-terminated, un-ended, made anew from the surprising consequences of interactivity. Think of it as like the feeling of wonder that we sometimes come away with after a good conversation—a sense of something new and unanticipated that leaves us changed and invigorated.[13] The dialogue of ecology is like that, at least potentially—full of moments of creativity and unpredictability. This constant capacity for re-ordering and de-ordering, which stems from the interactiveness of that which is always at least partially un-ordered, is what dialogics terms *unfinalizability.*

Environmental sociology does have to enter the dialogue somewhere, though. In order to be able to understand the interactiveness of the world, you have to be able to distinguish that which is interacting. In order to understand connection, overlap, and mutual constitution, you must be able to see *difference*. Otherwise there will be nothing to connect, overlap, and mutually constitute. The same can be said of disconnection, for not everything is, nor perhaps should be, connected. (I am happy to be disconnected from the smallpox virus.) It, too, depends on imagining difference. In this part of the book, we enter the ecological dialogue from the material side of things. In the next part, we will enter from the ideal side. Although we enter the dialogue from these different sides, these different moments, in each chapter of each part we will be inevitably drawn over from the material to the ideal and from the ideal back to the material. But we have to start somewhere.

The Hierarchy of Needs

One of the most famous explanations of the relationship between the material and the ideal is psychologist Abraham Maslow's theory of the *hierarchy of needs*. This hierarchy, Maslow argued in a 1943 paper, is relatively fixed and universal across cultures.[14] At the bottom of the hierarchy are the "basic needs," beginning with a foundation in physiological needs and moving up to needs for safety, belongingness and love, esteem, and self-actualization. Above the basic needs are additional needs for knowledge and understanding and for aesthetic satisfaction (see Figure 2.1).

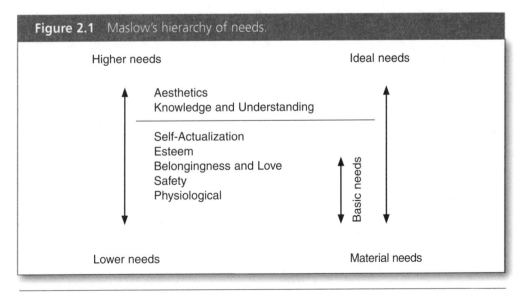

Figure 2.1 Maslow's hierarchy of needs.

Source: Author.

Why this hierarchy? Isn't knowledge and understanding, for example, a pretty basic need? Maslow argues that some needs have to be satisfied before we will direct our efforts toward others and that these are therefore the most basic. When a person's "belly is chronically full," wrote Maslow, "at once other (and higher) needs emerge and these, rather than physiological hungers, dominate the organism. And when these in turn are satisfied, again new (and still higher) needs emerge, and so on."[15]

Maslow made several key observations about the hierarchy of needs:

- Our efforts to gratify the hierarchy of needs are both conscious and unconscious, but more often unconscious.
- Any one action may be directed at several needs simultaneously.
- Most people's needs are never more than partially satisfied, but their lower needs have to be relatively satisfied before they can move on to the higher ones.
- We sometimes undervalue a lower need after it has been met for a while, but we typically can do so only temporarily. Maslow gave the example of a person who quits a job rather than lose self-respect—and yet after 6 months of starving is willing to take the job back, even if it means losing self-respect.
- Those who gain the opportunity to work on the higher needs tend to engage in more socially beneficial behavior. As Maslow put it, "The higher the need the less selfish it must be. Hunger is highly egocentric . . . but the search for love and respect necessarily involves other people."[16]

The hierarchy of needs is an intuitively compelling theory. Think how hard it is to study on an empty stomach. If you are not in good shape, it isn't sensible to try to climb a mountain to enjoy the view. Place a book and a piece of bread in front of a starving person and there is little doubt which the person will reach for. The intuitive appeal of Maslow's theory suggests that it speaks to a widely shared understanding of human motivation, which probably accounts for much of the theory's popularity: It feels right to many of us.

The theory is not without significant problems, though. First, a case can be made that we do not experience our needs as a hierarchy. Rather, we tend to act on whatever need is currently least well met or most under threat and that we are in a position to do something about.[17] Threatened with failure, students may study so hard that they forget to eat. Eager for the pleasure of a good view, people in poor shape sometimes climb mountains anyway. There are starving people—for example, those fasting for a political cause—who will refuse the bread and take the book. Similarly, we may find that the higher needs, like the lower, can be only temporarily ignored and undervalued. That person who took back a demeaning job after 6 months of starving may well quit once again to regain self-respect, even if it means more starving.

Second, the theory lacks a sense of dialogical interplay between the more material and the more ideal needs. If, because of my aesthetic judgment, I do not like the meal before me, I might not eat very much, even if I was initially hungry. I may "lose my appetite," as we sometimes say. Similarly, there have been cases where starving people have refused the food sent by relief agencies because the food was unappetizing to them. Thus my aesthetic, ideal need can lessen the strength of my physiological, material need. Material needs also shape our aesthetic ideals, however. If I am very hungry, I am more likely to decide that unappealing food really does not taste so bad after all. Think of how good everything tastes when you are camping or just in from the cold. In other words, there is a dialogic interaction between my state of hunger and my aesthetic sensibilities concerning what is good food.

Third, Maslow's theory could be seen as condescending to non-Westerners. The higher needs—self-actualization, the pursuit of knowledge, aesthetics—are also the achievements we stereotypically associate with the degree of "civilization" and "development" of a country or a people. Poor people in "developing" countries who must concern themselves more with the lower needs lead a lower form of existence, the theory could imply. The hierarchy of needs is therefore flattering to Westerners, who typically see themselves as being more developed and more civilized than non-Westerners. Part of the appeal of Maslow's theory, then, may be Western hubris.

And fourth, perhaps most important for this chapter, Maslow's theory cannot account for why we consume more material stuff than we need. According to the hierarchy of needs, no one should overconsume. Because their "lower" needs are satisfied, all wealthy and well-fed people should be composing symphonies, writing poetry, and volunteering for Oxfam or Habitat for Humanity. Sometimes, of course, wealthy and well-fed people do these "higher" things. But very often they buy more shoes, more clothes, or another TV—and sometimes a home Ferris wheel.

The Original Affluent Society

In fact, we moderns are the real materialists, perennially concerned with the "lower" needs, argued the anthropologist Marshall Sahlins in his classic 1972 essay, "The Original Affluent Society."[18] We who are so rich are far more preoccupied with material things than any previous society. Thus, in a way, maybe we're not particularly rich, suggests Sahlins, even in comparison to hunter-gatherers, who are usually regarded as the poorest of the poor. In his words, "By common understanding, an affluent society is one in which all the people's material needs are easily satisfied."[19] Sahlins argues that hunter-gatherers eat well, work little, and have lots of leisure time, despite living on far less than modern peoples. And not only are the material needs of hunter-gatherers easily satisfied by their manner of living, their material wants are easily satisfied as well—because they don't want much. Hunter-gatherers are thus the world's original rich, for they are rich in terms of meeting their "lower" needs—and, as I'll come to, their "higher" ones, too.

Time allocation studies conducted by anthropologists with the few remaining (and fast-disappearing) hunter-gatherer societies show that it does not take long to gather and hunt. Typical adult hunter-gatherers work 2 to 5 hours a day. In that time, they secure a diet that compares very well with our own in calories, protein, and other nutrients. The rest of the day, the typical hunter-gatherer has lots of time open for hanging around the campfire chatting and visiting, singing songs, exchanging information, telling stories, making art objects—thus satisfying the "higher" needs for belongingness, love, esteem, self-actualization, knowledge, and aesthetics. Nearly everyone in hunter-gatherer societies is some kind of artist. Anthropologists have long been impressed by the incredible richness of "primitive" sculpture, costumes, music, folktales, and religion and have filled many a museum and library shelf with evidence of these. But we moderns are mostly much too busy to develop our artistic sides, and work away our lives. As Sahlins put it, the biblical "sentence of 'life at hard labor' was passed uniquely upon us."[20]

Another distinctive feature of hunter-gatherers' lives is that they have to keep moving house, usually several times a year, when the local hunting and gathering gets thin. Many early observers, particularly missionaries, took this constant movement as a sign of shiftiness, as well as taking the hunter-gatherers' great amount of leisure time as a sign of laziness. They thought hunter-gatherers should settle down and do something productive, like raise crops. But here's what one hunter-gatherer told a visiting anthropologist: "Why should we plant when there are so many mongomongo nuts in the world?"[21]

Perhaps most confounding to modern observers is the apparent disregard in which hunter-gatherers hold material goods, despite their seeming poverty. Here's how one anthropologist described the attitude toward possessions of the Yahgan Indians of South America:

> Actually, no one clings to his few goods and chattels which, as it is, are often and easily lost, but just as easily replaced. . . . A European is likely to shake his head at the boundless indifference of these people who drag brand-new objects, precious clothing, fresh provisions, and valuable items through thick mud, or abandon them to their swift destruction by children and dogs. . . . Expensive things that are given them are treasured for a few hours out of curiosity; after that they thoughtlessly let everything deteriorate in the mud and wet.[22]

Why don't the Yahgan bother to put things away in clean and dry places? Not because they have no clean and dry places or have no time to construct them. Rather, hunter-gatherers have their own version of the law of diminishing returns, says Sahlins. The returns from hunting and gathering dwindle after a while as nearby game, roots, and berries are gradually harvested. So the group has to up and move, carrying with them what they want to keep. But if you can remake what you need wherever you go, and if the pace of your work life is gentle enough that you have plenty of time to do this remaking, and if you enjoy this remaking as a communal and artistic activity anyway, why trouble yourself to haul it? And why care much about what happens to things after they are used?

Consequently, hunter-gatherers do not bother with an institution like private property. Individuals do not amass goods and commodities, and what goods and commodities there are in the community are equitably distributed and communally held. If you make a particularly nice bow and your neighbor breaks it hunting or your neighbor's child breaks it playing, there's no need to fuss. Now you have an excuse to sit around the fire and make another one, even nicer. We who worry ceaselessly about goods and commodities live by the great economic motto of "Waste not, want not." The great economic motto of the hunter-gatherer is, as Sahlins wrote, "Want not, lack not."[23]

But the hunter-gatherer life is not an Eden that the rest of us have been thrown out of. It has its struggles. Hunter-gatherers do have to move often, and they also have to keep their population low. Numbers of infants beyond what the group can feed and those too old or too sick to make it to the next place must be killed or must go off alone into the bush to die so as not to harm the group. It's their "cost of living well," as Sahlins says. This trade-off may sound grim, but compare it to the many tragedies of our own lifestyle: stress, repetitive and meaningless work, far less leisure, individualistic isolation bred by economic competitiveness, and other tragedies that we each might list.

The hunter-gatherer's cost of living may well still sound grim. But Sahlins's point remains: You do not have to have a lot of money or goods to live well—to lead a life that affords a focus on the needs Maslow termed "higher"—and thus to be rich. As Sahlins says, "The world's most primitive people have few possessions, but they are not poor. . . . Poverty is a social status. As such it is the invention of civilization."[24]

Using Maslow's hierarchy as a measure, we moderns are the ones who may be the poorer, for we must spend so much more of our time in work, securing our physiological and safety needs. Wealth, that most basic measure of material well-being, depends on how you look at it. Being rich isn't having a lot of stuff. It's having everything you want. If you don't want much, you won't want for much.

Consumption, Modern Style

One of the great modern sins is being late. Reputations are sullied, grades sunk, and jobs lost through lateness. Consequently, lateness is something that modern people are perpetually anxious about. Most of us wear watches or carry a cell phone, constantly check them throughout the day, and regularly synchronize our readings with the community time standard that is broadcast on TV and radio and displayed on the clock on the wall and on the computer. And we often inquire of others what their watches or cell phones read. "What time do you have?" we ask, meaning the reading on the person's watch or cell phone. Being moderns, we can easily guess at another meaning—how *much* time the person has: very little.

But if time-keeping devices can be found pretty much wherever we go, why bother to bring one along ourselves? (I usually don't, in fact.) If you are in any social setting where precise timing needs to be adhered to, chances are a timepiece of some sort is there already.

About 30 years ago, I learned why people have personal timepieces nevertheless. (At the time, that meant a watch and not a cell phone.) I was working as a geologist in the Talamanca Mountains of Costa Rica in the middle of a dense rainforest some 3 days' walk from the nearest road. I was mapping the rock formations for an American mining company, a job I later came to regret because of the environmental and social implications of what I was doing—but that's a story for another time. Our supply helicopter had suffered a minor crash that decommissioned it for awhile, and we had to trade for food with the local Bribri Indians. The Bribri in that area, and at that time, maintained a mixed economy of hunting, gathering, and limited agriculture. The missionaries had been through, so local people owned radios and wore modern-style clothes, and a few times a year they mounted trading and shopping expeditions to town. They knew what money was and much of what it does.

One local man who traded with us even sported a watch. I wore a watch too in those days, and he would often ask me what time it was and then check his wrist to compare. After a few days of these exchanges, thinking to make a little small talk, I asked to have a look at his watch, mentioning that it appeared to me to be quite a nice one. I was shocked to discover that in fact it was broken, missing a hand. It clearly had been inoperable for some time.

"How ignorant these people are!" a coworker exclaimed when I told him the story later. But rather than being a sign of his ignorance, this man's broken watch was a sign of his sophistication. He understood perfectly well what a watch was really for: status. He could tell time just fine without one—that is, he was completely competent to temporally coordinate his activities as well as his local community required. Living at a latitude where the day length and path of the sun hardly change throughout the year, he had only to check the sky. Although he had no need for a functioning watch, his sophistication had given him another need. It had given him an awareness of something he never knew before: that he was "poor" and that others might consider him even poorer unless he took some conspicuous measures to give a different impression.[25]

The Leisure Class

A century ago, Thorstein Veblen argued that this form of sophistication is what lies behind modern materialism. In his 1899 book, *The Theory of the Leisure Class,* Veblen wrote that most of modern culture revolves around attempts to signal our comparative degree of social power through what he famously termed *conspicuous consumption,* as well as through *conspicuous leisure* and *conspicuous waste.*[26] It is not enough merely to be socially powerful. We have to display it. Power in itself is not easy to see. We consume, we engage in leisure, and we waste in conspicuous ways to demonstrate to others our comparative power.

In part, this conspicuous display is about showing off our wealth. But there is an important environmental connection here. Veblen argued that conspicuous consumption, leisure, and waste are convincing statements of power because they show that someone is above being constrained by the brute necessities of material life and the environment. Because of your wealth and position, you do not have to engage in productive activities yourself. You can command the environment through your command of other people, a command made possible by wealth and social position. What we might term *environmental power* thus demonstrates social power, and vice versa.

Let me give a few examples of each form of environmental power Veblen identified.[27] By *conspicuous consumption,* Veblen had in mind visible displays of wealth, such as expensive homes, cars, clothes, computers, boats, and the like, as well as sheer volume of consumption. The material visibility of these displays shows one's social ability to command a steady flow of material goods from the environment. Such display is, I think, well-known to all of us.

Conspicuous leisure is often more subtle. By this term, Veblen meant the nonproductive consumption of time, an indication of distance from environmental needs—from productive needs—and thus a sign of power. The most obvious example is a long vacation to a faraway and expensive place, 2 weeks at a Club Med hotel, say. But Veblen also had in mind social refinements, like good table manners, which require sufficient time free from productive activities to master. Maintaining a pristinely clean home is a similar demonstration of time free from productive necessities. Wearing the clothing of leisured pursuits—sports shirts, jeans, running shoes, backpacks—as daily wear is another form of conspicuous leisure, for such clothes suggest that a person regularly engages in nonproductive activities. Choosing forms of employment that are far removed from environmental production, such as being a lawyer or a corporate manager, is a particularly important form of conspicuous leisure, Veblen suggested. He noted that most high-status and well-paid jobs are far removed from environmental production, which is why he referred to the wealthy as the "leisure class"—not just because the wealthy have more leisure time.

By *conspicuous waste,* Veblen meant using excessive amounts of goods or discarding something rather than reusing it or repairing it. Examples might be buying the latest model of a consumer item, running a gas-guzzling power boat at top speed, or routinely leaving food on your plate. Those who can afford to waste in these (and countless other) ways thus demonstrate their elevation above material concerns.

Veblen also pointed out that the leisure class engages in *vicarious consumption, vicarious leisure,* and *vicarious waste*—that is, consumption, leisure, and waste that others engage in because of your environmental power. The vicarious can be a highly effective statement. Veblen had in mind here everything from parents who dress their children in easily spoiled clothes and send them off to a leisurely pursuit like college, with a new car and credit card, to male business executives who insist that their wives refrain from productive employment.

Veblen's terms overlap (as do probably all categorical distinctions about social life). For example, wearing leisure clothes can be simultaneously a form of conspicuous consumption and conspicuous leisure. Designer jeans and name-brand running shoes are far from cheap, and everyone knows that; plus, they are emblems of a life of leisure. The occasional fashion of ripped jeans, which emerged for a while in the late 1960s, reemerged for a while in the early 1990s, and seems to be coming back again now, is a form of conspicuous waste: Who but the environmentally powerful could afford to deliberately rip their clothes and be so confident about their social status as to wear them in public? Thus, the same pair of designer jeans could be a form of conspicuous consumption, conspicuous leisure, and conspicuous waste.

In a way, all modern materialism can be reduced to just one of these: conspicuous waste. Conspicuous consumption is wasteful, and leisure is a waste of time. As Veblen put it, modern society is guided by "the great economic law of wasted effort"—a theoretical, and satirical, dig at utilitarian economics and its idea that modern life is guided by ever-increasing efficiency.[28]

I'd like to highlight two important ecological implications of Veblen's analysis. The first stems from the relative subtlety of conspicuous leisure. Consumption and waste are generally much more conspicuous than leisure. It is hard to show to others all the nonproductive hours you put into cultivating your good table manners, cleaning your home, or perfecting your personal website. Nor can friends and associates see you when you are away on your winter holiday.

Perhaps that is part of the reason why tourists are so fond of bringing back photos, knickknacks, and a tan. These visible symbols of a vacation trip allow you to turn inconspicuous leisure into conspicuous consumption and conspicuous waste.

The greater visibility of consumption and waste is ecologically significant because leisure is potentially less environmentally damaging. Spending time with family and friends, reading books (particularly books borrowed from family, friends, or the library), taking a walk, riding a bike—leisure activities like these consume fewer resources than spending your salary on the latest bit of loud and colorful plastic. Leisure is not always less environmentally damaging, however. It depends on how you engage in it. For example, travel—particularly travel by air and automobile—consumes energy and creates pollution. But in general, consuming or wasting time is less environmentally damaging than consuming or wasting things.

The second and more important environmental implication of Veblen's work is the competitive and comparative character of conspicuous consumption, leisure, and waste. To be conspicuous, you have to exceed the prevailing community norm. As long as others are also attempting to signal their social power through conspicuous consumption, leisure, and waste, the levels required to make a conspicuous statement of power continually rise. Therefore, the environmental impacts from conspicuous consumption, leisure, and waste also continually rise.

To summarize Veblen, through wealth we signal our environmental power and thus our social power. When we can engage in conspicuous consumption, leisure, and waste, we feel socially powerful. But as this is a comparative and competitive matter, we must continually up the ante of conspicuous signals of wealth and power. As a result, we moderns are motivated in our environmental relations not by a hierarchy of needs, but by a hierarchy of wants, guided by the hierarchy of society.

Positional Goods

Although Veblen's theories aren't perfect, as I'll come to, let us first explore the ideas of a more recent scholar whose work bears a close affinity with Veblen's: the economist Fred Hirsch.

Why do we experience so much scarcity in the world? Hirsch argues that scarcity is due not only to physical limits in the supply of goods, but also to social limits. Conventional economics tells us that when many people want something of which there is not very much, shortages are likely when demand exceeds supply. Hirsch, on the other hand, argues that people often want something precisely *because* it is in short supply. The supply and demand of a good are not independent phenomena, with the price set at the point where they meet. Rather, there can be important interactions between supply and demand. Hirsch avers that for some goods, demand will go up as supply goes down, and demand will go down as supply goes up. As Hirsch puts it, "An increase in physical availability of these goods . . . changes their characteristics in such a way that a given amount of use yields less satisfaction."[29]

The point is, scarce goods create an opportunity for conferring status and prestige upon those who gain access to or possession of them. As Mark Twain wrote, describing the famous white-washing scene in *The Adventures of Tom Sawyer,* "[Tom] had discovered a great law of human action, without knowing it—namely, that in order to make a man or boy covet a thing, it is only necessary to make that thing difficult to attain."[30] Hirsch terms such difficult-to-attain things *positional goods,* goods whose desirability is predicated at least in part on short supplies, limited access, higher prices, and consequent social honor or position.

The notion of positional goods helps us understand why some goods and not others become the objects of conspicuous consumption. Goods that are in short supply, or can be made to be in short supply, are most likely to take on positional importance. One example of such a good is lakefront property. The amount of shoreline on lakes is limited by the physical landscape and the expense of reengineering that landscape. Hirsch suggests that, although we rarely admit it to ourselves, part of the desirability of such property is the fact that there is so little of it—particularly in places that have a good climate and are relatively close to cities. (There are plenty of lakes up north in the tundra.) The same is true of having a place in the country. If everyone lived in the countryside, it would not be so desirable (nor would it be the countryside).

Moreover, says Hirsch, the owners of positional goods may deliberately attempt to limit access to these goods, increasing their own positional advantage. An illustration of what Hirsch has in mind are some political movements to preserve the countryside, such as the 2-acre minimum lot sizes that many American exurban communities instituted in the 1970s and 1980s. Local proponents argued that 2-acre lots would limit development and thereby protect wildlife habitat and preserve rural character. Hirsch would say that such zoning provisions provided a means for exurban residents to pull up the drawbridge behind them, even though they might not admit such a motivation to others or even to themselves. I might add that, in fact, 2-acre zoning accelerates the deterioration of habitat and rural character by increasing the amount of land consumed by any new development. It is environmentally far more effective to concentrate development in a few areas and leave the bulk of it open.[31] Two-acre zoning does, however, ensure that only those wealthy enough to afford a lot of that size will move in, protecting the social honor of the exurban landscape.

The concept of positional goods also helps us understand the social pressures that sometimes result in the extinction of valued species of plants and animals. If a species is particularly valued, one might expect that those who appreciate it would do everything possible to protect it. The reverse is very often the case.

For example, on August 7, 2003, the Australian customs ship *Southern Supporter* spotted the Uruguayan fishing boat *Viarsa* in the territorial waters of Australia, just above the Antarctic Circle. The *Viarsa* was fishing for the rare and sought-after Patagonian toothfish—also known as the Chilean sea bass—whose oil-rich flesh is much prized in sushi restaurants in Japan and the United States. A single fish, which can weigh over 400 pounds, can fetch $1,000. Consequently, toothfish stocks are, at this writing, virtually in collapse.[32] Since 1988, the toothfish has been protected under an international conservation agreement, called the Convention for the Conservation of Antarctic Marine Living Resources, which most nations have signed. Australia is among those, and the *Southern Supporter* gave chase to the *Viarsa*—a 7,000-kilometer chase, in fact, across the southern oceans. Three weeks later, with the help of ships from Britain and South Africa, the *Southern Supporter* caught up with the *Viarsa*, and in 20-foot seas arrested its crew and took control of the 150 tons of toothfish on board, worth $1.5 million.[33]

Why do people prize toothfish so much? In part for its taste, no doubt, but also for the positional value of eating such rare fish. (Indeed, when one considers the wide variability in what peoples of the world consider good-tasting food, one has to wonder if part of the very taste of all food is in part its local status value.) As the supply of toothfish goes down, the enjoyment some find in this taste will only go up. The price will go up, too, making the poaching of toothfish all the more attractive an endeavor for boats like the *Viarsa*.

The same can be said of the trade in rhino horn, which is a favored ingredient in some Asian traditional medicines, particularly as an aphrodisiac, and is also a symbol of masculine pride for trophy collectors, especially in Vietnam. The world population of rhinos as of 2007 was approximately 25,000 animals in the wild for all five surviving species.[34] But the positional value of rhino horn for trophies and aphrodisiacs is such that it sells for over $50,000 a kilo, or about $150,000 per horn.[35] The result has been vastly accelerated rhino poaching in recent years.[36] There are even break-ins at hunting lodges, where the horns are sawed off the stuffed heads on the walls.[37]

It can only be seen as a sad irony that the establishment of conventions and reserves to protect rare and endangered species contributes to their positional value by decreasing their supply on world markets. The more we protect them, the more they become sought after. Which doesn't mean protecting them doesn't work. For example, rhino numbers are up considerably since the mid-1990s because of conservation efforts.[38] But the positional consequences of protecting rare and endangered species greatly complicate conservation work.

Sadly, this is an old story—at least as old as the Roman Empire. When the Greeks established the colony of Cyrene in present-day Libya in the seventh century BCE, someone figured out that a local member of the fennel family of plants had a number of remarkable properties. You could eat the stalks and leaves. It was a delightful spice (somewhat garlicky). It treated fevers and sore throats. But most remarkable of all, it was a highly effective contraceptive and morning-after treatment, according to several ancient sources. Soon women throughout the ancient world were using it. *Silphium,* they called this wonder plant. Cyrene grew famously rich on its trade. The trouble was that it grew in only a narrow range of climate, on the Mediterranean slopes of Libya's mountains in a zone about 30 miles wide. Plus, no one could figure out how to cultivate it, and silphium began to become very scarce.[39]

The Cyrene government passed many laws to regulate the harvest of silphium, which helped some. People recognized that it was a very useful plant, after all. But as the supply dwindled, its positional value soared. Pliny the Elder reported in his *Natural History,* which dates from the middle of the first century CE, that juice from it "was being sold at the same rate as silver." Eventually, like the "truffula trees" of Dr. Seuss's classic *The Lorax,* there was only one plant left. What do you do with the last of the species? Everything possible to let it spread and replenish itself? Apparently not, at least if you are living in the Roman Empire. You get out your "super-axe-whacker" (Dr. Seuss's memorable phrase), chop it off, and send it to the person with the greatest position of all. As Pliny reports, the last stalk of silphium "within memory of the present generation . . . was sent as a curiosity to the Emperor Nero" for his dining pleasure. Silphium has never been seen again.[40]

Hirsch's analysis of positional goods can also be extended to our concepts of beauty. If scarcity makes something desirable, then, in a way, scarcity makes something beautiful. The most scenic countryside is rarely the most ordinary. The most cherished forms of wildlife are rarely the most common. The most delicious foods are rarely the easiest to get. This suggests to me a tragic point: Destroying some of the environment can sometimes make the rest of it seem more beautiful.

Goods and Sentiments

A Veblenesque portrait of social motivation is a familiar form of social critique in modern life, and it can have a certain intuitive appeal, particularly when one is in a cynical frame of mind. Most of us have, I imagine, leveled Veblenesque charges at those around us—at least in our minds—reducing the behavior of others to greedy, competitive, self-serving display. We can easily imagine others having these motivations because, I believe, we have often sensed them in ourselves.

But there is more to people than a will to gain power and to show off. Many sociologists have long contended that ascribing all human motivation to *interest,* the desire to achieve self-regarding ends, is too narrow a view. As humans, we are equally motivated by *sentiment,* the desire to achieve other-regarding ends revolving around our norms and social ties.[41] It is wise, though, to maintain a critical outlook on the sentimental side of human motivation, lest we be seduced by the potential flattery of such an interpretation of social behavior. I try to maintain such a critical outlook in the pages that follow.

The Reality of Sentiments

We can get a handle on the sentimental side of motivation for acquiring goods by looking at the way we "cultivate" meaning in objects, to use language suggested by Eugene Rochberg-Halton.[42] One of the principal sources of the meanings we cultivate is the network of our social ties. The goods we surround ourselves with show not only how we set ourselves apart from others—Veblen's point—but also how we connect ourselves to others. They are talismans of community.

We all can give many personal examples. Here is one of mine. For many years, until it wore out completely, one of my favorite T-shirts was from a softball team I used to play on, the Slough Creek Toughs. We were all geology students, my undergraduate major, and the team name came from the name of a rock formation, the Slough Creek Tuff. (A tuff is a kind of volcanic rock.) It was neither a particularly clever name nor a particularly nicely designed shirt, but the shirt brought back pleasant memories of good times with friends from long ago. This shirt was not an emblem of the "old boy" network of my student days, for I have since changed professions from geology to sociology, and it has been years since I saw any of the team's former members. Perhaps there is some social advantage in the conspicuous leisure of wearing a T-shirt, but it is hard for me to see what social advantage I gained in wearing that particular—rather plain and obscure—T-shirt. I think what I gained was not advantage but a chance to express a sentimental connection to others, and mainly to express it to myself.

Not all social scientists would agree that such an interpretation is justified. An important theoretical tradition, long established in economics but also in other social sciences, argues that all we ever do is act on our interests, as best we understand them and the possibilities of achieving the desires that stem from them. This is often called the theory of *rational choice*. Veblen's theory of the leisure class is an important forerunner of this tradition. A close parallel is the theory of the selfish gene in evolutionary biology. Rational choice theory would argue that I had lots of self-serving reasons to keep wearing that old T-shirt. For example, I might wear it because one of my interests is to have a good opinion of myself, and one route to such a good opinion is the self-flattery of believing myself to be motivated by more than self-interests. In this view, sentiments are thus a self-serving fiction.

The rational choice perspective yields many fruitful insights. But it is also extraordinarily materialist in approach. As I argue throughout this book, theories of social life that emphasize either the material or the ideal generally turn out, on closer inspection, to be unbalanced. Although we cannot discount the accuracy of a purely materialist or purely idealist perspective out of hand, a case can usually be made for the equal importance of the other side of the dialogue.

Let me try to make such a case here.[43] Consider the rational choice view that sentiments are a self-serving fiction. Now it may be true that sentiments always have self-interest behind them, but that is not how you and I experience sentiments, at least not always. We experience our sentiments as sentiments: as feelings of concern, affection, empathy, and affection for others; as commitments to common values and norms; and frequently as the lack of these feelings and commitments (such lacks being equally manifestations of our sentiments). Thus, we may at times give up or refuse material gain because we experience concern for, and commitment to, the interests of others. For example, someone might refuse a raise, or ask for a small one, in order to make space in a budget for someone else to get a bit more. Or someone might make a donation to charity. We may also hold dislikes for others, and their values, that run contrary to our potential for material gain—refusing a high-paying job with an unpleasant boss in an environmentally damaging industry, for example.

Now perhaps there is always some hidden agenda of self-interest behind such refusals of personal material gain. But if we do not consciously experience interests behind our sentiments, we must then be basing our conscious decision making at least in part on other criteria. In other words, as long as the agenda of self-interest really is truly hidden—hidden even from the self—sentiments will be important sources on their own of what it pleases us to do.

In any event, as I have elsewhere written, there is simply "no way of knowing that that hidden agenda always exists, for, after all, if it does exist, it is often hidden."[44] But what we can know is what experience tells us does exist: that we have conscious orientations toward material goods that are both self-regarding and other-regarding, both materialist and idealist, both interested and sentimental.

Hau: **The Spirit of Goods**

One aspect of the sentimental experience of material goods is what the Maori people of New Zealand traditionally called the *hau,* the social spirit that attaches to gifts.[45] A Maori wise man, Tamati Ranapiri, once explained the *hau* to a visiting anthropologist this way:

> Let me speak to you about the *hau.* . . . Let us suppose that you possess a certain article and that you give me this article. You give it to me without setting a price on it. We strike no bargain about it. Now, I give this article to a third person who, after a certain lapse of time, decides to give me some things as a payment in return. . . . It would not be fair on my part to keep these gifts for myself, whether they were desirable or undesirable. I must give them to you because they are a *hau* of the gifts that you gave me. If I kept these other gifts for myself, serious harm might befall me, even death. This is the nature of the *hau,* the *hau* of personal property, the *hau* of the gift, the *hau* of the forest. But enough on this subject.[46]

Hau is the Maori word for both "wind" and "spirit," much as the Latin word *spiritus* means both "wind" and "spirit," and it is probably not accidental that two such widely separated cultures should have such a parallel. All peoples recognize that there can be a kind of palpable, yet intangible presence in things. For Tamati Ranapiri, that presence was the interconnected *hau* of personal property, gifts, and the life-giving forest. This interconnected *hau* watched over the movement of goods through the community to make sure each gift was reciprocated. In a way, the soul of the person who gave the gift—that soul being connected to the wider soul of the forest—lingered on within the gift, even after it had been subsequently given to someone else. As the anthropologist Marcell Mauss observed, "This represents an intermingling. Souls are mixed with things; things are mixed with Souls."[47] Through this mixing, as Mauss also noted, the tangibility of gifts brought a Maori group together, causing them to recognize and to celebrate something intangible: their sentimental connections.

Think about it. Articles we receive as gifts have a very different influence on our behavior from articles we buy for ourselves. The same is true for any good that we come to appreciate not just for its material purpose (if it even has one), but for the association we make between that good and a person or persons. We moderns still mix souls with things.

The wedding ring I wear on the fourth finger of my left hand is an example. It has no material purpose, and only modest material value. And yet if some experimentally inclined social scientist were to offer me an absolutely identical ring, plus $100, I would refuse the trade. I would refuse it for $1,000, and probably for $10,000—although at this level, I am less sure! But even if I weaken at such a figure, my ring, I believe, remains more than a cold, material object to me.

What makes my wedding ring more than a material object to me is my sentimental sense that it has a *hau,* in this case the *hau* of two souls joined in marriage. Similarly, my old Slough Creek Toughs T-shirt contains the *hau* of that softball team. Although I could go to the store and buy an end table for my study that is not as stained and wobbly as the one my grandfather made, no store sells end tables that embody the *hau* of my grandfather. All of us, I imagine, could point to similar spirited articles of social sentiment among our own possessions.

Goods are thus not merely objects of social competitiveness and social interest. They are also the means by which we remind ourselves, and indeed even create, the web of sentimental ties that helps support our feelings of social communion. In the words of Mary Douglas and Baron Isherwood, material possessions serve our interests, "but at the same time it is apparent that the goods have another important use: they also make and maintain social relationships."[48]

Sentiments and Advertising

Most goods today, however, do not have a *hau*. In this age of the global economy and the shopping mall, most of the goods that surround us were made by people we will never meet, bought from people we do not know, and chosen primarily because of price and convenience. We care about these purchased goods because of how they serve our interests. They are socially empty, or nearly so.

Companies routinely try to persuade us that their wares are not socially empty, however. Advertisers routinely appeal to sentiments to pitch products, and it is instructive to examine the techniques they use. The agenda behind sentimental appeals is very often not so hidden.

The principal form of advertising is price advertising. With so many purchasing options available, it is not easy to persuade consumers to spend their cash on a particular product. Moreover, most people do have considerable resistance to ads. They know that ads are manipulative propaganda.[49] From the extravagant "blowout clearance sale with unbelievable values" to the simple statement of object and cost that is the norm on eBay and Craigslist, advertisers have long found that price is the single most effective means of generating sales.[50] In part, buyers want to retain as much of their purchasing power as possible, perhaps with Veblenesque ends in mind. But also, given all the propaganda involved in selling, price is the one comparative aspect of a product about which a buyer can be relatively (excepting the prevalence of hidden fees) sure.

Price advertising does have limits to its effectiveness, though, particularly when a rival is advertising a similarly low price. In addition, emphasizing the monetary aspect of a transaction reminds the potential purchaser that the principal interest of the seller is similarly financial, undermining consumer trust in the quality of the good. So, in order to divert attention from the fact that, in reality, the seller is out for our money, advertisements routinely try to appeal to our sentiments.

One popular technique is to claim that a product is being offered for sale out of concern for you. Here's a sample newspaper ad in this "you" genre:

Truly Exceptional Service Starts With Careful Listening

It is why your Republic Account Officer makes sure to obtain a precise picture of your financial goals, time frame, risk acceptance, and other key factors. He keeps these constantly in mind as he looks after your interests. So year after year, you can count on us for the exceptionally complete, timely and personalized service that makes Republic truly unique.[51]

A rather dull ad. Yet note how, after the banner, the words *you* or *yours* appear in every sentence. But whose interests does a Republic Account Officer really look after?

In addition to service, advertisements attempt to bestow a feeling of concern for you through claims to offer choice and individualized products. "Double the size and even more choice," reads an ad for a new branch of Marks and Spencer, a large British retailer. "Have it your way" is the corporate motto of Burger King—a remarkable claim for a company that sells food made on an assembly line. Specialty shops make a related pitch by seeking the business of only a select group of customers with particular needs. "The choice for big or tall men" runs the slogan for High and Mighty, a European chain of men's clothing stores. Ads like these evade the mass-produced, machine-made, *hau*-less origin of modern goods by saying, "See, we care enough about you to provide just your size and color and taste. Mom herself couldn't have done it better at her sewing machine at home."

"You" advertising may exhibit supposed sentiments on the part of merchants, but, of course, it also appeals directly to the buyer's status and power interests. Statements about the merchant's commitment to *your* every wish entice the buyer with a romance of the customer's high status and power over the merchant. Indeed, after price, status is likely the principal theme of advertisements. "The finest watch in the world will only be worn by exceptional people," runs an ad for Audemars Piguet. "Compromise shouldn't enter your vocabulary, let alone your garage," puffs an ad for Volvo. A store name like High and Mighty is another effort to cash in on status. Status advertising is particularly characteristic of ads for clothing and luxury goods. Similarly, companies often boast of their own reputations as the "best in the business," not only to proclaim the quality of their products but also to establish a status enticement to "shop with the best."

Few of us like to feel that we are mainly motivated by a desire for social status, however. Whether or not we are only fooling ourselves about our sentimental concerns, thinking purely in terms of status and interest does leave one feeling a bit hollow, and probably thinking as much of the advertiser. So ads very often try to convince us that the product on offer is not just for status display, but for displaying love, too.

Christmas advertising is notorious for this. Dad (it is usually Dad) hands his college-age child the keys to the gleaming car in the background, as the music swells. Happy children play with this year's trendy toy while Mom (it is usually Mom) looks on contentedly. Similar "hooks" appear in other seasonal ads, particularly those for the Hallmark holidays invented by the advertisers, and most popular in the United States: Mother's Day, Father's Day, Secretary's Day, and the like. Veblen would argue that the real motive behind buying your child the latest piece of expensive junk is vicarious consumption, and I believe it would be hard to deny the common existence of such a desire. Yet vicarious consumption is far more likely to be psychologically palatable if we experience it as, at least in part, an expression of sentiment. The dialogical converse also applies: We are more likely to consume vicariously through those for whom we have strong sentimental ties.

Ads often portray a more generalized sentimentalism, too, placing a kind of good-for-the-world, friendly, family values frame around the item on offer. "Get Together," proclaims an ad for Nokia cellular phones opposite a photograph of a seven-hands handshake. "It's nice to meet you," runs an ad for LG Electronics, along with a photograph of some of the company's chip designers. "It took you a long time before you could walk. Air France will save you some when you want to fly," reads the caption for a photo of a father's hand helpfully reaching down to a toddler. Here as well, "you" is a prominent theme in the advertising copy.

Green Advertising

A new form of sentimental hook is green consumerism. Companies like The Body Shop (cosmetics), Ben and Jerry's (ice cream), and Whole Foods (an American organic supermarket chain) seek to demonstrate through the environmental and social good works they support that they are concerned about more than profit. They also make a sentimental appeal to the guilt we feel over our own consumptive habits.

The online catalog of Gaiam—an American "lifestyle media company" whose mission is to provide "the resources you need to achieve self-growth, live a healthy lifestyle or to positively transform your life"—is an example. It is full of expensive "you" products, personal care items that could hardly be deemed essential but are made with organic cotton, recycled rubber, and the like. Environmentalists might be pleased to see that Gaiam promotes the use of organic and recycled products. But there must be some disappointment in seeing them pitch items like a $49 bath mat hand-woven from vetiver root ("as seen in *House Beautiful*," according to the company website), a $79 water hyacinth meditation tray ("add stylish texture, durability and practicality to your living space," counsels the description), $219 for a pillow made out of something called CleanDown™ ("each luxury-quality CleanDown™ pillow," plumps the caption, "is made with exacting environmental and purity standards for an exquisite sleeping experience"), and a set of king-sized organic sheets for $419 (the full-sized set is just $319).[52] The message is that you can consume conspicuously and still be an environmentalist. Indeed, you can be conspicuous *about* your environmental consumerism when friends and family visit your home. Sentiment itself becomes display.

The environment is also a common theme in the ad campaigns of major corporations, particularly oil companies, automobile manufacturers, pesticide firms, and other industries with spotty environmental records. "Green-washing" is what critics call it, and perhaps with some justification.[53] General Motors is not on very many people's list of environmental nice guys, but did have a whole line of cars, the Saturn, which showed a kinder side. "It's nice to know the environment also impacts the auto industry," an ad for Saturn used to comfort, before GM retired the brand in 2010. ExxonMobil could hardly be considered a paragon of environmental virtue, given the *Exxon Valdez* oil spill on March 24, 1989, and the way it quietly spent $16 million between 1998 and 2005 to pump up the coffers of global warming skeptics, according to a Union of Concerned Scientists report.[54] It's the sort of company that regularly shows up on lists of companies with the worst environmental records.[55] But ExxonMobil has been very loud about the $15 million it has spent on its "Save the Tiger Fund," supporting groups like the World Wildlife Fund and Conservation International, which has the nice side benefit of promoting its corporate mascot, a cartoon tiger.[56] BP for years has also been making regular appearances on lists of corporate environmental villains. So some people were not surprised that BP was the company who brought us the *Deepwater Horizon* blowout and resulting oil spill in the Gulf of Mexico in 2010, a spill even larger than the *Exxon Valdez* disaster. But take a critical look at BP's new corporate logo, which dates from 2000. Instead of the old shield with BP in the center, it now tries to soothe us with a green and yellow sunflower-like image—which, incidentally, bears more than a passing resemblance to the logo of the Canadian Green Party.[57] At the same time, BP introduced a new corporate slogan, "Beyond Petroleum," while proceeding to spend very little on alternative energy projects, compared with its oil and gas sectors.[58]

In this age of widespread support for environmental concerns, a green halo is good for a corporation's image. To be sure, some of the things that corporations do to acquire that halo truly may help to resolve environmental problems. It is a good thing that ExxonMobil contributes money to the World Wildlife Fund; that The Body Shop allows customers to recycle their bottles; that Ben and Jerry's supports the local dairy farms of Vermont; and that corporations in general are learning that, in fact, it is possible to make money in more benign ways. We should not lose sight of the environmental significance of green business practices, as Chapter 11 discusses in some detail. But we should also recognize that good environmental citizenship makes a corporation's sentimental appeal to consume all the more potent.

Goods and Community

Why are appeals to our sentiments so potent? Social scientists have long observed the way the individualism of modern life has weakened the ties of community. We feel the lack of these ties, even though we may not consciously recognize it. So we try to buy community, the psychologist Paul Wachtel has argued. We try to buy a feeling of community in the goods we purchase for ourselves and the goods we buy for others, and we try to gain status within our community through the goods we display. Goods, then, are a substitute for social needs. In Wachtel's words,

> Faced with the loneliness and vulnerability that come with the deprivation of a securely encompassing community, we have sought to quell the vulnerability through our possessions. When we can buy nice things, and we can look around and see our homes well stocked and well equipped, we feel strong and expansive rather than small and endangered.[59]

Moreover, suggests Wachtel, a vicious circle is in operation. The more we lose community, the more we seek to find it through goods. And the more we seek the wealth to attain these goods, the more we immerse ourselves in the competitive individualism of the modern economy, thus undermining community. Meanwhile, the environment is undermined as well.

The Time Crunch

Part of the problem is simply a lack of time. The hunter-gatherer life of leisure is lost to us, and perhaps to our overall benefit. But with it went the hunter-gatherer's abundant opportunity for interaction with family and community.

Loss of time got really out of hand in the early years of the Industrial Revolution. In 1840, the average worker in the United Kingdom (the first country to experience widespread industrialization) put in 69 hours a week.[60] By the 1960s, in response to widespread protest over these conditions, the average workweek had fallen to half that throughout the industrialized world. Many observers foresaw the coming of a society that had hardly any need for work as improved machines replaced the drudgery of early industrialism and a new sense of a social contract between workers and employers ensured that time demands remained reasonable.

Yet in the United States, the length of the workweek is now back on the upswing, and the trend seems to be spreading throughout Europe.[61] Rising competitiveness in a globalizing economy, low job security, declining government controls, and weak (or even nonexistent) unions together encourage workers to acquiesce to employers' demands for long hours. The economist and sociologist Juliet Schor estimates that the average American worker in 2006 was putting in 204 more hours of work in a year than in 1973, the equivalent of five 40-hour workweeks.[62] Simultaneously, paid time off—vacation, holidays, sick leave—had slipped back by several days, despite already being at levels far below those of most European countries, where paid vacation alone is commonly 4 to 6 weeks by law, even for low-paid workers with just a year of service in a company.[63] In the United States, the norm is now just 8.9 days of vacation after a year's service, and there is no legal mandate for even this much (see Figure 2.2). The percentage of U.S. companies that offer any paid vacation time at all dropped from 99 percent in 1980 to 77 percent in 2010.[64]

Figure 2.2   Variations in legally mandated minimum vacation time across the world for employees with 1 year of service. Note that the United States is the only country in this group for which there is no mandated minimum. Some 77 percent of U.S. companies do offer paid vacation nonetheless, but an average of only 8 days per year. Plus, this figure has dropped from 99 percent in 1980.

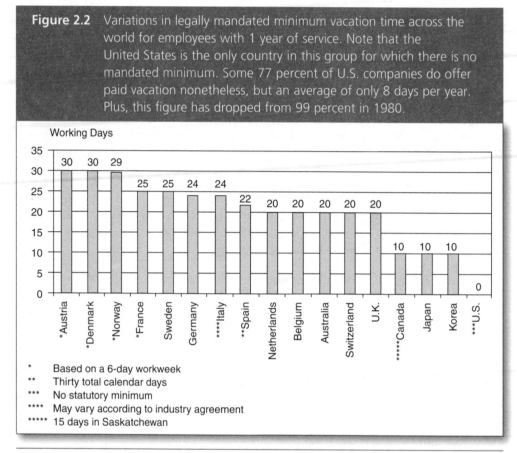

*   Based on a 6-day workweek
**   Thirty total calendar days
***   No statutory minimum
****   May vary according to industry agreement
*****   15 days in Saskatchewan

Source: Hewitt Associates and Bureau of Labor Statistics (1981 and 2007). Used with permission of Hewitt Associates.

Schor suggests that this time crunch may be much of the reason that consumerism is so pronounced in the United States. Americans are trapped in what Schor calls the cycle of *work-and-spend*: They must maintain a highly consumptive lifestyle in order to be able to put in all those hours at work. The clothes to wear to work, the several cars most households need to get there, the time-saving home conveniences and prepared food—these are the unavoidable costs of holding down a job, or the two or three jobs many Americans now work. This work-related consumption, in turn, increases the amount of time spent shopping. The result is even less time for other pursuits—like spending time with family, eating meals together, visiting with friends and neighbors, joining clubs, and participating in local voluntary organizations, all activities that surveys show have fallen off in the United States since the 1950s (see Figure 2.3).[65] With everyone doing so much consuming, the pace of competitive display ratchets up, leading to yet more need for work. More than the people in other wealthy countries, Americans find themselves working and spending, spending and working, rather than enjoying the vacation time they don't have anyway. Indeed, very often American workers don't even take the limited vacation time they do have, so caught up are they in the culture of work-and-spend. Plus, the most detailed international comparison we have—something called the Multinational Time Use Study—found that Americans had the least free time of any country in the survey.[66]

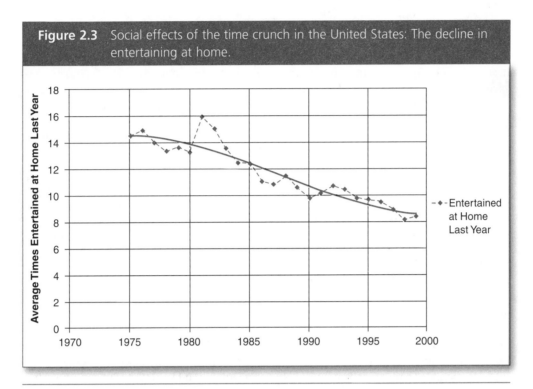

Figure 2.3 Social effects of the time crunch in the United States: The decline in entertaining at home.

Source: Reprinted with the permission of Simon & Schuster Adult Publishing Group from *Bowling Alone: The Collapse and Revival of American Community*, by Robert D. Putnam, p. 99. Copyright © 2000 by Robert D. Putnam.

The time crunch propels environmental damage along with consumerism. The raw materials to support this high level of consumerism have to come from somewhere. Moreover, with so little vacation time and free time, Americans have less opportunity to use leisure as a status symbol, generally a less environmentally damaging form of conspicuous display than consumption and waste.

Similar trends are underway elsewhere—for example, in Britain, where work hours and shopping hours are up and participation in voluntary organizations is down.[67] Between 1961 and 2001, time spent shopping more than doubled in Britain—from 25 minutes a day to 52 minutes a day for the average adult.[68] Although consumerism is most pronounced in the United States, other wealthy countries are not far behind. Indeed, as several social critics have commented, it appears that shopping is now the industrialized world's leading recreational activity—especially when we consider that television watching and Internet use are themselves increasingly shopping activities, in view of the volume of ads and the growth of credit card purchases through websites and shoppers' television networks.

Increases in productivity per worker have been such that people in all the wealthy countries could be working far, far less. For example, Schor has calculated that in the United States, everyone could work a 4-hour day, or only 6 months a year, and still have the same standard of living that prevailed in 1948.[69] But no population in any country has chosen such a course. Perhaps we have not been allowed to choose such a course, and perhaps we have not allowed ourselves to choose it. Probably, and dialogically, both are true. Nevertheless, we work and we spend, and we simultaneously drift further away from one another as we increase our rate of environmental consumption.

Consumption and the Building of Community

The rhetoric on consumerism and its relationship to the loss of community can easily get over-heated, though. To begin with, although we are frequently critical of grabs for social status and social power, few would deny the central importance of status and power to one's social-psychological health. All of us need some status and power within our communities. Indeed, we are critical of unequal distributions of status and power because these needs are important to everyone. The granting of a degree of status and the power with which it is closely associated is something we expect from our communities. Moreover, that expectation helps build our commitment to our communities.

Consumption can also enhance community, despite all the competitive individualism it can promote. Consumption can have a kind of festive air about it. Christmas gatherings, wedding ceremonies, and birthday parties have some of the hallmarks of potlatch—festivals of community in which people strengthen social ties through gift exchange—as anthropologists have argued.[70] Some *hau* exists even in the consumer society. And in addition to circulating *hau* through gifts, we make community through the goods we consume in common. Reciprocal exchanges and what Douglas and Isherwood called "consumption matching" can bring us together.

But Douglas and Isherwood carried this argument too far when they wrote that "consuming at the same level as one's friends should not carry derogatory meaning. How else should one relate to the Joneses if not by keeping up with them?"[71]

Must we really keep up materially with someone to relate to him or her socially? This is, no doubt, quite a common approach to fellowship.[72] But such an attitude quickly divides a society into class-bounded patterns of community. Note that Douglas and Isherwood did not suggest that one should lower one's consumption level to that of, say, the Collinses as a way to find community. When we engage in consumption matching, we nearly always attempt to match those above us in status. In other words, consumption matching is rarely only about building community.

Just as it would not be accurate to ascribe all consumption to competitive display, it is not accurate to ascribe it all to reciprocity and fellowship, as Douglas and Isherwood do. Both motivations can exist together. The consumption of goods generally represents a double message, a complex mix of competition and community, interest and sentiment. This complexity leads to considerable ambiguity in the meaning one person can read from another person's consumptive act. This ambiguity, to be frank, is often socially useful when we engage in a little "you" advertising of our own.

The Treadmill of Consumption

Meanwhile, the cycle of competitive and communal consumption accelerates. As one tries to keep up with the Joneses, the Joneses are trying to keep up with the neighbor on the other side, and up the line to Michael Jackson, the Rockefellers, the Walton family, Queen Elizabeth, the Sultan of Brunei, Bill Gates, and Carlos Slim Helú. And those up the line are constantly looking back over their shoulders.

Although the desire for more—more money, more stuff—is pervasive, one's level of wealth has little to do with a sense of happiness, at least beyond a certain minimum. A 1982 study in Britain found that unskilled and partly skilled workers, the bottom of the pay scale, were indeed less happy than others (measured by asking if a respondent was "very pleased with things yesterday"). But skilled manual workers from the lower middle of the pay scale were actually slightly happier than better-paid, nonmanual, professional workers.[73] Several American studies have found that the poor are least happy but the wealthy are only slightly more satisfied with their standard of living than are others.[74]

A cross-national comparison from 2002 shows the same weak relationship between wealth and happiness—what is sometimes called the *Easterlin paradox* after the economist Richard Easterlin, who has been forcefully arguing this point for many years (see Figure 2.4).[75] Ghana was one of the 3 poorest among the 68 countries surveyed, yet it had one of the highest levels of reported happiness. Only eight countries reported greater happiness than Ghana. Colombia was tied for happiest, yet was the 25th poorest. The wealthiest countries do tend to be among the happiest. But among the wealthy nations, greater riches do not lead to more satisfaction with life. Austria, France, and Japan are some of the world's richest countries, yet they reported happiness at levels lower than about half the countries in the survey. The United States, the second-richest country in the survey, registered the same happiness levels as El Salvador, the 51st-richest (or 18th-poorest) country in the survey. A 2008 survey by the Gallup Poll shows much the same results.[76] However, taken as a whole, the poorer countries in such surveys show a far wider range of happiness, and the least happy countries are all poor ones.

There is also no certain link between economic growth and increasing happiness. For example, the percentage of Americans who report themselves as "very happy" peaked in 1957 at 53 percent and has not recovered that height, despite continuous national economic growth since that time.[77] In fact, in a 2006 survey by Pew Research, only 34 percent of American adults described themselves as "very happy."[78] Which suggests the following about the human condition: If you are rich and happy or rich and unhappy, more money will probably do little to change your mood—because you are already rich.

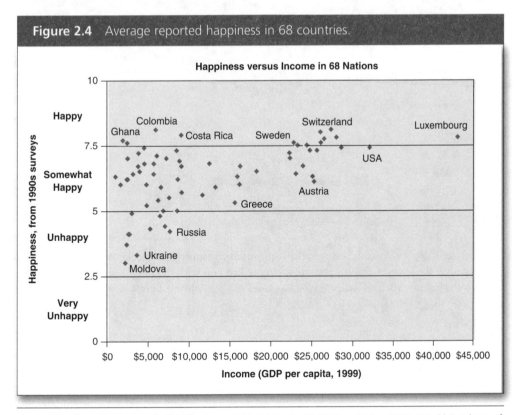

Figure 2.4 Average reported happiness in 68 countries.

Source: Based on Veenhoven, R. (2002). *Average Happiness in 68 Nations in the 1990s.* World Database of Happiness, Rank Report 2002/1. United Nations Development Programme (UNDP). (2000). *Human Development Report 2000.* New York and Oxford: Oxford University Press.

Moreover, getting richer can make you poorer—a paradox of the positional economy. Levels of consumption are constantly devalued as more people attain them through general economic growth. Because this constant devaluing affects everyone on the economic ladder equally, no one is made happier than anyone else by moving up a rung, except perhaps for the very poor. Consequently, economic advancement in an expanding economy has little to do with differences in overall happiness. You gain in this comparative game only when you are fortunate enough to advance in comparison to others, and by definition that advance must be limited to a few. We cannot all be like Carlos Slim Helú, the current richest person. Perhaps he is a supremely happy person. I don't know. But by definition, only one person can be the world's richest.

Yet still we try. We could all be like the Collinses, but although consumption may be an act of community, there is more to it than that. So we try to be like Carlos Slim Helú and other wealthy people instead. Meanwhile, the wealthy have gotten even richer. The result is no end to our wants and little improvement, if any, in our satisfaction—despite increased consumption of goods. This whole process of moving materially ahead without making any real gain in satisfaction can be termed the *treadmill of consumption* (see Figure 2.5).[79] I discuss two parallel treadmills, the treadmills of production and of underproduction, in the next chapter.

Figure 2.5 The treadmill of consumption: Even as consumption increases, satisfaction often remains elusive.

Source: Matthew Robinson and the author.

Moreover, at one level or another, most of us are aware that money and goods are not happiness—even in the United States, that paragon of consumerism. A 1989 Gallup Poll asked Americans to rank what was most important to them, and "having a nice home, car, and other belongings" came out last among nine options.[80] A 1995 survey I helped conduct among Iowa farmers (who, as a group, are sometimes accused of being concerned more about having a big new tractor than about the social and environmental implications of their farming practices) showed a similar result.[81] These kinds of responses, which are common throughout the world, have been shaped by cultural values every bit as deep as consumerism. Virtually all the world's major philosophical traditions counsel that money is the route neither to happiness nor to saintliness, as Figure 2.6 shows.

Figure 2.6	The teachings of world religions and major cultures on consumption and materialism.
Religion or Culture	*Teaching and Source*
American Indian	"Miserable as we seem in the eyes, we consider ourselves . . . much happier than thou, in this that we are very content with the little that we have." (Micmac chief)
Buddhist	"Whoever in this world overcomes his selfish cravings, his sorrows fall away from him, like drops of water from a lotus flower." (Dhammapada, 336)
Christian	"It is easier for a camel to go through the eye of a needle than for a rich man to enter into the kingdom of God." (Matthew 19:24)
Confucian	"Excess and deficiency are equally at fault." (Confucius, XI. 15)
Ancient Greek	"Nothing in Excess." (Inscribed at Oracle of Delphi)
Hindu	"That person who lives completely free from desires, without longing . . . attains peace." (Bhagavad-Gita, II. 71)
Islamic	"Poverty is my pride." (Muhammad)
Jewish	"Give me neither poverty nor riches." (Proverbs 30:8)
Taoist	"He who knows he has enough is rich." (Tao Te Ching)

Source: From Durning, Alan T. (1992). *How Much Is Enough? The Consumer Society and the Future of the Earth.* New York: Norton. Used with permission of the Worldwatch Institute.

Sometimes, however, it takes a shock to put things in proper perspective. On a late flight across the country a few years ago, I got to talking with a flight attendant about a recent tragedy in her life: Six months previously, her home had burned to the ground. Although no one was hurt, the house was a total loss. "How awful!" I exclaimed.

"No, not really," she replied. "Friends and family immediately came to my husband and me, giving us a place to stay, helping us clean up and salvage what we could, loaning us money, getting us back going again." She paused, and reflected. "It taught me what is really important, and that's people. It's not your things."

Similar reactions are well-known to social scientists who study the social consequences of disasters.[82] Major disasters in which many of one's friends and family suffer or even die are deeply troubling to the survivors, but disasters that take property and not lives—as can happen, for example, in floods that rise slowly enough for residents to evacuate—can actually leave people feeling better some months after the event.[83] The cleanup efforts bring people from behind the doors that usually separate them and join everyone together in a common endeavor. Unburdened of their possessions, they rediscover their community, without the ambiguity of the double message of goods.

Which suggests something significant, I think: Not only do we consume more than we need, we also consume more than we want.

CHAPTER 3

Money and Machines

We are becoming the servants in thought, as in action, of the machine we have created to serve us.

—John Kenneth Galbraith, 1958

A n ordinary day, nothing special. You roll over and glare at the clock radio. Wave-Rave, reads the embossing on the front. (Sweet machine.) You flick the radio switch for a weather forecast. A fatuously excited announcer is exclaiming about the fabulous once-in-a-lifetime sale this weekend at Loony Lucy's, a local discount store. Hit the switch. Where's the cell? Could try that new weather app. You reach for the bedside table and pick up your GOLDcell, with its little gold icon. (Nice phone.) The battery's flat. "Damn, I depend on that thing," you silently curse. Meant to charge it last night. Maybe the weather channel on cable would be better. Where's the remote? The screen comes to life just as the forecaster finishes with the segment on world weather, and the picture fades to a shot of a lush tropical beach, empty except for a lone sun umbrella emblazoned with a commercial logo you don't recognize. "Have you ever . . . ?" the voice-over begins. Who cares? Click. You look out the window to gauge for yourself if it's going to rain today, and you notice the company van from your neighbor's carpet cleaning business in her driveway. Her car must still be in the shop. "Our machines mean clean," the van's side panel promises. Jeans are fine for today, you decide. You reach for the unfolded pair from the pile of clean clothes you haven't put away yet. "Statue" reads the label on the rear. (Hot jeans.)

Six ads already, and you haven't even made it out of the bedroom.

The previous chapter described how the materialism of modern life is propelled by social psychological desires for status and social connection. But these desires are not fixed, and they can be manifested in many different ways. The organization of social life shapes the desires we experience in daily living and their manifestations. This chapter explores the economic and technologic basis of social organization and how, through interaction with culture and social psychology, economic and technologic factors help create the material motivations of modern life.

The inescapable presence of advertising, even in our bedrooms, is an obvious example of how technology and economics shape our motivations. Yet there is a need here for dialogical caution. Economics and technology are often construed as imperatives. As Stephen Hill has written, "The *experience* of technology is the experience of apparent inevitability," a statement that applies equally well to the experience of the economy.[1] With both, their materiality gives us the feeling that there is little we can do about them. Contrary to that material experience, a constant theme of this chapter is that technology and the economy are not imperatives. We do have control over them, even as they have control over us. We can resist their influence and frequently do. We can resist the economics and technologies of advertising.[2] We can switch off the radio and the TV. We can get by just fine without a cell. (Really. I do, at least, except when I'm travelling.) We can lobby for social controls on the amount and form of advertising. We can buy something else, or even nothing at all. But simply by being present as something to resist or to succumb to, ads—and technology and the economy more generally—nevertheless unavoidably shape the conditions in which we form our desires and motivations.

The shaping of motivation by economics and technology is of particular importance in understanding why we consume so much more than we need. The current arrangement of the economy and the current use to which we put our technologies tend to encourage not only consumption but also growth in consumption and thus in the economy. Economic growth, as many have observed, has become the most widely considered thermometer of economic health and even the social and political health of nations. As David Korten has written, "Perhaps no single idea is more deeply embedded in modern political culture than the belief that economic growth is the key to meeting most important human needs."[3] A rise or fall in economic growth is always headline news, and politicians constantly gear their efforts toward encouraging growth—in part because growth serves many powerful economic and technologic interests, but also because growth has become a central cultural value. Economic growth is a dialogical matter of both material interests and cultural ideals.

It is important to recognize that economic growth (and consumption too, for that matter) does not in itself lead to environmental damage. As the environmental economist Michael Jacobs notes, certain forms of economic growth, such as investment in companies that promote the use of green technologies, may in fact be beneficial to the environment.[4] However, the structure of economic interests does tend to overwhelm efforts to direct economic growth in ways that do not damage the environment. The pressure for economic growth tends to shift environmental considerations off our collective and individual agendas. Economic growth can take such a hold over our lives that issues of sustainability, environmental justice, and beauty of ecology fade from concern. As well, economic and technologic conditions may shove us into a motivational hole, consciously digging deeper into the firmament of ecology, even when we know we should not.

Large among those conditions is economic inequality. Without economic inequality, motivations for keeping up with the Joneses through conspicuous consumption disappear, because we are already all equal to the Joneses. Understanding the origins of economic inequality is thus a—and perhaps *the*—central problem of social organization for environmental sociology to consider.

The Needs of Money

In most of the world, money is now the principal means of arranging for one's material needs and wants, and in much of the world it is virtually the only means. Thus, the prudent person maintains a good supply of money, if possible. Money is useful stuff. Money is power. Because money is so powerfully useful, most of us wish for more of it rather than less.

In order to have a money economy, though, we have to keep money moving—to keep it mobile through the economy. Capitalism (that's what we are talking about here) always faces a basic conundrum. To make a profit, workers have to be paid less than the price of what they make, which makes it hard for workers to buy all they make. And who but workers is going to buy all that stuff? So how do capitalists sell enough to pay workers to begin with?

I like to describe this ever-present *wage–price gap* with a little parable I call the *problem of the original capitalist*.[5] Imagine you are the original capitalist, the very first one, and you hire me to be the original worker. You offer to pay me 4 shekels to produce a 2-cubit high pile of carrots. (Shekels and cubits? It was a long time ago, after all.) But your brave new little plan is to charge more than 4 shekels for the carrots so you can both pay me and keep something for yourself. In fact, since it was your idea, you think you should make more money than me, even though I'm doing the work. So you intend to charge 10 shekels for the carrots, keeping 6 for yourself. And that's where your little plan falls apart. The only people with shekels are you and—when and if I get paid by you—me. (Remember, you're the original capitalist and I'm the original worker.) Even at a price of 5 shekels, I won't be able to buy all the carrots, because you will have only paid me 4. You could buy the difference, but then you would only be buying from yourself, which wouldn't be buying at all. No money to be made there. Or anywhere.

Of course, capitalism doesn't work this way or it wouldn't work at all. Fortunately for capitalism, there are lots of workers and lots of capitalists to pay them. But even if we consider the whole capitalist world, there is still a major issue to contend with: The sum of all workers everywhere must buy more than what the sum of all workers everywhere were paid, if there is to be profit and some way to pay workers to begin with. (It might help to read that last sentence over again a time or two.) So how does it work? By keeping money ever moving in space and across time, through loans and bills and wages not yet paid, so when the loans, bills, and wages need to be paid in one place there is another place to pay them from.

That makes for a pretty delicate balancing act, though. If doubt festers about the economic promises people have made to each other (for that's what loans, bills, and wages are: promises), the house of cards collapses into recession or even depression. This is perhaps the most funda-mental reason why economic growth is so central to capitalist economies. If there is simply more money around by the time our various economic promises have to be redeemed, it makes it much, much easier to deal with a few little imbalances here and there and keep everyone's confidence going. The capitalist doesn't have to always be holding his or her breath, lest the card house tumbles in the exhale.

Not just capitalists, though: We all find ourselves hoping for some economic growth, for a variety of reasons. One is inflation. Because of it, money left by itself decreases in value. Stuffing your wad in a mattress, even a very secure one, is a rather shortsighted practice. But even without inflation, the uncertainties of the market lead us to nurture our money so that we are not left short. For instance, the possibility that employers may decide to downsize keeps nearly everyone at some degree of economic risk. Another source of risk is the business cycle, the seemingly inevitable tendency of any nation's economy to have a few cards fall, increasing unemployment and other forms of economic stress. Consequently, almost everyone with any accumulation of money seeks to build with it a cushion that will at the very least retain, if not increase, its size and comfort. For these reasons—putting aside for the moment the common desire for simply having more—those with money invest it, preferably in a way that offers high returns.

The inclination to seek high returns resonates throughout the economy. A bank can offer good interest rates only because of its own success in investing its depositors' money in loans, stocks, bonds, and other monetary instruments with strong returns. If the bank fails to offer competitive interest rates, it will lose depositors and possibly be forced to close. We often choose to improve on the rate of return banks offer us by investing (if we have the time and sufficient capital) in stocks, bonds, and other financial markets. Managers of mutual funds and individual investors, like banks, seek a high rate of return for these investments. Mutual fund managers want to keep customers, and individual investors want to make it worth their time to play the market themselves. Even if an individual invests in socially and environmentally responsible stocks, bonds, and mutual funds, the tendency is still to seek the highest rate of return possible from those opportunities.

Those of us with no savings—perhaps most especially those of us with no savings—also usually look to increase our stock of money. We sell our labor power, auctioning it to the high-est bidder. We seek to purchase goods—even positional goods—at low prices. And sometimes we trade work and goods informally to avoid depleting our scarce cash stocks.

In other words, nearly all of us find ourselves continually seeking more money—even to hold our economic place—which almost unavoidably leads us to assume the motives of the market: Seek the highest returns for labor and capital, and minimize costs. Sell high, buy low. The needs of money become our needs, too.

This *generalization of the market* has the important consequence of promoting political interest in economic growth. It's not just a thing for capitalists. (Or maybe in this sense, we are all capitalists.) If everyone is to have more wealth at the end of the year than at the beginning of the year, the economy must grow. This is simple math. Most politicians therefore see eco-nomic growth as a potential way to maximize the number of people who have been made wealthier—and thus, they hope, to maximize votes. Consequently, virtually all modern govern-ments try to promote economic growth.

But there are no guarantees that growth will increase everyone's wealth or that it will increase everyone's wealth equally. Either way, growth in overall wealth may be accompanied by growth in inequality. Indeed, without mechanisms for continually re-leveling the playing field, an increase in inequality is probably an unavoidable consequence of economic growth. Even if everyone gains at least some wealth, the trend toward inequality will persist, since the wealthy can almost always take better advantage of investments, labor auctions, and other economic opportunities—an economic version of what the sociologist Robert Merton once called the "accumulation of advantage."[6] And if the gap grows too wide, the problem of the original capitalist will stretch everyone's economic promises to the breaking point. Workers won't be able to pay their loans and buy things like houses and widescreen TVs. And when workers can't buy enough, capitalists can't pay enough, so workers buy even less and capitalists pay even less—until the economic house has very few cards left standing. This is not a happy thing.

Thus, although in the short term economic growth may help resolve political conflict, as many politicians hope, in the long term economic growth may only exacerbate it. Nevertheless, country after country continues to seek political salvation in economic growth, often with little regard for economic equality. Environmental concern usually remains on the political sidelines, bumped aside by the political momentum for economic growth, and economic inequality compounds as countries increasingly bump re-leveling mechanisms to the sidelines as well.

The Treadmill of Production

The momentum toward economic growth, economic inequality, and environmental sidelining is greatly heightened by the competitive pressure for production faced by firms and by the disadvantaged position of workers in this competition.

To maximize profits—a common desire and, because investors have to be repaid and retained, a common need of business—each firm tries to produce more goods more cheaply than the others. Merely making a profit isn't good enough. A firm continually needs to maximize its profits or investors will withdraw their support and put their resources in a firm that does. As the environmental economist Richard Douthwaite has written, "It is not just that firms like growth because it makes them more profitable: they positively need it if they are to survive."[7] Employee-owned and privately owned firms can often shelter themselves from the maximizing pressures of investors. But even they will sometimes have to go looking for outside capital. Repaying and retaining investors requires economic success on a scale of months and quarters, diverting attention from longer-run issues like the environment.

Firms are interested in more than just repaying and holding onto their investors, however. They are also interested in profit for themselves. Just like our original capitalist, owners and management usually decide to put as much profit as they can in their own pockets, after paying their debts, paying for needed reinvestment in the business, and paying employees enough to keep them coming to work. Thus, it is a virtually universal pattern, although by no means economically necessary, that employers take more home than employees. (Indeed, it is a virtually universal pattern in both nonprofit and for-profit institutions that those who have the most control over budgets are generally the highest paid. The best predictor of pay is not how hard you work—the hardest work, such as manual labor, is often the worst paid—but the power of your organizational position.)

In sum, the tendency of unfettered market forces is for increased growth, increased production, increased environmental consequences, and increased inequality.

A Widget Treadmill

Say I'm a widget maker. To set myself up in widget making, I probably had to borrow quite a bit of money. Machinery, land, buildings, labor—all these have to be paid for, often before the business is generating sufficient returns to cover these costs. Over time, the loans have to be repaid, which requires paying back more than the value of the loan. (Creditors want their profit, too.) If I raised capital through selling stock, those investors have to be repaid as well, and at a level of return that attracts their investment in the first place. In short, I'm under a lot of pressure to make a profit.

Let's say I am a good widget maker, though, and I have managed to figure out a way to keep my production high enough, my costs low enough, and my market big enough that the books are balanced. And let's say that I am doing well enough to take home a good profit for myself, not just for my creditors and investors.

But I am not the only widget maker. You make widgets, too. Like me, you are also interested in profit, perhaps even a bit more interested in it. Maybe your house is not as big as mine and your car is not as fancy. Maybe your workers have asked for a raise and you do not want to take it out of your own returns, however large they may be. Maybe your suppliers have just raised their prices. Maybe your shareholders are putting pressure on you, threatening to dump your stock in favor of a more profitable enterprise. Maybe all of these things are true. So you try a bold move. You agree to pay your workers a bit more, but only if they agree to a new kind of widget-making machine that requires fewer workers to run it and yet increases your factory's output considerably. You work out a deal whereby the workforce shrinks over time through retirement. No current employees lose their jobs, you are paying higher wages, and still your labor bills fall over time. Prices drop as your new widgets flood the market, but because your output is now so high, your profits go up nevertheless. Sounds great for your company.

Meanwhile, I've got to find a way to deal with these lower prices. My investors are increasingly dissatisfied, but I'm stuck with my old widget-making machine and my old rate of output. Your new widget-making machine is patented, and you've put a very high price on the rights to the technology.

So I call up some local government officials and politicians and ask to have some labor and environmental laws relaxed. Otherwise, I tell them, I will have to close my plant or move it elsewhere. I have always been a good contributor to the political party in power. Besides, the region needs jobs. The politicians agree and relax the laws. Meanwhile, I pressure my workers into accepting a pay cut in order to keep the plant open. Because the new labor laws have undermined their bargaining position, they accept.

This solution works fine for me until the government in your area also relaxes labor and environmental laws. (You are a good campaign contributor as well.) Now I have to try something else. I have already pinched my workers as far as they will go, and they are ready to strike. The government is cracking down on illegal foreign workers, and I can't find cheaper labor anywhere else. I have also pinched my creditors as far as they are willing to go before they call in their loans and shut me down. My mood is grim.

Then one of my engineers calls to say that she thinks she has a way to increase production even more, if I can just acquire the capital to put into place the even better widget-making machine that she has just dreamed up. Although the new machine will use a toxic chemical, that should be okay under the newly loosened environmental laws. The idea is intriguing, but I fear that this plan will so flood the market with cheap widgets that prices will drop even further. I still will not climb out on top.

Then I hear that the national government is mounting a summit on free trade with a formerly hostile power, and I sign on as an industry representative. The deal concludes favorably, and I suddenly have a new market for widgets. I install my engineer's new widget makers in all the production lines, having used the promise of a new market to attract the necessary capital, and things are looking great. Pollution has increased quite a bit, and the workers are still upset about their low pay, but I am turning a profit once again.

Then the widget makers in that formerly hostile country start feeling pinched, and they pass the pinch on to their own national government and their own engineers and their own laborers. They also send industrial spies into my plant to figure out what I am doing. Since free trade works both ways (if it really is free trade), pretty soon their widgets are coming over here. My profit drops, but I am still alright because I was able to pay off some loans before the foreign widgets started coming in. I consider reinvesting in a still-faster version of my new widget-making machine, but I decide to sit tight to see what you do.

You are desperate. You threaten your workers with a pay cut, and they counter-threaten with a strike. You try boosting prices and mounting an ad campaign that promotes the supposedly higher quality of your widgets. But the campaign is a bust, orders drop, and some of the backlog of widgets in your warehouse develops rust. Your reputation for quality, in fact, goes down.

So you are faced with two general choices. One is to cut your losses and shut down your factory. This move would lower the overall output of widgets and raise prices, to the benefit of your competitors—like me—who are still making widgets. This would be emotionally hard for you to take, though. The other choice is to try to attract new creditors, to come up with a new invention, to win more concessions from your workers, to discover another new market, to talk to your local government about loosening more laws or about giving you a tax break, or to find some other means of keeping the money coming in and the products going out. But if you choose the latter course, the foreign competitors and I will probably try to respond in kind, which will increase overall production, lower prices, and raise economic pressures once again—all the while diverting attention from environmental considerations.

In broad outline, this is a familiar and endlessly repeated story in industry after industry. Through competition and increased production, returns to capital—profit—decline over time, creating what the environmental sociologist Alan Schnaiberg and others have termed the *treadmill of production* (see Figure 3.1).[8] It's a process of mutual economic pinching that gets everyone running faster but advancing only a little, if at all, and always tending to increase production and to sideline the environment.

Figure 3.1 The treadmill of production: Mutual economic striving keeps us always struggling to increase production, often with little regard for social and environmental consequences.

Source: Matthew Robinson and the author.

The Struggle to Stay on the Treadmill

And yet there are limits to how fast people—and competing companies—can run. You can only work so hard. One adjustment mechanism is for someone to be forced off the treadmill—to accept the first option mentioned above—lowering production to some kind of equilibrium with costs and prices. This is an outcome that everyone on the treadmill resists, however. The second option, making competitive adjustments, tends to be favored, at least initially. Eventually, someone does get forced off the treadmill, though. But that does not necessarily lower overall production. The struggle to find new customers that goes on before someone gets forced off usually means that the market expands in the process. The common result is fewer and bigger businesses—monopolization—and a higher level of overall production, often much higher. Meanwhile, those who are forced out of the market often switch to making different products or delivering different services, increasing the overall level of production of the economy by creating new treadmills.

As firms struggle to stay on the treadmill, they cut back where they can. The usual results are job losses and increased economic inequality as well as further disregard for environmental consequences. The social conditions of industrialism have generally given workers less control than management, owners, and shareholders over cutbacks. The worldwide democratic revolution of the past couple of centuries still pretty much stops at the boardroom door. By constantly pointing to the threats posed by the treadmill of production, and by pointing to the hungry pool of unemployed people, management is able to win the consent of workers to take home less than an equal division of a company's profits.

Not only do firms struggle to stay on the treadmill, but people do, too. Those who lose their jobs may one day find new employment, but they are often forced to accept positions lower on the economic ladder than they previously enjoyed, if they are to have a position at all. The unemployment caused by the treadmill of production may be temporary (or may not be), but it also provides another opportunity for maintaining unequal distribution of income across a corporation, and thus across society at large.

In recent years, the ability of workers to bargain for more of the treadmill's proceeds has slipped even further, leading to declining incomes despite overall economic growth.[9] Consequently, rates of corporate profit have regularly exceeded the economic growth rate. Typical rates of corporate profit are currently in the range of 2 to 12 percent, while typical rates of economic growth are in the range of 1 to 3 percent.[10] This differential is possible only if some are getting less than others. It is thus a measure of the extent to which the pressures of the production treadmill have been turned into an opportunity for the rich to get richer.

Meanwhile, the environment continues to be largely ignored.

Development and the Growth Machine

The dynamics of the treadmill also have an important spatial dimension, which leads to the constant conflicts over development that are familiar to local communities everywhere.

As the sociologists John Logan and Harvey Molotch have observed, local businesses have an interest in local economic growth: Investment is often relatively fixed in space. Buildings, land, machines, and a well-trained workforce are hard to move around, and firms try to create as much economic activity as possible for these fixed investments. Consequently, business leaders almost universally advocate pro-growth policies that increase the circulation of capital through their local areas. Although local business leaders are often in competition with one another, one thing they can usually agree on is increasing the size of the local economic pie. They band together into a variety of alliances that Logan and Molotch call "local growth coalitions." To the extent that these coalitions can persuade local government of the importance of increasing local economic activity, growth becomes a leading cause of political leaders as well, making them a part of coalitions for growth.[11]

The result is that a city or a town or a state acts as what Logan and Molotch termed a *growth machine*, dedicated to encouraging almost any kind of economic development—frequently with little regard for environmental consequences or the wishes of affected neighborhoods. Local people have spatially fixed investments of a different sort, and because of them, conflicts with growth coalitions often arise. Logan and Molotch call these local investments the *use values* of a place: homes, strong neighborhoods, supportive networks of friends and family, feelings of identification with the local landscape, aesthetic appeal, a clean and secure environment. Business, on the other hand, is interested in the *exchange values* of places, the ways that places can be used to make money. The use values that local people gain from a place are often incompatible with the exchange values business can gain. Maintaining open land for a park versus using that land for a housing development is an example from a conflict in an Iowa community where I used to live.[12]

And when there is conflict, the pro-growth business interests typically win. Neighborhood groups tend to be far less organized than local growth coalitions and are usually less able to influence the political process. Moreover, neighborhoods may feel divided allegiances between what is happening in some other neighborhood and their own economic interests, sometimes leading to not-in-my-backyard politics as opposed to not-in-anybody's-backyard politics. In the face of such divided interests, local governments tend to follow the pro-growth policies of the more united, better-organized business community.

A further result of the politics of the growth machine is an inherent contradiction in the tasks we set for local government. On the one hand, we expect government to promote economic growth; on the other hand, we expect government to monitor and regulate environmental impacts.[13] Sometimes this dual role results in well-thought-out development projects that strike a good balance and promote economic growth without compromising the local environment. Yet, depending on the outcome of the political process, it can also turn government into an economic fox that guards the environmental chicken coop.

The "Invisible Elbow"

Adam Smith, the eighteenth-century founder of modern economic theory, envisioned that individual competitive decisions would guide us all toward prosperity by increasing production and efficiency. He suggested the famous image of the "invisible hand" to describe this process. The treadmill of production, however, makes the economy act like what Michael Jacobs has described as the "invisible elbow." Even if the goal is merely to hold one's place on the treadmill, economic actors are involved in a constant jostle. Although this jostling is often unintentional—Jacobs says it is usually unintentional—both people and the environment get compromised in the process. "Elbows are sometimes used to push people aside in the desire to get ahead," Jacobs writes.

> But more often elbows are not used deliberately at all; they knock things over inadvertently. Market forces cause environmental degradation by both methods. Sometimes there is deliberate and intended destruction, the foreseen cost of ruthless consumption. But more usually degradation occurs by mistake, the unwitting result of other, smaller decisions.[14]

The elbowing effect that Jacobs describes is more technically described by economists as *externalities*, economic effects not taken into account in the decision making in a market. Externalities may be divided into two broad types. Increased inequality and pollution are examples of *negative externalities*, costs not included in economic decision making and generally borne by those who did not make the decision. There may also be *positive externalities*, benefits that were not taken into account in an economic decision and may have wide utility, such as more efficient production or, conceivably, an economic arrangement in which individual economic choices promote greater equality, less pollution, and other public goods. The problem with Smith's image of the invisible hand is its rosy suggestion that the treadmill's market competition leads only to positive externalities. Jacobs's invisible elbow points out that negative externalities are also common consequences of the treadmill.

Externalities are not necessarily invisible, though, as Jacobs also points out. As we rush along on the treadmill, we may be well aware of some of the consequences of flying elbows. The *visibility of externalities* is enormously significant for our social decision making. The ability to see and appreciate an externality, whether positive or negative, is the first step toward creating the social conditions that promote the former over the latter. If we are unaware of something, we are certainly unlikely to direct our actions with that something in mind. Creating this visibility is a political act of considerable social and environmental importance.

The Politics of Treadmills

Visibly and invisibly, the elbows really fly as people come to recognize the unequal outcomes of life on the treadmill and the unequal outcomes for life. To be caught on a treadmill, or to be caught up on one, is to be caught in struggle.

Take the expansion of "factory farms" for livestock across the landscape, what the government regulators call CAFOs, for "confined animal feeding operations." The livestock may be confined, but the consequences are not. Next time you stick a fork into a succulent slice of pork roast, consider the red meat political conflict that it took for a corporation like Smithfield Foods—or Groupe Smithfield, as it is called in Europe—to establish a 10,000-head facility in some rural community (see Figure 3.2). A facility with 10,000 hogs means a lot of hog food, a lot of hog manure, and a lot of hog manure stink. Hogs are large animals, and they do little but eat and excrete in these facilities. But the consumer in the city doesn't smell the sometimes overpowering stench that drives longtime rural people from their homes and makes selling their properties sometimes impossible. The consumer doesn't see the manure spills from the often-leaky manure "lagoons" companies erect to contain the hog effluent—spills that killed 5.7 million fish in 152 incidents in Iowa alone between 1996 and 2002.[15]

These elbows are plenty visible to local rural people, however, and across the country many rural communities have fought hard to keep CAFOs away. They have tried to make the elbows of the treadmill more visible, building coalitions with environmental groups concerned about the water and fish, animal welfare groups concerned that the hogs are too packed in and never see the light of day, and economic justice groups concerned about the pay and conditions for the mostly migrant workers, as well as the ways CAFOs knock family farms off the treadmill. And they have joined together to lobby regulators and politicians. Meanwhile, the CAFOs have put together their own coalitions of interests eager to keep the exchange valves flowing. The grain farmers eager for a little boost in price and sales. The meat processors (who often own the CAFOs to begin with) happy to expand their trade. The politicians anxious to generate tax revenue and perhaps a few donations for their reelection funds. Most decisive, of course, is which side the politicians come down on in a treadmill struggle. In the case of CAFOs, few readers will be surprised to hear that in most cases—whether the facility produces hogs, cattle (including dairy), or poultry—industrial interests have mainly won. The bigger the coop, the happier the fox usually is, at least thus far in the political debate. Consequently, in the United States, local state governments have often taken away the right of localities to enact anti-CAFO zoning, and have banned "nuisance lawsuits" against them. Plus, they have put into place generally mild environmental regulations, often

Figure 3.2   Inside a large-scale hog confinement facility or CAFO, near Waucoma, Iowa. Such "factory farms" for livestock are becoming increasingly common throughout the industrialized world, despite concerns about their implications for human health, animal health, economic justice, and the environment.

Source: Corbis, used by permission.

hand in hand with special government funds to pay the new costs associated with the regulations, such as money to build stronger manure lagoons. Which is a good thing, say advocates, but mainly a bit of political window dressing, say the critics. And as the treadmill continues, so does the political debate.

The Treadmill of Underproduction

Even with political interests lined up to smooth laws, quell opposition, and build markets, producers still have their challenges. The environmental sociologist James O'Connor argues that two of these challenges are especially fundamental.[16] The first is what O'Connor calls the *crisis of overproduction,* which is pretty much what the treadmill of production theory describes. O'Connor argues that this crisis stems from the "contradiction" in the structure of the economy that producers never truly resolve. Despite any current success they may have achieved in technology, markets, labor costs, and politics, the pinching and elbowing still go on, as overproduction lowers profit margins and spins the treadmill of production ever faster.

The second challenge is what O'Connor calls the *crisis of underproduction,* and it stems from another "contradiction"—that sometimes producers find themselves undermining the same social and environmental relations that make their production possible. The crisis of overproduction points to social and environmental consequences that are mainly external to each producer, such as the negative externalities of CAFOs: air and water pollution, economic inequality, and flooded markets. For the adept producer, with big elbows, these are other people's problems. But the crisis of underproduction is a more direct and fundamental threat, as the declining capacity of environment and labor to keep output going undermines the very possibility of production.

The environmental sociologist John Bellamy Foster, along with his colleagues Brett Clark and Richard York, make a related point.[17] They argue that human society's interaction with the earth can best be understood as a process of metabolism, in which we gain the nutrients necessary to our sustenance and later return the products of our metabolism back to the Earth, yielding more sustenance. Think of it as a kind of farming of the whole Earth—but a regenerative kind of farming that gives back to, and does not rob from, the land. However, our economic structures, Foster and his colleagues hold, discourage this regenerative interaction. Instead of cycling back the nutrients of human existence, our economy imposes what they suggest calling a *metabolic rift* in our lives. As a result, producers wind up robbing both from society and from the land.

I like to think of it as another treadmill, the *treadmill of underproduction,* in which the efforts of producers to respond to declining production lead to even greater production declines through the destruction of overworked productive capacity. Fishing is a vivid example. In fishery after fishery around the world, fish harvests have been so complete that insufficient spawning stock remained to repopulate the fish, which in turn led to more harvesting that cut further into the spawning stock, until the stocks collapsed, making further production impossible. The cod stocks of the Grand Banks, once considered inexhaustible, virtually disappeared, compelling the Canadian government in 1992 to declare a moratorium on further cod fishing there. Cod lingers on in the Georges Bank and the North Sea, where the fishery remains legal, but stocks are dangerously depleted. The U.S. National Oceanic and Atmospheric Administration estimates that, as of 2010, the Georges Bank cod stock is 10 percent of what it should be.[18] Meanwhile, the orange roughy, one of the species the world turned to in the 1970s when cod went into decline, has now in many places (especially New Zealand and Australia) already collapsed as well.[19] Swordfish, toothfish, Atlantic bluefin tuna, plaice, anchovy, sand eel, sole, blue whiting, and more: The list of fully exploited, overexploited, or depleted fish now totals some 77 percent of world marine fish stocks.[20] And did I mention whales?

It is not uncommon, though, for a producer to be found with one foot on both treadmills, facing declining profit margins through production increases at the same time that it takes greater and greater effort even to maintain the same level of production. In fishing, this scrambling dance of two treadmills has often led to a few huge and highly mechanized boats that harvest way more fish per vessel and yet fewer overall. Our ordinary understanding of economics would tell us that at least the price of the fish should rise, due to the lower supply, reversing the problem of declining profits. But the *Jevons paradox*—named after William Stanley Jevons, the nineteenth-century British economist who first observed it with regard to coal—points out a dilemma.[21] The implementation of more efficient means of production can lower costs so much that people demand the product more. Let's hear from Jevons on the subject:

> It is the very economy of [coal's] use which leads to its extensive consumption. . . . Economy multiplies the value and efficiency of our chief material; it indefinitely increases our wealth and means of subsistence, and leads to an extension of our population, works, and commerce, which is gratifying in the present, but must lead to an earlier end. Economical inventions are what I should look forward to as likely to continue our rate of increasing consumption. . . . But the end would only thus be hastened—the exhaustion of our seams more rapidly carried out.[22]

In other words, more efficient production can actually increase resource use until in the end there is simply nothing left.

The Social Creation of Treadmills

These kinds of arguments can make for pretty gloomy reading. Purely materialist and organizational explanations like the treadmills of production and underproduction, and economic theories in general, tend to give off an air of inevitability that makes one just want to go back to bed, with a pillow over the head. It all feels so external, so out of our control.

So here's some good news—at least sort of good news. The environmental sociologists William Freudenburg, Peter Nowak, Dana Fisher, and their colleagues point out that the environmental impact of different economic actors is highly variable.[23] Some industries are far more polluting than others. Think of oil refineries versus software developers. And some companies within the same industry are far more polluting than others. Think of big hog confinements versus organic farmers that raise small numbers of hogs outdoors on pastureland. In fact, it turns out that most of the damage comes from a relatively small number of economic actors. Freudenburg found, for example, that 46 percent of toxic releases in the United States come from the chemical industry, and yet the chemical industry is only 2.9 percent of the GNP of the United States.[24] He then looked within one industry, the primary nonferrous metals sector—processors of primary ores (i.e., not recycled) other than iron. Here Freudenburg found that a single company, the Magnesium Corporation of America, accounted for 95 percent of all toxic releases in this industrial category, according to government records.

In other words, a few "bad apples" are mainly responsible for spoiling the whole barrel of ecology.[25] *Disproportionality* is what environmental sociologists call this finding. The next question, then, is how do those bad apples get away with it? Freudenburg argues that this privilege to pollute mainly comes about by political arguments that divert attention from disproportionality.[26] For example, a common argument against regulating polluters is that it would drive an industry either out of business or out of the country to a nation that doesn't regulate it. This claim implies that all companies in that industry would be affected the same by the regulation. But if most of the pollution is coming from only a few companies in an industry, that means the rest of them shouldn't be bothered much by regulations that force them to clean up. Why? Because they are already pretty clean. It also means that it is possible to be pretty clean—that alternatives are achievable.

This doesn't mean that all economic actors will utilize those alternatives, though. Economics is fundamentally a social process—the outcome of the interests of various social actors, the power that those actors have, and the level of their concern for the interests of others. How an actor responds to the treadmills of the economy is a people thing, not just a money thing.

The Treadmills Inside

In other words, how we respond to economic treadmills also depends upon how we understand our interests—that is, upon the ideas and sentiments we bring to bear on our lives. There is an important ideal dimension to treadmills of production and underproduction.

For example, our productivist mentality depends in part on our sense that hard work is a moral virtue—a historically recent notion of virtue, in fact. (Chapter 6 considers the origin of this notion of virtue in some detail when we explore the theories of the sociologist Max Weber.) We usually regard laziness as somehow immoral, even when working harder will only secure far more wealth than we physically need. In other words, we work hard not only to maintain our footing on economic treadmills but also because we have internalized the notion that hard work is virtuous.

The idea that hard work is virtuous is particularly pronounced in the United States and Japan, as many have commented (see Figure 2.2 in Chapter 2). The prevalence of this attitude is likely part of the reason vacation time is typically about 2 weeks a year in these two countries. Japanese and American workers have not fought harder for longer vacations in part because doing so would seem culturally inappropriate, which in turn perpetuates the cultural inappropriateness of long vacations. Meanwhile, both countries fall ever deeper into the cycle of "work-and-spend" discussed in Chapter 2.

In other words, economic treadmills are not just external pressures that we must conform to; they are also due to pressures that come from within. At some level, most of us actually want to work hard—which we think reflects well on us—although probably not as hard as economic treadmills often require.

Given the speed increasingly required to stay on the production and underproduction treadmills, it is probably materially advantageous that most of us are committed to the idea that hard work is a virtue. This advantage does not mean, however, that external forces determine internal ones. The relationship between external forces and internal forces is a dialogical one. True, hard work helps one stay on a treadmill, but hard work also leads to an acceleration of a treadmill—thereby creating conditions that encourage people to work even harder. Internal social factors such as a person's values thus support and are supported by external social factors like the economic speed of a treadmill. The treadmill within and the treadmills outside create each other.

Also, John Bellamy Foster points out that the goal of an economic treadmill isn't production (and it certainly isn't underproduction, either).[27] Rather, what keeps people going on it is the desire for *accumulation*—more money, more stuff, and the sense of greater environmental and social power that comes with more of both. It's about more, stupid. Production is just the way to get it. And as Chapter 2 discussed, the desire for more is not some kind of unchanging and universal inner hunger, even though it often appears that way from the perspective of our own time and place. The treadmills of production and underproduction are as much cultural matters as economic matters, as much effects of our ideas as of our material conditions. They are dialogically created, even as they present themselves to us with monological force.

The Dialogue of Production and Consumption

In other words, the twin treadmills of production and underproduction interlock their whizzing movements with the treadmill of consumption. Increasing the pace of the consumption treadmill increases the pace of the production twins, and vice versa. But also, the desire to increase the pace of one treadmill depends in part upon the social conditions created by another (see Figure 3.3).

For example, the desire for *more* is propelled by both the treadmill of consumption and the twin production treadmills through the relationship between *more* and profit. Because you need profit to survive on the production and underproduction treadmills, they encourage people to regard profit as a virtue. But in order to demonstrate your success in attaining the virtue of profit, you need to consume *more*. You need to buy a BMW or its equally conspicuous equivalent. Of course, consuming *more* is a desire that, because of the consumption treadmill, you likely already had. We want profit to consume, just as we consume to demonstrate profit. This, too, is a dialogue—although perhaps not one well calculated to support sustainability, environmental justice, or the beauty of ecology.

Figure 3.3   Interconnected treadmills: The treadmills of production and consumption each propel the other along, and their social and environmental effects.

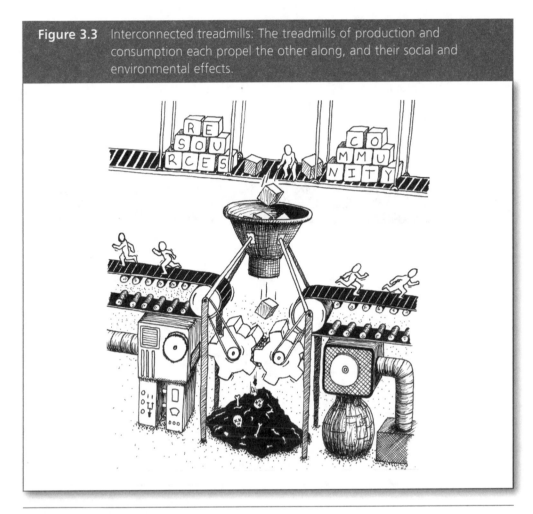

Source: Matthew Robinson and the author.

Hard work is another internal treadmill that is propelled by both production and consumption. Hard work keeps us going on the twin production treadmills, as the previous section describes. But it also keeps us going on the consumption treadmill by helping us to rationalize inequalities in consumption. "God helps those who help themselves," we often hear. Perhaps the real theological help is moral: relieving guilty consciences through the message that it's alright to consume *more* as long as you worked hard for it.

The internal treadmills of *more* and of hard work, then, dialogically connect the treadmills of production and consumption. Of course, you cannot have production without consumption, and vice versa—at least not for long—as classical economics readily recognizes. But classical economics typically treats production and consumption as if they were separately generated phenomena, balanced by price. That balancing is often the only acknowledgment of dialogical interconnections. But what a fully dialogical approach suggests is that production and consumption do not merely balance each other; they create each other.

Moreover, the treadmills of production and consumption do not just mutually condition internal values like hard work and the desire for more. They also dialogically organize the external circumstances in which we have these values. Here is where economic inequality plays its most important dialogical role. The inequality created by the treadmills of production and underproduction helps create the conditions under which the Joneses are constantly rising above us, fueling the treadmill of consumption. That fueling, in turn, creates the opportunity for further acceleration of the twin production treadmills and their frequently unequal economic outcomes. It's a vicious dialogical circle: Production treadmills create inequality, which creates the consumption treadmill, which creates more inequality and a further speeding of the twin production treadmills, thus keeping the whole cycle whirling ever faster (see Figure 3.4).

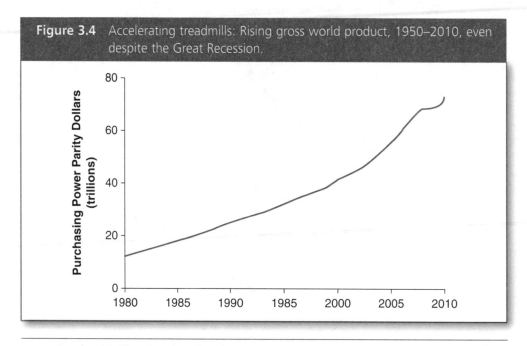

Figure 3.4 Accelerating treadmills: Rising gross world product, 1950–2010, even despite the Great Recession.

Source: Based on Worldwatch Institute (2010) and the International Monetary Fund World Economic Outlook Database.

But because this circle is dialogical, its vicious outcome is not inevitable. For one thing, there is also a potential twin to the treadmill of consumption: a *treadmill of underconsumption.* I don't mean competitive starvation here. Rather, I mean what Veblen might have called "conspicuous nonconsumption," and its moral siblings: "conspicuous non-leisure" and "conspicuous non-waste." If your goal is to show off—and why not, at least a little—a reversal of the now-common Veblenesque logic will sometimes do very nicely. Consider monks, nuns, and the Amish in their plain dress, sworn to a life of simplicity and poverty. Or consider your friend the eco-freak, decked out in patched and recycled garb from the thrift shop. Their clothing may not be high fashion, but it usually stands out just the same, affording a sense of confident identity. For these and other forms of conspicuous nonconsumption definitely come with a strong statement of environmental power: I am so secure in my self and status that I don't have to overconsume to demonstrate my worthiness. There can even be a little competitiveness at work here, both with conspicuous consumers and with other conspicuous nonconsumers, with ever-newer standards of what counts as conspicuous nonconsumption, ever-newer fashions of non-fashion.

Of course, we don't see a lot of conspicuous nonconsumption among the wealthy countries and wealthy classes today. But it is certainly imaginable, and within our powers to bring about. And why not—at least a little?

The Dialogue of State and Market

Economics, in other words, is not something that, like some force of nature, just happens. Economics is something we do. A vital place to recognize the social origin of economic forces is in the generally misunderstood dialogic relationship between the state and the market. Markets function best when the state intervenes least, we are often told. The glory of the West is its "free market" economy, as opposed to the failed centrally planned economies of the old Soviet bloc. Here's the classic definition of that glory, from an introductory economics textbook of a few years ago: "Markets in which governments do not intervene are called *free markets*."[28]

But recall the way government is hooked on the treadmill, too. Western politicians know that their political futures hang from the thread of economic growth, as this chapter has discussed. And recall the way that businesses encourage politicians to keep their focus here through lobbying and political contributions. That's because both politicians and businesses recognize that state intervention is, in fact, as central to capitalist "free market" economies as it is to a "centrally planned" economy. A heck of a lot of central planning, and central control, goes on in both. Where would a market economy be without a police force and a court system to enforce contracts, limit fraud, regulate trade, and establish product standards? Where would it be without a trustworthy money supply? Or without trade agreements with other nations and the political muscle (and sometimes military muscle) to back them up? Or without an independent body to enforce property rights? Otherwise you might try to steal from me, and I from you, without fear of retribution from anyone but each other. And no market society could last long today if it ran so roughshod over workers and the environment that its political legitimacy was undermined. Plus, there are all the other "free" services that government provides the market—like roads, an educated workforce, and a healthy and productive ecosystem. Moreover, in terms of sheer dollars spent, government is about a half to a third of the total economy in most Western nations—including the United States, that paragon of the free market. In other words, the treadmill is hooked on government as much as the other way around.

But isn't this heavy role of government the main problem facing Western economies today? Sure, maybe we can't get rid of government entirely, but can't we get rid of all the red tape and regulations, or at least most of it? Can't we at least have a *freer* market?

The answer to this question is to ask another question: Freer for whom?[29] Let's take a typical case of government intervention to enhance a "free market," like the banning of nuisance lawsuits against CAFOs and the prevention of local control over where CAFOs can be placed. Is there any less regulation as a result? It is true that CAFO owners are freer to do what they want, without fear of retribution, but those who object to CAFOs are now a lot less free. It's much harder to sue CAFOs now in states that have passed such laws. If you meet an unsympathetic judge who deems your suit a "nuisance," you'll have to pay the legal fees of the other side—a strong restraint on advancing any suit at all, no matter how strong you think your case is. And local people in these states can't use local zoning ordinances to control CAFOs either. Or take the "clear skies" initiative advanced by the U.S. government in the early years of George W. Bush's administration. Under it, industrial facilities would not have had to upgrade pollution equipment when they upgraded plants, unless the upgrade was more than 20 percent of the cost of that aspect of the facility. Factories would have been freer to do what they wanted to do, but everyone else would have been a lot less free to pressure factories to clean up their pollution.

Freedom can be like that. Freedom for one person may well entail constraint for someone else. As an old English proverb puts it, "Freedom for the pike is death for the minnow." This is why the philosopher Isaiah Berlin, from whom I get this proverb, used to speak of two kinds of freedom: *freedom-from* and *freedom-to,* the freedom from others stopping you from doing what you want and the freedom to take agency over the conditions of your life, which is to say other people. Berlin liked to call freedom-from *negative liberty* and freedom-to *positive liberty.* The pike eating minnows experiences negative liberty from anyone stopping it from gorging at the minnows' expense. If minnows could stop the pike, perhaps by enlisting the help of some piscine police force, they would experience positive liberty to take action against being gorged on.[30]

Which makes it sound like free markets are about negative liberty and the absence of regulations while positive liberty is what we get through regulation. So if you want more freedom-from—more negative liberty—you need deregulation. Many have read the implications of Berlin's argument this way.[31]

But that is to see negative liberty in purely individualistic terms. For you to have freedom-from, someone else may very well have to be denied it. For CAFOs to have freedom from complaint, others must be denied freedom to complain. In other words, freedom-from requires as much regulation as freedom-to. It's just a different kind of regulation. The "deregulation" that many call for in support of the negative freedom of the "free market" or a "freer market" is what we might call *negative regulation,* regulation-*from*—that is, regulation from interference. The kind of regulation we are used to calling regulation is *positive regulation,* regulation-to—that is, regulation to interfere.[32]

And both forms of regulation require copious lines of legal code in the statute books of every government. Take the U.S. Code of Federal Regulations. In 1980, it took 164 volumes to hold its 102,195 pages. By 2002, it had risen to 207 volumes and 145,099 pages—a 42 percent increase in pages—despite years of efforts at "deregulation" and "reinventing government."[33] But there was no deregulation here. Rather, it was mainly *re*-regulation of the negative sort, to the advantage of some and the disadvantage of others.

Freedom doesn't have to be *zero-sum,* as game theorists call situations in which one person's advantage depends upon another's disadvantage. Wise and just policy finds ways to help everyone so that there is more freedom for all. But in practice in economic matters, things don't work out in that happy way. In situations that encourage thinking about competitive self-interest, social actors typically have little regard for the interests of others. Indeed, the comparative disadvantage of others is generally exactly what they are after, by whatever means possible. Consequently, a "free market" is certainly no freer of politics than any other form of market.

The Social Creation of Economics

We can summarize the usual results of these dialogues as follows:

- A wage–price gap
- Decline in rates of economic return
- Bargaining with labor
- Bargaining with the state
- Search for new markets
- Fewer workers and more investment per unit of production
- Monopolization
- Increased production
- Increased undermining of productive capacity

- Increased state intervention and regulation
- Social inequality
- Reinforcement of the desire for more and the virtue of hard work
- Reinforcement of the treadmill of consumption
- Conflicts over development
- An obsession with economic growth
- Sidelining of environmental concerns

None of these outcomes is inevitable. As Alan Schnaiberg, John Logan, Harvey Molotch, and William Freudenburg have reminded us, and as I have tried to, treadmills are political. It all comes back to politics—to what sociologists often call *political economy,* the interplay of politics and economics, another dialogue. That is, treadmills result from the actions of human agents pursuing what they take to be their interests and sentiments. If we do not like the result, it must be because we have not fully understood what our interests and sentiments are or because the current distribution of power has prevented us from attaining our true interests and sentiments.

To achieve the outcomes we desire, we must first recognize the *social creation of economics.* The pinching pressures of treadmills encourage us to think of the economy as something outside of us, over which we have little control, as an external structure to which we must submit. And true, the economy has power over us. But we also have power over it. The economy is a result of countless individual decisions, as classical economics has long taught, with effects that present themselves as external structures. Yet it is precisely the fact that the economy begins with real human agents that makes it dialogically possible to direct the economy—by changing the circumstances in which we make our individual decisions.

Those economic circumstances are the result of bargaining among social actors such as the state, labor, and management. They are the result of legal precedents, of moral judgments, and of power relations within society. They are the result of the invisibility of externalities and the current limits to our imagination. They are the result of negative regulation and positive regulation, as well as negative liberty and positive liberty. Economies create societies, but societies create economies. Through bargaining between competing interests, through the selective involvement of the state, through the dynamics of the distribution of social power, through the moral visions of those involved, we shape the economic structures that shape us.

The Needs of Technology

"Not another call! I'll never get this lecture written!" I groaned as I reached for my ringing office phone. "Hello. This is Mike Bell," I said, hoping my impatience wouldn't show in the tone of my voice.

"I'm glad I reached you, Professor Bell," came a pleasant middle-aged male voice.

"Well, I hope I can be of some help," I replied, trying to sound interested.

"I hope so too," he quipped, and we both laughed. "You were recommended to me as a speaker for a conference I'm organizing on how local communities can adapt to technological change."

I was warming up to him. Still, he said "*can* adapt," I noted to myself. I wonder if what he really means is *should* adapt.

"That sounds interesting," I cautiously replied. "Who is sponsoring the conference and whom do you expect to attend?"

"I'm the development coordinator for several towns in the northern part of the state, and we're trying to put together an evening program for local people about hog lots. Are you familiar with the issue?"

"Oh sure," I said. "Of course." Hog lots are CAFOs for pigs. At the time, I lived in Iowa, the leading U.S. pork producer, and you'd have to have been a complete recluse to avoid hearing about the controversy.

"Then you'll appreciate the importance of the conference," my caller continued. "The problem is, people don't like change. But hog lots are coming, and coming fast. So we'd like you to talk about how communities can adapt to them."

I thought for a moment. It did sound like he meant *should*, not *can*. Finally, I replied, "This sounds like an important conference. I'd be happy to speak. But I'd like to speak not just about how communities can adapt to technological change. I'd also want to discuss how technologies can adapt to people. After all, it is people that invent technology."

Now it was his turn to say, "That sounds interesting." Then he added, "I'll have to discuss it with our planning board, though. I'll get back to you."

A week later he called again, a bit more curt. "The board decided that it would like to keep the focus on how communities should adapt," he said. "Thanks very much for your willingness to participate, but . . ."

Can adapt had indeed become *should* adapt. He said it himself.

Technology as a Dialogue

I tell the above story to introduce another of the central dialogues of environmental sociology, the *dialogue of technology*.[34] People often point to technology as one of the great motors of social change. One of the truisms of modern life is how much we have been changed by our technologies. The automobile, the tractor, the airplane, electronic media, birth control, the computer, the atomic bomb—these are all frequently cited examples of the role of technology as a social actor, as an independent agent of social change. Technology is also often seen as central to the pinching demands of the twin production treadmills. Indeed, we are often asked to accept the fact that more change will be coming in our lives because of technology. As my caller suggested, we had better be prepared for it, like it or not.

Our cultural training as Western modernists may make it particularly hard to see technology as an ecological dialogue, as a practical interplay between the material and the ideal, the external and the internal—as something that conditions our lives at the same time that we condition it. The metaphor of the machine, with its vision of sequential and linear causality, has become our master metaphor of technology. The word *technology* itself immediately conjures up images of machines. But a dialogical conception more accurately describes what technology is and does. A dialogical conception helps us to see that, like the economy, technology is not a mechanical imperative—unless we allow it to become one. Technology, too, is a social creation as well as a force that shapes the forms our social creativity takes.

For critics and supporters alike, there is great rhetorical attractiveness in a deterministic view of technology. In an attempt to energize us into action against a form of technology, critics sometimes portray it as a grim juggernaut that is rolling over our lives.[35] On the other hand, supporters like my phone caller often claim that there is little we can do once the Pandora's box of technology has been opened and that we had better make way for the changes that are bound to result. "Get big or get out" was the dictum of U.S. Secretary of Agriculture Earl Butz in the 1970s, explaining why American farmers had no choice but to buy bigger tractors and bigger farms.

Thus, fear mongering is used by both sides. The logic of neither argument holds up, however. If technology is an uncontrollable juggernaut, then there is no point in asking us to try to resist it: We can't. If technology is an uncloseable Pandora's box, then there is no need to ask us to make way for it: It is coming anyhow. Both arguments, in fact, logically depend on our having at least some control over technology.

As sociologists Keith Warner and Lynn England argue, technology is more than mere machines: Technology is all the techniques we have for gaining our desired ends. The knowledge of how to work a computer is just as much technology as the computer itself. Technology is the "how-to" of life, Warner and England write.[36] As such, it is not the sole product of external forces: of juggernauts and Pandora's boxes. Technology is something that human agents create. Plenty of human choice is involved.

But once we have made those choices, we will find that our future options have gained not only new possibilities but new limits as well. In other words, technology is not only a *how-to*, it is also a *have-to*. As the French environmental sociologist Bruno Latour and his colleagues have argued, humans make technology, and technology makes humans. Each acts on the other, linking together in a kind of network of mutual consequence, what Latour calls an "actor network"—a theoretical approach that goes by the happy acronym ANT for *actor network theory*.[37] (For more on ANT, see Chapter 8.) In this way, technology shapes the conditions of our lives and thereby helps direct the kind of choices we will feel compelled or inclined to make in the future.

Technology does indeed structure our lives, but we also make that structure and pattern our lives accordingly. The way we interact with technology changes what its consequences are and thus what it is. For as pragmatist philosophy teaches, a thing is what it does.[38] Sit on a table and it becomes a chair. Sit on the floor and lay out your lunch on a chair and it becomes a table. Make way for hog lots and then you'd better make way for hog lots. *Technology is political.*[39] Technology is thus not a mechanical structure, but a social structure, a form of social organization that we control as it controls us, and that sometimes we use to control each other.

Technology as a Social Structure

The automobile is an increasingly prevalent example of technology as a social structure. The automobile industry produces about 50 to 60 million cars every year, and the world fleet of cars and trucks stands at about 1 billion, as of 2010.[40] Nearly every household in Canada, the United States, and Australia has at least one, and the average for the United States is almost two (1.9, to be precise).[41] The United States now has more vehicles for personal use than it has drivers (about 256 million vehicles for 202 million drivers) and about 840 cars for every 1,000 people.[42] Vehicle-to-people ratios are lower in other countries, but rising rapidly. Only 20 percent of Japanese households had cars in 1970, but 72 percent did in 1988. Comparable figures apply to most of Europe. In France, over 30 percent of households now have two or more cars.[43] Western Europe now has about 590 vehicles for every 1,000 people. Figures are much lower in poorer regions. The Far Eastern countries have about 54 vehicles for every 1,000 people, and Africa has just 27 per 1,000.[44] But many poorer countries are also seeing big increases. In 1979, China had only 150,000 private cars; by 1996, it had 2.7 million; by 2003, more than 10 million; and by 2008, it had 24 million.[45] Car sales in India in 2001 were about 700,000 per year, but as of 2010 are approaching 2 million per year.[46] One industry forecast predicts that, as a result of the above, the world car fleet will hit 3 billion by the year 2035.[47]

That is a frightening thought, I think, for the consequences of the current 1 billion cars and trucks are already truly catastrophic (see Figure 3.5). More than 30,000 people die in traffic accidents each year in the United States, far more than die of AIDS each year.[48] For those over the age of 1 and under the age of 35, traffic accidents are the leading cause of death in the United States. Some 59 percent of all deaths for this age group are from traffic accidents.[49] The 5.5 million annual car crashes in the United States injure 2.2 million people, meaning that almost 1 out of 150 people will be injured in any given year.[50] There is thus a very high likelihood that lifelong residents of the United States will be injured in an automobile accident at some point in their lives.

Figure 3.5 "Cross With Caution" is right. Some 5,000 pedestrians die in automobile accidents in the United States each year, and some 60,000 are injured (NHTSA 2007) Worldwide, likely hundreds of thousands of pedestrians die each year (WHO 2004). Speed is a major factor, as the probability of pedestrian death from a motor vehicle impact rises rapidly above 20 miles an hour—from about a 1 in 10 chance to about a 9 in 10 chance at 35 miles an hour (WHO 2004).

Source: Author.

The death rate per capita from automobile accidents is even higher in Portugal, Greece, Estonia, and Latvia. Still, U.S. figures are among the highest in the developed world. U.S. death rates per mile driven are relatively low, but Americans drive a huge amount.[51] Worldwide, about 1.2 million people die each year in traffic accidents, and 50 million more are injured.[52] Most of these are in the less developed nations, where road conditions are often very poor, despite fewer cars. And the trend is up. The World Health Organization (WHO) reports that as of 1998, road accidents were the ninth-leading cause of death injury worldwide, and projects that if current trends hold it will be the third-leading cause by 2020.[53]

Millions of other animals also die each year because of traffic. In Britain, one study found that roads claim the lives of tens of millions of birds a year.[54] Comparable losses no doubt occur in all automobile-dominated countries.

Cars and trucks remain serious polluters, despite efforts to clean up emissions. Growth in the use of cars and trucks has wiped out much of the gain from emission controls, fulfilling the Jevons paradox, and most large cities remain enveloped in smog and fine particulates. In Britain, according to a government study, automobile exhaust kills 10,000 people a year.[55] A 2000 study published in the leading British medical journal, *The Lancet,* found that in France, Switzerland, and Austria, 40,000 people die every year from causes that can be linked back to air pollution, about half of it from cars.[56] The figure for the United States is 64,000 per year, according to a 1996 study, and likely half of that also from cars.[57] The figure for the world, according to WHO, is 865,000 per year, again with perhaps half due to vehicle exhaust.[58] In other words, the worst damage that cars and trucks cause may be effects that we rarely think about and connect to them.

Cars and trucks are also important contributors to global warming and acid rain. In the United States, 27 percent of carbon dioxide emissions come from motor vehicle tailpipes.[59] Motor vehicles also have substantial indirect environmental impacts through the mining required to supply vehicle manufacturers with raw materials, the oil spills and other toxic waste disasters associated with keeping the fuel coming, and the consumptive patterns of land use with which cars and trucks are associated. Road noise can also affect the reproductive success of many wildlife species.[60]

Traffic has a huge impact on the quality of places—on the beauty of ecology. Traffic is noisy and dirty and turns even quiet streets into potential sources of death for us and our children. Cars bring out the worst emotions in drivers. Inside their wheeled cocoons, drivers commonly experience fury over slight infringements on the social decorum of traffic. "Road rage" is what the media call it. Drivers cut each other off, zoom close past bicycles, and accelerate right up to jaywalking pedestrians, routinely threatening all with death and maiming. Streets are a major means by which we encounter the wider community. Through their danger, noise, and dirt, automobiles have made our daily encounters with one another hazardous and unpleasant. By terrorizing public life in these ways, cars contribute to the erosion of social commitment.

Why, then, are cars so popular? The standard answer is because cars are so convenient. Cars vastly increase our personal mobility, it is often said. They are an incredibly flexible and relatively inexpensive form of transportation. They protect travelers from inclement weather. They reduce physical labor. Cars, people often observe, are simply a better way of getting around. To be sure, traffic is sometimes a problem, and so is parking. But, goes the standard answer, bigger roads and parking lots can take care of these annoyances. And of course, we also gain a certain amount of personal pride and romance from owning an automobile, leading to what Americans term their "love affair" with the "dream machine." The pride and romance are understandable when one considers the superiority of automotive transport. Right?

The Social Organization of Convenience

At the center of this familiar argument about the benefits of cars is the image of technological choice, of opportunity, of convenience, and of cars as a better how-to. Taking a bus or a train is so inconvenient. It is no wonder that people across the world are adopting the car as soon as they have the opportunity, we often hear.

But the convenience of cars is not a mere matter of a machine that makes our lives easier. A dialogical understanding of technology points to the *social organization of convenience*—the way we often set up our lives around a particular way of doing things so that it becomes difficult to do things any other way (see Figure 3.6). A dialogical understanding of technology thus points to the common transformation of a "how-to" into a "have-to." If we allow the alternatives to disappear, it indeed becomes hard to do things any other way—which is precisely what has happened with cars.

However, the transformation of a how-to into a have-to is by no means inevitable, as cars also show. It depends on how we spend our money and allocate our resources. Supporters of the car often point to the substantial government subsidies given to public transportation.

Private transportation, the argument goes, pays its own way through gasoline taxes and highway tolls. But many studies have challenged this view by pointing to the hidden subsidies that cars receive—costs that are paid out of general revenues, not automobile-specific taxes.[61] In the United States alone, according to a study by the World Resources Institute, such subsidies total $300 billion every year.[62]

Figure 3.6 The social organization of convenience.

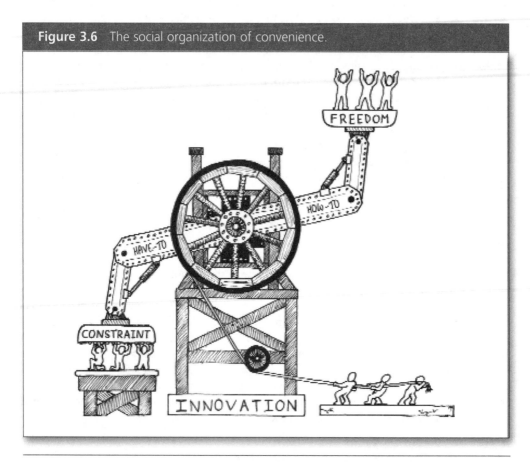

Source: Matthew Robinson and the author.

Some $68 billion of the hidden subsidies for cars goes to road and highway services. Highway patrols, parking enforcement, police work in tracking down stolen vehicles and responding to accidents, and street maintenance—these are very expensive car-related services. Another $13 billion goes to road and bridge construction and repair not covered by gas taxes and other user fees. Road maintenance—snow removal, patching, pavement marking, litter removal, grass cutting on the sides of highways—adds another $12 billion not covered by user fees.

One of the largest hidden subsidies is for parking. Approximately 87 percent of trips in the United States are by car, and 90 percent of these travelers find a free parking space when they get there.[63] The average annual cost to business of one free parking space is $1,000 per year, taking into account what it costs to build and maintain a parking lot and the lost opportunity of developing that land for something else. The same is true for the "free" parking spaces provided by shops. These costs are passed along to customers through higher prices. Of course, drivers help pay for those prices when they shop, and all of a company's employees who drive to work get the same subsidy for parking. But anyone who arrives at work or at a store by some other means is also paying for those spaces and is being denied what amounts to a salary bonus and a price discount for the car drivers—to the tune of $85 billion per year.

Motor vehicle accidents cost the United States $358 billion a year, according to the World Resources Institute study. Most of this cost is borne by drivers through their own insurance. But some $55 billion is not, largely because of the medical costs incurred by the pedestrians and bicyclists who are involved in accidents.

Additional hidden costs include the price of military intervention to keep oil flowing from places like the Middle East and the cost of air pollution on health, crops, and buildings. These figures are hard to pin down, but the World Resources Institute study estimated them at roughly $60 billion a year. Some other studies put these numbers far higher. For example, the United States' tab for the Iraq War alone has been estimated at $3 trillion—although the extent to which the Iraq War was about oil is a matter of political judgment.[64]

The $300 billion in hidden subsidies for cars and trucks works out to 15 cents for each of the 2 trillion miles Americans drove each year at the time. (It is now up to about 3 trillion miles per year.[65]) If a vehicle gets 25 miles to the gallon, the annual subsidy works out to $3.75 per gallon. Per vehicle, the subsidy is $1,579 a year.[66] Another study—which included higher figures for the health costs of air pollution, pollution cleanup, and a figure for the economic inefficiencies of automobile dependence—came up with a whopping 9 bucks a gallon and $5,000 per car per year.[67] No wonder Americans drive so much.

The hidden subsidies for the automobile are classic examples of negative externalities, costs that the actor does not directly bear. Hidden subsidies are more than economic matters, though. They are also political matters. Powerful lobbies—such as the American Automobile Manufacturers Association, the American Automobile Association, the American Petroleum Institute, and the various road-building associations—do what they can to keep these subsidies flowing in ways that benefit their industries.

During the 1930s and 1940s, General Motors (GM) went one step further. Using a company called National City Lines (NCL), a firm it set up for this purpose, GM went around the United States buying up the country's vast network of electric streetcar lines. GM then proceeded to convert these lines to buses, figuring that buses would not be as effective as streetcars at countering demand for automobiles and that GM could sell buses in the meantime. Firestone Tire and Rubber, Phillips Petroleum, Standard Oil, and Mack Truck agreed with this strategy. These companies purchased much of the stock of NCL and helped finance its bus lines with sweetheart supplier contracts.[68]

In the 1970s, the U.S. government successfully brought suit against NCL for antitrust violations. But the damage had been done. The streetcar rails had been ripped up, and roadways had been expanded in their place. As a lawyer representing NCL at the trial put it, "This was all part of a very reasonable corporate strategy to develop a market."[69]

The history of cars thus shows how we often socially organize (and sometimes socially manipulate) the convenience of a technology so that it comes to seem the most appropriate way to do things. The pattern of our lives, from where we live and work and shop to how we organize our families, from the kinds of things we expect from government to the kinds of things we expect of our neighbors, all come to revolve around and thus reinforce the convenience of the car. In this way, the convenience of a car comes to be what Robert K. Merton termed a "self-fulfilling prophecy."[70] As Merton observed, if we decide a thing is true and plan accordingly, very often it turns out just that way.

The Constraints of Convenience

But convenience also constrains. To say something is convenient is to say that we find our lives constrained so that some other thing is not convenient.

Again consider the car. We drive because we have allowed the location of common destinations to become decentralized, on the assumption that most people own cars. We drive because the bus now comes only every 30 minutes and the train once an hour, and maybe not at all. We drive because even if the schedules for the bus or the train fit our own, they probably do not go where we need to go, given that land use has become decentralized. We drive because little provision for bicycles has been built into the streets, because sidewalks are nonexistent or unpleasantly close to a thundering stream of traffic, and because the places we now have to go are so far away. There is no freedom in a car. We drive because this how-to really has become a have-to.

My point is not that we would have more freedom with more buses and trains, trams and light rail, bicycle lanes and sidewalks. Freedom always exists in a dialogue with constraint. If we were to organize our lives so that these technologies were the most convenient ways to get places (as they in fact still are in parts of some cities, even in the United States), we would still find our lives constrained. We would, for example, find it most inconvenient to use a car. Roads would be narrower, far less parking would be available, and land use would reflect the denser and more centralized patterns that make walking and public transport "convenient."

But although such a reorganization might not bring us any more freedom and convenience, it would not necessarily bring us any less, either. And the pattern of convenience that public transportation, walking, and traditional town planning brings could in fact cost far, far less—less pollution, less land and energy consumption, less community alienation, and less loss of life.

Technological Somnambulism

Why, then, do we accept the increasing dominance of cars, despite their considerable social and environmental impacts, when more benign alternatives are readily at hand? The same question could be asked of other technologies and habits of doing things. "The interesting puzzle in our times," Langdon Winner has written, "is that we so willingly sleepwalk through the process of reconstituting the conditions of human existence"—a phenomenon Winner called *technological somnambulism*.[71]

In this section, I briefly take up three answers to this puzzle: the phenomenology of technology, the culture of technology, and the politics of technology.

Phenomenology

Phenomenology means the manner in which we experience the phenomena of everyday life, and one of the central experiences that phenomenologists have long emphasized is *routinization*. If you stop to think about it, even the simplest tasks of everyday life are extraordinarily complex. We need ways to simplify what we do, and making routines is a common way.

Alfred Schutz, the founding figure in this area of sociological research, liked to use the example of walking out to the street corner to place a letter in a mailbox.[72] This simple task is filled with presumptions about how the world works. The postal service will pick up the letter. Postal workers know how to read in the language I have used. The person I am sending the letter to knows how to read this language. There is such a thing as letter writing, and the post office and the person to whom I am writing understand that. There is such a thing as a postal service. There is such a thing as a mailbox, and there is no hungry monster at the bottom of the mailbox, waiting to eat the letter. In order to get to the mailbox, I need to get up, put my feet forward, cross the room, open the door, walk down the hallway, get my coat, put it on, open the outer door, step outside before I close it, close it, walk to the street, and so on.

If one did stop to think about all the presumptions involved every step of the way, one would likely never reach the door. Instead, through accumulated experience, we establish a series of little routines—*recipes of understanding*—that we can call upon without having to think the whole thing through each time.[73]

The metaphor of recipes is quite appropriate; routinization is exactly what a cook's recipe allows. Say I have just found out that a friend is having a birthday and I decide to bake a cake as a surprise. It is 4:00 in the afternoon and the cake needs to be ready by 7:00 if the surprise is going to work out. But how do I bake a chocolate cake, and what kind of chocolate cake should I bake? I certainly do not have time to experiment much, so I reach for my family's favorite recipe for chocolate cake, a recipe I have followed many times before and I know is likely to please. I bake the cake in time, and it is a big hit—because, through my use of a recipe of understanding, I had routinized the process of baking a chocolate cake.

Making use of any form of technology—and mailing letters and baking cakes are both forms of technology—requires a similar process. I know a little bit about how my computer works, although far from everything. But if I stop to think about all that I know every time I urge my fingers to strike the keyboard, it would take me all day to write this sentence, if not longer. I know that my computer works and, given the time constraints I have, I must usually be content to proceed with no more than that tacit confidence—technological sleepwalking.

But suppose a lightning bolt were to strike the building I'm in and send a charge through the electric lines that melts some essential bit of my computer's innards? (In fact, that happened to me once.) How could I keep on writing? I'm quite good about backing up computer files, but many of my files are stored only electronically. Until I got my own machine working again, or bought a new one, I would have to borrow a computer from someone else to read my backup files. If one were available, I might have to contend with software I'm not quite used to. Still, despite these annoyances and economic barriers, I would only reluctantly resort to pen and paper, because doing so would force a major reorganization of my work habits.

The point I'm trying to establish is that our technological routines tend to lock us into continuing those routines and into trying only new routines that mesh well with our older ones. If something disturbs our technological sleepwalking, we will likely do all we can to walk around the disturbance, re-close our eyes, and return to pleasant walking slumber. And if this tactic fails, we may suddenly awake in a startled panic, having lost our confidence in the secure ordering of our experience.

The threat of broken technological routine was widely experienced in the United States during the gasoline shortages of the 1970s. The Arab oil-producing states of the time instituted a cartel to limit production, resulting in much higher prices and widespread gasoline shortages. Local governments instituted various rationing strategies, such as alternating "odd" and "even" days for buying gas, depending on whether a car's license plate ended in an odd or even number. In the Washington, D.C., area, motorists were prohibited from buying less than 5 gallons of gas, to prevent hoarding through topping off tanks. In a car-dominated society, this rule produced near hysteria. A neighbor of my mother-in-law, who lives near Washington, D.C., drove around the block for an hour one day, trying to empty her tank enough to be able to buy five gallons of gas.

The result of this panic attack in the middle of the technological night was not a United States with more public transportation and more centralized land use. Rather, the result was a United States that, critics contend, will go to war to keep the oil flowing and the cars rolling, as the two Gulf Wars later demonstrated. Routinization follows the patterns of technology we have socially organized as the convenient way to do things. Once these routines are in place, they reinforce the same patterns of social organization that gave rise to them in the first place.

Culture

The same kind of dialogical momentum underlies the relationship between technology and culture.

We perceive all technological knowledge through the cultural lenses we use to make sense of our world. These lenses provide our technological means with technological meaning. Without culture, without meaning, we would not know what to do with our technological means. Yet technological means help shape technological meaning. The means available to us mold our contours of choice and thus our sense of the possible. By structuring the character of experience, means affect and sometimes even justify our cultural ends. We love cars in part because cars are now the main way we have to get around. Means become meaning.

Not only does technology shape culture, however; culture also shapes technology. Should we come to believe that cars are ugly, dangerous, and environmentally and socially destructive (a change I think may finally be underway), we may decide to reorganize our lives so that cars are less necessary. We may invent new technology to fit our new cultural aims. Meaning becomes means. Should we maintain a cultural commitment to what we see as the beauty, freedom, and pleasure of cars, we will be unlikely to seek alternatives—an outcome that equally represents the cultural shaping of technology.

Technological progress is one widespread example of the dialogue of means and meaning. On the one hand is our cultural faith in progress: We have gained so much in organizing our lives around technological change, who could dispute its value? We therefore ask for more new and wondrous technological means. On the other hand is our continuing effort to discover such new and wondrous means: We have given up so much in organizing our lives around technological change, how could we go back now? We therefore find ourselves compelled to entrust our lives to the future benefits of further technological change. To do otherwise would take an enormous realignment of our cultural commitments, a rude awaking from our technological sleepwalk.

Signs of such a realignment of meaning are, however, increasingly apparent—in part because of problems with the means. Pollution, danger, increasing social inequality, and the loss of traditional forms of beauty were never mentioned in the technological promise, and yet they have all occurred. But as long as technology keeps delivering on its promise of ever more wonders. . . .

Fears that technology might not even do that may have been behind the quasi-religious reactions to the infamous "Y2K bug." Remember that night? Hardly anyone mentions it now, perhaps out of embarrassment. But the predictions of what would happen when computers cashed in their chips at midnight on New Year's Eve, 2000, are just astonishing to recall. In case you don't recall what all the fuss was about, computer engineers had designed the calendars inside most computers, particularly big mainframe ones, to count only the last two digits in the number of a year and to assume the 1 and the 9. Consequently, when the year 2000 (2K in computer speak) came around, all these computers were expected to reset their clocks to 1900, tangling the subtle dynamics of programming. Normally sober publications like the *New York Times* and *Scientific American* published astoundingly dire predictions of what might happen. Phone lines and electric transmission lines would go dead. Computer-controlled factory production lines wouldn't function. Air traffic control computers would have to be shut down. Automobiles with computerized engine management systems wouldn't start. Some fire engines wouldn't start. Automotive air bags would be spontaneously activated. Land mines would explode. Digital watches would go blank. Medical equipment would lose calibration and give faulty readings. Pharmacies wouldn't dole out refills because prescriptions would show up on the screen as expired for a century. Elevators would stop. Pacemakers would stop. Fire sprinklers would turn on and wouldn't turn off. Nuclear power plants would explode. Nuclear weapons would be automatically launched. Stores would run out of food as refrigeration equipment switched off, processed food production lines shut down, oil refineries stop producing fuel for transporting produce, crops fail because of fertilizer shortages, and people everywhere strip store shelves bare in a massive spree of panic buying. Lawless looting would take over our cities as desperate people struggled to survive. Vigilantism would become the only form of law and order as police forces and the military became incapacitated by the collapse of communication and transportation systems.

Talk about waking up the sleepwalkers. Millions of people took this very seriously. The web was full of sites advising people how best to stock up for the collapse of commerce and how to defend their homes from looters. Probably most people in the computer-dependent countries at least had a look in their kitchen cupboards to make sure there was plenty of canned food and pasta on hand, as well as making sure their wallets were full in case ATM machines really did stop working, as many predicted would be the case for at least a few days. A friend of a friend of mine even flew down to Fiji for the coming of Y2K, in case of Armageddon in the United States—"bugging out," as it was called at the time.

Government and industry took it all very seriously, too. "Y2K readiness" and "Y2K disclosure" became significant factors in investors' evaluation of stock values. In October 1998, President Clinton signed into law the Year 2000 Information and Readiness Disclosure Act during the midst of National Y2K Action Week, and he told government agencies to spend whatever was necessary to set their computers right. Other wealthy nations followed suit, and the World Bank set up the Year 2000 Initiative to help poorer nations debug their computers. According to some estimates, hundreds of billions of dollars were spent worldwide.[74]

And then, virtually nothing happened, even in places like Russia, where computers were old and more likely to have the bug, and where almost nothing was spent on "Y2K readiness," due to the post-Soviet disarray of its government and economy. Y2K quickly became "Yawn2K," and the Y2K bug became the "Y2K bust," as the media smugly put it afterward. But as smug as some tried to be, it was but a thin mask over a red-faced industrial world.

How could we have gotten it so wrong? Probably because of the threat Y2K presented to the quasi-religious character of what is rightly called our "technological faith." As Lewis Mumford, probably the greatest critical scholar of technology, put it, "If anything was unconditionally believed in and worshipped during the last two centuries, at least by the leaders and masters of society, it was the machine."[75] The computer is undoubtedly the central contemporary icon in the religion of the machine. And to have this religious machine proclaim that the end is nigh, precisely on the date that some Christian traditions had long warned of, was to touch deeply into our most ancient fears.

Yet after that moment of worldwide collective doubt, faith was restored, and the computer took up once again its continual advance into the network of our dependencies.

Sure, computers really are amazing. Amazement is the essence of the miracle. But my point is that cultural faith in technological progress dialogically propels us to reorganize ourselves around the computer as much as the existence of the computer alerts us to the possibility of such reorganization and cultivates our faith. And who speaks of Y2K now?

Politics

Technology is also political, as I earlier noted. The political character of technology is made plain both by instances of active manipulation of the market through lobbying for subsidies and by instances of popular resistance to technological structures such as the environmental justice movement. Technological politics can also be passive and tacit, though. Technology is political even when we are sleepwalking. Because of routinization and romance—because of the phenomenology and culture of technology—we tend to uncritically promote particular technological structures and their social interests, sleepwalking a political order into place.

A decision to organize your life around a certain technology draws you into becoming one of its political supporters. Once you have bought your house in the car-dependent and car-worshipping suburbs, you have organized your life in such a way that you are likely to promote continued car use. If gas prices rise or your car becomes old and unreliable, you cannot easily switch to another means of transport. You will likely find yourself paying the higher gas prices and buying a new car—as well as clamoring for better parking and new highways to bypass the traffic caused by all your neighbors' cars. (Your own car never causes traffic and parking problems.) You will likely find yourself maintaining a place in the line of social interests that leads to the subsidies for the automobile, as well as the automobile's social and environmental consequences, a line that stretches from you to the local garage to the automobile manufacturers to the road builders to the oil companies to—say some—the troops patrolling the streets of faraway cities.

Technological somnambulism is technological politics. Because of it, we may ironically find ourselves in the frequent position of promoting technological interests we wish we did not have.

The Needs of Neither

People have a common tendency to externalize the economy and technology—to treat them both almost as forces of nature. We speak of economic and technical changes as being driven by efficiency, subtly implying that these changes follow the thermodynamic laws of physics regarding the conservation of energy. It would be unnatural to reject them, we seem to be telling ourselves. In a parallel way, we speak of these changes as being driven by convenience, suggesting again a kind of external objectivity to the course of technical and economic development. We must therefore adapt to these imperatives, we probably all find ourselves thinking at least occasionally, and live as best we can with their social and environmental consequences.

But neither the economy nor technology is an imperative. Neither has needs. They often appear to us as external structures, as forces over which we have little control, as ends to which we must give way. But for all their objective status as facts of the market and of science, money and machines are our own creations, a part of human culture. It is we who have the needs. What Lewis Mumford said of technology applies equally to the economy: "The machine itself makes no demands and keeps no promises: it is the human spirit that makes demands and keeps promises."[76] Technology and the economy take the shape that they do only in response to human demands, promises, and broken promises. There are no imperatives but our own.

Once set into motion, however, the mills of technology and the treadmills of production and consumption become social structures that shape our needs and interests, just as those needs and interests dialogically shape the mills and treadmills to begin with. We are quickly lulled into the routines and cultural desires organized by these structures. We find ourselves dropping off into both technological and economic somnambulism.

Yet the dialogical influence that the economy and technology have over us should not be a cause for despair. In fact, in that influence lies hope. By changing the conditions under which we find ourselves making the decisions that we do, we can reinforce our movement toward the ends we truly desire. All we need to do is wake up and decide where we really want to go.

CHAPTER 4

Population and Development

God forbid that India should ever take to industrialization after the manner of the West. The economic imperialism of a single tiny island kingdom [England] is today keeping the world in chains. If an entire nation of 300 million [India] took to similar economic exploitation, it would strip the world bare like locusts.

—Mahatma Gandhi, 1928

Thomas Malthus's *An Essay on the Principle of Population*, originally published in 1798, is assuredly one of the most controversial works of all time. The book's basic argument is that unless checked in some way, population growth tends to continue until it runs up against environmental limits, causing poverty, hunger, misery, and resource scarcity. The eventual result is a population crash.[1]

Although Malthus's ideas seem like common sense, you could etch glass with some of the critics' reactions. Friedrich Engels termed Malthus's theory a "vile and infamous doctrine," a "repulsive blasphemy against man and nature," for it implied that the poor, through their alleged inability to control their reproduction, were to blame for their own poverty.[2] Others have called Malthus's theory "racist," for when applied to today it seemingly places the bulk of blame for environmental problems on poor countries, where population growth is the fastest, and thus on people of color.[3] Still others, most notably economists confident about the ability of human ingenuity in a free market economy to overcome just about any problem, have labeled Malthus's views "nonsense."[4]

There truly is a lot to object to in Malthus's ideas, as well as in many of the later applications of his argument. Yet his book continues to be the point of departure for discussions of the relationships among population, development, and the environment. Two hundred years later, we are still arguing about Malthus because, despite the inadequacies of his book, it forcefully states a basic incontrovertible truth: The world is only so big. It offers only so much room for development and only so many resources for people to consume.

The idea that we might face limits does not mesh comfortably with the modern worship of constant growth. It represents a direct challenge to resolving the problem of the original capitalist and the economic treadmills around which we have organized so much of our lives. Much of the objection to Malthus can be understood as ideological reactions to this challenge.

But Malthus also presented his theory of population and its environmental limits in a deterministic way—as something that is inevitable, beyond our control. In the previous chapter, I argued against a deterministic view of the economy and of technology. Malthus's critics are right that we should not accept a deterministic view of population and the environment, either.

Neither should we ignore the influence that the environment does have on society, however. The relationship between society and the environment is a dialogue. Each shapes, but does not determine, the other. We need to avoid both the deterministic outlook of crude Malthusian arguments and the anything-goes outlook of crude anti-Malthusian arguments. The goal of this chapter is to tread our way through the problems of Malthusianism and anti-Malthusianism and to come to a balanced understanding of the dialogic relationships of population, development, and the environment.

The Malthusian Argument

Many of the world's leading environmental thinkers argue that despite his errors and overstatements, we still need to pay attention to the basic point of Malthus's thesis: the environmental and social threats posed by *exponential growth* in population, or growth that is constantly compounding.[5] Malthusians—those who agree with at least this much of Malthus—point out that because of compounding, a constant rate of population growth leads to accelerating effects. Since the annual rate of growth also applies to the additional increments of previous years, not only will the number of people added each year go up, but the rate of the rise will constantly accelerate.

It's like a savings account. A deposit of $100 at a constant 10 percent annual interest rate will increase year by year, first $10, then $11, $12.10, $13.31, $14.64, and so on. Each annual increment is larger by a wider margin than the one before because the interest is applied to an ever-increasing total. Because of compounding, the size of the increments can go up even if the growth rate falls

somewhat, although the increments will eventually begin to drop if the growth rate falls sufficiently. Compounding is great news when you're trying to make money, but the situation is rather
gloomier, say Malthusians, when we consider the appetite of a growing population in a finite
world. As a result, writes William Catton, one of the founders of the discipline of environmental
sociology, "Human life is now being lived in an era of deepening carrying capacity deficit."[6]

The rate of world population growth is, in fact, falling. The peak came in the early 1960s,
when the world population was growing at an annual rate of 2.2 percent.[7] By 2011, the rate had
fallen to 1.08 percent.[8] We therefore have little cause for alarm, say some.

Not so, respond Malthusians. Because of compounding, the size of new increments to the
world's population is still higher than it was in the 1960s, even though the growth rate has
declined. Today's slower growth rate is applied to a bigger population, so the world's average
annual population increase has nevertheless jumped from around 50 million a year in the
1950s to around 70 million in the 1960s to approximately 85 million a year in the late 1980s,
although it has now, as of 2010, fallen to about 76 million a year.[9] The time it took to add a
billion people to the world therefore also sped up. It took 14 years—from 1959 to 1974—to go
from 3 billion to 4 billion people. It took only 13 years to get to 5 billion in 1987, and 12 years
to get to 6 billion in 1999.[10]

More than 90 percent of the world's current population growth is taking place in poor countries.[11] Growth in rich and middle-income countries is far slower. In over 30 countries, including Japan, South Africa, and most of the former European Bloc countries, the population
growth rate as of 2010 is zero or below.[12] In 59 countries, however, population growth is running at 2 percent or more per year—and these are almost all countries with high poverty rates
and low levels of human development.[13] A dozen of them are growing at 3 percent or more.[14]
These may not seem like substantially higher rates, but the power of compounding can be surprising. At a 2 percent growth rate, a country's population would double in 34 years. In 100
years, it would rise sevenfold. A country that maintained a 3 percent growth rate would double
in size in 23 years, and after 100 years would see its population increase 19-fold.[15]

Here are the numbers for a few individual countries. Niger, currently the world's fastest
growing country, has 15.9 million people as of 2010 but is growing at 3.66 percent a year, which
if unchecked would give it a population of 617.9 million in 2110—twice as large as the United
States today. The size of the population base matters a lot, though. Ethiopia has the seventh
fastest growth rate at 3.20 percent per year but a current population of 88 million. If this growth
rate doesn't change, in a hundred years Ethiopia would have a population of about 2.2 *billion*,
larger than any country today. India, now the world's second-largest country, has a current
population of about 1.2 billion, but a growth rate much lower than Ethiopia's or Niger's. If India
continues as its current rate of 1.38 percent growth per year, it would have a population of
4.7 billion in 2010. Finally, let's look at China's 1.3 billion people, the world's largest population
today, but who now have a fairly low annual growth rate of 0.49 percent, due to the country's
severe population policies. At current growth rates, in a hundred years China's population
would be the same size as Ethiopia's: also 2.2 billion.[16]

It is important to distinguish between a population's rate of growth and its level of fertility,
the average number of children born to women in a population. Between 1970 and 2010, the
world fertility rate dropped 60 percent, from 4.3 children to 2.6 children. In poor countries, the
rate dropped from 6.7 to 4.4, and in rich countries, it dropped from 2.2 to just 1.6—less than
what a couple would need to replace themselves.[17] And yet the number of people added to the
world in 1970 was 78 million, only a small amount larger than the 76 million added in 2010.[18]
But a drop in fertility does not necessarily translate into fewer numbers of children born each
year. Earlier high levels of growth often result in a population with a large proportion of young
adults in the prime of their childbearing years. Lower fertility levels may thus result in just as
many, or even more, births—an effect demographers term *population momentum*. Because of

population momentum, it can be many years before a decline in fertility in a country leads to a slowdown in its population growth.

Nevertheless, fertility declines have finally begun to reverse the growth in annual additions to the world population. Some of the recent fall is attributable not to fertility declines, but to increased warfare, to high rates of AIDS deaths in sub-Saharan Africa, and to sharp declines in life expectancy in the former Eastern Bloc due to economic dislocation. But the time it takes to add another billion people is increasing independently of these problems. Demographers now predict that we will reach 7 billion sometime in 2011 or early in 2012—once again 12 or 13 years after the previous billion, but on a bigger population base, indicating that the rate of growth is continuing to slow. They further predict that rates will slow considerably after that, as population momentum lessens, leading to a world population of about 9.2 billion in 2050, with the world growing at about 41 million people a year—about a new billion every 24 years (see Figure 4.1).[19]

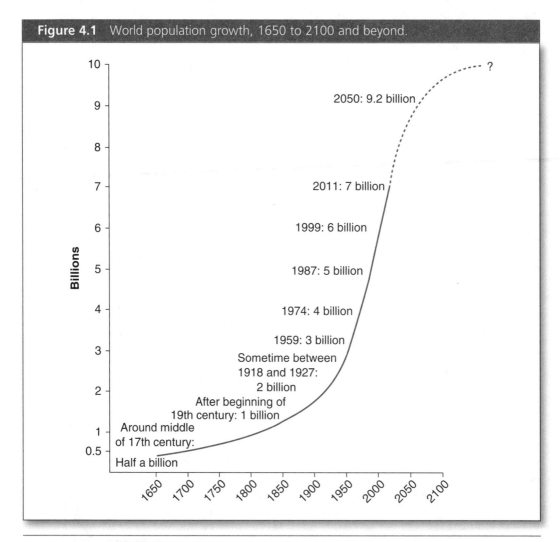

Figure 4.1 World population growth, 1650 to 2100 and beyond.

Sources: Based on United Nations Population Division. (2009). *World Population Prospects: The 2008 Revision* and the midyear estimates of the US Census Bureau *International Data Base* (2010).

Many Malthusians worry that we will still have too many people even if population growth levels out at that 9.2 billion figure, let alone continues to grow by a few billion more. Because of economic treadmills, a constant population may still have an increasing appetite for material resources. Moreover, the same logic of compounding applies to consumption and production: A constant rate of growth means not only increasing effects, but also accelerating effects. Currently, population growth compounds this compounding. Increases in population come on top of accelerating per capita consumption and production. And even if world population eventually stabilizes, environmental impacts may continue to compound because of our growing appetite for resources—perhaps, argue some, leading to a general condition of underproduction and lessened overall carrying capacity. *Overshoot* is William Catton's term for it, a very undesirable situation.[20]

The implications of compounding growth need to be considered carefully. On the one hand, growth in consumption, production, and population does not necessarily degrade the environment—at least theoretically. In fact, population growth itself has no environmental consequence at all. Any environmental impacts depend on what is being consumed and what is being produced by those increased numbers of people and on how they go about their consuming and producing.[21] Improved technology and social organization could possibly compensate for any potential impacts and even leave the environment in better shape than it was to begin with. We could even decide to send a good bit of the human population to another planet someday, if we gain the necessary technological means.

On the other hand, although there are no theoretically necessary environmental outcomes, there are some theoretically likely environmental outcomes of our current forms of consumption, production, and population growth. Consider climate change, air and water pollution, deforestation, loss of habitat and biodiversity, soil erosion, soil impoverishment, per capita declines in food availability, shrinking water supplies, and social differences in the distribution of environmental goods and bads. Challenges to the three central environmental issues—sustainability, environmental justice, and the beauty of ecology—are already well underway. Moving to another planet to escape the pollution and depletion of ours is not currently a realistic option (and in my judgment would be a sad reason for planetary pioneering even if it were). We need, therefore, to be "taking population seriously," according to Frances Moore Lappé, a longtime critic of Malthusian arguments, who might not be expected to hold such a view.[22] It is indeed a small planet.

In the remainder of this chapter, I review the three main critiques of Malthusian arguments: the inequality critique, the technologic critique, and the demographic critique. Along the way, I point out the environmental implications of these arguments in preparation for a final section on the dialogic role of the environment as a social actor. Because the relationships among population, development, and environment are dialogic and not deterministic, I argue that it is possible to change those relationships should we see problems in the current situation.

But before we wade into all that, we should consider the cultural factors that make population a perennially contentious and emotional issue.

Population as Culture

Population is more than just numbers. "Population means *people*," as the Independent Commission on Population and Quality of Life, which conducted a 5-year international study of population issues, observed in its 1996 final report. A South African witness in the hearings convened by the commission across the world put it this way: "Women have children; they do not have population."[23] Children are one of the most important sources of meaning and purpose in human life; they are central to our cultural values.

Population issues can fundamentally challenge these central values. Let's consider six ways.

First, many people sense a hint of misanthropy in Malthusian arguments. People have inherent value, inherent rights of existence, most of us believe. People are also the source of most of our central interests in life. To say that we should have fewer people sounds to some like people hating.

Second, population issues rapidly connect to issues of racial, ethnic, local, national, and religious pride. For some people, controlling population means diminishing the group. If our population shrinks, the feeling sometimes goes, our people will become less important in the world. And within a country, subgroups sometimes feel a sense of what the psychologist Dorothy Stein terms "demographic competition."[24]

Third, the techniques of population control confront some religious values. Conception and birth are processes that even in a scientific age still evoke awe and a feeling of mystery. Many people look to religion for a framework for understanding these mysteries. All the major world religions emphasize the sanctity of life, of course, but they interpret this sanctity in various ways. In some cases, their interpretations conflict with particular means of population control. Most notable is the contemporary Catholic Church's stated rejection of any method of contraception other than "natural" family planning—avoiding intercourse during the fertile period of a woman's menstrual cycle—as a means of population control. Although this teaching is widely disregarded by the Catholic laity, the Vatican reiterated its firm stance against "artificial" contraception in 2008 in the *Dignitas Personae*, a recent set of doctrinal directives. (Note that the Catholic Church rejects contraception it regards as artificial, not population control itself.) Moreover, population control threatens the sheer numbers of the faithful, a fact that many religious leaders have no doubt pondered.

Fourth, population issues are fundamentally related to sexual activity. To talk about population control is to talk about sex and sometimes to talk about controlling sex. You do not have to be a deep student of social life to recognize that issues relating to sexuality will likely be highly controversial. Moreover, because sex remains a subject many find embarrassing or even immoral to talk about openly, even the mere discussion of contraception and other issues relating to population control can make some people anxious.

Fifth, population issues are necessarily gender issues. Reproduction lies at the center of ideas of appropriate gender relations, with all the implications that these relations have for social power and social motivation. Population control can threaten social power dynamics built around gender by undermining the basis on which many people legitimize them, such as the still-common idea that because of child rearing, a woman's place is in the home.

Sixth and finally, family is a central source of social identity and feelings of transcendence. It is from family relations, in part, that we understand who we are. Most of us live through most of our adult years as parents, and we thereby gain a deep sense of who we are and who we should be. Without someone to fill the role of a child, you cannot be a "mom" or a "dad." Through children, we also gain a sense of transcendence over the confines of our own individual lives, a kind of immortality. Although we have other sources of transcendental values—such as religion, art, and the sense of having made a contribution to society through our work—children are a particularly direct and accessible source. For many people, to suggest controlling population is to infringe on their principal solution to the problem of mortality: family.

These are common reactions, but they are not cultural necessities. Rather than misanthropy, population control could be seen as an enlightened form of being pro-people, for it may improve the quality of human life. Instead of diminishing the group, population control could be seen as a way to strengthen the group by making it more ecologically secure. As for religious values, our various traditions have a wide range of responses to population issues. For example, some accept "artificial" contraception, while others do not. Our level of comfort with sexuality is equally variable. Plenty of people feel little difficulty in openly discussing sexual matters. As for gender relations, changes in the relations of power between women and men can be seen as liberating rather than threatening. And instead of compromising family identity and transcendence, many see population control as a surer route to improving the life chances of all our families and children.

I raise these counterpoints to illustrate the diversity of possible cultural responses to the issues often raised by population control. Significantly, all these responses have moral implications. It therefore seems unlikely that any of us can evaluate population issues in a morally neutral way. Nor should we try to, for it is these implications that in large part make population issues so significant for us. But even though moral neutrality is not possible, we can still evaluate population issues in a reasoned way.

Being conscious of our own moral values seems a surer route to reason than falsely assuming that we have unbiased perspectives. With our moral passions in mind—including any opinions we may have concerning sustainability, environmental justice, and the beauty of ecology—let us now evaluate Malthusianism and the critiques of it.

The Inequality Critique of Malthusianism

Looking around the world, we can see an indisputable association between high levels of population growth and poverty, a wide range of environmental problems, and declines in per capita food production. For example, Africa has the highest overall rate of population growth of any continent and is simultaneously experiencing substantial declines in food production per capita as well as overgrazing, desertification, shortening fallow periods, and deforestation. Between 1960 and 1964, sub-Saharan Africa produced 147.8 kilos of grain per person per year, which was about half of the world average; by 2000 to 2004, the region's annual production was down even further to 121.2 kilos per person.[25] The treadmill of underproduction, apparently, is not limited to industrialized settings.

The question is why. Malthus would likely have seen diagnostic evidence in Africa for his view that population growth eventually overwhelms the productive capacity of the environment, leading to poverty and hunger. He would have probably argued that, fundamentally, economic treadmills gain their jostling whirl from the steady increase in the number of people trying to find a footing on them. But the story is, at the very least, more complex.

A long tradition of scholars has even argued that the direction of causality should be reversed. It is not population growth that causes environmental decline and ultimately poverty and hunger. Rather, it is poverty and hunger that cause environmental decline and population growth, as the poor struggle to gain their living. To understand the origins of poverty, we should look not to population pressure, goes this counterargument, but rather to the history of international development and the social and economic inequality it has fostered between (and within) the countries of the world.

The Development of Underdevelopment

We sometimes forget, or at least overlook, that a scant century ago, most of the poor countries of today were the colonial possessions of most of the countries that are rich today. As recently as 1950, only three countries in Africa—Egypt, Ethiopia, and Liberia—could claim full independence (see Figure 4.2).[26] A century ago, all of southern Asia, most of southeastern Asia and the Pacific Islands, and a substantial part of the Caribbean were also under the control of various empires. Much of the rest of the world was divided into spheres of influence that achieved a similar political result. Although colonization brought some benefits to the affected regions, Western Europe, Japan, the United States, and the former Soviet Union gained the most from these relationships, growing in wealth while their colonial possessions, for the most part, languished in poverty.

The period following World War II, however, saw a new global commitment to resolving these inequalities. The lessons of the war brought about a new global consciousness—a sense that we are all in this together, that the rich countries should help the poor, and that the right of self-determination applies to all countries. The United Nations was one product of this consciousness. Partly in response to this new sense of global commitment, rich countries gave up their empires. New states sprang up everywhere as colonial powers pulled out. And in part,

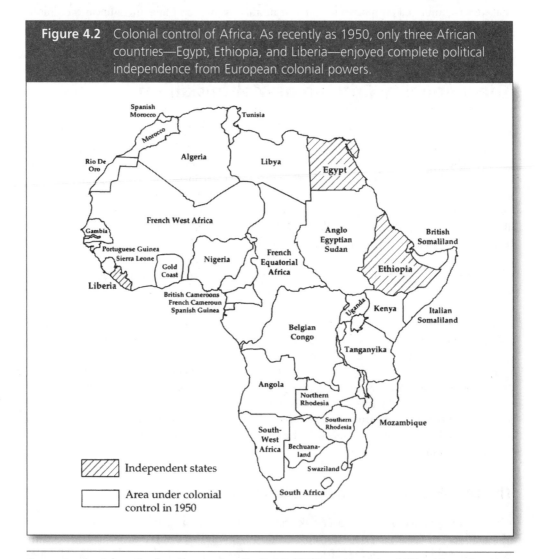

Figure 4.2 Colonial control of Africa. As recently as 1950, only three African countries—Egypt, Ethiopia, and Liberia—enjoyed complete political independence from European colonial powers.

Source: Based on information from Freeman-Grenville, G. S. P. (1991). *The New Atlas of African History.* New York: Prentice Hall.

it must be said, the old empires folded up because the devastations of world war left few imperial countries with the financial or military means to maintain direct control over their former colonies.

But the new commitment to helping poor countries was, at least partially, heartfelt.[27] Many believed that poor countries could modernize just as the rich countries had done, once the rich freed the poor from colonial control. With education, industrial infrastructure, industrialization of agriculture, modern political institutions, and lots of exports and imports to connect these countries into the increasingly global economy, the poor countries could soon join the rich at the table of modern luxury and avoid the dismal prospect of Malthusian decline.

In a word, these poor countries lacked only one thing: *development.* Modernization would bring it. This is the basic tenet of an influential perspective known as *modernization theory,* an idea associated with many thinkers in many fields; within sociology, Talcott Parsons and Seymour Martin Lipset remain the best-known advocates.[28]

Perhaps President Harry S. Truman best stated the underlying spirit of modernization. As he proposed in his inaugural address on January 20, 1949,

> We must embark on a bold new program for making the benefits of our scientific advances and industrial progress available for the improvement and growth of underdeveloped areas. The old imperialism—exploitation for foreign profit—has no place in our plans. What we envision is a program of development based on the concepts of democratic fair dealing.[29]

Truman was articulating a growing consensus among world leaders. Just a few years earlier, in July of 1944, delegates from 44 countries got together in a quiet country resort in a place called Bretton Woods, New Hampshire, to chart the course of the world economy after the war. By that time, World War II seemed on its way to an inevitable close. Led by John Maynard Keynes, probably the most influential economist of the twentieth century, the Bretton Woods delegates sought to use the conclusion of World War II as an opportunity to fundamentally reshape global society and economy. Their stated intent was to broaden the benefits of modernity, through development, and thereby make sure that such a calamity might never again occur. But as we shall see, grounds exist for a more cynical interpretation of the result, if not the intent.

Whatever the reasons were, the delegates set up two key institutions that have had an enormous influence on the subsequent course of economic development: the International Monetary Fund (IMF) and the International Bank for Reconstruction and Development. The latter evolved into the World Bank, which is composed of two arms: the International Development Association, which focuses on the very poorest countries, and the International Bank for Reconstruction and Development, which now focuses on middle-income nations.

Both the IMF and the World Bank make loans to countries for development purposes. The IMF gives mainly short-term loans to help countries balance their national budgets. The World Bank emphasizes loans for major infrastructure projects intended to have long-term effects, like dams, roads, irrigation canals, and schools. Some of the money loaned by the "Bretton Woods institutions," as they are often called, comes from capital subscriptions from member countries, but most comes from the sale of bonds.

Since the Bretton Woods conference, several other international or "multilateral" banks have been founded, most with a regional focus, such as the Inter-American Development Bank and the Asian Development Bank. Most of the capital for international development projects comes from private foreign banks, generally in the range of 75 to 85 percent of the total, depending on the year.[30] But loans from multilateral banks are usually at least a part of most major development projects. Countries use the seal of approval of a multilateral loan to leverage development funds from private sources, both domestic and foreign. In addition to these sources, there has been an enormous proliferation of private international aid organizations and governmental aid agencies, which, unlike the big development banks, usually make gifts and not loans.

International development has become a major human activity, involving movements of capital equivalent to hundreds of billions of U.S. dollars annually. In the world's poorest countries, as of 2008, development monies from abroad typically amount to some 16 percent of GNI (Gross National Income).[31] The result of a half century of international development and modernization, however, is continued disparities in wealth among countries. In fact, the gap between the rich and the poor has widened substantially.[32]

Between 1970 and 2008, per capita GDP in the wealthiest countries grew by 2.4 percent, but per capita GDP in what the United Nations Development Programme (UNDP) calls the "low human development" countries fell by 0.4 percent.[33] India and China are important exceptions, however. During this same period, India's per capita GDP grew 3.6 percent, and in China per capita GDP grew 7.9 percent.[34] Considering that these are by far the two largest poor nations (and by far the two largest of all nations), this is significant growth.

Nevertheless, a longer look shows an overall widening gap even with respect to India and China. For example, because of the United States' huge economic head start, the difference in per capita GDP between the United States and China increased by 92 percent—almost doubling—between 1970 and 2010, despite slower economic growth in the United States.[35] The accumulation of advantage, a process discussed in Chapter 3, also works at the level of nations.

Understanding Underdevelopment

The reason for the ever-widening gap between rich and poor lies in the structure of the world economy, say a number of social scientists. The treadmill-driven tendency to solve the problem of the original capitalist by seeking new markets and new places for investment has sent the capital of wealthy nations overseas. Poor nations have usually welcomed this investment, but most development funds have come in the form of loans, not gifts. Although aid and charity account for a significant portion of the capital flow, the rest has been sent with the expectation that it would be returned, with a comfortable margin for profit. The World Bank, for example, is a highly profitable institution. Its bonds are considered unusually secure investments, and it has cleared billions in profit for its investors.[36]

Meanwhile, poor nations have become mired in debt. Although the debt crisis faced by developing countries has received much international attention, debt continues to rise. As of 2008, the total external debt of developing countries was $3.7 trillion dollars, or 22.1 percent of their GNI.[37]

To put the level of international debt in perspective, economists often cite the ratio of a country's annual external debt payments to the value of its annual exports and international income from other sources, such as the money migrant workers send back home. These are the sources of the funds needed to pay off foreign creditors. In developing nations today, that ratio averages 9.5 percent—a considerable improvement from the 17.1 percent of 1995, yet still dire enough.[38] In several countries, it rises to more than 30 percent. In Kazakhstan, it is 41.8 percent; in Liberia, it is a whopping 131.3 percent.[39] These countries therefore need to maintain at least a 9.5 percent margin on their international income, and return none of that 9.5 percent margin to investors within their own borders, in order to pay off their debt. This is a highly unlikely scenario. The typical rate of profit for a corporation ranges only from 2 to 12 percent, as Chapter 3 discussed, so these countries will have to run for many years at the high end of capitalist profitability to get ahead, given their current debt load. Could they do it through economic growth? That is not likely either. The average economic growth rate in the world was only about 2.1 percent over the 1970 to 2008 period, so these countries will also have to beat world averages by a considerable margin.[40] In fact, as noted above, the poorest countries saw their economies contract during this period. Even China's staggeringly high growth rate of 7.9 percent during this period falls short of the 9.5 margin these countries need as a whole to maintain to repay their debts, let alone grow their economies. In short, these countries owe far more than they can comfortably pay back without impoverishing themselves still further.

Data like these support the world systems theory advanced by Andre Gundar Frank, Immanuel Wallerstein, and others.[41] *World systems theory* sees the process of development as inherently unequal, dividing the world into core regions and periphery regions. Because of differences in political and economic power, wealth tends to flow to the core regions from the peripheral ones, feeding the former and bleeding the latter. Thus, over time, development tends to exacerbate economic differences instead of leveling them. World systems theorists also sometimes point to regions that have some of the features of both core and periphery, what they term semi-periphery regions. Examples of core countries would be the United States, Japan, and the wealthier nations of Europe. Peripheral countries include Uganda, Zaire, Vietnam, Bangladesh, Ecuador, and Panama. Countries like Costa Rica, the Slovak Republic, and Turkey would be semi-periphery, and China and India are fast joining this group. According to world systems theorists, core, periphery, and semi-periphery relations can emerge not only between countries but within them as well.

Figures on comparative income bear out the notion that despite their commitment to international development, the core nations have received the most benefit. Development sociologists Joseph Cohen and Miguel Centano use a simple yet clarifying measure of changes in comparative income: what they call the DFR, or "distance from rich," ratio, which uses a country like the United States as a benchmark of what constitutes being rich. They find that the world as a whole had a DFR ratio of 0.32 in 1980, which fell to 0.25 in 1990, and then a bit more to 0.24 in 2000. These figures include all the countries of the world in comparison with the United States, including the other rich countries and the new semi-periphery countries of China and India, where the DFR has changed little or even gone up. If we look elsewhere, we see the same pattern and often significantly lower ratios throughout these years—such as sub-Saharan Africa's DFR figures of 0.09, 0.07, and 0.06 for 1980, 1990, and 2000, respectively.[42]

There are grounds to think that the situation is now improving. Debt relief has become a major cause on the world stage, and quite a bit of international debt has been forgiven already. But the dominant trend over the half century since the Bretton Woods conference has been much the same as it was during the period that President Truman termed the "old imperialism."

The Structural Adjustment Trap

The continued poverty of poor countries was much exacerbated by a World Bank and IMF policy known as *structural adjustment*, a term coined by Robert McNamara, president of the World Bank from 1969 to 1981. (McNamara was also U.S. Secretary of Defense during the beginning of the Vietnam War.) Structural adjustment referred to a comprehensive program of radical "free market" changes, such as reducing public services, liberalizing trade, emphasizing export crops, eliminating subsidies, and curbing inflation through high interest rates and reduced wages. The latter part of the program—expensive money and low pay—is what was euphemistically referred to as "demand management." Structural adjustment programs were instituted in dozens of countries from the late 1970s onward, across Africa, Asia, the Americas, and the former Soviet Union. The idea was cooked up in the West, though, with the United States leading the way through its traditional role of picking the World Bank president, who has always been an American. But other Western countries largely agreed with this U.S.-led approach, which is why structural adjustment policies are also often called the "Washington Consensus."

The idea of structural adjustment was to help—some say to force—countries to reshape their economies so that they could pay off their mounting debts. Private banks often reschedule and in some cases write off loans, but the World Bank and the IMF have been reluctant to do so out of fear of undermining their high bond ratings and thus being forced to raise the interest rates they offer investors.[43] So they encouraged—again, some say forced—debtor countries to adopt structural adjustment programs instead. Whether one calls it help, encouragement, or force, the two Bretton Woods institutions certainly gave poor countries a compelling incentive to adopt structural adjustment: no more World Bank and IMF loans unless they did, which also meant a greatly reduced ability to attract private foreign capital.

Free market policies may not seem like such a bad idea (leaving aside here the question of whether a "free market" can be said to exist, as discussed in Chapter 3). After all, these are the same policies most Western governments advocate for themselves these days. But the free market policies increasingly adopted by rich countries have been put into practice without anything like the severity and inflexibility the World Bank and IMF have imposed on poor countries throughout the world. "Shock therapy" is what World Bank and IMF officials used to call the regimen. The result has been devastating for the marginal peoples of nearly every country that has taken this stern medicine. Typically, under structural adjustment, basic social services such as education and health care have been sharply cut, and the price of formerly subsidized food-stuffs like bread has gone sky high, leaving the poor in often desperate circumstances.

Throughout the 1980s, the heyday of structural adjustment, "IMF food riots" plagued cities across Africa as starving people took to the streets, sometimes toppling governments.[44] And as investors bought up farmland to produce export crops for the newly liberalized export trade, displaced peoples moved into more marginal lands, promoting deforestation and land degradation. With less productive land growing food for local consumption, poor countries have found themselves more dependent than ever on imports to meet their basic food needs. Moreover, once displaced from their land, people in poor countries were less able to compensate for the newly increased food prices by growing their own. When you are poor, it is very risky to be dependent on money to get enough to eat.

Another result of structural adjustment is that many poor countries are in the unenviable position of exporting raw materials elsewhere, only to buy these materials back later as finished goods. Value is added elsewhere, and these poor countries must pay for that added value out of their scant stocks of foreign exchange funds, further crimping their ability to pay off foreign debts. The trade liberalization features of structural adjustment and a number of recent international treaties have also enabled companies from rich countries to come into poor countries and set up low-wage factories, later exporting the goods to places where people have enough money to buy them. The governments of poor countries, eager to attract the foreign investment needed to fulfill structural adjustment plans, often allow these companies to evade the environmental and labor laws they would have to contend with in their own countries. During the mid-1990s, for example, many lines of Nike sneakers were being assembled in Vietnam by women working 12-hour days at 20 cents an hour to make shoes they could never afford themselves—to the embarrassment of a number of celebrities who were paid millions to endorse Nike products.[45]

Under these conditions, poor countries got the short end of the economic stick of structural adjustment. As a result, value is added in peripheral areas, but the peripheral areas do not get to keep much of that value. Local workers are commonly paid almost nothing, and somebody else owns the products of their labor. By opening up poor nations for increased foreign investment under such unequal terms of trade, structural adjustment assisted in draining value away from the periphery and moving it toward the core.

Structural adjustment has been widely criticized, and the prestige of both Bretton Woods institutions has been severely undermined. The World Bank itself is now backing away from the policy. The World Bank's former chief economist, Joseph Stiglitz, has written widely on its inadequacies.[46] One doesn't usually hear the phrase "shock therapy" in development circles anymore, and even the phrase "structural adjustment" has largely disappeared from the policy statements and progress reports that continually stream out of the Bretton Woods institutions. Plus, the debtor nations increasingly take a dim view of the radical "free market" agenda of the Washington Consensus, soured by its heavy and uneven-handed approach. Particularly in Latin America, debtor nations have been responding with a different radicalism, re-embracing social-ist ideas while retaining (some say perfecting) many market institutions—a hybrid sometimes called "market socialism."[47] The evidence suggests few successes that the Washington Consensus can point to among poor countries. Aside from China, Taiwan, South Korea, and Singapore—the "Asian tigers," as development officials call them—structural adjustment led to increased inequality and lowered standards of living in the poor countries, according to one recent com-prehensive review of the economic performance of 100 countries between 1980 and 2000.[48] A major rethink of development policies now appears to be underway, but the human and envi-ronmental damage has been done, reinforcing centuries of unequal exchanges between the rich and poor countries of the world.

The anti-Malthusian moral is this: The poverty experienced throughout the world is not just a population issue. (Some argue it is not a population issue at all.) Poverty cannot be under-stood apart from the history of development, a history that has favored some regions over oth-ers. Any claim concerning the relationship between the environment and poverty needs to take this history into account.

Food for All

Malthusians point not only to poverty but also to the 925 million people who suffer from not enough to eat, as an indication of population pressures on the land. Anti-Malthusian critics of this position argue that, on the contrary, the world has plenty of food for everyone. The prob-lem of food shortages, they contend, is really a problem of *access* to food and of overconsump-tion by those who do have access. The principal names associated with this counterargument are Frances Moore Lappé and the Nobel Prize–winning economist Amartya Sen.

One of the points that Lappé forcefully raises in a number of books is that there is little correlation, if any, between population density and hunger or between the amount of cropland per person in a country and hunger. The obvious example is Europe, which has some of the most densely populated countries in the world and yet very little hunger. Africa, which suffers from a far greater percentage of hunger, nevertheless has a far lower population density. Of course, much of Africa is desert and cannot be farmed. But even when considering the amount of cropland per person, Lappé finds no particular relationship with hunger. Japan has about 10 people for every acre of cropland and very little hunger. Tiny Singapore has 143 people for every acre of cropland and very little hunger. But Chad has 1.68 acres of cropland for every person—17 times as much as Japan and 240 times as much as Singapore—and experiences quite extensive hunger.[49]

Lappé argues that the world has plenty of food, even at current population levels. In 2009, the world produced 328.8 kilograms of grain per person, which is 1.98 pounds per day, or about 3,145 calories per day.[50] Except for an exceptionally active person, that would be an ample amount—even without other calorie sources. (The recommended daily calorie intake for adult males in the United States is 2,000 to 2,200.) However, 37 to 40 percent of that grain, depending on the year, is fed to livestock. As many have pointed out, much of grain's food value to humans is lost because farmers feed their animals not only so they put on mass but also to give them energy to stay alive. Various kinds of livestock take from 2 to 7 pounds of grain to produce a pound of meat.[51] With their high-meat diet, Americans consume (directly and indirectly) approximately 800 kilograms of grain each year per person. In India, which has a diet low in meat, people consume (again, directly and indirectly) about 200 kilograms of grain per person per year.[52] Changes in the diet of the wealthy and a different distribution of food could raise the figure for India considerably.

In place of Malthusianism, Lappé and her colleague, the environmental sociologist Rachel Schurman, advocate a *power structures* perspective on food and population. Even though Japan and Singapore have little cropland per capita, they are wealthy countries; Chad and India have more cropland, but they are poor. And people who are wealthy have a lot more power—a lot more ability to gain the food they require and a lot more "say in the decisions that shape their lives," as Lappé and Schurman point out.[53]

Lappé and Schurman argue that a power structures perspective explains not only inequalities in access to food, but also high rates of population growth. A lack of control over their lives leads poor people to regard children as an economic resource. "Living at the economic margin," Lappé and Schurman observe, "many poor parents perceive their children's labor as necessary to augment meager family income. By working in the fields and around the home, children also free up adults and elder siblings to earn outside income."[54]

Given their lack of options, poor parents thus have a strong economic incentive to have lots of children, the inverse of Malthus's argument that population growth, through environmental decline, leads to poverty. Rather, Lappé and Schurman respond, poverty and environmental decline lead to population growth by decreasing the power people have over their lives, leading them out of desperation to seek economic security through having large families.

The Politics of Famine

Amartya Sen makes a similar argument concerning famine.[55] Famines, says Sen, are caused not by a lack of availability of food, but rather by a lack of access to food. All societies have social systems for what Sen terms "entitlements" to food and other goods, such as the distribution of ownership of land to grow food and the ability to acquire food through trade, usually through the medium of money. Entitlements allow people to gain command of food and other goods. Breakdowns in *entitlement systems* are what cause famines, Sen argues, not environmental decline.

Sen makes his case by analyzing four major famines in the twentieth century: the Great Bengal Famine of 1943, the Sahel famine of the 1970s, the Ethiopian famine of 1973–1974, and the Bangladesh famine of 1974. He argues that in each instance, sufficient food to feed everyone was on hand in the affected countries. The problem was that people could not get access to the food.

For example, in the 1974 Bangladesh famine, a series of summer floods on the Brahmaputra River largely wiped out one of the three annual rice crops and damaged a second one. But food imports and stocks of rice remaining from earlier harvests provided plenty of food throughout the crisis.[56] The real problem was that the flood threw a lot of farmers and agricultural laborers out of work. With nothing to harvest from one rice crop and no chance to plant the next because of the continuing floods, laborers could not find paid work. Farmers didn't have much money, either, because they had nothing to sell. Consequently, these laborers and farmers couldn't afford to buy much rice. Also, the United States chose this moment to cut off its normal food aid, because Bangladesh was exporting jute to Cuba. In anticipation of shortages due to the flooding and the loss of U.S. food aid, the rice market went haywire. Prices for rice in Bangladesh jumped by 18 to 24 percent at a time when many people had little money on hand to make up the difference. Because of fluctuations in the labor market and the rice market, they had lost their entitlement to food. Somewhere between 26,000 and 100,000 people died of starvation and malnutrition within 3 months. The tragic irony was that, in the country's warehouses, there was more than enough unsold rice to feed everyone.

Similar kinds of arguments have been applied to other famines, such as the infamous Irish Potato Famine of the 1840s. While millions starved, Ireland continued to export large quantities of wheat to England (some 800 boatloads in all) because of earlier export contracts.[57] In the recent famines in war-torn Somalia, Rwanda, and Burundi, there may not have been sufficient food from local farms nor much exporting of food. But high population density relative to cropland was still not the cause of the starvation, some argue. Rather, because of the war, people were not given access to food, nor could they plant in order to feed themselves. War broke down their food entitlement systems.

Sen's argument has an important practical (and political) implication. If there is food available even in the midst of most famines, then the long-term solution to hunger in the world is not the importation of more food. In the short term, in the midst of a crisis when people are dying, food imports are frequently necessary. But even if the long-term problem is the distribution of food, the long-term solution is not redistribution of food. Rather, it is redistribution of *access* to food, as Lappé and Schurman also argue. "What is needed is not ensuring food availability," says Sen, "but guaranteeing food entitlement."[58] In other words, don't give the poor food (except when they are starving). Rather, give them farms, give them jobs, and give them democracy.

Limits of the Inequality Perspective

The inequality perspective makes an important case for the significance of the social origins of poverty, population growth, and hunger. A purely Malthusian perspective, as nearly all scholars now agree, is clearly inadequate.[59]

Yet there is much that the inequality perspectives of Sen, Lappé, and Schurman cannot explain about hunger. Although Chad has a higher ratio of cropland per person than Japan and Singapore and a ratio roughly equivalent to many European countries, not all cropland is equal in its productivity. The climate in Chad is quite dry, and production per acre is quite low. Much more significant than cropland per capita is annual grain production per capita, which, as we have seen, runs at about 120 kilograms per person in sub-Saharan Africa. In contrast, Europe produces about 500 kilograms per person per year.[60] The power structures perspective of Lappé and Schurman needs to take into account the spatial distribution of environmental productivity, not merely wave it aside. Environmental productivity is itself a source of power—a source of environmental power. The power that comes from environmental relations should be part of a power structures perspective.

Nor does entitlement breakdown seem sufficient to explain all famines, as a number of Sen's critics have argued.[61] Six years of warfare in Europe between 1939 and 1945 severely disrupted systems of entitlement, as did the strife in Yugoslavia in the 1990s. But the disruptions of war in Europe did not result in the widespread starvation that Rwanda, Burundi, and Somalia suffered during their recent wars, nor the massive food imports that the war in Iraq necessitated. Europe has long had far more food production per capita. It is also striking to consider the relatively minor disruptions that led to the Bangladesh famine. At the peak of the famine, the price of rice rose only 18 to 24 percent. Bangladesh, however, is very poor, both financially and in terms of the per capita productive capacity of its environment. With a population of some 160 million in a region the size of Greece (which has a population one fifteenth the size), the country finds itself compelled to import much of its food, leaving its people dependent on something most of them have in short supply—money—in order to eat. Thus, most of the people of Bangladesh live very close to the margin they need to survive.

When you live close to the edge, both environmentally and economically, even a minor disruption can have a big impact. Bangladesh and all the countries that have experienced famine in the twentieth century are poor countries with unfavorable levels of grain production per capita. They have very marginal systems of both food entitlement and food production.[62] This is a dangerous combination. When a country is too poor to easily command food imports and when it doesn't have much local food production to begin with, it will have less food around for its people to be entitled to—even in good times.

The Technologic Critique of Malthusianism

In October 1990, the anti-Malthusian economist Julian Simon won a much-discussed bet with Paul Ehrlich, a biologist and a prominent figure in the Malthusian tradition. Ehrlich is the author of the 1968 bestseller *The Population Bomb,* which predicted widespread famine and starvation within 10 years, and Simon and Ehrlich published a series of counterattacks on each other from the early 1980s until Simon's death in 1998.[63] Simon bet that the price of five metals of Ehrlich's choosing—chrome, copper, nickel, tin, and tungsten—would fall during the next 10 years, as opposed to rising in the face of Malthusian scarcity. Ten years later, the price of all five metals had actually dropped, after taking inflation into account. (They had agreed to pay each other the difference in value accrued by the market movement of $200 worth of each metal over the 10-year period. Simon would pay for all the metals that went up in price, and Ehrlich would pay for all the metals that went down. Ehrlich quite honorably sent Simon a check for the $576.07 difference.)[64]

It was a foolish bet for Ehrlich to make, even from a Malthusian point of view. Market forces are complex and reflect environmental conditions crudely at best. Many of the costs of natural resource production are externalized, disguising their true environmental (and social and economic) significance. The prevalence of the Jevons paradox means that the crisis of underproduction may set in, even when prices fall. Also, the prices of these particular metals have little to do with the resource scarcities—land, water, food—that would be significant to the poor and marginalized, those who are most likely to experience a Malthusian crisis, if anyone will. Even if the prices of the metals had gone up, the bet would have said little about Malthusian shortages.

A Cornucopian World?

Although the bet proved nothing, it did serve to highlight the debate between Malthusian arguments and a kind of anti-Malthusian argument often called *cornucopian,* of which Simon was the most prominent proponent. Simon controversially claimed that the solution to resource scarcity is actually to increase population. People, said Simon, are the "ultimate resource." A larger number of people means more brainpower and labor to work out technological solutions to scarcity, Simon argued. When confronted with scarcity, we apply our collective brainpower and find new sources of formerly scarce resources and new techniques for extracting them. In some cases, new technology will allow us to substitute different materials for ones that have become scarce, what Simon called the principle of *substitutability.*[65] Because of substitutability, we will never be long on the treadmill of underproduction, Simon argued.

Simon cited a variety of evidence to support his arguments. Population has successfully continued to increase, and increase rapidly, for the roughly 200 years since Malthus first published his book, Simon noted. As well, life expectancy has leapt to unprecedented levels, while infant mortality has declined considerably. Standards of living for many of us today are astonishing when compared with living standards of the past, which suggested to Simon that Malthusian limits are far from inescapable. The prices of most basic commodities have actually dropped over the decades, in line with the results of Simon's bet with Ehrlich, and Simon found no paradox here. Simon also disputed the significance of acid rain, global warming, the ozone hole, and species loss, arguing that these issues either have been exaggerated by environmentalists or that they represent challenges that future technological innovation will overcome. He argued that, in fact, the state of the environment is now much improved, pointing to the drop in air pollution emissions in the United States and the many advances in public health.[66]

But Simon had—and has—many critics, and rightly so. One point that is often raised is Simon's neglect of social inequality. Although the lives of many have improved, the percentage of the world's people who live in poverty, facing hunger and malnutrition throughout their lives, hasn't declined all that much in the last 50 years. The sheer number living in poverty has doubled.[67] It is true that even the desperately poor are generally living longer, in part because of medical and other technological improvements, but life expectancy is still very uneven across the world. Simon paints an overly rosy picture of the world. We can do better.

Simon's argument that more people means more brainpower to work out problems is also rather dubious. Sure, two people may come up with more ideas than one person (although they may also come up with the same ideas). But by Simon's argument, the Roman Empire should still be with us, continuing to expand, ever increasing the number of people enlisted into the task of solving the empire's problems. It is clearly not the sheer size of a society that makes it innovative. Innovativeness depends on social circumstances that encourage creative thinking, such as democratic discussion and a good educational system, not mere numbers of people. In fact, greater numbers of people may only increase a society's stock of misguided ideas if that society is set up in a way that stamps everyone in the same mold or clouds their creativity. Also, the kinds of improvements that Simon looked to are mainly high-tech. But the bulk of population growth currently is taking place among those who do not have the educational background to contribute to high-tech solutions.

Critics also doubt Simon's optimism about technology. With every advance in technological how-to, as Chapter 3 discusses, generally comes an equal measure of technological have-to. Technological freedom's relation to technological constraint is another of the dialogues of social life. Moreover, there are limits to what technology can do. Technology has indeed made possible substantial substitutions in the resources we depend on, often in the face of scarcity, such as the techniques for the use of fossil fuel that resolved the fuel wood and water power shortages of early industrialism. But will technology always come to the rescue in time to prevent serious problems? This question is particularly germane as we encounter limits in resources that seem less substitutable, such as fresh water, clean air, land for agriculture, and habitat for biodiversity.

Even if we come up with an innovation, new technology can bring with it unfortunate unintended consequences, such as the substitution of HCFCs (hydrochlorofluorocarbons), a potent greenhouse gas, for ozone-depleting CFCs (chlorofluorocarbons). Solving one problem often contributes to another. Besides, waiting for a shortage to stimulate innovation and substitution could put humanity on a path of crisis management in which we struggle to solve difficult problems instead of avoiding them to begin with. This is a risky strategy, especially for the world's poor.[68]

Take, for example, increasing food production. During the 1960s and 1970s, intensification of agriculture swept through the developing countries, a transformation often called the "green revolution." Mechanization, irrigation, pesticides, a 10-fold increase in fertilizer use, the introduction of high-yielding hybrids, rural road construction to open up new areas for clearing and cultivation—these practices allowed world grain production to increase 2.6-fold between 1950 and 1984.[69] The per capita world grain harvest rose by 40 percent.[70] Much of the gain was in rice, the centerpiece of the diet of billions, and much of that due to a wonder variety known widely as "miracle rice," developed by the International Rice Research Institute.

But since then, the tale is not as pretty, and especially regarding rice. Because crop yields fluctuate with the weather, let's compare the 5 most recent years of world rice production data with their counterparts 20 years ago. Between the 5-year periods 2004 to 2008 and 1984 to 1988, global rice production rose an average of 34.2 percent—just barely ahead of the 32.5 percent rise in world population over the same period.[71] That made for a 1.3 percent increase in per capita rice yields, a far cry from the wonders of miracle rice. Plus, the world's fastest population growth rates are generally in rice-dependent regions. Concern for the future of rice led the United Nations to declare 2004 the International Year of Rice, in hopes of stimulating yields. Meanwhile, per capita yield of all grain worldwide has been basically flat since the mid-1980s (see Figure 4.3).[72] Researchers at the Worldwatch Institute suggest that the declining responsiveness of crops to further fertilization, as well as "soil erosion, the conversion of grainland to nonfarm uses, and spreading water scarcity," is limiting the growth of grain production.[73]

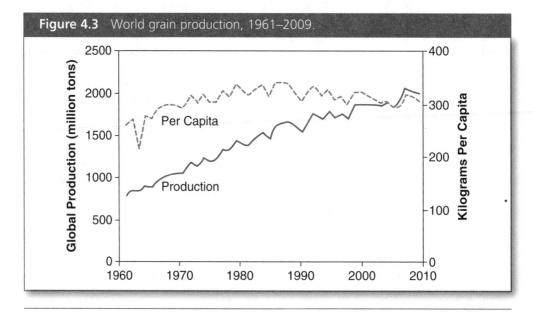

Figure 4.3   World grain production, 1961–2009.

Source: Worldwatch Institute (2010), *Vital Signs, 2010,* New York: Norton; and FAO (2010, *Food Outlook 2010,* Rome: FAO.

Will new biotech varieties like "golden rice" and "Bt rice" eventually bring yield boosts of rice and other grains back to at least matching population growth? Can sustainable methods provide us a way to boost yields without the negative effects of the green revolution techniques?[74] Well, I guess we have to hope so.

The Boserup Effect

Ester Boserup offered a closely related but more temperate argument for how population growth can, in some circumstances, stimulate technological change. In a famous study of agricultural development, Boserup suggested that population pressure is a primary factor stimulating the adoption of more productive farming practices.[75] Malthus held that population pressure reduces food availability, but Boserup's view was that population can increase food production by giving people an incentive to switch to more intensive farming methods.

Consider a low–population-density farming practice like shifting cultivation, in which farmers cultivate a particular parcel of land for a few years and then let it lie fallow for 20 or 30 years before cultivating it again. In the interim, trees and other vegetation grow back, restoring the fertility of the soil and breaking the life cycle of crop pests. This is an effective and relatively low-labor method of farming. But it supports few people per acre. So as population grows, rural people shorten the fallow periods—already a technological change—until problems with fertility and crop pests increase and population pressures rise even more. At that point, and usually reluctantly, villagers begin cropping fields annually and using small plows, fertilizers, and pesticides to maintain fertility and control pests. Where possible, and if pressures remain high, they may eventually irrigate their fields and purchase high-yielding hybrids rather than saving their own seed. In most cases, these more intensive practices require more labor and higher cash outlays, making villagers reluctant to switch to them. But eventually people do switch, if they can, as demonstrated by the dramatic increases in food output in developing countries in recent decades.[76] In short, it is possible to step off a treadmill of underproduction like shortening fallow cycles. Indeed, the very existence of a treadmill provides a huge incentive for doing so.

But Boserup was no starry-eyed optimist. She identified many qualifications to this process, which is now sometimes called the *Boserup effect*. First, technological change is not the same as innovation. Population pressures provide the incentive to adopt technologies that have been invented elsewhere but that may not prove attractive until population pressures override labor and financial costs. Such pressures also encourage innovation, she suggested, but the principle at work is necessity, not Simonian collective brainpower. In any case, "Societies have most often advanced technologically by introducing technologies already in use in other societies," Boserup wrote.[77]

Issues of inequality can also limit the influence of a Boserup effect. The investments required to increase the intensity of production may not be available in developing areas. And if farmers do attain sufficient capital to intensify their operations through mechanization, they may also put farmworkers out of work, increasing poverty and inequality.[78]

Boserup further noted that in conditions of *rapid population growth*, economic development may be severely limited.[79] Population density needs to be considered separately from the rate of population growth.[80] Whereas population density may provide the incentive to intensify production, rapid population growth may overwhelm the economic and social resources that are essential to intensification. Governments and local communities can be left constantly scrambling to provide a burgeoning population with education, health care, poverty relief, and infrastructure improvements like roads and irrigation. Rapid population growth also leads to a population with a high percentage of children requiring schooling and caregiving and thus competing for scarce funds and adult labor.

The problem of rapid population growth can be particularly pronounced in urban areas. When the population is generally poor, taxation does not yield sufficient funds to keep investing in new roads, public transportation, sewage lines, clean-water supplies, school buildings, hospitals, phone lines, and power generation. Nor do people have enough money to attract much private investment to provide these services. Government officials and police receive low pay and turn to corruption to maintain their incomes, making it even harder to coordinate rapid growth. Kickbacks increase the cost of providing infrastructure, and polluters avoid regulations through payoffs to officials.

Even when the population is wealthier, rapid growth presents a serious organizational problem. Mexico City, capital of a country that actually has relatively high per capita income, is often pointed to as an example of the difficulty of planning in the midst of rapid expansion. Despite the horrendous traffic in the city, residents increasingly turn to cars as an alternative to inadequate public transportation, only making matters worse. Public transit services simply can't keep up with the rapidly increasing demand of the rapidly increasing population. Because of these planning difficulties, the likelihood of corruption increases. Thus, as industries and car owners bribe their way around regulations limiting polluting emissions, the 21.2 million (as of 2009) people of Mexico City experience the worst air quality of any city in the world. In 2006, ozone levels exceeded the World Health Organization's (WHO) standard on 209 days.[81] A 2004 study estimated that 4,000 premature deaths each year can be attributed to Mexico City's smog, as well as 2.5 million lost days of work.[82] And yet this represents a considerable improvement over the situation a decade ago.[83]

The water shortages in Atlanta, Georgia, one of the fastest-growing cities in the United States, may be an example of the environmental challenges of rapid population growth in one of the world's richest countries. Between 2000 and 2006, Atlanta added more new residents than any other U.S. city, and the city's infrastructure simply can't cope—particularly in the face of global climate change and the ensuing drought in the southeastern United States.[84] The troubles there are so severe, and so beyond what human planning seems capable of handling, that on November 14, 2007, Governor Sonny Perdue even held a public prayer session to pray for rain. "Oh Father, we acknowledge our wastefulness," Perdue said from the steps of the Georgia Capitol, his head bowed.[85]

We need to make another important qualification to the Boserup effect. Like Simon, Boserup did not allow a big role for the environment in her theories of technological change. But the environment can significantly limit the potential of a Boserup effect. The problem of the unintended environmental consequences of technological change, mentioned with regard to Simon's theory of technological substitution, also applies to Boserup's theory. New production strategies bring new consequences. Equally important, because of the pressure to increase environmental yields quickly, more intensive production often proceeds by increasing the overall level of resource use rather than by increasing the efficiency of resource use. Sometimes efficiency even declines, resulting in soil erosion, soil degradation, deforestation, and water shortages. Moreover, efficiency gains can set off a Jevons paradox, so that people wind up running faster on a treadmill of underproduction, not stepping off it.

The Demographic Critique of Malthusianism

Let's turn now to the third set of critiques of Malthusianism, beginning with the ideas of the demographer Frank W. Notestein. In 1945, Notestein offered a simple model of population that has since become central to the debate over population growth.[86] Looking over European history, Notestein suggested that development and modernization initially raise population growth rates but eventually lead to a return to a stable population, albeit at a higher level. Notestein distinguished three stages (see Figure 4.4).

Stage 1: In premodern times, countries experience high birthrates and mortality rates that roughly cancel each other out. Because of disease, malnutrition, and accidents, average life expectancy is about 35 years. Children and infants are particularly hard hit because of the diseases and fragility of childhood. Parents compensate by having a lot of children, a practice supported by pro-natal social norms and social institutions as well as the common use of children as a source of household and agricultural labor.

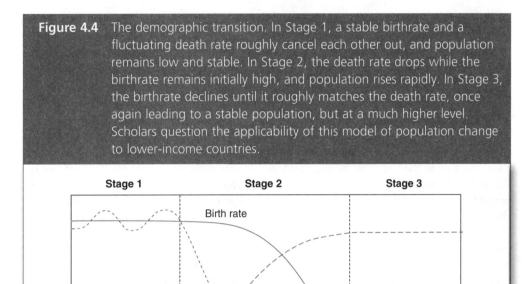

Figure 4.4 The demographic transition. In Stage 1, a stable birthrate and a fluctuating death rate roughly cancel each other out, and population remains low and stable. In Stage 2, the death rate drops while the birthrate remains initially high, and population rises rapidly. In Stage 3, the birthrate declines until it roughly matches the death rate, once again leading to a stable population, but at a much higher level. Scholars question the applicability of this model of population change to lower-income countries.

Source: Based on information from Sarre, Philip, & Blunden, John. (1995). *An Overcrowded World? Population, Resources, and the Environment.* Oxford and New York: Oxford University Press and the Open University.

Stage 2: With the beginning of modernization, new scientific discoveries lead to improved health and an increased food supply, and industrialization increases wealth. Mortality levels fall, but pro-natal social norms continue to promote a high birthrate. Population growth rates climb, eventually to high levels.

Stage 3: Finally, social norms and social institutions catch up with the fact that children are quite likely to survive, leading to a fall in the birthrate. Increased urbanization promotes a lower birthrate because children are no longer seen as a labor supply for the farm. And with the coming of universal schooling, children become an economic burden rather than an economic resource. Changes in social structures associated with modernism lead parents to invest more in each child. Having fewer, better-educated, better-financed children becomes more attractive than spreading a family's investment over many children.[87] Also, as a country gains wealth through industrialization, it can more easily afford social security, pensions, health care, and other social benefits, reducing the tendency for parents to see their children as their future caregivers in old age.

Notestein called this sequence the *demographic transition*, a theory that is closely related to modernization theory. Demographic transition theory is an implicit critique of Malthusianism.

It suggests that, rather than expanding until environmental constraints cause it to collapse, population growth eventually levels off of its own accord. The same factors that lead to population growth—scientific advances, industrialism, and modernization—are also those that eventually lead to a return to population stability, once societies adjust to their newfound social and technological circumstances.

A New Demographic Transition?

Notestein based his model on the European experience. But scholars and development organizations seized on the idea that the less-developed countries might also eventually go through a demographic transition. If such an extension is correct, the solution to population problems would be for poor nations to continue down the path of modern development. Eventually, they theorized, the less-developed countries will grow out of rapid growth and the planning difficulties it causes. As a slogan that came out of the 1974 World Population Conference in Bucharest put it, "Development is the best contraceptive."[88]

The theory of demographic transition fits European history reasonably well (although there are plenty of exceptions, such as the drop in the birthrate experienced in many regions of France before mortality declined).[89] Yet there are substantial reasons to doubt the applicability of the demographic transition to the less-developed countries. A number of important conditions seem to set the contemporary less-developed countries apart from nineteenth-century Europe.

First, the rate of population growth currently experienced in most less-developed countries is considerably higher than the growth rate in nineteenth-century Europe. Since contemporary less-developed countries have been able to import already-established medical technologies, mortality rates have declined much faster. At the same time, birthrates are far higher. Many European cultures of the nineteenth century practiced late marriage and frequent non-marriage, as opposed to the practice of early and nearly universal marriage in most contemporary less-developed countries. High fertility and low mortality has, in turn, led to a high proportion of young people throughout the less-developed world, leading to population momentum as younger generations reach their reproductive years.[90]

Second, the poverty that leads children to be regarded as a labor supply may be difficult to change in the face of contemporary forces of global inequality. Economic structures that channel wealth toward core countries may limit the extent to which less-developed countries will ever develop, or at least may very much retard the spread of development's benefits.

Third, some critics doubt the very desirability of some of those benefits. The principal doubt centers on the homogenizing tendency of development and the charge that development proceeds from a "West knows best" point of view. Critics see development not only as economic imperialism but as cultural imperialism as well. In the words of Wolfgang Sachs, "From the start, development's hidden agenda was nothing else than the Westernization of the world."[91] Hidden in the idea of "development" is the presumption that those who are "underdeveloped" or "less developed" or "developing" are missing something—that there is something inadequate about them. Western education and Western values do not liberate traditional peoples, argues Helena Norberg-Hodge, another critic of West-knows-best development.[92] Rather, they deny traditional peoples the cultural tools to function in anything but a Western economy, trapping them instead. Given that unequal relationships of economic exchange may prevent poor countries from becoming developed or may delay development for generations, the trap is particularly tight.

Fourth, there is the environmental critique. Levels and inefficiencies of consumption associated with the European demographic transition may simply be unsustainable and as well may compromise environmental justice and the beauty of ecology. Even though only the wealthy few currently enjoy them, such consumption levels already seem to be compromising the environment—the situation may become far worse when the whole world tries to keep up with the Joneses. Although we must avoid the simplicity of crude Malthusianism, we must also avoid crude anti-Malthusianism, social environmentalists like Jules Pretty and Lester Brown argue.[93] If the rest of the world is to attain development, then the meaning and the techniques of development must change.

In light of these critiques, there is increasing agreement among population specialists that if the poor countries of the world are to achieve a demographic transition, they will have to do so by different paths than European countries took.[94]

Women and Development

One vision of those different paths sees women having a key role in development. Scholars and development organizations now see improving the status of women both as an end in itself and also as one of the most significant means of reducing rapid population growth and improving the life chances of poor children.

The first two decades of development efforts, 1950 to 1970, gave scarcely any consideration to the importance of gender. Women were largely invisible, both as actors in and potential beneficiaries (or victims) of the development process. In another path-breaking book, her 1970 *Women's Role in Economic Development,* Ester Boserup was the first development scholar to highlight women, and it came as something of an academic bolt of lightning. Boserup pointed out that women in less-developed countries make vital contributions not only in the domestic sphere of reproduction but also in economic production.

In retrospect, Boserup's finding seems obvious, but at the time the official economic statistics of countries around the world consistently underestimated women's nondomestic work. For example, Egypt's national statistics for 1970 listed only 3.6 percent of the agricultural workforce as female. In-depth studies revealed quite a different picture. Half of women participated in plowing and leveling land, and three-quarters participated in dairy and poultry production.[95] A 1972 census in Peru registered only 2.6 percent of the agricultural workforce as female, whereas an interview study found 86 percent of women participating in fieldwork.[96]

Boserup's survey found that although there are substantial variations by region and by level of agricultural intensity, women do as much work in agriculture as men, if not more.[97] The kinds of agricultural work women and men perform do differ, however. Men tend to be more involved in the mechanized and animal-assisted aspects of production, such as running harvesters and cultivators, and women tend to be more involved with hand operations, such as sowing seed and hoeing weeds, which are generally more laborious. Contrary to stereotypes about the greater physical capabilities of men, women in fact do the bulk of the world's physical work: sowing, weeding, washing, and carrying—lots and lots of carrying, of water, wood, and children, plus goods to and from the local market.

Indeed, women around the world, in country after country, rich and poor, work more than men do. Much more of women's work is outside of the market, but if we sum up all hours of work in and out of the market, women work more than men do. In the wealthy countries, it averages about 20 minutes more a day, according to the UNDP. In developing countries, women average almost an hour more work a day—57 minutes, to be exact.[98] These may not seem like big differences. But at the end of the day, those minutes of lost leisure are precious indeed. There is still truth to the old saying that men work from sun to sun, but a woman's work is never done.

Development efforts, however, were ignoring the implications of development policies for the kind of work women do and for women's status. In many rural regions in less-developed countries, particularly in Africa, women collect the bulk of fuel wood essential for cooking food—another activity dominated by women. Development policies stressing exports encouraged poor countries to convert forest lands to timber and crop production, taking little notice of their importance as a source of fuel wood. Women soon found themselves walking miles and miles each day to gather fuel. In general, the kinds of economic activity stressed by development projects emphasized men's work, underestimated the agricultural contributions of women, and almost completely ignored domestic work, seeing what was outside the money economy as not really economic activity at all.

The status of women in households and communities, as well as in politics and the economy, was not seen as a development issue at the time Boserup wrote. In the years afterward, however, issues of women's status and gender relations came closer to the center of the development debate. Scholars came to recognize that women were disproportionately represented in the ranks of the poor. The United Nations proclaimed 1976 to 1985 the Decade for Women, and today few development projects go ahead without some explicit attention to women, even if only cursory.

One reason why "women in development" (or WID, as it is often called by development specialists) has captured so much interest is the increasing recognition of the importance of women in population issues. Demographic studies find that the status of women, measured through their education and participation in the paid economy, is the most consistent factor in fertility reduction.[99] Women frequently want to reduce fertility rates, sometimes in contrast to their male sexual partners.[100] When men come to see women as economic equals, they tend to see them more as social equals as well, and women gain more say in family planning and other family decisions. Education gives women, as well as men, a broader understanding of possibilities, eroding fatalism and building a sense of empowerment. The greater economic standing of women in paid work in an increasingly monetarized world also means that childbirth and child care can become more of an economic burden than an economic opportunity for families. As opposed to general economic development of the structural adjustment and modernization variety, improving women's standing may be one of the principal paths to a demographic transition for less-developed countries, and perhaps the most important.

Some feminist critics are suspicious of this approach to development, however, as it seems to view improving women's status as a means to the end of population stabilization, not as a moral end in itself. The emphasis should be on ending patriarchy, not on furthering women's economic development, argues Sylvia Walby.[101] Patriarchy is a system of social organization in which women consistently hold lower status and less social power than men—a system that, most scholars agree, still characterizes virtually all human societies. Emphasizing women's economic development may be misplacing priorities.

Part of the reason for this doubt about women's economic development is the tendency to relegate women to lower-paid work. As less-developed countries have tried to build their exports in order to reduce debt and comply with structural adjustment plans, they have promoted cheap factory work, generally performed by women. As Valentine Moghadam has put it, women are the "new proletariat worldwide."[102]

There is considerable controversy among scholars about this phenomenon, often termed the "feminization of labor."[103] Does it represent the continued subordination of women in a new form, or does it represent an opportunity for poor women to gain better lives for themselves and their families through one of the few means available to them? Is it empowerment or continued disempowerment?

I strongly suspect the answer is both. As other writers have argued, improving women's status is not simply a means to population stabilization and increased exports.[104] However it may be seen by the governments and development agencies involved, improved status is good for both women and their families. If nothing else, reproduction should be seen as a women's health issue.

Some 358,000 women die each year from pregnancy-related causes—through childbirth, unsafe and illegal abortions, and post-birth infections and other illnesses. Ninety-nine percent of these deaths occur in less-developed countries. The good news is that maternal mortality has declined by a third since 1990.[105] But we still have a long way to go to ensure reproductive health for women. Better women's health also means better health for their children.

The persistence of patriarchy, despite improvements in women's health and economic and social status, seems undeniable. Eliminating patriarchal social relations is ultimately the only way to achieve equal status for women. But the fact that attention is finally being given to women in development (although perhaps not yet with sufficient sensitivity and commitment) should not be seen merely as a patriarchal ploy. Rather, it may be a sign that the world is beginning to acknowledge that improving the status of women is good not only for women. It is good for everyone.

Family Planning and Birth Control

Another controversial aspect of population is the use of birth control in family planning. The controversy stems partly from the coercive way that birth control has been applied in some instances, partly from moral judgments concerning some forms of birth control, and partly from questions about the significance of birth control technologies in reducing fertility.

The promotion of birth control has some serious black marks on its record. One of the worst instances was India's National Population Policy of 1976, initiated during the 18-month period between June 1975 and January 1977, when Prime Minister Indira Gandhi ruled as a dictator. Prime Minister Gandhi had been found guilty of election fraud, and in order to hold onto power, she declared a national state of emergency. The press was censored, dissidents jailed, and civil liberties curtailed. In this climate of extreme state control, the government put forth the National Population Policy under the direction of Mrs. Gandhi's son, Sanjay Gandhi. The policy emphasized sterilization—as well as health care, nutrition, and education for girls. Sterilization plans went quickly ahead, but the other aspects of the policy were more long-term and were for the most part ignored.

Most Indian states set bureaucratic quotas to monitor the "performance," as it was called, of the policy. Although people were paid for being sterilized, there was much coercive abuse as government officials in this strikingly undemocratic period in India's history struggled to meet their quotas. Near the capital, Delhi, the government set up vasectomy booths. People were harassed, threatened, and bribed. In about 6 months, some 8 million sterilizations were performed, mainly on the poor, who were vulnerable to the fees, harassment, and threats. Hundreds died in the riots that broke out in protest, as well as through infections caused by the sterilization procedures. When Mrs. Gandhi finally lifted the national state of emergency, the program was quickly dropped.[106]

Instances such as this, or the sterilization of Native Americans that was carried out on some U.S. reservations, are intolerable. They can also lead people, in anger and suspicion, to associate all advocacy for population control with oppression. A number of critics have seen the concern about population as part of, to quote one author, a "racist eugenic and patriarchal tradition"—the fears of the rich and White about a rising darker-skinned horde, as well as an effort to control women's bodies.[107] Critics have had particular concern about the single-minded attention that some Malthusians have given to birth control as a means for reducing population growth, given that most contemporary population growth is outside the West.

Paul and Anne Ehrlich's 1990 book, *The Population Explosion,* may be a case in point. They predicted that "the population explosion will come to an end before very long. The only remaining question is whether it will be halted through the humane method of birth control, or by nature wiping out the surplus."[108] There is nothing explicitly, and perhaps not even implicitly, racist about such a statement. Nevertheless, critics have argued that placing all the emphasis on birth control as a solution to population growth leaves intact the social inequality that is the primary cause of population growth.[109] Whatever the explicit intent of the Ehrlichs' position (and I believe the Ehrlichs are, in fact, strongly committed to social equality), critics suggest that the implicit effect would be the continuance of social inequality of race, class, and gender.

But just because racism, classism, and sexism have been a dimension of some birth control policies, and possibly some theories, this does not mean that birth control is necessarily racist, classist, or sexist. Indeed, preventing people from controlling births can be just as racist, classist, and sexist as any policy that seeks to control births. Reproduction, I believe most people would agree, is a basic human right. But so too is the right *not* to reproduce. Most couples around the world voluntarily seek to control and regulate—to plan—their reproduction. Limiting their ability to do so can be coercive, too.

One example of a policy of coerced reproduction took place in the late 1960s in Romania under the regime of Nicolae Ceauçescu, one of the most iron-fisted dictators of the twentieth century. In 1966, Ceauçescu suddenly declared any form of birth control, as well as abortion, illegal. Women had to undergo gynecological exams every 3 months to determine whether they were complying with the new law. As a result, birthrates doubled, at least initially.

Maternal mortality doubled too, with about 85 percent of these deaths due to botched abortions, illegally performed. Women across Romania also began avoiding gynecologists as much as possible, skipping appointments and failing to sign up for them, even for routine gynecological checkups. The result was that Romania for a time suffered from Europe's highest rate of death due to cervical cancer. Infant mortality also went up considerably (by one-third) as parents neglected, abused, and even abandoned unwanted babies.[110]

Granting a right to control and plan births is not the same as approving of all forms of birth control and all national birth control policies. There is certainly extensive disagreement on the morality of some forms of birth control, particularly abortion. But one can disapprove of abortion and still support other means of controlling births.

The question remains, though, whether modern birth control technologies are effective means of reducing population growth. In detailed historical studies, scholars have noted that in Europe, fertility decline generally began before modern birth control technologies were widely available. Indeed, in some places, fertility declined even before industrialization began.

The point is, there is nothing new about family planning. People have been using, and continue to use, many family planning techniques other than the pill, the diaphragm, the condom, the sponge, and other modern birth control technologies. Practices such as late marriage, extended nursing, abstinence, rhythm, withdrawal, and polyandry, among others, can be and have been effective forms of family planning.

But no doubt modern methods can be even more effective, which is one of the main reasons why so many couples across the world choose them when they are available. The commitment to plan births is absolutely essential to the success of any family planning practice, however. If social conditions are such that people are unable or unwilling to make such a commitment, no technique can be effective. In other words, birth control and greater social equality can be complementary, rather than contradictory, social policies.

The Environment as a Social Actor

Malthus went too far. It is clearly incorrect to adopt a position of *environmental determinism*—the view that the environment controls our lives and that there is little we can do about it. Human population has certainly increased to unprecedented levels, despite environmental limits, and in many wealthy countries population levels have returned to stability for reasons other than environmental scarcity. Technological and social change has allowed societies across the world to increase the production of food and other resources. Although the numbers of the poor grew dramatically in the last half of the twentieth century, particularly in areas with rapid population growth, their poverty cannot be understood apart from the dynamics of the world economy. The too-simple claim that population growth inevitably leads to poverty through environmental scarcity is just that: too simple.

But Malthus was not entirely wrong. Access to food depends not only on systems of entitlement; it also depends upon the environmental availability of food. Some resources seem hard to substitute with something else, even with the highest of technologies. And too often the risk inherent in some technologies puts the poor and marginal most in danger. Rapid population growth is also an enormous problem in itself, apart from any environmental implications, both for organizing social benefits, such as schools and a coordinated economy, and for safeguarding health. Rapid growth can indeed cause poverty, just as it is itself a product of poverty.

Moreover, because of the compounding effects of population with consumption and production, the question of growth and development is not merely one of finding enough food to feed everyone. It is also a question of whether we will ultimately be able to sustain everyone—humans and other creatures alike—if the competitive consumption and production levels of the world's rich become the ever-escalating norm.

Accepting a degree of *environmental agency,* accepting the environment's causal power in social life, is not the same as accepting environmental determinism. The role that the environment plays in our lives depends upon our interactions with it. The environment is not a given. We shape the significance it has. The French sociologist Bruno Latour, whom Chapter 3 introduced and Chapter 8 discusses in more detail, makes this point well. The environment is, in effect, a different place depending upon how we wish to use it and how we envision what it is. An environmental resource is only a resource if, because of our technical and social relations and because of our ideas, we find it to be a resource. It is we who make resources as much as it is the environment that provides them to us. It is we who make the environment as much as it is the environment that makes us.

The sociologist Fred Cottrel put it this way: The environment limits what we can do and influences what we will do.[111] Take, for example, the process of technological change, which is often presented as constrained only by our imagination and not by the environment. Even the cornucopian vision of technological change presented by Julian Simon implicitly grants a considerable degree of agency to the environment. For Simon, environmental scarcity prompts technical innovation. Thus, the environment helps guide the directions in which we exercise our powers of imagination.

And we should not forget that the consumption and production in which the human population engages are aspects of the environment in their own right. They are not external forces that may or may not impact the environment. All human activities are part of ecological dialogue.

One of the most important lessons to draw from that dialogue, I have tried to argue, is that population growth is a real issue in the conversation. But it is one that must always be understood within the context of consumption, production, and social inequality. Although birthrates among the wealthy are lower than rates among the poor, the consumption and production levels of the wealthy are far higher, and so are their per person environmental consequences, given current technological conditions. Social inequality—by region, class, race, ethnicity, and gender—is also in itself a principal factor in population growth, as well as in growth in consumption and production. But population growth does have environmental implications. And since we are a part of the environment, those are necessarily implications for us and how we may live.

Yet perhaps the most important dialogical lesson is that we can change the ecological dialogue. Changing the current dialogue of population growth seems to me to be a very good idea. Maybe we will be able to cope with the outcome of that dialogue in a way that provides general and sustainable well-being. Maybe. We're certainly not doing a great job of it now. But I think there are still reasonable grounds for hope. In order to do a better job, we need a dialogical understanding of the situation—an understanding that recognizes the inequalities, complexities, and interactions of social and ecologic life. For the problem of population is not just one of "too many people." Rather, it is also a problem of too many people with too much and too many people with not enough.

CHAPTER 5

Body and Health

There is a kinship between the being of the earth and that of my body.

—Maurice Merleau-Ponty, 1970

D ecember 2, 1984, was the date. The people who lived in the shadow of Union Carbide's pesticide plant in Bhopal, the capital of the Indian state of Madhya Pradesh, knew it wasn't the best place to call home. The main product of the Bhopal plant was the insecticide Sevin. One stage of making Sevin requires the production of methyl isocyanate, or MIC, a highly toxic chemical related to the nerve gas phosgene, which has the unfortunate property of reacting very strongly with a very common substance: water. So it has to be handled with unusual care. Even moisture in the air can be a problem. MIC is not the kind of thing that recommends itself to people looking for an area in which to settle down and raise a family.

But the people of the Jaiprakash Nagar neighborhood, 100 yards from the plant, were poor and didn't have much choice in the matter. At least they had roofs over the heads of their families—if only tiny, ramshackle ones propped up by thin and shaky walls. As the residents drifted off to sleep that cool evening, the late-night voices of the neighborhood came filtering through those thin walls, as they did every night, a comforting music of place to those accustomed to it. Someone laughing in the distance. Someone comforting a crying baby. Someone rummaging around in the dark.[1]

On the other side of the chain-link fence separating the plant from the neighborhood, however, there was mounting panic. Production at the plant had been shut down for a month for maintenance. Workers were beginning the complex series of operations to get it up and running again. About 9:30 p.m., they started washing out a few lines with water, downstream in the production process from the MIC storage area, which should have been safe enough. But there followed a whole series of troubles, individually minor and collectively disastrous. A clogged valve. A line left open. A few standard safety procedures not followed. A dysfunctional safety mechanism—the burner that was supposed to scrub any gases venting from the system. A recent, poorly thought-out modification of the plant's initial design that, in fact, connected the MIC storage area with the lines the workers were washing out. A recent reduction in the size of the work crew from 12 to 6. An inexperienced supervisor.

The last may have been the most crucial. Around 11:30 p.m., workers detected an MIC leak in the way they usually did: a burning in their eyes and throats. (This was far from the first MIC leak the plant had had. Although Union Carbide claimed that the Bhopal plant was the twin of a trouble-free one in West Virginia, it had been built without several of the safety features of its supposed twin.) The workers reported the problem to the supervisor, who shortly called a tea break, with plans to attend to the leak afterward, once fortified with caffeine. By 12:30 a.m. (now on December 3), the reaction of MIC with water became too much for the system to contain. A major leak began as MIC from tank E610 started to rush past the dysfunctional burner and out into the atmosphere above the plant. At 12:40 a.m., burning eyes and throats ended the tea break. At 12:50 a.m., workers pulled a general alarm at the plant when they discovered they were unable to get the burner working. At 2:00 a.m., the leak petered out as tank E610 reached empty. At 2:15 a.m., workers pulled the public alarm siren and walked over to a nearby police control room to report that the "leak has been plugged" and to give the first public admission that there had been a leak at all.[2]

By then, thousands had already died. In the coming hours, days, and years, thousands more would die. Some 5,000 to 15,000 in all would lose their lives, including a quarter of the residents of Jaiprakash Nagar and two other shantytown neighborhoods close to the plant. Some say 30,000 died.[3] No one knows for sure how many.[4] The residents of the worst affected neighborhoods were not the sort of people whose troubles the local government pays much attention to or who take their troubles (including their dead) to the government for help. Many died in their sleep, and there may have been some luck in that. Others awoke, breathless, coughing, with burning sensations, vomiting blood and frothing at the mouth, and rushed out of their homes in agony, right into the depths of the chemical fog, before collapsing in the street. Tens of thousands of cattle died as well. The stench of death was everywhere. At least another 500,000 people were injured.[5]

Here are some reports from the local papers the next day:[6]

> Jaiprakash Nagar, a sleepy locality of Old Bhopal, is today a ghost colony. Every second house in the locality has lost at least one family member in yesterday's night of horror.

> This correspondent who went round the locality early morning found more than fifty dead bodies lying unattended and unnoticed. . . . The dead included mostly children below ten years of age.

> The scene was so gruesome that it was difficult for survivors to identify their own dead family members. The neighbors were not willing to tell anything to anybody. They just sat glassy eyed, dumb-founded.

The tragedy continues today among the survivors. Numbness. Trembling. Polluted breast milk. Monstrous birth defects. Memory problems. Breathing problems. Immune system problems. Psychological problems. Plus, it turned out that Union Carbide had been dumping large quantities of toxic waste on the site for years, polluting the land and the water below. Cleanup efforts are underway, but 20,000 people still live in the immediate vicinity of the plant and had been drinking water from local wells for years before authorities started providing them with clean water in September 2000 (see Figure 5.1.).[7]

Figure 5.1   A protestor outside the fence at the former Union Carbide pesticide plant in Bhopal, India.

Source: © Chris Rainier/Corbis.

Survivors are still struggling for just compensation, and there were worldwide hunger strikes on their behalf in 2003, 2006, and 2008. In 1989, Union Carbide did offer the Indian government a $470 million settlement. As this amounted to less than $1,000 per victim, survivors were furious and demanded much more. The Indian government has repeatedly requested the extradition of Warren Anderson, Union Carbide's CEO at the time, to face trial for culpable homicide, but the United States has not complied, and does not seem likely to.[8] Indeed, in August of 2010, the U.S. State Department declared the Bhopal case "closed."[9] Plus, U.S. courts have thrown out several class action lawsuits. In June of 2010, seven Indian employees of Union Carbide's Indian subsidiary at the time of the tragedy were sentenced to 2 years in jail each, although they were later released on bail pending appeal. At this writing, several more cases and lawsuits are still pending in both Indian and U.S. courts.

Enough. I know this is a grueling story to read about. I know because it was grueling to write about. But I tell it to remind us in a forceful way of a central implication of environmental questions: the health of our bodies. I also tell it to point out, in what I hope is an equally forceful way, that these questions of body, health, and environment are sociological ones as well. As Eric Klinenberg noted with regard to the hundreds of deaths in the Chicago heat wave of 1995, "We have collectively created the conditions that made it possible for so many . . . to die."[10] In other words, the tragedy of Bhopal was a social tragedy as much as anything else, patterned by factors by now familiar to a reader of this book. The patterns of our economy. The patterns of our technology. The patterns of our politics. The patterns of our distribution of environmental goods and environmental bads—of environmental justice. And, perhaps less obviously at first look, the patterns of our ideas and their mutually constituting interaction with our material conditions.

In this chapter, we explore these patterns with respect to the *environmental sociology of the body*. True, the Bhopal tragedy is one of the worst industrial accidents ever, with maybe only Chernobyl as a rival. It is hardly representative. But the cumulative impact of the countless ways, large and small, seen and unseen, that our bodies are affected by the technologies of our environmental interaction connects us all to these patterns.

Connects us all—the environmental sociology of the body is a way to explore ecological issues at the most personal of levels. Ecology, recall from Chapter 1, is the study of natural communities. But literally it is the study of *ecos*, which is ancient Greek for "home." Ecology is thus the study of natural home as community. Environmental sociology, recall as well, is the study of the biggest community of all—and thus the biggest home of all: the abode we share with everyone, human and nonhuman alike. A body is an abode, too. As we will discover—perhaps surprisingly, given its constitution of each of us as individuals—a body, too, is an abode we share with everyone.

Welcome to the Invironment

It may come as another surprise to learn that the body and health have not always had an easy and welcome place within environmental discussions. Indeed, the body and health have often been understood as diametrically opposed to the concerns of environmentalism.

Take health. Why do we alter the environment by draining swamps, dousing our crops with pesticides, and burning fossil fuels? To eliminate the habitat for insects that carry disease. To compete better with the creatures that would deny us a portion of that essential substance of health: food. To make our lives less grindingly arduous through heating, cooling, lighting, and mechanized transportation. Seen from this view, the environment seems a threat to health, not an aspect of it. Thus, concerns for public health, for eliminating hunger, and for human comfort have often promoted transformations of the environment, on behalf of our bodies, that run afoul of at least some conceptions of sustainability and the beauty of ecology, and even those of environmental justice. As a result, historically, the public health movement and efforts to end hunger have had surprisingly little to do with the environmental movement. While efforts to clean air, water, and land of pollution are certainly central environmental concerns, they have often taken a rhetorical backseat to efforts like wilderness and biodiversity protection. And many proponents of efforts to improve human comfort have long regarded the environment as something of an enemy.

Take the body. The very meaning of the word *environment* has a connotation of what is around us. The environment is our environs, our ecological neighborhood, not us. A body may have an environment, but it is separate from it, our normal usage seems to imply. Lurking inside this opposition may be something of a disgust for the body, at least in the minds of some. If you think through what it takes to maintain a body as an ongoing entity, you are pretty much inevitably led to our links to the environment, and thus to the recognition that we are animals with animal needs. By keeping the body conceptually separate from the environment, perhaps we are unconsciously trying to ignore this embarrassing "low" reality of human life.[11]

But these oppositions in our thought are lessening. Increasingly, environmental concerns routinely embrace health and the body as environmental issues, and thus the body as intimately involved with the environment—a part of, not apart from, indeed its most intimate part.

In order to help that lessening along, I have a little term to offer, one that my Dutch colleague Kris van Koppen and I came up with some years ago.[12] We could continue using the terms body and environment, as traditionally has been done in the West, and try constantly to remind ourselves not to regard them as oppositional. But by establishing an initial separation, these terms force us to undertake an extra intellectual step to recognize their interconnections, their dialogue. So rather than always speaking of the body and the environment, it might help sometimes to speak of the *in*vironment and the *en*vironment—where the *invironment* refers to the inner zone of the environment, where we find the body in perpetual dialogue with the environment. Invironmental issues, then, would be issues that concern the dynamics of that inner zone of dialogue, with health being perhaps the prime example.

In the usage I'm suggesting, then, environment is a more encompassing term than invironment. Some environmental issues—like global warming or species loss—are not, in the first instance, invironmental issues. But as we consider environmental issues more closely, we will likely discover that they all have invironmental dimensions. Global warming has implications for food production, for water supplies, for the spread of disease, and more, including the sheer level of warmth with which bodies must contend in the summer. Species loss has implications for the loss of potential medicines and crop varieties that might help relieve concerns for health and hunger.

Species loss is also an invironmental issue of immediate concern for the nonhuman bodies involved—the bodies of the nonhuman animals who are losing habitat, being hunted, or facing competition from recent migrants to their ecological niches. In this chapter, I take an almost exclusively human-oriented stance because I have more than enough to say about the human invironment on its own. But the difference between the human and nonhuman invironments is another potential opposition we would do well to be wary of.

In sum, the environment is not only something "out there." It is also something "in here."

Food: The Taste of the Invironment

What is that substance there on your plate, just beneath your hovering fork? Food, yes: Something nourishing that you soon intend to put into your mouth, and then to swallow deeper within and later disperse throughout your body. Some of it will pass through—perhaps rather quickly, if I may say so. But some of it will stay awhile in one form or another, perhaps even for the rest of your life. If you are what you eat, that is your future body there beneath your fork.

But that substance is also a lot more than that. It is ecology in motion, moving to you and soon to move through you. You may be thinking about it as a simple, light lunch: pasta tossed with olive oil, a crushed garlic clove, sautéed mushrooms, some grated cheese, and a bit of pepper and basil. But it likely contains atoms from the soil, air, and water of many continents, assembled into tasty unions by plants and sunlight in interaction with bacteria, fungi, and animals.

The wheat in the pasta probably didn't come from anywhere near you. Pasta makers favor an especially hardy variety of wheat with a double load of chromosomes called "durum" wheat, which doesn't like a lot of rain. Farmers grow a lot of it in North Dakota, Saskatchewan, Spain, northern Italy, Turkey, Syria, Northwest Africa, India, Australia, and a few other places. But durum wheat dries well, stores well, and pours well into containers, and thus is fairly easy to send flowing throughout the world, including to you.

Unless you are lucky enough to live in a Mediterranean climate, the olives that yielded the olive oil probably didn't grow near you either. Olive trees like their summers hot and dry and their winters mild and wet. But olive oil also stores well and pours well, and is easy to send flowing from afar—probably from California or a Mediterranean country.

If that's black pepper on your pasta, it assuredly came from a tropical country, most likely Vietnam, currently the world's leading exporter. So if you're from higher latitudes and western longitudes like me, that likely represents yet another continent on your plate.

If you are having lunch at my house, though, everything else could have come from much closer by, including my own backyard here in Wisconsin. We don't grow mushrooms at home, but lots of farmers in Wisconsin do—which is nice, because I like mushrooms even though they are hard to grow and are fairly perishable. Mushrooms aren't like wheat. They store badly and pour badly. Local is definitely better with mushrooms. I don't have a cow either, but we sure have a lot of dairy farms in Wisconsin, so getting the cheese is not a worry. It's probably not a worry for you either, even if you don't live in Wisconsin, California, France, Britain, Italy, or Switzerland. Milk pours well but stores badly. Cheese pours badly but fortunately stores pretty well, and it's packable enough to make up for bad pouring—as long as you can charge a good amount for it. So one can often find tangy bits of Wisconsin in much of the world. But the garlic and the basil I can assure you came from our vegetable garden, 30 feet away. And if instead of black pepper that's red pepper on your pasta, some years we grow that here at home, too.

Quite conceivably you have on the plate before you durum wheat from Australia; olive oil from Italy; black pepper from Vietnam; and cheese, mushrooms, basil, and garlic from Wisconsin, the latter two from my backyard. And maybe we'll finish up with some chocolate blended from varieties grown in Africa and South America, as is often the case with fine chocolate. That would be six of the seven continents. (Not much grows in Antarctica.) That would also be four of the six kingdoms of life: cheese from animals and eubacteria; wheat, olive oil, basil, garlic, and pepper from plants; and mushrooms from the fungi. (People do not, as yet, cultivate archeobacteria for food. In some cuisines, there are some protists, mainly in the form of algae, as in the nori in sushi, but people don't eat a lot of protists either.)

That's a lot of ecology. You are thus an intersection point in a vast flowage of ecological relations. In that sense, you eat what you are.

You eat what you are in another sense, too. What we eat is more than merely material. Our food also embodies social relations. Our sense of identity suffuses what we judge as worth eating. Don't like cheese? Then you are probably from a nondairy culture, like most of China, where cheese can seem as alien as tofu can feel to others. Don't care for garlic or pasta? Then perhaps you are the Queen of England, who is said to regard these as foreign intrusions on British culture. Absolutely adore that cake recipe that has been handed down in your family from your mother's mother, or that stew recipe from your father's father? And do you adore it, if truth be said, in part because you know these recipes are family traditions? Then you are like everyone else who senses in food the presence of those who are not physically there, what we might term *food ghosts* that give meaning to what we eat.[13]

Plus, there likely are also social presences in your food that you don't immediately recognize, but probably would allow are there to some degree if someone raised the matter. The farmers and farm workers who raised the food. The processors who got it into a form that enabled it to be mobile. The distributors that moved it. The retail clerks who put it on a store shelf and made you pay for it. The factory workers who made the tools that all of the preceding used. The miners and foresters who provided raw materials to the factory workers. The teachers who educated them. The doctors and nurses who kept them healthy. The financial officers who provided loans to get and keep all this going. The government bureaucrats who monitored the financial officers (or tried to) to ensure they did everything on the up and up. The politicians who kept an eye on the bureaucrats (or were supposed to) to ensure they did their work well, too. And so on. And so on. And so on.

In this way, food makes farmers of us all, cultivating our environment as we are assimilating our invironment.

A Moving World

All of the connections between bodies and environments that I've just described are obvious enough, once pointed out. The trouble is, we rarely do point them out, says a new direction in environmental sociological theory. Our usual ways of thinking about the world separate and disassociate one thing, place, and living thing from another, holding them steady in a fixed stare of reductionism. Under such a mind-set, food is just fuel, just something to buy and then eat. When Aldo Leopold warned about "the danger of supposing that breakfast comes from the grocery," he was making the same point.[14] The division of labor that is so central to the contemporary economy gives us a divided way of thinking about our lives. We see dead commodities in packages, not the living beings and places that created them and sent them our way.

But the British theorist John Urry and the Spanish theorist Manuel Castells invite us to make motion and interaction our usual ways of thinking about the world. And they suggest metaphors drawn from liquids to help us with this re-envisioning. Urry suggests that we think of "global fluids" of peoples, ideas, and materials, streaming across the planet, connecting localities and bodies across the globe.[15] Some recent developments in technology and society have accelerated and enlarged what Castells calls the "space of flows," but the world has always been on the move.[16] The Dutch environmental sociologists Gert Spaargaren and Arthur Mol, along with the American environmental sociologist Fred Buttel, suggest the concept of *environmental flows* as a way to describe this age-old social ecology of motion.[17] The overall term for this line of thinking is *mobilities theory*, and it is getting a lot of attention from scholars right now.[18]

Of course, everything everywhere is not always and forever moving and intermingling. Were that so, we would soon have no difference between one place, person, or thing and another, as everything passed through some kind of great blender and wound up a single homogenous mix. We want, need, and have lots of distinctions and differences, what mobilities theorists call not separations but *stabilities* or *moorings*.[19] Our bodies are an example of a stability. But our bodies also need mobilities to remain stabilities. It's not as paradoxical as it sounds. It just means we need to eat to live—and to breathe and drink and move about to gain our food, air, and water and to secure the social and ecological relations that allow us to gain them.

There are also mobilities that we don't want—things that move and connect that we wish would stay more apart. We could take that sentence as the mobilities theory definition of pollution. Again, this is all obvious enough once pointed out. But in modern, industrial life, with its reductionist and disassociating ways of thinking, we have often assumed that bodies and places—invironments and environments—were separate that were not. Or, maybe more accurately, we often haven't even bothered to check.

Living Downstream: Justice and Our Threatened Invironment

The ecologist and author Sandra Steingraber wants us to check. She offers another fluid metaphor for alerting us to the connections between environment and invironment: "Living downstream," she calls it in her book by that title.[20] Because of the body's perpetual dialogue with an environment on the move, we are all always living downstream of what goes on around us. We may wish sometimes that we were separate. But we're not. We just aren't. Moreover, it turns out that, in matters of who gets what is coming downstream, we are neither separate nor equal. Which suggests a reformulation of the golden rule: In the words of Wendell Berry, "Do unto those downstream as you would have those upstream do unto you."[21]

Mercury and the People of Grassy Narrows

In 1970, a small band of Ojibwa Indians living in the Grassy Narrows reservation in remote northwest Ontario learned that they were literally living downstream. In that year, government authorities realized that some 20,000 pounds of mercury had been released over a 10-year period from a paper mill into the Wabigoon River, 80 miles upstream from Grassy Narrows. There, in the river, ecological processes converted the mercury to methyl mercury, one of the most toxic substances to be found in any chemistry book. It steals a person's vision, hearing, agility, ability to feel, memory, emotional control, and eventually a person's life. The affected walk with a kind of stagger, a glazed and glassy expression on their faces.

Methyl mercury is a sly and crafty toxin. It has no taste and no smell. It can't be seen in the water or in the fish. It can't be felt either. As an Ojibwa elder described to a visiting journalist, "But you know it's there. You know it can hurt you, make your limbs go numb, make your spirit sick. But I don't understand it. I don't understand how the land can turn against us."[22]

This sense of everything, even the land, turning against you can be one of the worst effects of methyl mercury and other toxins that are largely invisible to our senses, as the sociologist Kai Erikson has observed.[23] As Erikson puts it, methyl mercury poisoned the minds of the Ojibwa of Grassy Narrows with "a pervasive fear that the world of nature and the world of human beings cannot be trusted in the old way."[24]

Not without reason. There is frequently much empirical justification for a feeling of persecution in invironmental issues. Although we all "live downstream," there are definite patterns in the social characteristics of those who find themselves living closest to the outfall pipes of our industrial economy. The Ojibwa people of Grassy Narrows are poor and disenfranchised, and that is no sociological surprise. Study after study has documented a persistent finding: As Chapter 1 discusses, those who get more of the bads are disproportionately those who receive less of the goods.[25] Those bads are very often matters that affect us intimately, right in our own bodies. In these cases, issues of environmental justice are also issues of the justice of our invironment. (We could call this "invironmental justice," and in earlier editions of this book I did. But let's not add too much to this world's confusion of terms.)

The connection between violations of environmental justice and economic justice is often an interactive one, as the people of Grassy Narrows discovered. Once it became known that the river on which they had long relied for food and income was polluted with methyl mercury, the Canadian government banned fishing there. It had to be banned, of course, but that threw an already poor people out of much of the little work they had.

It also threw them, in a way, out of their culture. For the work we do is more than a source of income and sustenance. It is a source of pride and purpose, of self and the embedding of self in the lives of others. "We are now a people with a broken culture," is how Simon Fobister, chief of the Grassy Narrows band, put it.[26]

When your culture is broken, people often look for purpose in drink. At least so it was in Grassy Narrows, where after the closing of the river to fishing, alcohol abuse skyrocketed. Mortality rates skyrocketed, too, as alcohol abuse led to violence, accidents, and health troubles. As Erikson notes, out of roughly 400 members of the Grassy Narrows band, 35 persons—about 9 percent—died between 1974 and 1978. This is a huge mortality rate. Some 80 percent of these deaths were either directly or indirectly related to alcohol: suicides, murders, drownings, alcohol poisoning, and heart failure from excessive drink.[27]

Which is not to transfer the blame from methyl mercury to alcohol abuse. It is methyl mercury's poisoning of their minds, breaking their culture, that led to alcoholism problems among the Objiwa of Grassy Narrows. Plus, there is a further cruel feature of methyl mercury's poisoning of the flesh: the way its effects mimic those of alcohol abuse. As Erikson observes, considering the physical symptoms of methyl mercury poisoning,

> Now if you were asked where one might find a group of persons with slurred speech and difficulty in focusing, with a lumbering gait and uncertain coordination, with a glazed and numbed look about them interrupted at times by violent outbursts of temper, what might you suggest?[28]

Thus, when alcohol abuse is also present, it can be a hard matter to ascertain with any surety that methyl mercury's effects are being directly manifested. It's a long, slow poison.

But either directly or indirectly, slow or fast, the result is the same. As another Grassy Narrows elder put it, "Now we have nothing. Not the old. Not the new."[29]

A Factory Explosion and the People of Toulouse

On September 21, 2001, at 10:15 on a Friday morning, a huge explosion ripped through the AZF (Azote de France) fertilizer factory on the outskirts of Toulouse, France. Something touched off a silo with 300 tons of ammonium nitrate, a chemical used both in fertilizer production and in explosives. It was like an earthquake. Some 27,000 nearby homes were damaged, 11,000 seriously, with crumbled walls and missing roofs. Windows were shattered in stores 2.5 miles away.[30] Eighty schools had to be closed for repairs. It left a crater 165 feet across and 33 feet deep. Miraculously, only 30 people died, mostly plant workers, although more than 2,000 people were injured.

Coming a scant 10 days after the September 11, 2001 tragedy, the AZF explosion hardly registered in the world's news. (I didn't hear of it at the time myself.) But it was one of the biggest stories of the year in France, in part because some worried that it might be connected with the events of September 11. There was much buzz in the French media about whether one of the workers who died, a Tunisian Muslim with reputed radical Islamist leanings, might have set off the explosion as an act of terrorism. But investigators eventually dismissed that theory as groundless. Although a definite cause has not been established, investigators for the French judiciary believe that the cause was likely a reaction with some of the other chemicals used at the plant, notably, sulfuric acid, lime, and soda. In other words, sloppy procedures by the company were likely at fault, according to the investigators.

Much of the debate since then has been over corporate culpability in the explosion. In the now-common tangled chain of ownership that vertical integration and increasing scale have led to in industry after industry, the AZF brand is owned by GPN, which was formerly called Grande Paroisse, a large French fertilizer and environmental engineering company. Back when it was called Grande Paroisse, GPN was owned by Atofina, which in 2004 became Arkema, the fifth-largest chemical company in the world, and just one branch of an outfit now calling itself Total—which is itself the result of a series of amalgamations that have now made it one of the world's largest oil companies. Confused? You're not the only one. The executives buying and selling chunks of business like so many deeds in a game of "Monopoly" often have little idea what is going on down the chain of ownership, which cannot be good for long-term investments in plant safety, critics argue.

Not that there should have been much doubt the AZF plant could be trouble someday. The European Union had some time ago given the plant a "Seveso" designation as a high-risk facility—a designation whose name derives from the Italian village of Seveso, where, in 1976, an industrial accident at a pharmaceutical company released a toxic cloud of dioxin. A Seveso designation means a plant must be subjected to unusually strict safety procedures, including notification of nearby residents of the dangers of the facility. But Seveso guidelines were not being carefully followed at the AZF plant or in the surrounding neighborhoods. Nor were the guidelines being carefully followed at many of the other 371 Seveso-designated facilities in France—a point that had led the European Commission a few weeks before the blast to announce that it intended to take France to court to get it to comply with the Seveso directive.[31]

In fact, there wasn't much doubt locally about the dangers. It is no simple accident of blind planning that the residential area closest to the plant is one of the poorest districts in Toulouse. Across a highway from the plant, many residents live in a series of cheaply made apartment towers—*banlieues* is the French term for them, which sounds much nicer to the English speaker's ear than "housing projects" (in the United States) or "council estates" (in the United Kingdom)—a number of which were seriously damaged in the blast. The AZF plant was established in 1924 in what was then open countryside. Toulouse soon grew out that far, though, and when the suburban tide reached the factory zone, it was not the homes of the rich planners and developers that were located there.[32]

In other words, the residents of the *banlieues* discovered for themselves the negative association between who gets the bads and who gets the goods of the environment. Those who get the bads typically are those who don't get much of the goods, like decent housing. Why do people put up with this association? Because they often fear that what homes and jobs they do have depend on it. After all, living near a fertilizer plant is not exactly a sought-after situation for those with more choice in the market. Nor is working there. When your choices are limited, it's hard to make a good one.

As the mayor of Toulouse, Philippe Douste-Blazy, observed in the immediate aftermath of the explosion, "This kind of incident should date from another era. It's time to change. We must stop asking our citizens to choose between their work and their lives."[33]

Pesticides and the People of Everywhere

Currently, the people of the world apply some 5.4 billion pounds of pesticides to the land every year, mostly for agricultural uses, including both chemical pesticides like atrazine and biological control agents like *Bacillus thuringiensis*.[34] In all, some 881 pesticides were in world use as of

2006.[35] And they're mostly toxic, some highly toxic. That's why we use them, after all: to kill things. Of that 881, the World Health Organization (WHO) classifies 28 as "extremely hazardous," 56 as "highly hazardous," 107 as "moderately hazardous," and 119 as "slightly hazardous." Eliminating the 322 pesticides that WHO did not include in its classification of hazard, that gives a figure of 55 percent of pesticides as being some degree of hazardous (see Figure 5.2).[36]

Figure 5.2 World Health Organization rating of pesticide hazards, 2006.		
WHO rating	*Number*	*Percentage of all classified*
extremely hazardous	28	5.0
highly hazardous	56	10.0
moderately hazardous	107	19.1
slightly hazardous	119	21.3
not hazardous	249	44.5
not classified	322	NA

Source: World Health Organization. (2006b). *The WHO Recommended Classification of Pesticides by Hazard and Guidelines to Classification: 2004.* Corrigenda of June 28, 2006. Geneva, Switzerland.

WHO has some good reason to label half of pesticides as hazardous, according to toxicological and epidemiological studies. Take research linking pesticide use to birth defects, for example. A Finnish study found that women who worked during the first trimester of a pregnancy in agricultural occupations that used pesticides had double the normal rate of cleft lips and palates among their newborns. A similar study in Spain found that the rate was 3 times the normal. A study of 700 women in California found that those who lived nearest to crop areas where particular pesticides were in use had higher rates of fetal death due to developmental defects. In Iowa, a study found that rates of birth defects and congenital heart problems were elevated in communities whose water supplies were contaminated with atrazine, a common pesticide in the United States.[37]

Studies even show a correlation between time of pesticide application and birth defects. A 1996 examination of birth defects in Minnesota compared western Minnesota—where most of the state's agriculture is—with the eastern side. This study found not only that farm families from western Minnesota had higher rates of birth defects, but that nonfarming families on the western side did, too. Plus, the defects on the western side of the state followed the seasonality of pesticide application. Most pesticides are applied in the spring, and birth defects were highest for western Minnesota babies conceived in springtime.[38]

A more comprehensive 2009 study showed the same effect nationally.[39] Researchers looked at the U.S. Geological Survey's monthly National Water Quality Assessment data. Over a 6-year period, from 1996 to 2002, they found that pesticide concentrations in the nation's water supply were highest from April to July, which is just when farmers apply pesticides the most. And they used data from the Centers for Disease Control to track conception dates and birth defects. Babies conceived between April and July were significantly more likely to have birth defects, including Down syndrome, club foot, cleft lip, and other troubles. There was a particularly striking association with concentrations of atrazine in the nation's water (see Figure 5.3.).

Figure 5.3   The association between birth defects, mother's last menstrual period (LMP), and levels of the herbicide atrazine in US surface water, 1996–2002.

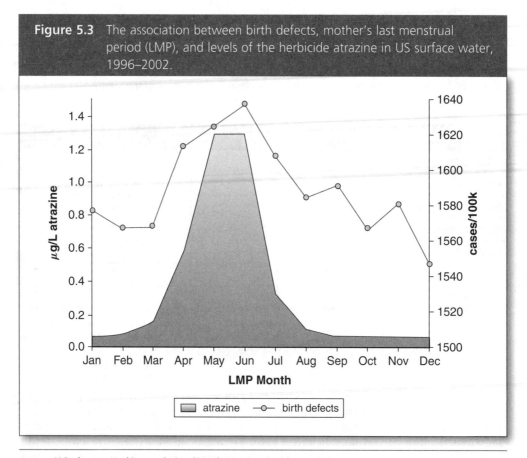

Source: Winchester, Huskins, and Ying (2009). Reprinted with permission.

Or take the evidence linking pesticides to fertility problems for both men and women. There's the 2001 study of 225 men in Argentina who were experiencing fertility problems. Here, researchers found an association between pesticide exposure, low sperm counts, and elevated levels of female sex hormones in their bodies.[40] Then there's a 2003 study from Missouri— similar findings.[41] And there's the 1994 study of 30 Danish organic farmers, published in the journal *The Lancet*. This one found that organic farmers (who use no pesticides on their farms) had a sperm count twice as high as Danish blue-collar workers.[42] As for women's infertility, a 2003 National Institute for Occupational Safety and Health study of 644 Wisconsin women found that those who had experienced fertility problems—defined as a year's unsuccessful effort to become pregnant—were 27 times more likely to have been involved in mixing and applying pesticides in the 2 previous years.[43]

Stick with me while I run through a few more facts and figures from studies of the health effects of pesticides. Here are a few on neurological and mental problems associated with pesticide use. A 2007 British study of 959 people with Parkinson's disease and another 1,989 without, across five European countries, showed "significantly increased odds" of pesticide exposure among the Parkinson's patients.[44] A 2000 Dutch study of 830 people found that those who had been regularly exposed to pesticides in their work were 5 times more likely to suffer "mild cognitive dysfunction"—which meant that they had trouble recognizing words, colors, and numbers and had trouble speaking.[45] Earlier, a study from the mid-1990s of farm families in

Colorado found that exposure to certain pesticides was associated with levels of depression 6 times the rate of the nonexposed.[46]

And here are some results from studies on breast cancer and prostate cancer, two of cancer's most pervasive forms. A 2003 Belgian study found that women with breast cancer were 5 times more likely to have pesticides in their blood.[47] This is in line with concerns that a number of pesticides are endocrine disrupters—that is, they mimic and interfere with hormonal activity. Endocrine disruption is, in turn, suspected to cause cancer in some cases. As for prostate cancer, a 2003 study of farmers and nursery workers in Iowa and North Carolina found elevated incidence rates.[48] A 2002 study found that workers at Sygenta Corporation's atrazine plant in St. Gabriel, Louisiana, are coming down with prostate cancer at 3 and a half times the expected rate for Louisiana.[49]

Or take this sampling of studies, from just the last year as I write these lines in February of 2011. A North Carolina study found that levels of cholinesterase, which is central to proper nerve function, goes up in and down in the blood of Latino farmworkers in time with periods of pesticide application and pesticide levels in their blood.[50] A study of greenhouse workers in Ecuador found that prenatal exposure to pesticides is associated with neurobehavioral problems in their children.[51] A study of licensed pesticide applicators in the United States found an association of skin cancer rates with occupational exposure to several pesticides.[52] Another study found an association of pesticide exposure with risk of developing type II diabetes by looking at blood samples collected 18 years earlier.[53]

I could go on, and were this a book on environmental toxicology I would. But I think the point should be clear by now: There is a good deal of very worrisome evidence about the invironmental effects of pesticides, effects that concern the health of all of us.

There is also evidence that disagrees with some of these findings. For example, a 1999 Yale University study of 1,000 women in Connecticut did not show a link between breast cancer and exposure to pesticides.[54] And the Environmental Protection Agency (EPA) argues that the elevated prostate cancer rates at Syngenta's St. Gabriel plant are due to a sampling effect. Most prostate cancer is never detected, according to autopsy studies, so the research at St. Gabriel may just be showing the prostate cancer that would be expected for the general population, were a similarly comprehensive study of non-plant workers to be done.[55] Research is like that, which gives ground for much political debate as to whether particular pesticides should be banned.

Take atrazine. In addition to birth defects, studies have implicated atrazine in cancer, menstrual problems, miscarriages, low birth weight, sexual development disorders, retinal damage, and muscle wasting in humans, as well as hermaphroditism in frogs. It is now banned in Europe, while the EPA in the United States is waiting for more studies and conducting a review of the matter, as of the winter of 2010.

Many do not want to wait for all the pesticide studies to be done, however. That's in part because all the studies never will be done. This, too, is what research is like. So in the meantime, many are turning to organic food, which in recent years has become one of the fastest-growing sectors in food retailing. Growth in consumption of organics has been running at 15 to 20 percent a year in the United States, although it declined to 5 percent in 2009 during the height of the Great Recession. But even in 2009, organic food sales rose at 3 times the rate of nonorganic food sales.[56] Organic sales in Europe have also been seeing double-digit annual growth, with a similar dip to 5 percent growth in 2009.[57] In the United States, half of consumers now use organic products at least occasionally.[58] In most cities, even conventional supermarkets have put in organic product lines (see Figure 5.4). Most of the major food corporations are busily buying up the smaller organic companies, in an effort to cash in on this booming market and citizens' increasing concerns with the health effects of pesticides.

Figure 5.4 Just a few of the many organic certification labels that national governments have established.

New Zealand   France   United Kingdom   Sweden

United States   Germany   Japan

Source: Author.

Pesticides and Environmental Justice

But not everyone is buying organic food yet, including many who would like to. Lack of availability is indeed part of the problem. Price is the big barrier, though, due to market factors that go beyond the constraints of supply. "Yuppie chow" is what critics sometimes call organic food, and not unreasonably so in many circumstances. In an organic supermarket a mile from my house, part of a national chain of such supermarkets, tomatoes were selling for $5 a pound recently. True, these were specialty tomatoes—an heirloom variety raised by a local farm. But they don't have to sell them for so much, another local organic farmer told me at the farmer's market that is also near my home. She used to sell her own heirloom tomatoes to this store for a fraction of that $5 price. When she saw what the store was selling them for, she refused to supply the store any longer. (Now she sells them on her own at the farmer's market for $1.50 to $2 a pound, more than the wholesale price she had been getting, she informed me.) In other words, organic food often becomes vanity food—a positional good, as Chapter 2 termed such an item, limited in supply and thus a ready source for social display for those in a position to pay for it.

The use of organic food as a positional good raises issues of environmental justice. While the environmental effects of pesticides have consequences for everyone, some people are better able than others to avoid these effects. It would be hard to escape all effects of pesticides. Even the wealthy have neighbors who douse their lawns with the latest chemical wonders, and even the wealthy sometimes find themselves in situations where it is difficult to eat as one might like. The wealthy have a considerable advantage in avoiding pesticides, though. For one thing, they rarely earn their living as farmworkers and lawn care workers, the people with the most direct contact with pesticides.

There are other social inequalities in pesticide exposure as well. Certain of us have increased vulnerabilities to pesticides and other toxins, especially at certain stages in the life of our bodies. Children and pregnant and nursing women have particular sensitivities that mean the same level of exposure can lead to outsized effects. When the body's cells are reproducing and growing rapidly, chances are particularly high that chemical pollutants can disrupt the body's development. Plus, children and many pregnant and nursing women eat more in relation to their body weight, increasing their level of exposure. They may also eat more of foods high in pesticide residues, such as many of the apples that go into apple juice, a staple of young children's diets in many countries.[59] And young children can increase their exposure through eating what they shouldn't and through playing with what they shouldn't—such as dirt from a lawn treated with pesticides, or pesticides stored at home. Even playing on the floor of a home with a lawn and garden treated with pesticides can increase children's exposure, because of chemicals tracked in on shoes, notes the EPA.[60] The U.S. 1996 Food Quality Protection Act tries to take children's sensitivities and exposure into account in the setting of pesticide residue guidelines, but few parents consider these matters in making food choices for their children.

Another source of children's exposure to pesticides and other toxins is from their own mothers. Breast milk turns out to be, unfortunately, remarkably sensitive to chemical pollutants. Some 200 different chemical pollutants have been detected in the breast milk of American mothers, according to a 1981 tabulation.[61] If a similar tabulation were done today, it would undoubtedly yield a substantially higher number. A particular problem is persistent organic pollutants, or POPs—chemicals like dioxin, PCBs, and the once-popular pesticide DDT (the latter usually in the form of DDE, a metabolic byproduct of DDT). As Sandra Steingraber observes, "Prevailing levels of chemical contaminants in human milk often exceed legally allowable limits in commercial foodstuffs."[62] Consequently, breast-fed babies consume 50 times more dioxin and PCBs than do adults, according to recent studies, and a German study from 1998 found that breast-fed babies have 10 to 15 times the level of POPs in their bodies that formula-fed babies do.[63]

These are tragic findings. "On the one hand we have the chemical adulteration of human milk," notes Steingraber. "On the other is the bodily sacrament between mother and child. Can we speak of them both in the same breath? Can we look at one without turning away from the other?"[64]

That parents could be placed in such a quandary is also an issue of environmental justice and its invironmental reach into all our bodies.

The Sociology of Environmental Justice

These are uncertain matters, though. I don't just mean the inescapable uncertainty of science, particularly when dealing with largely invisible matters, like how much of a particular chemical has made it into our bodies and the time it takes for any health consequences to be manifested. I mean also the uncertainty of whether those consequences will matter for a particular person like you or me, or for particular peoples like them or us. Chapter 9, on the subject of risk, explores the sociological implications of these uncertainties in more detail, but here I want to raise an important issue of justice and how we think about it and act on it.

Let us say, for instance, that the risk of getting Parkinson's disease from pesticide exposure is on the order of 1,000,000 to 1. I don't know what the actual number is, and I'm not even sure that it could be known. But it is common enough to think of such an issue in such a way. Or let us say that the chances of being caught in the blast wave of a factory explosion are also 1,000,000 to 1. It might be. I don't know for sure, but certainly it is not a very regular occurrence. It's never happened to me or any of my friends, at any rate.

So maybe neither of these problems is something I should worry about. Besides, we need food if we are to solve the environmental challenges of hunger. The pesticides produced at Bhopal and the fertilizer that Toulouse plant produced are very helpful in our battle with pests and with the Earth itself, it might be argued, and indeed often is argued. On the whole, pesticides and fertilizers are for the greater good of everyone, it is frequently said.

Of everyone? Really? Such might be the response to the view in the preceding paragraph. If a hazard has a 1 in 1,000,000 chance, that means in a country the current size of the United States, 300 people will suffer from it, even die from it. You probably won't be affected, and the same is true for almost everyone else. But there are 300 people out there that the hazard will indeed affect. So it would not be for the greater good of *everyone*. Likely most agree that it is not right for the majority to suffer for the benefit of a few. Yet is it right for a few to suffer for the benefit of the majority? Certainly, if you knew that you were going to be among that few, you would be unlikely to agree to such an arrangement.

The debate I have just sketched here is the essence of recent discussions among political philosophers about the meaning of justice and related ideas like human rights. This debate is not only an academic one, however. The social dynamics of environmental justice to a large extent revolve around the way the debate plays out in contemporary society. If we are to understand the sociology of justice and its implications for the invironment, we should inspect this philosophical question with some care.

What Is Justice?

The view that if something is for the greater good then it must be okay is technically referred to as *utilitarianism,* and it is an old and pervasive idea. The classic formulation of utilitarianism is that we should strive for actions whose tendency "to augment the happiness of the community is greater than any it has to diminish it," as the philosopher Jeremy Bentham put it in 1779.[65] Another standard phrase to describe the goals of utilitarianism is that we should promote "the greatest good for the greatest number."[66] It's called "utilitarianism" because the good is defined as "utility," or the degree to which an action promotes happiness. Central to utilitarianism is the principle that everyone's happiness has equal moral standing, no matter what the person's social position. Also, most utilitarianism is strongly focused on results rather than the way we get there, a philosophical position known as *consequentialism.*

Utilitarianism can be found in much contemporary thought. Many democratic principles, such as the will of the majority, resonate strongly with utilitarianism. We can also see utilitarian ideals enshrined in, for example, the focus on growth in gross domestic product (GDP) as a mark of social improvement, what we might call economic utilitarianism. In this case, GDP is taken as a direct measure of the greatest source of utility in modern economies: money. If there is more of it around in the economy, people on the whole must be able to do more of what they want to do, which can only be for the greater good, the argument goes. Similarly, if the use of pesticides and other toxins helps feed the world and keep us in comfort, that must be for the greater good as well.

This deceptively simple and goodhearted notion has a few stingers, though. Most prominent is that utilitarianism (at least as conventionally understood) tolerates inequality in the distribution of the good. The greatest good for the greatest number, sure. But what if the greatest number's good comes at the expense of a smaller number of people? Moreover, there is nothing inherent in utilitarianism that requires the greatest number to be a majority of the population, or even any more than a small minority of it—as long as the good applies to more than it might otherwise. All other things being equal, utilitarianism prefers to have the greatest good spread around. But the greatest good comes first. So if GDP for the country goes up, then utility has gone up, too, even if all the increase went to just a few, or even just to one. Besides, isn't a flourishing economy good for everyone in it? And isn't an abundant food supply good for everyone, too, even if many of us face greater cancer risks from the pesticides we use to get that abundance? After all, as the saying goes, sometimes somebody has to "take one for the team." Perhaps "taking one for the team" could even be extended into murder, if the murder of one enhanced the average happiness of most others.

Another trouble with utilitarianism is what the philosopher Derek Parfit calls the "repugnant conclusion."[67] If we accept that we want to maximize happiness, surely there would be more happiness to add up if there were simply more people around, as long as they were all happy. Parfit points out that, in fact, all these additional people would not have to be very happy at all. Imagine 1 billion people with 10 units of happiness each, adding up to 10 billion units of happiness in total. Now imagine 10 billion and 1 people with 1 unit of happiness each, adding up to 10 billion and 1 units of happiness in total. By this logic, having lots of barely happy people is better than having fewer quite happy people—meaning that we should encourage the human population to continually rise. Few environmentally minded people would welcome such a prospect.

It may seem absurd to interpret the phrase "greatest good for the greatest number" in these ways. Most contemporary utilitarian philosophers, in fact, agree it is absurd to promote great good for a small number, or small happiness for a great number. For example, they argue that simply raising GDP per capita does not necessarily result in the greatest possible increase in total happiness if that rise in GDP per capita is badly distributed, since the happiness increase you get from more money goes down the richer you are. A thousand dollars means a lot more to a poor person than it does to a rich one. Thus, bad distribution of GDP will not constitute the *greatest* good for the greatest number. However, all utilitarians are united by the idea that sometimes the minority must accept an unpleasant sacrifice to benefit the majority.[68]

Moreover, the logic of the primacy of economic growth as the be-all and end-all of the political good takes precisely this absurd form, as it does not take into account distribution. And yet this is a pervasive form of utilitarian thinking in contemporary society and politics, even if most philosophers—including most utilitarian philosophers—reject it. Economic growth is supposed to be a sign of great good for most of us, but is it always? Similarly for technology. New advances are supposed to enhance the lives of most people, but is that always so? This form of utilitarianism—what we might call *economic utilitarianism*—can easily lead to the perverse result of benefiting only a minority while professing to do the opposite. And even if there is benefit for the majority, should we be concerned about the minority who are disadvantaged?

John Rawls and Justice as Fairness

The philosopher John Rawls said that we should be concerned, and for a very selfish reason: We might ourselves wind up among the disadvantaged.[69] In his much discussed 1971 book, *A Theory of Justice*, Rawls asked that we stop for a moment and try to figure out what our principles of justice would be if none of us had any idea of where we will likely wind up in life. Put on an imaginary "veil of ignorance," as Rawls termed it, about your life chances and sit down with everyone else, similarly garbed. From this "original position," asked Rawls, what principles of justice would people come up with?

Rawls's answer is that we would all commit to two basic principles, and I'll quote him verbatim on this:

1. Each person is to have an equal right to the most extensive basic liberty compatible with a similar liberty for others.

2. Social and economic inequalities are to be arranged so that they are both (a) reasonably expected to be to everyone's advantage, and (b) attached to positions and offices open to all.[70]

In other words, Rawls says that if we didn't know where we would personally get to in life, we would want the greatest good for *everyone*, not just for the greatest number. We would also recognize that our own good must be realized within the context of others similarly, and justly, pursuing their own good. Liberty has constraints, most notably the liberty of other people. And we would recognize that we would not want others to seek advantage over us and they would not want us to seek advantage over them.

We would also recognize that justice does not depend upon everyone being exactly equal. There are times when some forms of inequality are advantageous to everyone, if properly handled. Children gain some advantage from the authority their parents—who are more experienced in the dangers of the world—have over them. Students similarly, we must hope, gain some advantage from the greater experience of their teachers in the topic at hand. Citizens gain some advantage through the coordination of social organization afforded by having police officers, mayors of cities, licensed medical experts such as nurses and doctors, and the like, all of whom we grant powers to that we may not have as individuals. But these offices must be advantageous to everyone, Rawls argued, not just to a majority, and they must be open to everyone as well, in part to ensure that they are indeed advantageous to everyone.

Rawls distilled his entire 607-page book into the following sentence:

All social values—liberty and opportunity, income and wealth, and bases of self-respect—are to be distributed equally unless an unequal distribution of any, or all, of these values is to everyone's advantage.[71]

He got it down even more tightly when he offered a definition of the opposite of justice: "Injustice, then, is simply inequalities that are not to the benefit of all."[72]

Rawls called his approach "justice as fairness," and it is a form of *egalitarianism*.[73] Utilitarianism may often feel egalitarian in its avowal of what promotes the greatest good for the greatest number. But utilitarianism can lead to outcomes that are not to everyone's advantage, as the previous section describes, even if the total happiness of the community concerned has gone up. Utilitarianism might well accept a technology that increases the risk of cancer by 1 in 1,000,000 if more happiness were created than lost in the process. But justice as fairness would not accept such a technology. Maximizing happiness isn't its point. Maximizing fairness is.

Speaking of fairness, I should point out that some forms of utilitarianism are more egalitarian than others—which is to say that some are sharply less so.[74] For an example of one that is less egalitarian, take the economic utilitarian notion of *Pareto optimality*, named for Vilfredo Pareto, an Italian economist from a century ago. Pareto counseled that we can allow some greater advantage to someone, or someones, as long as it does not make others worse off. That sounds pleasantly accommodating, but it immediately makes the nonadvantaged relatively worse off, even if not absolutely worse off. After a few relative nonadvantages accumulate, political disempowerment will, too. Thus, in practice, a Pareto optimal form of utilitarianism accepts inequality that does not advantage everyone. And that's not fair, said Rawls.

The publication of Rawls's book touched off a huge discussion in political philosophy that continues today, even after Rawls's death in 2003. Many philosophers had long been troubled by some of the implications of utilitarianism, but hadn't quite put their finger on a more compelling way to think about what is just. Many (but by no means all) now think Rawls's theory provides that. Rawls's theory also presented new questions, however, in part through the issues it did not resolve. Most prominently, readers of Rawls found themselves pondering the pluralism of social life and how to incorporate social difference into political theory. Rawls's theory was a universalistic one. He thought it applied to everyone, everywhere. Many of his readers were not so sure.

Justice and the Problem of Pluralism

One such reader is the economist Amartya Sen, who, while applauding much of what Rawls offered, has tried to build in a more pluralistic understanding of fairness.[75] Yes, says Sen, Rawls recognized that fairness didn't necessarily mean that everyone had to be equal in every regard, as long as that inequality was to everyone's advantage. But Rawls seemed to have in mind mostly unequal positions in social hierarchy that are helpful for social organization. What if not everyone wants the same "social values" that Rawls said should otherwise be distributed equally or does not want them in the same degree and amounts?

For example, should the illiterate be given subscriptions to the *New York Times* and the Book-of-the-Month Club? Not while they are still illiterate, Sen's arguments would suggest. Should Amazonian tribal people be air-dropped suitcases filled with $8,728 for each of them, that being the average gross national income per capita in the world as of 2009? Not if cash isn't something they need or that is relevant to their lives. Should all men—and I mean this quite seriously—be given appointments at gynecologists' offices, or conversely, should women be denied them if men can't have them? Not if men's bodies have little need for this medical specialty and women's bodies do.

If you read Rawls closely, his theory does not necessarily lead to any of these bizarre outcomes. But Sen wants to make the pluralism of people's needs and wants more explicit. So he suggests that we think of people as having *functionings,* beings and doings they have reason to value, in the language of Sen, and *capabilities,* freedoms for attaining these beings and doings. Justice for Sen is maximizing people's capabilities to achieve their functionings. Lack of justice is when we do not do what we can to give people these capabilities. So, too, is poverty. Rather than seeing poverty as a lack of money, as conventional economics does, Sen says poverty is *capability deprivation,* and our capabilities depend on a lot more than money. Ill health would also be a form of capability deprivation, a kind of poverty of the body that prevents one from attaining valued beings and doings.

Sen's work has been widely embraced, most especially by development agencies and scholars. The well-known Human Development Index (HDI) of the United Nations Development Programme (UNDP) is a direct application of Sen's ideas about poverty being more than a matter of money and a direct challenge to the standard GDP per capita approach. GDP per capita—a form of economic utilitarianism, as I discussed earlier—equates money with what we value, gives us no sense of inequality levels within a country, and doesn't consider the variability in what people want. The HDI does not reject the significance of money in an increasingly monetarized world. But the HDI combines GDP per capita with measures of two other widely valued beings and doings: health (using longevity rates) and knowledge access (using literacy and school enrollment rates). It gives us a window (albeit an imperfect one) on the problem of within-country inequality by looking at a couple of measures of how people are doing, aside from their average income. And the HDI gets a degree of pluralism into the mix by suggesting there is more than one route to increasing human well-being, other than money alone. While a country's world rank in GDP per capita is typically similar to its world rank in HDI, often there are wide disparities. For example, the countries of Chad and Kyrgyzstan have almost the same GDP per capita, but Kyrgyzstan has almost double the HDI figure—0.705 versus Chad's 0.368.[76]

The philosopher Ronald Dworkin is another fan of Rawls and of Sen's efforts to better deal with the pluralism of humanity. Dworkin argues that justice needs to be based on the recognition that we are all different and that across that difference we need to extend an equality of concern. He worries that we might choose to pursue our differences in ways that compromise the pursuits of others and thus compromise the first principle that Rawls said those wearing veils of ignorance about their life chances would agree to: that everyone's liberty has to be compatible with everyone else's liberty. He worries that Sen's notion of functionings doesn't take that necessary balancing sufficiently into account. Sen says we should give people the capabilities to achieve the functionings they have reason to value. But what if the functioning one person values conflicts with another's? For example, what if that one person values being a dictator? Dworkin also argues that if people do not apply themselves in the pursuit of what they individually value, equality of concern will be hard to sustain in the long term. (It's hard to maintain concern for someone who simply doesn't try.) Justice means the responsibility of concern for others and what they individually value, but it also means that we have responsibility for ourselves and what we individually value.[77]

Rawls came to have much sympathy for these and other efforts to combine fairness with difference, and much of his later writing concerned this subject.[78] The details of his responses need not concern us here. But one thing that Rawls remained firm about is that unless you have a pretty good idea that you're going to wind up in a privileged position, you wouldn't choose to live in a society organized around utilitarianism—especially economic utilitarianism. Rather, you'd want to live in a society organized around conceptions of justice that enhance everybody's needs and wants.

Power, Environmental Justice, and the Invironment

Every *body*'s needs and wants, that is: Everybody is a body, and that body is in constant dialogue with its environment and with other bodies, each with their own needs and wants. If we want a society organized around concern for every body, a society that strives for environmental justice for everyone's invironments, then we would do well to consider why we so manifestly do not live in such a society today.

To answer this question, we need to begin with the following fundamental sociological observation: We do not wear veils of ignorance about our life chances. No one knows for sure where she or he will wind up. You could get hit by a car tomorrow and be in a wheelchair for the rest of your life. You could indeed get cancer or Parkinson's or any of a host of potential maladies and calamities. But based on the reasonable presumption that where I'm headed today is where I'll be headed tomorrow, we all have some confidence about our likely future lot in life, whether good or bad.

This knowledge makes utilitarianism an attractive philosophy for those confident that they will indeed be on the upside tomorrow, and for tomorrows to come. Moreover, those who are likely to end up on the upside of, say, the pollution that might accompany some broadly beneficial technological process with uneven downstream consequences are also typically the same people who are advantaged in society's decision making. Were it otherwise, those advantaged in decision making would be unlikely to promote that technological process. This is one of the main reasons that we live in a society organized around utilitarianism far more than around justice as fairness, and the related notions of capabilities to function and equality of concern.

My point is not that those who are socially advantaged in decision making would reject justice as fairness out of hand. Rather, my point is that utilitarianism does not seem as personally threatening to the advantaged as to the disadvantaged. So the advantaged are less likely to easily envision its problems, and thus less likely to throw their social weight against its problems coming to pass. The origin of environmental injustices, then, may not lie in some widespread social rejection of justice as fairness as much as in the lack of rejection of utilitarianism by those in advantaged positions.

What I am describing are the invironmental consequences of inequalities in *social power*—people's scope for action in the social world.[79] Social power is not necessarily a bad thing, as

- utilitarian = unequal world; greatest good for greatest amount of people.

Chapter 2 discusses. It would be neither an enjoyable nor a just world if people were not able to exercise some control over their life situations. This is precisely why Sen emphasizes the importance of people's capabilities to achieve the functionings they value. Given that this control has to be realized within the context of others similarly seeking to achieve the functionings they value for themselves, justice requires some means of balancing these capabilities in ways that do not disadvantage anyone. However, that balance does not now exist in many places, if any.

Which is not to say that those who exercise control over the social world in ways that do not advantage everyone are necessarily evil and cruel. The image of the exercise of social power as some kind of mad organist playing at will on the keyboard of social life is rarely apt. Utilitarianism retains much of its influence because it usually feels democratic, not demonic.

For example, an action that appears to benefit the community at large surely feels more democratic than one that doesn't. And if that action has the perverse consequence of benefiting only a minority, the majority can exercise its democratic control and prevent it from happening. We're not stuck with it, necessarily. We can vote it out.

But from an equal-concern perspective, there are still some problems here with utilitarianism. First, a democratic response like taking a vote on something may then disadvantage the minority who voted the other way. This is the problem Alexis de Tocqueville long ago pointed out with America's democratic experiment: the potential in a democracy for a "tyranny of the majority."[80]

Second, there's something in the very way we understand democracy that can also make it hard to prevent the perverse outcomes of utilitarianism. We generally understand democracy as a matter of ensuring that everyone has equal political standing—that everyone can vote, can organize meetings, can express their views to elected officials, and can run for office themselves. These are all important ideals. However, we are unlikely to attain these ideals without confronting an important material reality: Those with material advantages are likely to be a lot more advantaged by these ideals. What I mean is that, while everyone can potentially vote in a democracy, there are great inequalities in people's abilities to influence what is voted on, and what those who are elected do with their offices. People with economic and other social advantages are far more likely to be able to organize effective social groups, to gain the ear of elected officials (as well as unelected officials), and to run successfully for office. The ideal of equality of political standing thus depends in part upon equality of material standing—and vice versa.

Take what happened in Bhopal, Toulouse, and Grassy Narrows. In each case, a large corporation was able to use its influence to establish industrial plants that horribly disadvantaged a minority, while contributing to GNP and to what could be argued were environmental advantages for the majority—pesticides and fertilizer for food production and, in Grassy Narrows, wood pulp for paper, one of the great conveniences of this bodily life, even in a digital age, one could argue. At least in Bhopal and Toulouse, lots of people were well aware of the potential dangers to local people. In Grassy Narrows, too, there was reason to be worried. The 1950s discovery in Minimata, Japan, of methyl mercury poisoning from a plastics factory should have been a warning. Many say the warning signs are already loud and clear about toxic wastes, air pollution, and pesticides. Using economic utilitarian arguments both to themselves and to others, these consequences were and are ignored by corporate executives and by government officials. What's good for General Motors is good for the nation, it is sometimes said, and by extension, what is good for any corporation is good for any nation. But from a justice-as-fairness point of view, the matter is far, far more complex.

Again, I want to emphasize that corporate executives and government officials are not necessarily being demonic when they violate fairness and equality of concern in favor of economic utilitarianism through the exercise of their greater political power using the normal and legal channels. Under our current dominantly utilitarian understandings of justice and political process, there is no reason why they should do otherwise. These are their rights, in the current way we have organized our democracies.

What we have in the world today are democracies of inequalities in which we grant, so we think, equality of political standing without addressing material inequalities. Without the latter we cannot attain the former, however. What Rawls, Sen, Dworkin, and the millions who call for environmental justice are asking for are not just democracies of inequalities, but rather democracies of equalities. They are asking for not just democracy, but for what we might term *isodemocracy*, democracy founded on equalities in both political and material standing—democracy in which the concerns of everybody and every body are the concerns of everybody and every body.[81]

Making Connections

Thus, we continue to grapple with the central issue of the body in environmental matters: making connections. We moderns tend to regard our bodies, and our selves, atomistically. When considering matters of justice and the environment, however, we quickly encounter the connectedness of our bodies with the world. Our bodies live in context. The interactiveness of ecological dialogue impels us to consider our connectedness with care. Our health, among other things, depends upon that careful consideration.

Such consideration depends in part upon the conceptions that we bring to bear. Although this chapter has concerned some basic material fundamentals—body and health—it has equally concerned some basic fundamentals in the realm of our ideas. "For meaning," David Abram has written, "remains rooted in the sensory life of the body—it cannot be completely cut off from the soil of direct, perceptual experience without withering and dying."[82] I did not highlight the importance of the ideal side of ecological dialogue as the chapter went along. But a moment's reflection should show it.

To begin with, the notion of an "invironment," that inner zone of the body's dialogue with the environment, represents the new conceptual place the body is gaining in environmental discussions. This is a recent change, as I noted. Earlier, issues of the body were often relegated to the sidelines of environmentalism, and environmental sociology as well. The body was always "there," of course, just as gravity was always there before the mythical apple conked Newton on the head. But the consequences of that *thereness* depend in part on how we understand it. The new connected understanding of the body's place in the environment is already leading to new environmental initiatives, such the rise of the "slow food" movement and its promotion of "good, clean, and fair food for all," as the group Slow Food International puts it in its mission statement.[83] By good food, slow food advocates mean food that is tasty and part of local culture; by clean food, they mean food that is good for the environment, animal welfare, and our health; and by fair food, they mean good and clean food that is accessible to everyone. Health is now an environmental issue. Food is now an environmental issue. The body is now an environmental issue.

As well, our ideas of justice are changing. The efforts of Rawls, Sen, Dworkin, and others to shift us away from utilitarian thinking are already having some material effects. The U.S. government's Food Quality Protection Act of 1996, with the special attention it gives to the developmental differences of children's bodies, is an example. Rather than using the utilitarian logic of greatest good for the greatest number, this portion of the act finds it fair to pay attention to the pluralism of our needs and to give equal concern to those different needs, even for the minority of the U.S. population who are children. Here again, our ideas are having material consequences.

In this light, it is worth inspecting the debates over what are sometimes called "NIMBY" or "Not-In-My-Back-Yard" responses to environmental issues. Often the term NIMBY is used as a utilitarian weapon. Why should a small group of people concerned about the local environmental consequences of some development proposal stand in the way of the greater good? They are just "nimbies," say utilitarian critics, individualistically out for themselves with a knee-jerk reaction against change. Similarly, with the new presence of the body in environmental issues, we are seeing the rise of what the environmental sociologist Melanie Dupuis has called "NIMB"—individualistic "Not-In-My-Body" responses that ignore questions of the greater good. Got a problem with eating food laced with pesticide residues? A NIMB response finds the solution in the organic produce section of the supermarket, not in the voting booth.[84] These conflicts make NIMBY and NIMB politics highly contentious. Justice as fairness gets us beyond this utilitarian turmoil by saying not-in-anybody's-back-yard and not-in-anybody's-body unless it is to everybody's and every body's advantage. Justice as fairness says that justice is not a zero-sum game. There really are ways to arrange our lives to the advantage of everyone, if we appropriately apply our minds to the task.

Speaking of fairness again, though, we must recognize that it is not easy to arrange our lives for everyone's advantage. I have been quite critical of utilitarianism in this chapter, and especially economic utilitarianism. But utilitarianism, for all its faults and uncomfortable moral implications, has one very great advantage: It is simply easier to work for the greater good than for everyone's good. That doesn't make utilitarianism right. But it does mean that utilitarianism may often be justifiable in the short term, while we're working on coming up with something that advantages everyone—only as long as we really do keep working on that fairer justice.

In other words, environmental justice demands our best ideas. Ideas may not be matter, but they do matter. And matter matters for our ideas. It's a dialogue, the most basic one of environmental sociology. It's a dialogue of connection and interconnection, of the unity of difference. This dialogue is no more evident than in that home, that abode, that we never leave as long as we live: our own bodies.

The Ideal

CHAPTER 6

The Ideology of Environmental Domination

No one seems to know how useful it is to be useless.

—Chuang Tzu, third century BCE

The view from Glacier Point in Yosemite National Park is one of the world's most famous. From this overlook, you can see a sweeping panorama of Yosemite, which many have called the most beautiful valley in America (see Figure 6.1). A number of years ago, my brother and my sister-in-law, Jon and Steph, were visiting her relatives in California, and they decided to take Steph's grandmother to see Yosemite, where she had never been. An elderly woman, she did not walk well, so they took her only to sites you can get to by car. You can drive right up to Glacier Point, and they did. As Jon later recounted the story to me, they helped Steph's grandmother up to the edge and stood there for a few minutes taking it all in. Then Jon turned and asked her, somewhat hopefully, "Well, what do you think?"

Figure 6.1 The view of Half-Dome and the Yosemite Valley from Glacier Point.

Source: © Michele Falzone/JAI/Corbis.

She considered the question carefully, and replied, "All that forest. What a waste. There should be people and houses down there."

When two people look out on a scene, a scene of any kind, they are unlikely to appreciate it in just the same way. Faced with the same material circumstances, we each see something different. Where my brother Jon saw the beauty of wild nature in that view from Glacier Point, Steph's grandmother saw wasted resources. Such differences are a part of our individuality. They also reflect social differences in the apparatus of understanding that we use to organize our experience. There are larger social and historical patterns in the distinctive mental apparatuses we each bring to bear on the world around us. In a word, there is *ideology* at work.

In this second part of the book, we take ideal factors as the point of entry into the ecological dialogue. As we saw in Part I, the other side of the dialogue is always close at hand, and we will find that here, too. Investigation of ideal factors inevitably leads back to material questions. But the emphasis in Chapters 6 through 9 will be on the form the environment takes in our minds.

The independent power of ideas in our lives is well illustrated by the history of environmental ideas. The material conditions we now regard as environmental problems have long historical precedents, yet few people in the first half of the twentieth century questioned the increasing per capita appetite for resources, the spread of the automobile and its sprawling land use, or the invention of yet another chemical or mechanical weapon for every instance of the environment's resistance to our desires. Early articles in *National Geographic,* for example, extolled the industrial might that spawned marvel after marvel, as their titles implied: "Synthetic Products: Chemists Make a New World," "Coal: Prodigious Worker for Man," "The Fire of Heaven: Electricity Revolutionizes the Modern World," "The Automobile Industry: An American Art That Has Revolutionized Methods in Manufacturing and Transformed Transportation."

In the decades from 1960 on, though, the ideological situation changed dramatically in country after country, as Chapter 7 discusses.[1] *National Geographic,* to continue with that barometer of Western cultural values, began running articles with titles like these: "Our Ecological Crisis," "African Wildlife: Man's Threatened Legacy," "Nature's Dwindling Treasures," "Pollution: Threat to Man's Only Home," "The Tallgrass Prairie: Can It Be Saved?" A different ideology had taken more general hold, at least among the writers and editors (and, we can presume, many of the readers) of this perennially popular magazine.

Scholars have studied the role of ideology in the dialogue of ecology in two broad ways, largely drawing on historical evidence. First, they have considered the ideological circumstances that make domination of the environment thinkable and tolerable, focusing on understanding Western cultural attitudes that support such a relationship to the environment. Second, scholars have considered the ideological circumstances that make such conditions and such domination increasingly unthinkable and intolerable, focusing on the social origins of the environmental movement.

This chapter considers that first role of ideology; Chapter 7 considers the second. In this chapter, then, we examine the ideological origins of the view that human beings can and should transform the environment for their own purposes. Scholars argue that three Western intellectual traditions—Christianity, individualism, and patriarchy—have in large part provided the ideological rationale for environmental domination. These ideologies of environmental domination are by no means exclusively Western, but they are certainly heavily present in the West, which may help account for the central role of Western institutions in the industrial transformation of the Earth. As well, all three of these ideologies of environmental domination have close links with ideas about hierarchy and inequality, suggesting an ideological connection between environmental domination and social domination, as we shall see.

Christianity and Environmental Domination

A common explanation for the modern urge to transform the Earth is the rise of the industrial economy. But the next question to ask is, *Where did the industrial economy come from?* As I suggested at various points in Part I of this book, the development of economics should not be seen in purely materialist terms. Ideas of work, leisure, social status, and community infuse the economy as much as the economy infuses those ideas.

A major source of those ideas in the West is Christianity. As Max Weber, one of the founding figures of sociology, argued in a famous 1905 book, *The Protestant Ethic and the Spirit of Capitalism,* Christian ideas—and, more specifically, Protestant ideas—form one of the great wellsprings of capitalist thought. It is more than accidental, said Weber, that the Protestant Reformation of the late sixteenth century immediately preceded the development of modern capitalism and the expansion of European economies all over the globe in the seventeenth, eighteenth, and nineteenth centuries. Capitalism is, in a way, a secular version of Protestantism.

The Moral Parallels of Protestantism and Capitalism

"A man does not 'by nature' wish to earn more and more money," Weber wrote, in the gendered phrasing of an earlier time, "but simply to live as he is accustomed to live and to earn as much as is necessary for that purpose."[2] So why do we work so hard to make more money than we need? A desire to maintain a place on the treadmills of consumption and production is part of it. But to leave the matter there does not answer the question of why we are on these treadmills to begin with.

The answer, suggested Weber, lies in the moral anxiety that early Protestantism inculcated in its followers. Medieval Catholicism was more forgiving, encouraging repentance and allowing last-minute, deathbed declarations of faith. If you were rich enough, you could literally buy your way into heaven by funding priests to say prayers for you and by purchasing "indulgences" from the church. But in 1517, a young professor of theology at the University of Wittenberg in Germany nailed a copy of a recent work of his to the door of a local church—a work he called *Disputation of Martin Luther on the Power and Efficacy of Indulgences.* Martin Luther's *Ninety-Five Theses,* as the work came to be known, touched off a storm of protest and a new religious tradition named after that storm: Protestantism. One of the early ideas of Protestantism was to replace indulgences with a kind of final weighing up of all the good and bad that a person had done in life. This final judgment view made it harder to overcome one's misdeeds and made entrance into heaven less ideologically certain.

A lot of the anxiety also stemmed from the doctrine of predestination—the idea that one is preordained either to go to hell or to be one of the "elect" who goes on to heaven. Predestination was a common belief of early Protestants, particularly early Calvinists, and it ratcheted up moral anxiety by several notches. On the face of it, predestination seems a lousy way to motivate people, for it suggests that how you act in life doesn't matter. You are still going to go where it has been preordained that you will go. So why not lead a carefree life of sin, laziness, and gluttony? But the trick about predestination was that no one knew for sure who had grace—who was one of the elect and who was not—except through a person's worldly deeds. Those who were good, moral, upright, and successful in this life must be the elect of the next life, early Protestant creeds such as Calvinism taught.

Thus, in order to convince themselves and the community that they were among the elect, early Calvinists became ascetics, denying themselves bodily pleasures like laziness and working incredibly hard to achieve the signs of success in this life. And they began to rationalize the work process, making work more orderly and efficient, in order to maximize their worldly signs of moral worth. Basically, said Weber, early Calvinism was a competitive cult of work, denial, and rationalization.

These same ideas still infuse capitalist economic life today, albeit without the religious framework (at least not explicitly). What has happened, Weber argued, is that we have secularized the idea that hard work and denial, rationally applied, are outward signs of how good and deserving one is. It remains one of the most basic assumptions of modern life that those who work hard are the most deserving, the most morally worthy of our admiration and of high salaries. Hard workers are the elect of the heaven of social esteem. They are the ones who have grace.

And now we have little choice but to be hardworking rational ascetics ourselves, even if (as is likely the case) we do not follow the religious tenets of early Calvinism. The anxiety of early Protestants produced huge accumulations of wealth. (If you work really hard and deny yourself, you are indeed more likely to be able to fill your wallet fuller. More likely: There is no firm correlation between hard work and wealth, as any coal miner or factory worker knows.) They reinvested this wealth, which led to even more wealth. And as each dedicated Protestant sought to increase his or her comparative success, the trend toward work, rationalization, and production accelerated. The treadmills of capitalism began turning ever faster. Soon one had to work hard, deny oneself, and rationalize one's life in order to attain any kind of economic foothold, for that was what everyone else was doing. Increasingly, people came to accept the idea that those who worked hard deserved to get more and to gain everyone's respect. Likewise, they came to accept a corollary: that those who had less must not have worked so hard and therefore deserved their fate. The Protestant ethic had become the spirit of capitalism.

The history of capitalist development provides some support for Weber's thesis. Modern capitalism arose first in the countries with a sizable population of Protestants: England, Scotland, the United States, and Germany. Within Europe even today, as of 2008, among the 10 wealthiest countries, 8 have a significant presence—at least 20 percent—of people who are at least nominally Protestants (as opposed to regular churchgoers, who are few and far between in most of Europe). The poorest European countries have comparatively few Protestants (see Figure 6.2). Outside of Europe, we can point to the continued high per capita incomes of Australia, Canada, New Zealand, and the United States. We can also speculate about South Korea, which has joined the ranks of the richest countries, and the 18 percent of its people who identify as Protestant. To be sure, there are some very wealthy Catholic countries, most notably Luxembourg and Austria. And one of the world's richest countries, Japan, is 95 percent Buddhist or Shintoist. But the association between Protestantism and income remains striking a century after Weber wrote.

Now modern capitalism has spread well beyond the confines of countries with many Protestants, and even beyond the dominantly Christian countries. Religion is no longer the driving force. The capitalist spirit steadily enfolds country after country into its secularized ethic of ascetic rationalism. Economic structures have taken over from Martin Luther and John Calvin in spreading this spirit, even as this spirit dialogically propels the structures, as in the way hard work speeds the treadmill faster and faster. Ascetic rationalism has become what Weber termed "an iron cage."[3] As Weber put it,

> This order is now bound to the technical and economic conditions of machine production which today determine the lives of all the individuals who are born into this mechanism, not only those directly concerned with economic acquisition, with irresistible force. Perhaps it will so determine them until the last ton of fossilized coal is burnt.[4]

In a way, we're all Calvinists now, burning up our lives and our planet as we race along the treadmills of capitalism.

The Moral Parallels of Christianity, Science, and Technology

Weber is not the only scholar who has traced a connection between Western religion and social developments that greatly impact the environment. In 1967, the historian Lynn White published a short essay that remains one of the most influential and widely read analyses of the environmental predicament: "The Historical Roots of Our Ecologic Crisis." White's basic argument was that environmental problems cannot be understood apart from the Western origins of modern

Figure 6.2 The relationship between Protestantism and per capita gross national income (GNI) in the European Union, plus Norway and Switzerland. Green highlight indicates countries with at least 20 percent nominal Protestants. Income data corrected for local purchasing power, or "purchasing power parity" (PPP).

Country	GNI/capita (PPP) (2008)	Dominant Religion
Norway	$59,250	Protestant
Luxembourg	$51,109	Catholic
Netherlands	$40,620	Cath/Prot
Switzerland	$39,210	Cath/Prot
Sweden	$37,780	Protestant
Denmark	$37,530	Protestant
Austria	$37,360	Catholic
United Kingdom	$36,240	Protestant
Germany	$35,950	Cath/Prot
Finland	$35,940	Protestant
Ireland	$35,710	Catholic
Belgium	$35,380	Catholic
France	$33,280	Catholic
Spain	$30,830	Catholic
Italy	$30,800	Catholic
Greece	$28,300	Orthodox
Slovenia	$25,857	Catholic
Czech Republic	$22,890	Catholic
Portugal	$22,330	Catholic
Cyprus	$21,962	Orthodox
Slovakia	$21,658	Catholic
Malta	$21,004	Catholic
Estonia	$19,320	Prot/Ortho
Hungary	$18,210	Catholic
Lithuania	$17,170	Catholic
Poland	$16,710	Catholic
Latvia	$16,010	Cath/Prot/Ortho
Romania	$12,844	Orthodox
Bulgaria	$11,370	Orthodox

Source: The author. Income data from United Nations Development Programme. (2010). *Human Development Report–2010.*

science and technology, which, in turn, derive from "distinctive attitudes toward nature that are deeply grounded in Christian dogma."[5] Not only does the economy of the West have religious origins, then, but Western science and technology do as well.

Many ancient cultures participated in laying the foundation stones of science—notably China and the Islamic world. Yet, White argued, "By the late thirteenth century Europe had seized global scientific leadership."[6] The achievements of Newton, Galileo, Copernicus, and other early scientists were accompanied by rapid advances in Western technology. White placed particular emphasis on the development of powered machines: the weight-driven clock, windmills, water-powered sawmills, and blast furnaces.

Even more significant, though, was the development of the moldboard plow in northern Europe during the latter part of the seventh century. The moldboard plow dramatically changed human attitudes toward the environment, said White. Previous plows had allowed farmers only to scratch at the ground. These shallow plows were adequate for the light soils of the Near East and the Mediterranean, although they restricted agriculture to being pretty much a subsistence affair, with little surplus for trade. The generally heavy soils of the North, on the other hand, required a stronger plow. The moldboard was invented to cut more deeply into the ground, loosening up the heavy northern soils. The difficult work of the moldboard plow normally took the pull of eight oxen, as opposed to the one or two used by earlier plows.

Thus, the moldboard plow was essentially a powered machine. In White's words, "Man's relation to the soil was profoundly changed. Formerly man had been part of nature; now he was the exploiter of nature."[7] Formerly, we had seen ourselves on a par with the natural world. Now we saw ourselves as standing above it, at least potentially.

Why this change? This exploitative and domineering attitude toward the environment, encompassing both unlettered farmers and scientific intellectuals, was so specific to one region that its origins must lie in a broad intellectual trend, White argued. The likely trend was one of the great intellectual revolutions of the Western tradition: the Christian ethic. For at roughly the same time that northern farmers were developing the moldboard plow to handle their heavy soils, White noted, they were also giving up paganism for Christianity.

For the pagan, the world is full of spirits. Every rock and tree is potentially animated by something. Nature is alive, organic, and magical. It is cyclical, and we are part of it. Early Christianity, on the other hand, building on Judaic philosophy, saw time as linear and nonrepeating, and it saw the environment as dead and inanimate, as separate from people. For early Christianity, the spirit world of God and the saints was not *immanent* in nature—that is, suffused throughout nature, making nature a direct embodiment of spirits—but rather *transcendent* above nature. God was up high in heaven, not an animating presence down here, on the Earth.

Moreover, early Christian doctrine taught that God gave the world to human beings to exploit, to change and re-create, much as God himself could do (which is why only human beings are made in God's image, some Christians believe). Changing nature was no longer a sacrilege. Indeed, all the Mosaic religions—Judaism, Islam, and Christianity—have commonly interpreted that it was God's will that we do so. In the words of Genesis 1:26,

And God said: Let us make man in our image, after our likeness; and let them have dominion over the fish of the sea, and over the fowl of the air, and over the cattle, and over all the earth, over every creeping thing that creepeth upon the earth.

By granting humans "dominion," Mosaic teachings thus gave us moral license to change the world as we see fit, White argued—a license gladly accepted and spread far and wide in Europe by Christianity. As White put it, "Christianity is the most anthropocentric religion the world has ever seen."[8]

The Greener Side of Christianity

The coincidence of the development of medieval technology and science alongside the spread of Christianity is intriguing and suggestive. The biblical license of dominion likely at least facilitated the development of technology and science. The association of the Protestant Reformation with the subsequent rise of modern capitalism and the striking parallels between contemporary secular morals and the ascetic rationalism of early Protestantism also suggest an important influence of religious ideas on our material conditions.

But we cannot conclude that Christianity unambiguously promotes science, technological progress, and capitalism at the expense of the environment. For one thing, Christians are not the only readers of the Bible, nor the first. The connection that White saw between Christianity and technology is based on the Old Testament, a work that is revered by Jews and Muslims, too. Thus, White should have been able to find a similar connection between technological advance and the spread of the Old Testament among the peoples of those faiths. Yet he made no such argument, and it is not immediately apparent that he could have. Moreover, Christianity is itself a geographically and ideologically diverse tradition. The Eastern Christianity of Constantinople, for example, was not linked to the development of science and technology to the degree that the Latin Christianity of Western Europe was. Why not? Surely Eastern Christians had environmental constraints of their own to contend with and therefore had equal incentive to develop science, technology, and a domineering attitude toward the environment.

A further problem with White's explanation is that Christianity has often been at odds with science. Consider the conflict between medieval scientists and the established church. The inquisition of Galileo for heresy is only the most well-known example. Far from welcoming science as a way of proving that, yes, God is indeed transcendent and that nature is an inanimate machine driven forward through linear time, the church found its authority threatened by the development of scientific thought. Even though almost all early scientists, including Galileo, presented their work as theological efforts to understand the true meaning of God, church authorities only grudgingly accepted the argument that science was about faith. And today, many Christian religious leaders object to a range of scientific techniques, such as genetic engineering. When Scottish scientists announced in 1997 that they had successfully cloned the sheep "Dolly," the news was greeted by many Christians as a blasphemy. The controversy over stem cell research is another example; this controversy has been almost exclusively confined to a mainly Christian and unusually religious country, the United States.

Another sign of Christianity's ambivalent views about environmental transformation is certain biblical passages. For example, right before the famous line in the Bible in which God tells Noah and his family to leave the ark and says, "be fruitful and multiply," which sounds rather domineering, there is a more ecological passage:

> And God spoke unto Noah, saying, Go forth from the ark, thou, and thy wife, and thy sons, and thy sons' wives with thee. Bring forth with thee every living thing that is with thee of all flesh, both fowl, and cattle, and every creeping thing that creepeth upon the earth; that they may swarm in the earth, and be fruitful and multiply upon the earth. (Genesis 8:15–17)

Note that in this passage, the animals, too, are given the right to "be fruitful and multiply"—in fact, even before people are given that right—and Noah is ordered to help make it happen. There is an even more ecological passage later on when God promises to establish a covenant both with Noah and with "every living creature," promising not to bring on another flood:

> And God said: This is the token of the covenant which I make between Me and you and every living creature that is with you, for perpetual generations: I have set my bow in the cloud, and it shall be a token of a covenant between Me and the earth. (Genesis 9:12–13)

This passage could be read as suggesting that humans are not the only beneficiaries in the rainbow covenant. The covenant includes "every living creature that is with you." And when the covenant is restated half a sentence later, human beings are not even specifically mentioned. The covenant is "between Me and the earth."

And indeed, many contemporary readers of the Bible take these lines in this more ecologically inclusive way.[9] From the recent ecological teachings of the Vatican to the "What would Jesus drive?" campaign of evangelical Protestants, environmental protection is now a religious calling for many Christians. Indeed, contend many, the interpretation that the Bible sanctions environmental domination is based in part on a long-held misreading of the word *dominion* in Genesis. To have dominion is to have responsibility, not a license for tyranny, just as a king or queen is expected to rule with the interests of the whole realm in mind. This implies a hierarchy, true enough. Still, it's a matter of good housekeeping, if one goes back to the Latin root *domus*, meaning "home," before *domus* developed a branch of meanings stemming from *dominus*, meaning "property," leading to the words *domination*, *domineering*, and *domain*. We should really understand dominion as meaning responsibility for one's home—not domination of it. Such an understanding is deeply ecological, argue green Christians, if we recall that the word *ecology*'s own root meaning is the "study of the home."[10]

Whether or not one finds theological instruction and comfort in biblical interpretations like these, green Christianity is clearly possible. Moreover, after something of a late start in comparison with other social institutions, green Christianity appears to be growing rapidly. New groups like the Green Christian Network and the Evangelical Climate Initiative have formed in the last few years. Thus, White's focus on Christianity as the origin of our environmental troubles may have been somewhat misplaced. The environmental ideas he discusses—linear time, an inanimate world, the dichotomy between people and nature, anthropocentrism—are certainly not explicit aspects of the Bible. They do not appear in the Ten Commandments nor the Sermon on the Mount, for example. And the Mosaic faiths, as we have seen, are neither exclusively Western nor unified in their teachings.

We might more accurately describe these ideas that support the domination and transformation of the environment as an underlying philosophy of the West, rather than of Christianity alone. This does not mean that religion has no role here, though. As the principal religious tradition of the West, Christianity must have been amenable to such ideas for them to become widespread. Indeed, any religious tradition capable of gathering such a wide range of cultures under its tent must be amenable to a similarly wide range of interpretations. The origin of modern ideas about the relationship between humans and the environment is therefore likely more than merely religious. There is no tradition without interpretation.

Non-Western Philosophies and the Environment

Although we cannot simply blame Christianity for ecological decline, non-Western philosophic and religious traditions do generally give recommendations for how humans ought to act toward the environment that are strikingly different from much Western thought. These traditions often promote a more egalitarian relationship with the Earth as well as an acceptance of the environment as it is.

Taoism, for example, advises *wu wei*, or "non-action," as the route to contentment. Non-action does not mean "non-doing." It means working with nature, instead of against it, by attempting to act without deliberate effort. (Translating Taoist ideas into Western terms is difficult, but "nature" is certainly close to what is meant here.)[11] Here is an explanation of wu wei from one of the great Taoist classics, *The Way of Chuang Tzu*, which dates from the third century BCE:

> *Fishes are born in water*
>
> *Man is born in Tao.*
>
> *If fishes, born in water,*

Seek the deep shadow

Of pond and pool,

All their needs

Are satisfied.

If man, born in Tao,

Sinks into the deep shadow

Of non-action

To forget aggression and concern,

He lacks nothing

His life is secure.

Moral: "All the fish needs

Is to get lost in water.

All man needs is to get lost

In Tao."[12]

Such a moral certainly does not appear to provide much license for transforming the Earth to suit human concerns. Rather, Taoism counsels us to forget human concerns so as to avoid the inevitable sorrow of materialism. When one "tries to extend his power over objects, those objects gain control of him," observes the *Chuang Tzu*.[13]

Yet as the geographer Yi-Fu Tuan has observed, China has long been one of the regions of the world most transformed by human action, despite the influence of Taoism and Buddhism.[14] The ancient Chinese canal system, the extensive clearing of the land for cultivation, the formal gardening style of Chinese parkland—all these represent considerable alteration of the environment. Such transformations continue today in huge projects such as the Three Gorges Dam, accelerating urbanization and industrialization, the mechanization of Chinese agriculture, the ready adoption of a consumer lifestyle by many of China's 1.3 billion inhabitants, and the increasingly perilous and deplorable condition of China's ecology.

Nor are asceticism and rationalism new to non-Western cultures. Rationalism built ancient China's canals, agricultural system, formal gardens, cities, centralized government, and complex philosophical systems. Ascetic denial was a part of the training of Japanese samurai warriors and has long been an important moral ideal in Japanese life.[15] The asceticism and rationalism of early Protestantism were not unique to the West.

None of this proves Weber and White fundamentally wrong. It just reins them in a bit. Medieval Christianity likely did play an important role in promoting acceptance of environmental transformation and exploitation, at least at the time. Early Protestantism similarly helped promote the train of reasoning that led to the rise of modern capitalism and the secular ideals of hard work and rationality now common throughout the West. But religion was not the only path that led to these increasingly global sensibilities.

Individualism and Environmental Domination

Another path that has led to environmental transformation is *individualism*, the emphasis on the self over the wider community that has long been a central dimension of the Western tradition. Individualism does not mentally prepare us to recognize how interconnected we all are with our wider surroundings, both social and environmental. With an individualistic frame of mind, we tend to ignore the consequences of our actions for those wider surroundings and therefore, because of our interconnections, sometimes for ourselves as well. Moreover, we in the West have understood our emphasis on the self in competitive and hierarchical ways. Thus, we sometimes pursue our individualistic ambitions not just with "invisible elbows" that jostle others accidentally but with elbows deliberately braced for bumping and shoving aside whoever, and whatever, stands in our way.

Individualism, the Body, and Ecology

One of the many scholars who has connected our Western sense of hierarchical individualism with environmental domination is Mikhail Bakhtin, a Russian social theorist. Bakhtin pointed out that individualism deeply influences the way we regard the main medium by which we are connected to the environment: our bodies. Individualism encourages us to see our bodies as sealed off from others and from the natural world, with a host of consequences for what we regard as dirty, as repulsive, as polite, as scary, and even what we regard as humorous. All of these cultural responses to how our bodies interact with the world have important environmental implications, as we shall see.[16]

Bakhtin based his argument on an unusual source: the quality of humor in the writings of the early French Renaissance writer, François Rabelais.[17] The novels of Rabelais are infamous for their scatological satire of French politics of the sixteenth century. They recount, in graphic detail, the outlandish and vulgar careers of Gargantua and his son Pantagruel, both fabulously obese giants. (The English word *gargantuan* derives from Rabelais's novels.) The two giants lead an outrageous life centered on feasting, drinking, excreting, copulating, and other earthy acts. Woven through the stories are references to the political figures of the day, who usually appear in unseemly and ridiculous situations.

Rabelais's novels, published together nowadays under the title *Gargantua and Pantagruel,* caused quite a stir when they first appeared. Rabelais was often in political trouble because of them. But he also found widespread favor, even among many of the political figures he lampooned, because even the king and his courtiers found the novels downright funny. Still, it was controversial stuff.

The political references in Rabelais's novels no longer mean much to readers. His writings remain controversial, though—but for a different reason than what caused Rabelais so much personal trouble: the style of the books, a style that many modern readers find distasteful and obscene.[18] Bakhtin sought to understand why it is the *style* of Rabelais's humor, rather than the *subject* of his humor, that is now so offensive.

Like Rabelais's novels, Bakhtin's answer caused quite a stir. His book on the subject, *Rabelais and His World,* could not be published until 1965, a full 25 years after it was written.[19] Writing during the height of Stalinist repression, Bakhtin, too, was often in trouble with the authorities. He was denied employment and eventually forced into exile in Kazakhstan during the 1930s. After World War II, he was able to regain the teaching job he had briefly held earlier at an obscure Russian university. Most other scholars thought him dead, though. Then, in the 1960s, some graduate students at Moscow's Gorky Institute rediscovered him. Now that Stalin was gone, *Rabelais and His World* was finally published, and Bakhtin's earlier works were reread and brought back into print. By the time Bakhtin died in 1975, his works were being read all over the world.

I tell the story of Bakhtin's career because it highlights the strong reactions that people often have to reminders that our own bodies perform the same basic functions as any other animal's body. Why should it be that references to the body and all its everyday—and biologically essential—activities should be considered dirty and indecent? What could be more commonplace than the body and its needs? So why is it usually considered a rude topic?

Bakhtin argued that people did not always react in this way. We moderns are offended because of a historical shift in our conceptions of the body, from what Bakhtin termed the "carnivalesque body" to the "classical body."

The *carnivalesque body* is a body of interconnections and exchanges with the social and natural environment. It is a body of openings and protrusions that connect us with other bodies and with the world around us: the mouth, the nose, the anus, the genitals, the stomach. Through these organs of connection, we exchange substances, some made by the body and some brought into the body from other bodies and from the surrounding world: air, smells, sounds, food, saliva, nasal mucus, urine, excrement, the various genital fluids, sweat, tears, mother's milk. The carnivalesque body is also one that relishes bodily acts and desires: eating, drinking, laziness, sleeping, snoring, sneezing, excreting, copulating, giving birth, nursing, kissing, hugging. The emphasis of the carnivalesque body is on what Bakhtin described as the body's "lower stratum." The carnivalesque body is also an ecological body, a body that is forever interacting and exchanging with the environment, bringing in solids and fluids at the top and sending them back out again at the bottom.

The *classical body,* on the other hand, is a body of separation from society and nature. Most of its orifices are hidden from view. Those that are not hidden are carefully controlled through rituals that de-emphasize their openness. Food is carefully introduced into the mouth with a fork, and the mouth is quickly closed again. The nose is blown into a Kleenex or handkerchief, and the mucus is carefully kept out of sight. The classical body does not belch, pass wind, cough or sneeze on others, eat with an open mouth, sweat, cry, or experience sexual desire. Excretory acts are kept strictly private. Openly discussing any of these activities is considered rude and immature, unless carried out under the strict linguistic supervision of "polite" language, such as I am using here. Emphasis is on the body's upper stratum. And the body's means of ecological connection become shameful.

The Carnivalesque Body

Bakhtin drew the term carnivalesque from the annual pre-Lenten festival of carnival, once one of the most important dates on the medieval calendar but which survives today in only a few places. Carnival traditionally was the people's holiday, often lasting for days. It was a time of merriment, feasting, parades, dancing, music, and general indulgence. It was a time for the outrageous.

But most important, carnival was a time of connection. In carnival, the community became all one flesh. (The *carn* in *carnival* means "flesh.") Everyone, high status and low, joined together in celebration. It was a time of social "uncrowning," as Bakhtin termed it, a time when the high and mighty were brought back down to Earth, the people's Earth. By dancing together; by celebrating the Earth's abundance with feasting and indulgence; and by joking together, often through references to the lower stratum of the body and to the substances that pass from and move through that lower stratum, people celebrated their connections with each other and the world. Through these constant references to the bodily connections we all share—the joy of food, the pleasures of leisure, the desires of the flesh, the necessity of excretion—even the famous and highly esteemed were brought down to a common level (see Figure 6.3).

Bakhtin makes a crucial distinction between the carnivalesque and bodily references that are merely gross and those that are degrading, however. In carnivalesque humor, the subject of the joke is lowered, but not lowered below the tellers of the joke. Rather, carnivalesque humor is egalitarian humor that seeks to unite everyone on the same earthy, bodily, social plane. We laugh not just at the subject of the joke but at ourselves, too. Carnivalesque humor is not mere mocking. It is, as Bakhtin put it, "also directed at those who laugh."[20] It is laughter that joins us all together in the joke, renewing community. Degrading jokes, on the other hand, create

Figure 6.3 This painting from 1498—Piero di Cosimo's *The Discovery of Honey*— celebrates the festive and open-mouthed character of what theorist Mikhail Bakhtin called the "carnivalesque body." As in di Cosimo's painting, such a body relishes exchanges and interactions with society and the natural world, rather than presenting itself as a sealed-off monad.

Source: Used with permission of the Worcester Art Museum, Worcester, Massachusetts.

hierarchy and separation. They seek to lower others without bringing them into the same common earthy community of bodily life.

Bakhtin wrote in defense of the carnivalesque. But he worried that bodily humor had become "nothing but senseless abuse. . . . Laughter [has been] cut down to cold humor, irony, sarcasm. It [has] ceased to be a joyful and triumphant hilarity."[21]

He also wrote to make a historical point. Why do we moderns have such trouble distinguishing between the carnivalesque and the merely gross? Why do we so often find any references to the body to be offensive and shameful? Because, Bakhtin argues, social mores have changed from medieval and early Renaissance times, in tandem with the modern rise of hierarchical individualism.

The Classical Body

A work like *Gargantua and Pantagruel* is generally offensive today, then, not because of its politics (what offended some early Renaissance readers), but because of its affront to bodily individualism (what virtually all early Renaissance readers found deliciously funny). Today, we find individualism a lot harder to laugh at. We are ashamed at references to our bodily connections with the world. The beauty of ecology has become offensive.

This change is evident not only in humor but also in modern codes of politeness, cleanliness, and privacy. Today we eat with cutlery, particularly in formal situations. Medieval people ate with their fingers. Today, we find it impolite to eat with an open mouth or with slurping noises. Medieval people were not so troubled. We have historically astonishing standards of cleanliness for our homes and bodies. We confine most bodily acts to the privacy of the bedroom and bathroom. In fact, the bathroom has become a kind of modern shrine to the individual, and expensive modern homes often include one for every member of the family, plus one for any guests—four- and five-bathroom homes have become standard in exclusive housing developments. And we medicalize birth, death, and all the stages in between of the body's growth and interactions with life. We keep the environment as much at a distance from our bodies as we can. Again, medieval people were not so troubled.

Why do we do all these things? Because, Bakhtin argues, these are symbols of social hierarchy. In order to be elite, you need to separate yourself from the common people. Separation from nature and bodily functioning is a particularly convincing way to make that distinction. As Thorstein Veblen noted, elites try to remove themselves from environmental concerns in part because doing so demonstrates their power, as Chapter 2 describes. It's a symbolic form of environmental power—power over others that we gain from the environment. Bakhtin would add that such environmental power also entails showing oneself to be above bodily concerns, not just the productive concerns Veblen discussed. It requires what Weber would recognize as a kind of asceticism, a denial of bodily existence. Having servants and machines to handle dirt, trash, and bodily excretions; being able to get through the day wearing the most impractical of clothes; traveling by means other than one's own bodily locomotion; maintaining impeccable standards of cleanliness for one's home and body; having a house and workplace big enough for separate rooms for private acts, and separate kinds of rooms for each kind of private act—to acquire these forms of ecological and social separation requires power. It requires money and status. Such separation is far harder for those without money and power, thus clearly establishing who is on top and who is on the bottom.

Our desires for social distinction are thus intimately connected with our desire to distance ourselves from the body, from the Earth, and from ecological reality. We cannot admit that we are connected to the Earth, for doing so would undermine the very feeling of separation and distinction that modern life seeks. Seeking to live the life of the high-status individual, we model ourselves after the classical image of the body and find references to carnivalesque connection dirty and threatening. We pretend that we have no need to heed nature's call.

Balancing the Ecological Self and the Ecological Community

As often happens when someone hits upon a new idea, Bakhtin probably overstated his case. His portrayal of medieval and early Renaissance life seems filtered through a romantic mist.[22] This period was not a golden age of unending feasting, merrymaking, and communalism. There was much hierarchy then, too, as well as grinding poverty, poor sanitation, and disease. Bodily connections with society and with the environment can be fatal, a point that surely was significant to medieval people. (But so, too, can be attempts to deny such connections.) Thus, we cannot pass off the modern interest in sanitation and medical intervention as merely the product of raging individualism. (But over-cleanliness can also be hazardous, and indeed is suspected by some researchers as being a factor in the dramatic rise in the incidence of allergies and asthma in the wealthy countries.)

We also need to be cautious about seeing the rise of a classical conception of the body and its implication of ecological separation as a purely Western phenomenon. Rather, it is characteristic of elites the world over. Nearly all elites adopt refined lifestyles that insulate them from the dirty, sweaty, smelly consequences of being a human animal. Bakhtin would have readily accepted this point, in fact. And he would have added that common people have long responded to the pretensions of the world's elites with carnivalesque humor. In Bakhtin's words, "Every act of world history was accompanied by a laughing chorus."[23]

Finally, we need to keep a sense of balance with respect to the carnivalesque and the classical. I for one am not prepared to lead a life of the purely carnivalesque. Besides, even during medieval times, carnival was not an everyday occurrence, although the spirit of carnival was no doubt a more regular presence in the lives of medieval people. Probably it ought to be in ours. But neither should we give up all forms of bodily individuality. A sense of our own difference is, after all, essential to a feeling of connection, for there must be something to connect from and with. It's another dialogue.

Yet we also need to balance a classical conception of our selves and our bodies with a carnivalesque understanding that we are part of ecology. Evidence suggests that we may be coming around to this point of view. The West has substantially changed its attitudes about the body in the 50 years since Bakhtin wrote *Rabelais and His World*. Thanks in large measure to the social changes and social movements of the 1960s, we are no longer so ashamed to speak of the body (although there are signs that such shame may be on the rise again). Hippie culture and the women's movement both emphasized the importance of being open about the body, its needs, its functions, and its realities. Hippies emphasized a more natural body style, breaking the taboos against long hair for men and leg hair and underarm hair for women, for example (although these taboos seem now to have returned, albeit in more muted form). Feminists helped break down the misconceptions and sense of shame long associated with women's bodies, perhaps most notably through the publication of the revolutionary book *Our Bodies, Ourselves*, now in its fourth edition.

These social changes suggest a connection between environmental awareness and bodily awareness. It may be no accident that the 1960s saw both an environmental movement and a body awareness movement. In other words, accepting the importance of environmental interactions may depend in part upon accepting a more ecological—and thus less hierarchical and more democratic—conception of the body. Our bodies, our ecologies.

Gender and Environmental Domination

Another source of our domineering attitudes toward the environment is gender relations. Consider, for example, the common metaphors we in the West use to describe the environment and our interactions with it, metaphors that are strikingly sexual and militaristic. The pioneers in North America "broke virgin land" and cleared "virgin forest." Farmers have long spoken of the "fertility" of the soil, and surveyors and military commanders assess the "lay of the land." Mariners sail on the "bosom of the deep." The environment in general is "Mother Nature." We speak of abuse of the environment as "raping the land," and we speak of civilization as the "conquest of nature." The sex of the environment in these examples, sometimes implied, sometimes overtly stated, is female.

In light of the violence of some of the imagery—the "breaking," "clearing," "rape," and "conquest" of female nature—these are disturbing metaphors. They suggest, along with a range of other evidence, that there is an ideological link between the domination of nature and the domination of women. If patriarchal ideas pervade our thinking about society, then they likely influence our thinking about the environment as well, for we use the same mind, the same culture, to understand both.

The Ecology of Patriarchy

Note the common Western tendency to consider women as being closer to nature than men. Not only is nature female, but females are more natural, our traditions often suggest. We tend to associate women with reproduction, broadly understood—with the natural necessities of giving birth, raising children, preparing food, healing the sick, cleaning, attending to emotional needs—as well as with the domestic sphere, the realm of the reproductive and the private. In contrast, we have conventionally associated men with production—with transforming nature so that it does what we want it to—and with the public sphere, the realm of rationality, civilization, government, and business.

These gendered associations imply a clear hierarchy, with men on top. Western thinkers have often considered women inferior because of their alleged animalistic closeness to nature and men as superior because of their allegedly greater skills in the allegedly higher aspects of human life. Edmund Burke, the late–eighteenth-century English philosopher, wrote that "a woman is but an animal and an animal not of the highest order." Hegel felt that "women are certainly capable of learning, but they are not made for the higher forms of science, such as philosophy and certain types of creative activities." Sigmund Freud mused that "women represent the interests of the family and sexual life; the work of civilization has become more and more men's business."[24] And here is Henry James Sr.—father of the philosopher William James and novelist Henry James, and himself a prolific author—writing in 1853 on the subject of "Woman and the 'Woman's Movement'": "Woman is "by nature inferior to man. She is inferior in passion, his inferior in intellect, and his inferior in physical strength." As he put it in another essay, discussing "The Marriage Question," a wife is her husband's "patient and unrepining drudge, his beast of burden, his toilsome ox, his dejected ass, his cook, his tailor, his own cheerful nurse and the sleepless guardian of his children."[25] These characteristics of women and their lives were not social inventions open to interrogation and change. For these men, and many others of their time, these were the writ of nature.

The social implications of such patriarchal presumptions are quite troubling, most would today agree. Many writers also argue that so are the environmental implications. By demeaning women for their stereotypical association with reproduction and with nature, we encourage both the domination of women and the domination of the environment.

Ecofeminism

The work of these writers comes out of a relatively new tradition of scholarly and philosophical inquiry, *ecofeminism*, which explores the links between the domination of women and the domination of the environment and argues that the domination of the environment originates together with social domination of all kinds—across not only gender but also race, ethnicity, class, age, and other forms of social difference treated as hierarchies.[26] It is common for socially dominated groups to be linked with nature, ecofeminists observe. People of color have often been associated with savagery. Lower classes have often been seen as primitive and as having inadequate control over their emotions, leading to a greater tendency toward violence and sexual licentiousness. And women have often been relegated to the realm of nature and its reproductive requirements, as opposed to reason and civilization.

It seems that when we think social hierarchy, we think ecological hierarchy—and probably vice versa, too. As the prominent ecofeminist Val Plumwood has written, the "human domination of nature wears a garment cut from the same cloth as intra-human domination, but one which, like each of the others, has a specific form and shape of its own."[27]

Environmental activists themselves have sometimes promoted the association of women with nature, for example, by using the image of "Mother Earth." An ever-popular environmental slogan is "Love your mother," referring to the Earth. In this case, nature is positively valued, and the activists who use the expression probably feel that it therefore positively values women as well, reversing the traditionally negative connotation of being associated with nature.

This is an ideologically dangerous strategy, say some ecofeminists. Read this statement from Charles Sitter, senior vice president of Exxon, who used the image of Mother Earth to minimize the significance of the infamous 1989 *Exxon Valdez* oil spill in Alaska's Prince William Sound: "I want to point out that water in the Sound replaces itself every twenty days. The Sound flushes itself out every twenty days. Mother Nature cleans up and does quite a cleaning job."[28]

During the discussion about the "cleanup" following the Gulf Oil Spill of 2010, many commentators used similar imagery. *National Geographic* reported that "Mother Nature has rolled up her sleeves" and helped with cleaning up methane released by the spill.[29] Ed Overton, a University of Louisiana professor who became something of a media star during the height of the spill, described the reabsorption of much of the oil into the environment as "mother nature doing her job."[30] Coast Guard Commander Randal Ogrydziak opined concerning storms that broke up oil patches on the surface that "Mother Nature is doing what she does best, putting things back in order."[31]

This "Mom will pick up after us" vision of the environment, as Joni Seager and Linda Weltner have termed it, is both ecologically problematic and sexist. As Weltner writes,

> Men are the ones who imagine that clean laundry gets into their drawers as if by magic, that muddy footprints evaporate into thin air, that toilet bowls are self-cleaning. It's these overindulged and over-aged boys who operate on the assumptions that disorder—spilled oil, radioactive wastes, plastic debris—is someone else's worry, whether that someone else is their mother, their wife, or Mother Earth herself.[32]

The point of ecofeminism is not to blame men for environmental problems. Nor are all ecofeminists women.[33] Ecofeminists, like other feminist scholars, are concerned about our patriarchal system of social organization, which is enacted by both men and women but results in the domination of women. What domination of women, you might say? Aren't we past all that, at least in the rich countries? Not yet, agree virtually all sociologists. Even in the rich countries, women are still paid some 15 to 25 percent less than men—23 percent less in the United States and 15 percent less in Europe—both because they are more likely to be consigned to lower-wage jobs and because they tend to receive less even when they hold the same job as men.[34] Many jobs and academic fields remain highly gender segregated. Women are less likely to hold political office, especially at the highest levels. Even when they are in paid employment as many hours as their partners, women still do the bulk of work at home—what the sociologist Arlie Hochschild calls "the second shift": cooking, cleaning, child rearing, paying bills, sending out thank you notes and holiday cards, and more.[35] Women still do the majority of all labor, paid and unpaid. But these persistent patterns of inequality are not men's fault alone. They are everyone's fault. We all enact them.

Ecofeminists add to feminist scholarship the notion that the domination of nature is linked to patriarchy and other forms of social domination, and vice versa. But ecofeminists observe that women, too, have been active agents in the domination of nature. As Plumwood points out,

> Western women may not have been in the forefront of the attack on nature, driving the bulldozers and operating the chainsaws, but many of them have been the support troops, or have been participants, often unwitting but still enthusiastic, in a modern consumer culture of which they are the main symbols, and which assaults nature in myriad direct and indirect ways daily.[36]

Patriarchal Dualisms

A key tenet of ecofeminism is that our cultural climate of domination has been built on dualisms—morally charged, oppositional categories with little gray area in between—that deny the dependency of each upon the other. Thus, man is man, and woman is woman. Nature is nature, and culture is culture. Our dualisms interlock into a larger cultural system of domination, ecofeminists such as Plumwood argue: culture versus nature, reason versus nature, male versus female, mind versus body, machine versus body, master versus slave, reason versus emotion, public versus private, self versus other.[37] In each dichotomy, the first member of each pair dominates over the second. The core dichotomy, Plumwood writes, "is the ideology of the control of reason over nature."[38] The dominating side in each pair is culturally linked to reason, and the dominated side is culturally linked to nature.

This tendency to separate the world into antagonistic pairs, Plumwood suggests, is a legacy of a Western us-versus-them logic of domination. Ecofeminists like Plumwood advocate a different form of logic, one that recognizes gray areas and interdependence and one that recognizes difference without making hierarchies. They want us to be able to make categorical distinctions that respect the diversity and interactiveness of the world and that do not rely on absolutist, mechanical, and hierarchical boundaries.

The Western logic of domination is not just an intellectual problem, argue ecofeminists. It has all-too-real material outcomes. Under Western rationality, the dominated and naturalized "other" does not receive fair environmental treatment. Women, people of color, people in lower socioeconomic groups, nonhuman animals, the land itself—all these groups tend to experience a lack of environmental justice because our cultural orientation is to regard them as generally less important and less deserving. Women, for example, are less likely than men to receive an even share of environmental goods, due to their lower average incomes. Worldwide, poverty rates are significantly higher for women—making women more susceptible to environmental bads as well.

But patriarchy also leads to the environmental oppression of men, even those from favored social groups. The patriarchal vision of masculinity leads men to take foolish risks with machines, chemicals, weather, and the land. Men often die as a result, or become maimed and diseased, which is some of the reason why men on the whole do not live as long as women. As Will Courtenay notes, in a review of gender and health practices, "men and boys are more likely than women and girls to engage in thirty controllable behaviors conclusively linked to a greater risk of disease, injury, and death."[39] Thus, all of us have an interest in changing the current social order.

Gender Differences in the Experience of Nature

The dualisms of patriarchal reasoning also affect the way women and men experience the environment. Although on the whole, Western women and men experience the environment quite similarly, some significant differences suggest that we have indeed internalized some of the patriarchal stereotypes. In the late 1980s, I conducted an ethnographic study of the experience of nature in an English exurban village. Although similarities far outweighed differences, village men described their natural experiences to me using significantly more aggressive, militaristic, and violent imagery. Village women emphasized a more domestic environmental vision based on their experience of nurturing in nature.[40] For example, men spoke of the pleasures of releasing their pent-up aggressive feelings through clearing brush and engaging in visceral rural sports such as "skirmish," a mock war game played in the woods with guns that shoot paintballs. As one village man described the game,

> I think when we were made, we were made with instincts to defend our tribe. . . . These instincts never get an airing. We sit in our office desks [isolated] from that danger, save-the-family type situation. . . . But when you go out there playing this game . . . it's like a dog that's been cooped up forever and then one day it's taken for a walk in the woods and

it sees a rabbit. It sniffs it and all its primitive instincts come alive. . . . It's quite exciting when a ton of people are coming at you with a gun.[41]

No village woman described such pleasures. Nor did any village man relate stories of nurturing in nature such as those told to me by several village women. One village woman, for example, told a story about a family cat that helped raise two ducklings, extending nurturing feelings even across the divide of predator and prey. She tells the story best, so here it is in her words:

> We had a cat [Suzy]. We always had lots of cats. And this particular time I went to Harchester, and there were two little ducklings in a pet shop window. And like a fool I thought, well, the kids will like them. And I brought them home, didn't I? And Suzy became a mother and she got kittens, at this particular time. And of course she took the two little ducklings over, didn't she? So wherever she went with the kittens, the ducklings followed. And they used to sleep together in this cardboard box. The cat and the ducklings! . . . It's completely true. She would wash and cuddle the ducklings, just like they were her own. It's the mothering instinct, I suppose.[42]

This is an incredible story, one that even got the family's picture in the paper, along with the cat and the ducklings. But significantly, this was a story that a woman told me. Her husband, whom I knew well, never mentioned it. This was her story, not his. Rather, he told me stories about rough weather and other hard environmental conditions and his feats of physical prowess and mental toughness in the face of these conditions. Perhaps village men and women told these different types of stories to conform to their expectations of what a male researcher should be told, and not to express their true feelings. Even so, it is significant that their expectations ran along such gendered lines.

I must emphasize once again, however, that the similarities between men's and women's stories far outweighed the differences. I must also emphasize that it is not helpful to blame men for experiencing nature in ways that I suspect most readers—both male and female—would regard as less laudable. The point of an ecofeminist perspective, as Joni Seager explains, "is not [to] reduc[e] environmental understanding to simplistic categories of 'wonderful women' and 'evil men.'"[43] Rather, the point is to highlight the environmental consequences for both women and men of patriarchal social structures and patterns of thinking, which both women and men bring into being.

The Controversy Over Ecofeminism

Ecofeminism remains a controversial viewpoint. Much of the debate has surrounded the attempt by some ecofeminist writers, mainly in ecofeminism's early days, to subvert Western patriarchy by reversing its moral polarity. These writers proposed that women and their associations with nature should be celebrated. Reproduction, nurturing, sensitivity to emotions, closeness to nature and the body—all these things are inherently good, the argument goes. Women should embrace these qualities, which one ecofeminist praised as the "feminine principle," not reject them.[44] It's the other side of patriarchy's dualisms—reason, civilization, machines—that has made such a mess of things.[45]

Critics both inside and outside of ecofeminism object that such a position reifies the very social order that needs to be changed. It perpetuates the dichotomy between men and women as well as the negative stereotypes of women as irrational, as controlled by their bodies, and as best suited for the domestic realm.[46] Critics also argue that this reification is alienating and fatalistic because it implies that biological differences between men and women are at the root of patriarchy. Such a position, suggests Deborah Slicer, is best termed "ecofeminine" and not "ecofeminist."[47]

There is also a spiritual and religious dimension to some ecofeminism associated with "goddess spirituality," Wicca, and neo-paganism. Spirituality and religiosity are, of course, important dimensions of human experience and are not in themselves necessarily problematic. Nor is there any reason in pluralistic societies to complain about the beliefs and practices of religions and spiritual perspectives that may differ from one's own. However, spirituality and religiosity are matters of faith, not social science, and should not be confused as such. So it is important that the spiritual strands of some ecofeminism be kept carefully separate from its social scientific claims. Many observers object that this separation has not always been maintained.

Another criticism is that a perspective like Plumwood's implies that the "logic of domination" is mainly a feature of Western thought. Are Eastern cultures less patriarchal than Western ones? The evidence suggests not. Also, Eastern cultures have shown themselves to be quite capable of dominating nature. Either the "logic of domination" that infuses both our social and our environmental actions must not be exclusively Western or the East must have its own logic of domination.

Also, in their effort to make clear the sexism that underlies some of our outlooks on the environment, ecofeminists have sometimes offered oversimplified arguments. For example, the patriarchal character of dualisms is not always so clear-cut. Consider the cultural association of women with nature and men with culture. In fact, the dualism often goes the other way, aligning women with culture and men with nature. Since Victorian times, one common stereotype of women has been that they are the bearers of culture and refinement and that they have responsibility for inculcating "civilization" in the next generation—and in men. One common current stereotype of men is that they are wild beasts driven by lust and violent passion, which women must tame for their own sake and for the sake of their children. Also, many of the spirits that various Western (and non-Western) traditions have sensed in the physical environment are characterized as male: Father Sky, the Greek sun god Apollo and ocean god Poseidon, and the notion of a "fatherland."

Indeed, it is an important feature of ecofeminist thought that we must recognize the gray areas and the interactiveness and interdependence of our categories. Unless we continually remind ourselves of the dialogics of categories, of the dialogue of difference and sameness, we easily slip into one-sided, deterministic, and hierarchical arguments. And, as ecofeminism also stresses, when you survey the world with a one-sided, deterministic, and hierarchical frame of mind to begin with, you are even more likely to slip in this way.[48] But ecofeminism has not always followed its own advice here as well as it might have.

In light of these controversial features of the ecofeminist debate, some social scientists have sought to find a different term to refer to explorations of the role of gender and patriarchy in social and environmental interrelations. "Environmental feminism" is what Michael Goldman and Rachel Schurman have suggested.[49] "Ecological feminism" is a similar phrase one increasingly encounters in social scientific literature. "Ecogender studies" is the term Damayanti Banerjee has offered.[50] "Feminist political ecology" is a term increasingly popular among geographers and others.[51] Time will tell if these terms prove analytically helpful. The good news is that social scientists are now investigating the effect of gender in environmental relations, no longer seeing such a concern as irrelevant or unscientific. Recent years have seen such diverse work as studies of the anti-environmental breast cancer movement, gender differences in environmental knowledge, rural masculinity, women's participation in forestry, men's lower overall concern for environmental risk (a subject Chapter 7 returns to), the significance of gender in watershed management, and so much more.[52] Studies of sexuality and environmental relations are also now beginning to appear.[53]

In any event, our environmental complaint with patriarchy should not be that it is wrong to create categories and draw distinctions. We need categories to recognize difference and thereby to build our theoretical and moral understanding of the world. (After all, ecofeminism itself represents a category—a category of thought.) But we also need better categories than the hierarchical, socially unjust, and environmentally destructive ones of patriarchy.

The Difference That Ideology Makes

These various theories of the environmental significance of religion, individualism, and patriarchy all have a common theme: the central roles of inequality and hierarchy in the way we think about the environment. Whether we are talking about the competitive desire to achieve grace through work, the notion that people and their God are above nature, the achievement of individual distinction through bodily distance from the world, or the dualistic thinking of patriarchy, social inequality influences our environmental relations.

I hope this chapter also makes it clear that social inequality has not only material but also ideological roots. This is another dialogue. Material factors structure our lives in unequal ways, leading to hierarchical visions of the world, just as ideological factors allow the material structures of inequality to develop and to persist.

Another common theme of this chapter is that, thus far, scholars have relied too much on the Western experience in formulating theories of the human transformation of the environment. Some of this neglect of the East has likely been due to a romantic view of the environmental sensitivity of that part of the world. But large-scale transformation of the environment in the East goes back thousands of years, just as it does in the West. Although this romantic view is flattering in some ways, it is also a backhanded insult, for it implies that the scientific and technological mind was beyond the ideological capabilities of the East. The view that the East was ecologically sensitive (until corrupted by the West) may thus perpetuate negative stereotypes of irrationality and backwardness.

Placing more emphasis on economic factors may help us understand how the ideology of transformation arose (as long as we also recall, with Weber, that any economic pattern is as much an ideological matter as a material one). The global spread of capitalism has been propelled by the accelerating treadmills of production and consumption, bringing with it social structures and ways of thinking that increase our orientation toward transforming the Earth.

But still the explanation is not complete. Environmental transformation was going o[n] capitalism arrived in both the East and West. Also, and perhaps even more important, w[e need] to remember that the socialist economies of the former Soviet Bloc and East Asia showe[d] as much tendency as capitalist economies to transform and dominate the Earth. We cann[ot] point our analytic finger at capitalism alone.

In short, we do not yet fully understand the ideological origins of the transformation and domination of the Earth. And it may be that even after we take into account both material and ideal factors, we still will not fully understand these origins. One implication of a dialogical view of causality is that complete explanations are rarely, if ever, possible. The spontaneous creativity that comes out of social interaction has effects that can never be completely predicted.[54]

Nevertheless, we should still pursue the analysis of social and environmental change. It is vitally important that we try to understand the material and ideal factors that dialogically shape, if not completely predict, our actions regarding the environment—particularly if we hope to guide those actions in a different direction.

CHAPTER 7

The Ideology of Environmental Concern

Rather than love, than money, than fame, give me truth.

—Henry David Thoreau, 1854

"It is our alarming misfortune," wrote Rachel Carson in 1962, describing the indiscriminate use of chemical pesticides, "that so primitive a science has armed itself with the most modern and terrible weapons, and that in turning them against the insects it has also turned them against the earth."[1] With these words, Carson concluded *Silent Spring*, a book that came like a thunderclap in a seemingly cloudless technological sky (see Figure 7.1). Because of chemical poisoning, argued Carson, it was a very real possibility—and indeed it had already happened in some areas—that a time could come when spring would arrive "unheralded by the return of the birds, and the early mornings are strangely silent where once they were filled with the beauty of bird song."[2] Carson carefully documented her claims with the results of hundreds of scientific studies, challenging science with science. Suddenly, the technological utopianism of the postwar period no longer seemed so utopian.

Figure 7.1 Rachel Carson, 1907–1964. A biologist for the U.S. Fish and Wildlife Service and a brilliant writer, Carson is widely credited with helping precipitate a great change in public attitudes toward the environment, particularly with her final book, *Silent Spring*.

Source: Photograph by Edwin Gray. Rachel Carson on the Dock at Woods Hole, 1950.

Of course, we cannot assign an absolute beginning to any historical trend; history always has precursors. But so dramatic were the subsequent shifts in public opinion that it has become conventional, with some justice, to date the start of the modern environmental movement from the publication of *Silent Spring*.[3] My own mother, who read extracts from the book in a popular magazine, recalled to me the heated discussions it touched off among her friends. "It really shocked a lot of people," she explained. "We didn't have any idea that pesticides could be so dangerous. Nobody used to question these things."

Today, however, millions—even billions—do. The domination of the Earth has become increasingly unthinkable to increasing numbers of people in the years since 1962. The third wave of the World Values Survey, which ran from 1999 to 2004, asked the people of 29 countries whether human beings should "master nature" or "coexist with nature." World opinion is overwhelmingly one-sided on this point: 78 percent of the 36,000 respondents said humans should coexist. In only one country, Saudi Arabia, did a clear majority—62 percent—feel otherwise.[4] In the International Millennium Survey of 60 nations from rich to poor conducted in 1999, a total of 65 percent said that their governments had "done too little" to protect the environment, and 57 percent found the state of their environment "mainly unsatisfactory" or "very unsatisfactory."[5] The 1990 first wave of the World Values Survey asked the people of 43 nations their view of the environmental movement. Ninety-six percent of the respondents said that they "approved" or "strongly approved." Most said they strongly approved.[6] There can be little else that so much of the world apparently agrees on.

Why this strong shift? Humans had been dramatically altering their environment for centuries without evoking a popular environmental movement. Yet even the most influential book can only crystallize concerns that must already have been held in dissolved suspension in the roiling sea of public opinion. How can we understand this ideological reorientation?

This chapter seeks sociological answers to this question. It does so through a sketch of the history of environmental concern and a review of the theories advanced by social scientists to explain the recent flowering of this concern into the modern environmental movement. We will see that, in the face of environmental domination, counter-ideas have always been around. These ideas became much more widely held in the latter half of the twentieth century for three primary reasons: the rediscovery of the moral attractiveness of nature, the increased scale of material alterations of the environment, and the spread of democratic attitudes and institutions. It is, yet again, a matter of both the material and the ideal.

Ancient Beginnings

Environmental concern has a long history—perhaps every bit as long as the history of conscious environmental transformation. Like other creatures, humans unavoidably influence their surroundings. But the first decisions to consciously tinker with the environment likely prompted some heated debate: Is this safe? Is it moral? Perhaps even, is it beautiful? And will the gods approve? At the very least, these kinds of debates have been with us since the time of the ancient Romans, Greeks, and Chinese.[7]

Rome

The poet Horace loved his country villa in the Sabine Hills above Rome. One day, in about 20 BCE, he took up his wax tablet and his reed stylus and scratched out the following lines to his friend Fuscus:

> *Fuscus, who lives in town and loves it, greetings from one who loves*
>
> *The country. . . .*
>
> *You stay in your nest, I sing my lovely rural*

Rivers, and trees, and moss-grown rocks. Why drag out

Our differences? I live here, I rule here, as soon as I leave

Those city pleasures celebrated with such noisy gabble:

Like a professional cake-taster I run looking for good plain bread,

Just crusty bread, no honeyed confections, dripping sweet!

If life in harmony with Nature is a primal law,

And we go looking for the land where we'll build our house, is anything

Better than the blissful country? Can you think of anything?

Where can we sleep, safer from biting envy?

Is grass less fragrant, less lovely, than your African tile?

Is your water as clear and sweet, there in its leaden pipes,

As here, tumbling, singing along hilly slopes?

Lord! You try to grow trees, there in your marble courtyards,

And you praise a house for its view of distant fields.

Push out Nature with a pitchfork, she'll always come back,

And our stupid contempt somehow falls on its face before her.

Live happy with what you have, Fuscus, and live well,

And never let me be busy gathering in more than I need,

Restlessly, endlessly: rap me on the knuckles, tell me the truth.

Piled-up gold can be master or slave, depending on its owner;

Never let it pull you along, like a goat on a rope.[8]

Astoundingly modern-sounding sentiments all. Like countless nature writers of the current day, Horace "sings" the beauty of the countryside, of rivers and trees and moss-grown rocks. And like many in recent decades who left the city for the country, Horace praises the simple life, close to nature. He has no need for the urban contrivances of "honeyed confections, dripping sweet." Just give Horace the plain crusty bread of country living. Since "life in harmony with Nature is a primal law," the country is the best place to live, he proclaims. After all, grass can be as beautiful to walk on as Fuscus's imported African tile. The water in the country is pure and sweet, Horace says, instead of the stale, piped-in stuff that Fuscus gets in town. (Although he mentions the lead in the pipes—Romans used lead extensively in their plumbing—Horace could not have know about the added danger of lead poisoning, as this hazard was unknown at the time.)

Horace praises not only the naturalness of rural living but also the social consequences. A country life frees one from the "biting envy" of the city. Horace doesn't want to live a life devoted to "gathering in more than I need," being pulled along "like a goat on a rope" by the pursuit of money and material possessions, and he warns Fuscus of these dangers. In the poem's most famous lines, Horace chastises those who contemptuously attempt to avoid these social and environmental truths by trying to "push out Nature with a pitchfork." Nature will "always come back," he warns.

That contempt was very evident in the Rome of 20 BCE. At that time, Rome was probably the largest city ever known, with close to a million inhabitants, the product of spectacular feats of technology and engineering. A vast system of aqueducts and pipes carried more than 200 million gallons (about a billion liters) of water a day in from the surrounding countryside. The resulting urban effluent poured into the Cloaca Maxima, an underground sewer large enough to accommodate a small sailboat, and thence into the badly abused River Tiber.[9] (The Cloaca Maxima, amazingly, is still in limited use today in Rome, a testament to the quality of Roman construction.) Wealthy Romans enjoyed running water in their homes, even showers, as well as central heating. Common people lived in *insulae,* apartment buildings the size of a full city block and sometimes as much as seven stories tall. Roman legionnaires had hot baths and flush toilets in their military camps. (Flush toilets are in fact even older; the Minoans had them at the Palace of Knossos a millennium earlier.) Showers, baths, flush toilets, and seven-story apartment buildings—and a technology, economy, and empire capable of supporting it all: mighty pitchforks against Nature (see Figure 7.2).

Figure 7.2 Roman central heating, Northumberland, England. The commander and his family lived in considerable environmental comfort at Cilurnam, a Roman fort along Hadrian's Wall, built around 122 CE. Even in this remote location, they enjoyed a *hypocaust,* a system of central heating in a which a furnace blew hot air through these passageways beneath the stone floor. Horace wrote his poetry about the importance of nature in reaction to such power to transform the environment.

Source: Author.

The sentiments that Horace expressed were, we cannot doubt, in some measure formed in reaction to these new environmental transformations—transformations that also had social meaning for him. It was a culture of money and power that produced the technological pitch-forks. The urban, commercial life of empire brought with it a widespread feeling that every-thing was becoming political—that social life was moved not by virtue but by self-serving desires for power, influence, and material possessions. Greed was overwhelming the Roman landscape and lifescape. And if all social motivations derived from the pursuit of interests, of materialist desire, where might one encounter an alternative?

For Horace, in nature. And what made nature so attractive to Horace still makes nature attractive today. Concern about nature cannot be separated from concern about social interests and how these shape our moral understandings.[10] Part of the attraction of nature stems from our struggles with the oldest of moral problems: the balance of power between us. And part of nature's attraction is its use in our struggles with the oldest of moral critiques: that interests underlie what we say, do, and believe. We look to nature for a moral base that lies outside our-selves, outside human power structures, and therefore outside the potential that we may have manipulated morality for our own ends. To experience nature is to experience an interest-free foundation upon which to build our motivations. To experience nature is to experience a point of rest from the constant charges that we act as we do because we seek power. To experience nature is thus to experience social innocence—or so we hope.

Horace felt that innocence in a country life. By living in harmony with nature, Horace believed, he could remain free of city dwellers' relentless pursuit of personal gain and their "bit-ing envy" over others' "piled-up gold." In nature, Horace felt he had discovered a moral realm that lies beyond the reach of the pollution of human interests and materialist desire. The search for this interest-free realm of innocence is a kind of conscience—what I like to term a *natural conscience*—and it is fundamental to our moral thought.[11]

Yet was Horace truly above the pursuit of self-interest himself? Had he really put "biting envy" behind him through his celebration of a natural life? One way to read his famous epistle to Fuscus is that Horace was trying his best to build a bit of biting envy in his friend—envy for Horace's lifestyle in his country villa. Wealthy Romans loved country living and established country villas for themselves across the empire. Horace is thought to have had quite a substan-tial one—about a 20,000-square-foot place, the ruins of which can still be visited. True, some archeologists say this villa was actually built later by the Emperor Vespasian and that Horace lived in the small farmhouse whose foundations were found beneath the villa in 2001.[12] No matter. Even a modest country house was already a positional good 2,000 years ago—which is perhaps why, despite his opening avowal, Horace's epistle goes on very forcefully to in fact "drag out our differences."

In other words, the discovery of a natural conscience does not necessarily mean that one has truly escaped the moral problem of interests. (Indeed, much philosophical and sociological work suggests such an escape is not possible.)[13] But it does mean that one is grappling with the issue.

Horace was not the first to grapple with this problem. The search for a natural realm of moral innocence has frequently accompanied the rise of a complex, urban-dominated political life and the growing wealth and social inequality that have so often been associated with such a life. For that, we must begin even earlier, with the ancient Greeks, and, in a few pages, with the ancient Chinese.

Greece

Nature is an old and powerful idea. Words are not the same as ideas, of course; the same boat can carry many different loads. But it is illuminating to trace the origin of the word *nature* and the historical sequence of conceptual loads it has been asked to carry.

The boat of nature was first loaded up in Greece. The English word *nature* is a rendering of the Latin *natura*, which first appeared in the third century BCE. But *natura* was itself a Roman translation of the older Greek word *physis*.[14] (Physis also makes its way into English, serving as the root for *physics, physician, physical, metaphysical*, and so on.) *Physis's* own roots are in the Greek *phy*, meaning simply "to be," and *phyein*, "to give birth." Although we can't be certain when *physis* was coined from *phy* and *phyein*, the word was nevertheless in use by the eighth century BCE to connote the permanent, essential aspects of an object by which it might be forever known—a meaning that was apparently derived from an earlier use of *physis* to mean "birthmark."[15]

By the end of the fifth century BCE, however, *physis* had come to take on a wider and more significant meaning. The fifth century was a period of enormous change in Greek society, a period of fantastic growth in the power, size, and wealth of the city-states. After 480 BCE, when the Greeks defeated the Persians, to whom they had previously paid tribute, their economy began to expand mightily. The economic expansion brought wealth and urban growth. Athens grew to a size never seen before. Perhaps as many as 275,000 people lived there in 431 BCE.[16]

In such a place, and at such a time, it was hard not to be impressed with an urban truth: Money and politics, not the gods and other lofty concerns, moved the world of everyday life. At least that was the message of the Sophists, a group of itinerate philosophers known for their cynical and relativistic teachings about the reach of self-interest into all human affairs and beliefs. "It is for themselves and their own advantage," declared the Sophist Callicles, "that they make their laws and distribute their praises and censures."[17] Not the gods, but "man is the measure of all things," claimed Protagoras. Human laws are "designed to serve the interest of the ruling class," Thrasymachus observed. In his view, "The actual ruler or governor thinks of his subjects as sheep [and] his chief occupation, day and night, is how he can best fleece them to his own benefit."[18]

What the Sophists were saying is that the moral order, including religion, is based upon mere convention—*nomos*, to use the Greek word—not some principle external to human interests, such as god, justice, or what we have come to call science. Many Sophists were famous for their rhetorical skills, which they used as proof of their view that morality is just a kind of con game. Indeed, their practice of lecturing only for a fee (which Plato complained about in *The Republic*) was itself a kind of demonstration of the ultimate Sophist point: Even truth has a price on it.

The Sophists are often considered the bad guys of ancient Greek philosophy because of their apparent anything-goes vision of morality. If morality is human derived, it follows that morality is whatever any human wants it to be, justice and inequality be damned. But most Sophists, in fact, were very concerned about social inequality and sought to expose hypocrisy. Their teachings fit well with the ambitions of the students who came to hear the latest Sophist to come to town. These were mainly young or disadvantaged citizens and free noncitizens eager to acquire the rhetorical skills essential for getting ahead in the city.[19] For them, the Sophist message was comforting and hopeful: The social order is not preordained.

The Sophist argument was ultimately circular, however, for if all morality is mere rhetoric and nomos, Sophism itself must be as well. There had to be somewhere else to stand. *Physis* suddenly sprang into widespread use as a word that might provide that foundation, a word that could root truth in something outside of human manipulation, rhetoric, bias, materialist desire, and self-interest—a word that could be a source of moral guidance in a jealous world of wealth and power. Hippocrates advised doctors that *physis* was the only true calling of medicine (thus the term *physician*).[20] In the middle of the fourth century, Aristotle wrote several books on *physis*, most notably *Physics*, and went on to use it as his proposed foundation for a just society in his work *Politics*.

Plato was a bit cautious about the word, though, perhaps because the Sophist opponents in his dialogues often based their arguments on *physis*. For example, Plato reports that the Sophist

Callicles believed might-makes-right was a truth "nature herself reveals."[21] Such a view really would make Sophists philosophical bad guys, but we only have Plato's word on this. Whether or not this was truly a common Sophist position, Plato had recognized that *physis* was a bendable enough concept that might-makes-right could be justified with it. So he sought a different solution than *physis*—but one that was equally an effort to discover a moral realm beyond interest. He agreed that the Greek pantheon of gods had become a philosophical shambles. So he proposed a new kind of god, a great ideal or "form" that he claimed governed the order of things: "the Good."

Plato argued that the Good was the divine agent of the world, what he called the "Demiurge." As pure goodness, the Demiurge could not suffer from a materialist sin like envy. And herein lies the origin of the world, as Plato explained in *Timaeus*:

> Let us therefore state the reason why the framer of this universe of change framed it at all. He was good, and what is good has no particle of envy in it; being therefore without envy he wished all things to be as like himself as possible.[22]

So the Demiurge created the material world, using as a blueprint "his" own ideal goodness. In other words, the primal act of the universe was the denial of envy, of materialist desire. As a consequence, the whole world is good, and goodness is the whole world.

Thus was born a new manner of god. Not the quarreling, querulous gods that populated the old Greek pantheon. Not the jealous and vengeful God of the Old Testament, constantly punishing those who don't believe in him. But a god who cannot sin and whose very power stems from being separated from the backstabbing ways and moral sleight of hand of human society.

Aristotle apparently thought the Demiurge story was a bit silly, and he did not repeat it. Aristotle's distinctive moral contribution was his effort to get around the might-makes-right problem and thereby to return to *physis* as a solution to the problem of justice. But he kept a bit of Plato in his argument and agreed that the ultimate motor of the world, the "final cause" and "unmoved mover" of *physis*, was "the Good."[23] Nature was not, as some Sophists had apparently concluded, a source of anything-goes morality. It was, in effect, itself the Demiurge, and thus following nature was following the Good. Now nature was cured of materialist desire and could become a source of moral guidance in a world of envy and greed. The Greeks had discovered the natural conscience.

China

The classical Greeks were not the first ancient people to critique materialist desire. *The Song of the Harper*, an Egyptian text, circa 2600 BCE, had this to say: "Remember it is not given to man to take his goods with him. No one goes away and then comes back."[24] Similarly, the Egyptian sage Ptahhotep admonished, "Beware an act of avarice; it is a bad and incurable disease."[25] The Hebrew Bible, most of which was written before the flowering of Greek philosophy, warned that "He who oppresses the poor to increase his own wealth, or gives to the rich, will only come to want."[26]

But few pursued this line of thinking with the philosophical vigor of the classical Greeks and another group of ancient thinkers: the Taoists. By the fifth century BCE, about the same time as the Greeks, Taoist philosophers began to take a rather skeptical view of the human world. In China, as in Greece, this was an age of commerce and empire in which power and self-interest seemed to be greater truths about the way the world worked than did earlier religious views. Taoist writings of the time abound in criticisms of the self-centered money interests overtaking ancient Chinese society. This critical impression grew into full-fledged doubt about the materialistic underpinnings of social life in general.

Lao Tzu (or rather the writer of the *Lao Tzu,* who is not known for certain) is famously direct on the topic: "There may be gold and jade to fill a hall, but there is none who can keep them." You can't take it with you—clearly a reminder that many through the ages have seen fit to give. "To be overbearing when one has wealth and position," the *Lao Tzu* goes on, "is to bring calamity upon oneself."[27] Lao Tzu taught that "the sage desires not to desire and does not value goods which are hard to come by."[28] "Is this not because," Lao Tzu asks in a different passage, "[the sage] does not wish to be considered a better man than others?"[29] The *Lao Tzu* makes the point most plainly in the allegory of the "uncarved block," the simple, unadorned state it counsels people to emulate: "The nameless uncarved block is but freedom from desire." The secret to contentment is to "have little thought of self and as few desires as possible."[30]

Similarly, the *Lieh-Tzu,* another early Taoist work, taught the uselessness of the fancy art objects made for the aristocracy. One story in this book is of a certain prince of Sung who commissioned an artisan to carve a morphologically correct leaf out of jade. When the artisan returned with the jade leaf after 3 years' labor, it was so perfect that no one could tell it from a real leaf, and the prince was overjoyed. But when Lieh Tzu heard about it, he replied, "If nature took 3 years to produce one leaf, there would be few trees with leaves on them!"[31]

These critiques had a deep resonance, and we find them in many later ancient Chinese writings. The *Lii-shih Ch'un Ch'iu,* a compendium of useful knowledge put together for a third-century BCE prime minister, portrayed commerce as a potentially corruptible tangent from the simple ways of agrarian society.[32] Writing about the same time, Chuang Tzu counseled against giving in to "desire," materialism, and ambition in "The Empty Boat." Whoever "can free himself from achievement and from fame . . . will flow like Tao." He who has no power and no reputation "is the perfect man: His boat is empty."[33]

Chuang Tzu also extolled the virtues of simplicity in the story of a Taoist sage, Hsu Yu, who, when offered the rulership of a kingdom, exclaimed,

> When the tailor-bird builds her nest in the deep wood, she uses no more than one branch. When the mole drinks at the river, he takes no more than a bellyful. . . . I have no use for the rulership of the world![34]

Materialist skepticism in ancient China, like Sophism in Greece, proceeded so far that many Taoists, particularly Lao Tzu, came to be deeply suspicious of all knowledge. The sage "learns to be without knowledge," wrote Lao Tzu.[35] "One who knows does not speak; one who speaks does not know."[36] Like the Sophists, the Taoists sharply criticized the deceptiveness of language. Here is Lao Tzu again: "Truthful words are not beautiful; beautiful words are not truthful. Good words are not persuasive; persuasive words are not good." The passage goes on to return to the problem of knowledge more generally: "He who knows has no wide learning; he who has wide learning does not know."[37]

But unlike the Greek Sophists, the Taoists had an alternative to self-interest, an alternative that provided a natural conscience they believed to be free from materialist ambition. The Taoists found their natural conscience in the *Tao,* or "the way," the principle that underlies things we in the West would term "natural"—in agrarian society, in the nests of tailor-birds, in moles and rivers, and in the leaves of trees. As the historian Fung Yu-Lan put it, Tao is "the unitary first 'that' from which all things in the universe come to be." It is "the all-embracing first principle of things," what Aristotle would have recognized as the "unmoved mover" of the world.[38] Tao is unaffected by human doings, but people do best when they allow themselves to be guided by it, and not by human desire.

Through the Tao, each thing obtains its *te,* its individual essential quality. Tao is within humans as well, and each human therefore has a *te.* We can allow *te* to express itself through *wu wei,* acting without deliberate effort.[39] Acting deliberately, what Taoists call *yu wei*, would inevitably lead one away from the Tao. Empty your boat, advise Taoists, and drift along with the natural conscience of the Tao.

The Moral Basis of Contemporary Environmental Concern

In the centuries that followed, nature remained a central pillar of the natural conscience. But nature was often tied in with other concepts—religion, science, conceptions of bodily difference such as race and gender, and many other ideas and institutions that similarly claimed at least a partial foundation in "nature." Such ideas and institutions have been strongly criticized in recent years. Critics have argued, like the Sophists once did, that human interests do indeed motivate these sources of moral judgment. In an era of unprecedented growth in wealth, inequality, environmental domination, and political conflict, it seems to many that rather than resting above power, our basic moral ideas—including those rooted in nature—are the products of power.

Consider race, once widely seen as a "natural," and therefore perfectly moral, basis on which to allocate social rewards and social position. Increasingly—and quite correctly—people see the idea of "race" as a way those in power have sought to stay that way. Race is rapidly losing its former status as an interest-free fact of "nature." (Chapter 8 explores this criticism in more detail.)

It is a common fate of every widely accepted formulation of a natural conscience that it is subsequently subjected to careful critical scrutiny to see if it really does rise above the problems of human interest and the balance of power. We should welcome such scrutiny, I believe, although we often do not. The scrutiny of various visions of the natural conscience have been unusually intense in recent decades, though, and it is the defining feature of a cultural trend often called *postmodernism*. Although skepticism about human motivations is certainly not wholly new, as we have seen, the moral dilemma is particularly strong today: If all that motivates us is power and self-interest, is there then no truly moral place to stand?

These questions had already begun to occur with some force to Henry David Thoreau in the middle of the nineteenth century, as he walked through the woods near Walden Pond. He was by no means the only one to whom such questions occurred at this time. Many people worried about the direction and motivations behind the social and environmental transformations brought about by the burgeoning Industrial Revolution. And the moral solution that many found was the same: a return to a purer vision of nature—nature as the wild; as woods and winds, farms and fields, grazing sheep and flitting butterflies; as the nonsocial world. A few decades earlier, Beethoven had penned the *Pastoral Symphony*. A few decades later, Bierdstadt would paint *The Last of the Buffalo*. Cities were establishing parks, zoos, and botanical gardens, and the rich were taking up second homes in the countryside. Natural historians were reveling in the wonders of a nature beyond society, and the theory of natural selection was slowly accumulating evidence on Darwin's desk, even as the coming of industrialism was radically transforming those wonders. But Thoreau was a leading exemplar of this broad cultural change (see Figure 7.3).

Thoreau in particular loved walking. Thoreau walked "four hours a day at least," he tells us in the essay "Walking."[40] And when he walked, he found himself inevitably drawn to the west. "Eastward I go only by force," he explained, "but westward I go free."[41]

From his cabin on Walden Pond, the west led Thoreau away from Boston and into the open countryside. The freedom he felt in this direction was a social freedom—that is, a freedom from the social. Thoreau exalted,

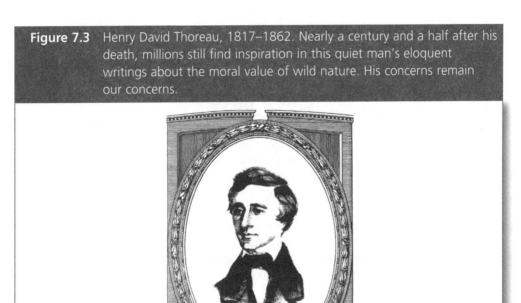

Figure 7.3 Henry David Thoreau, 1817–1862. Nearly a century and a half after his death, millions still find inspiration in this quiet man's eloquent writings about the moral value of wild nature. His concerns remain our concerns.

Man and his affairs, church and state and school, trade and commerce, and manufactures and agriculture, even politics, most alarming of them all—I am pleased to see how little space they occupy on the landscape. In one half-hour I can walk off to some portion of the earth's surface where a man does not stand from one year's end to another, and there, consequently, politics are not, for they are but as the cigar-smoke of a man.[42]

Instead of the cigar smoke of politics, to the west, Thoreau found the Wild. As he put it in the most famous line he ever wrote,

The west of which I speak is but another name for the Wild; and what I have been preparing to say is, that in Wildness is the preservation of the World.[43]

We can hear in Thoreau both a critique of power and interests and an alternative place to stand (or rather, walk). What alarms Thoreau about society and the east is the play of human politics—the interest-laden "affairs" of church, state, school, trade, commerce, manufactures, and agriculture. But to the west, Thoreau found release from power and self-interests. The vital serenity of the wild was, for Thoreau, the serenity of a realm without social conflict, for in the wild was no society with which to conflict. It was the serenity of what I have called in some of my research the *natural other*—a vision of an interest-free realm upon which to base a natural conscience.[44]

Like the Taoists and the Sophists, Thoreau carried his critique of material desire and self-interest into a suspicion of knowledge itself, especially the useful and arrogant knowledge of the technological ethos. Like the Taoists (and some Sophists, including the most famous Sophist of all: Socrates[45]), he praised ignorance and the recognition of how little we know:

> We have heard of a Society for the Diffusion of Useful Knowledge. It is said that knowledge is power; and the like. Methinks there is equal need of a Society for the Diffusion of Useful Ignorance, what we will call Beautiful Knowledge, a knowledge useful in a higher sense: for what is most of our boasted so-called knowledge but a conceit that we know something, which robs us of the advantage of our actual ignorance? . . . Which is the best man to deal with, he who knows nothing about a subject, and, what is extremely rare, knows that he knows nothing, or he who really knows something about it, but thinks that he knows all?[46]

It is the power behind knowledge that worried Thoreau the most, I believe. He saw in knowledge no authentic route to truth and self. That route was to be found in going west and opening oneself up to the wild, but without deliberate effort, again echoing Taoism. In Thoreau's words, "I believe that there is a subtile [sic] magnetism in Nature, which, if we unconsciously yield to it, will direct us aright."[47] This sounds to me like *wu wei*.

Through this unconscious yielding to the natural other of the wild, Thoreau counseled, we can find our real selves, and even our real names, as opposed to the "cheap and meaningless" names we receive from society. "It may be given to a savage who retains in secret his own wild title earned in the woods," thought Thoreau. "We have a wild savage in us, and a savage name is perchance somewhere recorded as ours."[48] In that wild savage within, and in that savage name, Thoreau found a *natural me*—the imagination of a truer, more authentic self that we believe the natural other sees in us, as opposed to the me that society sees.

The sense of a "natural me" is a social psychological invention, of course. Authenticity and truth are, in the end, matters of personal conviction, not eternal and external points of reference that we all agree upon. So, too, for the authenticity and truth of the natural me: It is something we formulate, not something we are given. But the human and social origin of the natural me does not make it any less significant for the human and the social. Indeed, the problem of the necessary invention of that which must be believed to be non-invented makes the achievement of a convincingly non-invented invention all the more compelling. Confusing? Contradictory? Paradoxical? Yes. But we must experience it otherwise for the natural conscience, with its natural other and natural me, to work morally and ideologically.

In any event, Thoreau's formulation of the natural conscience was very influential and is still a touchstone for environmental thought today. Thoreauvian *wu wei* can be found in the voluntary simplicity movement in all its many manifestations. His natural other can be found in the wilderness preservation movement and in the restful, yet exhilarating joy that so many find walking in their own Wild Wests, imagining freedom from social conflict, social constraint, and social power. His critique of materialist desire can be found in the environmentalist's condemnation of waste, pollution, corporate irresponsibility, state inaction, and the greed seen to underlie it all. Thoreau's natural me can be found in the environmentalists' self-conception that they are motivated by the interests of something outside themselves, and perhaps their self-conception that they are less artificial people.

If it is not Thoreau's own *wu wei*, natural other, natural me, and materialist critique that inform contemporary environmental concerns, then it is something of close resemblance, whether arrived at through direct moral descent from Thoreau or through moral convergence with Thoreau. A number of years ago, while doing research in England for a book about attitudes toward nature and the countryside, I had a memorably Thoreauvian conversation with a thoughtful, committed young man, very concerned about the environment. Nigel, as I'll call him, was also very concerned about the situation in Northern Ireland, and one evening we discussed the complexities and contradictions of that bitterly contested land. We found that we agreed that fault lay on both sides. There was a pause in the conversation, as we both let that thought sink in. And then Nigel said, "That's why I think I'm interested in the environment. You know what's right. It's clear where one should be standing. It's never that way with politics."[49]

I have no idea if Nigel had read Thoreau. Most of the 96 percent of respondents from that 1990–1993 world poll who supported the environmental movement almost certainly had not read Thoreau. Nor had they likely read the many environmental ethicists who followed Thoreau—such as David Abram, Mahatma Gandhi, Aldo Leopold, Carolyn Merchant, Mary Midgley, Lewis Mumford, Arne Naess, John Muir, Val Plumwood, Henry Salt, and Rudolf Steiner—let alone those who preceded him, like Horace, the Sophists, Aristotle, and Lao Tzu. They don't have to have read these classic works. The ideas they represent are part of the cultural waters in which we all swim.

Which doesn't mean, however, that we are all cultural clones of Thoreau. We reinvent the natural conscience as much as we borrow it from our cultural history. We do this because such an idea continues to help us contend with the ideological conditions of our lives. Thoreau is admired today because we see our ideas in him as much as (if not more than) we experience his ideas in us. The modern natural conscience is thus as much a new invention as it is a reinvention—as much a new discovery as it is a rediscovery.

The Extent of Contemporary Environmental Concern

The natural conscience is not all there is to environmental concern and its expansion into a broad popular movement in the last half of the twentieth century, however. In this section and the next, I argue that perceived environmental decline is an important factor in its own right. That is, there is an important material basis to the ideology of environmental concern.

This point may seem self-evident, but a number of scholars and critics of environmentalism have raised two potential counterarguments: first, that environmentalism is a passing flavor-of-the-month kind of issue and that people will rapidly lose interest, and second, that environmentalism is mainly an elite issue. Either counterargument, if true, would lessen the significance of the material origins of environmental concern.

The Persistence of Environmental Concern

In support of the flavor-of-the-month argument, it is important to note that the intensity of environmental concern varies considerably from person to person. This is something we all know from our personal interactions with others, and it also expresses itself in survey results.

Recall again the first wave of the World Values Survey, in which 96 percent of respondents from 43 countries expressed their approval of the environmental movement. That 96 percent figure needs to be interpreted with care. For example, in that same poll, only 65 percent indicated that they were willing to pay higher taxes to prevent environmental pollution. This still is a significant percentage, but the idea of having to pay to prevent pollution did dampen the enthusiasm of many. And when presented with the statement, "The government has to reduce pollution but it should not cost me any money," 55 percent of the sample agreed—meaning that only 45 percent were willing to pay to prevent pollution. Thus, the intensity of world support for environmental concerns dropped from 96 to 65 to 45 percent, depending upon how the pollsters phrased the question.[50]

Moreover, public concern for most issues follows a pattern that the political scientist Anthony Downs once called the "issue-attention cycle."[51] The vogue term for it these days is "compassion fatigue." Through the combined effects of boredom, the media's restless search for the new and novel, a realization that relieving the problem would entail significant costs, and maybe a sense that government must now be taking care of things, public interest in most issues tends to wane over time. Perhaps environmental concern, too, will drop.

And perhaps not. A 2007 survey of 45,000 people in 47 nations by the Pew Research Center found that environmental concern is rising worldwide. Among the 35 countries that Pew also surveyed in 2002, environmental concern had fallen in 3, stayed the same in 2, and risen in 30. The people of 14 of these countries rated environmental problems as one of the top two threats facing the world today. In a 2009 Pew survey of 25 countries, a majority in 23 of them agreed with the statement that we should "protect the environment even if it slows growth and costs jobs."[52] Not surprisingly, there are differences across the world, with concern lower in Africa and the Mideast and higher in the Americas, Europe, and Asia.[53]

Among industrial countries, the Pew polls recorded the highest levels of concern in Germany and the Scandinavian countries; moderate levels of concern in Britain, France, Spain, and Italy; and comparatively low levels of concern in Russia and the United States. But it also depends on how polls phrase their questions. In a 2007 Gallup poll of the G8 nations, Russians indicated the lowest level of satisfaction (and thus the highest concern) when asked, "in your country, are you satisfied or dissatisfied with efforts to preserve the environment?" Just 15 percent of Russians indicated satisfaction.[54] A 2006 Roper Center survey found that 83 percent of Americans completely or mostly agreed that "there need to be stricter laws and regulations to protect the environment."

Environmentalism has had both ups and downs, though (see Figure 7.4). Take the United States. After a dramatic upswing in the late 1960s, American environmental concern peaked shortly after the first Earth Day in 1970 and declined throughout the 1970s. As early as 1972, Anthony Downs was already predicting that interest in environmental problems would soon follow the issue-attention cycle into the shadows.[55] For a decade, it looked like he might be right. However, the 1980s brought a resurgence of concern, reaching the unprecedented levels recorded by surveys in the early 1990s. In 1990, an estimated 75 percent of Americans said the country isn't spending enough on "improving and protecting the environment," 23 percent higher than in 1980.[56] And then environmental concern underwent another steady decline, bottoming out in the early 2000s, just after the attacks of September 11, 2001, when the nation's focus shifted strongly to worries about terrorism. In 2002, the percentage who felt more needs to be spent on the environment had fallen to 62 percent, but by 2006, it had recovered to 69 percent, in line with other surveys showing a resurgent concern with the environment.[57] By 2007, a Yale University national poll even found that 63 percent agree the United States "is in as much danger from environmental hazards such as air pollution and global warming as it is from terrorists."[58] More recent polls using other measures show that Americans' environmental concern dropped once more at the onset of the Great Recession in 2008, but bounced back again after the Gulf Oil Spill in April of 2010, although it may have fallen again since.[59] Other countries where we have good polling data, such as Britain, show similar ups and downs in concern, but with much higher levels overall than before the 1960s.

Environmentalism has shown itself over 50 years to be more than a passing fad. Concern rose in the 1960s and the 1980s and 2000s, and fell during the 1970s and 1990s. We now appear to be in period of slackening concern in many countries, due to the Great Recession. Nevertheless, environmentalism seems likely to remain a persistent topic of worldwide public discussion for the foreseeable future. Scholars now generally agree that environmental concern is widespread and long lasting, although variable in intensity over time and from place to place and person to person.

Figure 7.4   The persistence of environmental concern in the United States, 1981–2007. Although there have been ups and downs, responses to the *New York Times*/CBS national polls have remained strong to the following statement: "Protecting the environment is so important that requirements and standards cannot be too high, and continuing environmental improvements must be made regardless of cost."

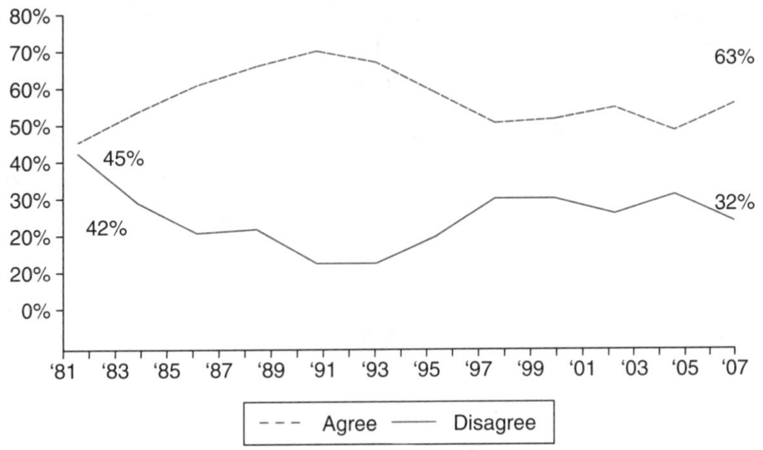

	Agree	Disagree
1981	45%	42%
1983	58	34
1986	66	27
1988	65	22
1989	74	18
1990	74	21
1992	67	29
1996	57	37
1997	57	36
2001	61	33
2002	56	39
2007	63	32

Source: Dunlap, Riley. (2002). "An Enduring Concern: Light Stays Green for Environmental Protection." *Public Perspective*, September/October, pp. 10–14; "Americans' Views on the Environment." (2007). *CBS News/New York Times Poll*, April 20–24, pp. 17.

Social Status and Environmental Concern

Now let's inspect the common suggestion that environmentalism is an elite movement of the liberal, the White, and the comfortably well-off.

Early public opinion polls in the West found that, in fact, these social factors had relatively little influence on the level of environmental concern. A 1979 British survey found that supporters of the environmental movement did tend to be drawn from middle-class service professions: teachers, social workers, doctors, and the like. But the strongest opponents of environmental views—business leaders—had even higher incomes on the whole. Moreover, many lower-income people and trade unionists showed strong support for environmentalism.[60]

A more comprehensive 1992 American study, drawing on 18 years of national surveys conducted between 1973 and 1990, showed similar results.[61] Income had only a minor effect on a person's level of environmental concern, even during recessions. People who were younger, liberal, members of the Democratic Party, well educated, urban, and did not work in primary industries like farming and forestry, did tend to be more supportive of environmentalism. But again, these effects were slight; plenty of older, conservative, Republican, less-educated, rural, and primary-industry–employed people professed environmental support.

Finally, a comparative study across 18 wealthy nations, based on a 1990 survey, found little influence of social status on levels of environmental concern. Level of wealth had no effect, and being younger and being female had very slight effects.[62]

But in recent years, social status has emerged as an important predictor of environmental concern, at least in the United States, and in completely the reverse direction of the old charge that environmentalism is an elite concern. People from privileged social groups now tend to have significantly less support for environmentalism than others do. In general, support for environmentalism is higher among women, people of color, and people with lower incomes. For example, in a 2000 poll of voting intentions in the presidential race, lower-income groups were far more likely to call protecting the environment "very important" in their voting decisions. Among those making less than $30,000 a year, 67 percent called environmental protection "very important," while just 40 percent of those making more than $75,000 a year did.[63]

Researchers have also recently noted what they term the "White male effect," in which White men tend to report considerably less concern with health risks, technological hazards, and environmental protection. In 1994 and 1996 national surveys, Whites showed significantly lower levels of environmental concern than other groups. Among Whites—and only among Whites— gender was also a significant factor, with men registering lower levels of concern than women.[64] A 2004 survey by a group led by Dan Kahan found that gender was also a significant factor among non-Whites and that four groups can be ranked by increasing environmental concern: White men, White women, non-White men, and non-White women (see Figure 7.5).[65]

Why? The evidence is at least coincident with the main conclusion of Chapter 6—that there is an ideological connection between social domination and environmental domination. As Linda Kalof and her colleagues note, "These anomalous attitudes of White men are likely the result of their historically privileged position regarding risk and power in society."[66] We may presume as much for the wealthy. Those with greater social power are less likely to suffer the consequences of environmental problems, and thus will be less concerned about them. Also, those with greater social power tend to have less egalitarian attitudes, and thus less concern about the consequences of environmental problems for others—both other humans and non-human environmental others.

Dan Kahan and his colleagues also find that the risk people perceive is not just material threat, but cultural threat as well. The anthropologist Mary Douglas argues that cultural values can be arranged on two broad axes: individualist to communitarian and egalitarian to hierarchical.[67] The Kahan study found that environmental concern rose as one moved from hierarchical individualists to hierarchical communitarians to egalitarian individualists, and, finally, to egalitarian communitarians (see Figure 7.6). Moreover, they found that White men were considerably more likely to be hierarchical individualists and that these men were the ones responsible for the overall "White male effect." White males with egalitarian and communitarian outlooks perceived risk similarly to other egalitarians and communitarians. Hierarchical individualists, argue Kahan and his colleagues, find that environmental issues challenge their cultural identity by implying that hierarchy and individualism are substantially at fault for environmental problems—as the sociological evidence in Chapter 6 suggests is indeed the case. In order to protect their identity, hierarchical individualists subconsciously downplay the significance of environmental concerns.[68]

As well, in recent years in the United States, environmental concern has become substantially more partisan, the environmental sociologist Riley Dunlap observes.[69] There is a partisan gap in other countries as well, but it is particularly strong in the United States—and quite recent. For example, 86 percent of Republicans in 1992 agreed that "there need to be stricter laws and regulations to protect the environment," and 93 percent of Democrats did—just a 7-point gap. By 2009, however, only 64 percent of Republicans agreed with this statement, while 94 percent of Democrats did—for a 30-point gap. (There was also a slight fall among independents over this time period, from 91 percent to 82 percent.)[70] The partisan gap is even wider with regard to climate change, Dunlap notes.[71] In 1998, an estimated 47 percent of Republicans felt the

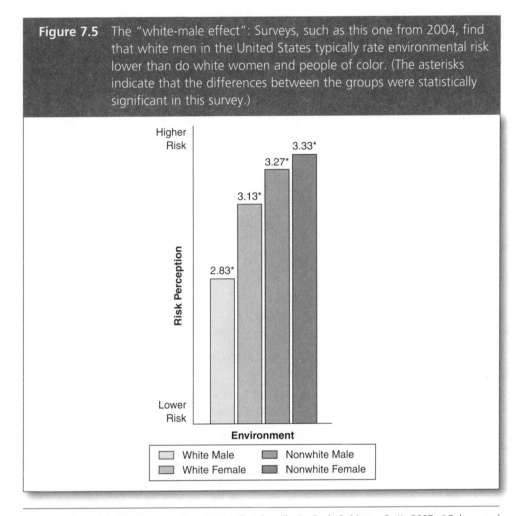

Figure 7.5 The "white-male effect": Surveys, such as this one from 2004, find that white men in the United States typically rate environmental risk lower than do white women and people of color. (The asterisks indicate that the differences between the groups were statistically significant in this survey.)

Source: Kahan, Dan M., Braman, Donald, Gastil, John, Slovic, Paul, & Mertz, C. K. 2007. "Culture and Identity- Protective Cognition: Explaining the White-Male Effect in Risk Perception." *Journal of Empirical Legal Studies* 4(3): 465–505.

effects of climate change had already begun, while 46 percent of Democrats did—which was a statistical tie. But by 2008, only 41 percent of Republicans agreed that effects had begun, while 76 percent of Democrats did—for a 35-point gap.

These are controversial findings and interpretations, and some may feel threatened or offended by them. So it is very important to remind ourselves that we are speaking here of tendencies and preponderances, not absolutes. Not all wealthy people, White people, male people, and conservative people are among the less environmentally concerned, and not all poor people, people of color, females, and liberals are among the more environmentally concerned. These are statistical matters and should not be allowed to become stereotypes.

But there may be some elitism in saying that environmentalism is an elite movement. It could be taken as implying that the poor are too ignorant or unsophisticated to comprehend the issues. It would also not be empirically accurate, and not only for the poor in a wealthy country like the United States. There is abundant evidence of considerable environmental concern among the

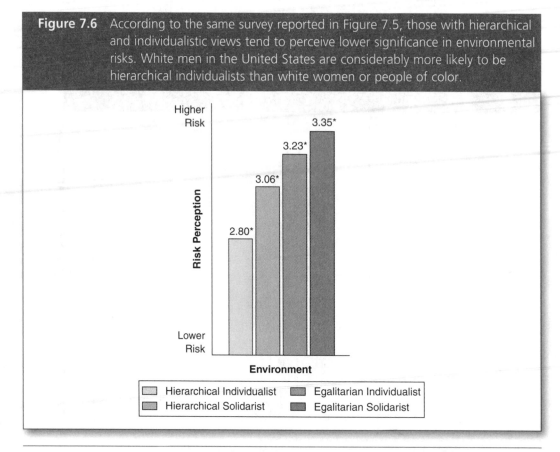

Figure 7.6   According to the same survey reported in Figure 7.5, those with hierarchical and individualistic views tend to perceive lower significance in environmental risks. White men in the United States are considerably more likely to be hierarchical individualists than white women or people of color.

Source: Kahan, Dan M., Braman, Donald, Gastil, John, Slovic, Paul, & Mertz, C. K. 2007. "Culture and Identity- Protective Cognition: Explaining the White-Male Effect in Risk Perception." *Journal of Empirical Legal Studies* 4(3):465–505.

poor elsewhere, as the Indian environmental sociologist Ramachandra Guha has forcefully pointed out. Consider the struggles of the Penan people of Malaysia fighting the logging of their forest home; the similar struggles of the Chipko ("hug the tree") movement of villagers in northern India; the hard-fought, and tragic, campaigns in Brazil and Nigeria that led to the murders of two of the world's most important environmental leaders, Chico Mendes and Ken Saro-Wiwa; the Amazonian native peoples of Ecuador protesting the despoiling of their forest home by fighting oil companies. Environmentalism has become a significant feature of political debate in the world's poorest countries, too.

The character of the "environmentalism of the poor," as Guha terms it, often differs significantly from the environmentalism of the rich. "First and foremost," Guha has written, "it combines a concern with the environment with an often more visible concern with social justice."[72] This heightened emphasis on environmental justice has also been central to the environmentalism of the poor in rich countries, as the thousands of community-level groups that have sprung up in the United States since the late 1980s demonstrate. These groups usually focus on the unequal distribution of environmental "bads," such as the location of toxic waste disposal sites and heavily polluting industrial facilities, blending environmental and social justice concerns, particularly over race and class, as Chapter 5 discussed.[73]

There is, then, little empirical support for the allegation of Lester Thurow and others over the years that, to quote Thurow, "poor countries and poor individuals simply aren't interested" in environmentalism.[74] In the early years of environmentalism, there may have been a grain of truth to this assertion, as access to education is of course higher for the wealthy. But the ideas of environmentalism have now spread well beyond the halls of the academy. It is no longer an elite concern, if it ever was. If anything, it is increasingly just the reverse.

Three Theories of Contemporary Environmental Concern

So how have these changes in environmental concern come about? Several theories suggest that the origin ultimately lies in changes in the material conditions of society—that increases in environmental concern began at a material moment in the ecological dialogue, moved to an ideal moment as understandings changed, and then has in turn led to attempts to change those material conditions. I'll compare three of these theories. First, I'll take up the political scientist Ronald Inglehart's theory of *postmaterialism*. Inglehart's analysis moves beyond the elitism critique of environmentalism, while at the same time returning to it in a way. Second, I'll consider the *paradigm shift* theory of Riley Dunlap (the sociologist we have just heard from about the recent partisan gap in environmental concern), William Catton, and their colleagues. This theory traces the rise of broad concern with evidence of environmental decline that may, in time, lead to a fundamental shift in our ideologies. Third, I'll discuss the *ecological modernization* theory of many scholars, but in environmental sociology most especially Arthur Mol and Gert Spaargaren. Ecological modernization argues that we have recently begun to overcome the opposition of economy and ecology that has so long prevented us from embracing environmental concern.

Material changes, as we'll see, play a central role in all three of these theories. There is also a fourth major theory of environmental concern, the *risk society* thesis of Ulrich Beck, in which material issues figure prominently. Beck's work is now most often applied in analyses of risk, so I take it up in Chapter 9, which focuses on risk.

Postmaterialism

Since the early 1970s, Inglehart has been documenting a broad intergenerational cultural shift in wealthy countries across the world, what he terms a turn from "materialist" to "postmaterialist" values. The word *materialist* can be confusing. Inglehart does not have in mind here materialist in the philosophical sense of explanations that emphasize economy, technology, biology, and the physical, which is the way that I usually use the term in this book. Nor does he mean "greediness." Along with his colleague Paul Abramson, Inglehart argues that younger generations are shifting away from "concerns about economic and physical security" (material values, in Inglehart's terms) and toward "a greater emphasis on freedom, self-expression, and the quality of life" (postmaterial values).[75]

Inglehart takes a leaf from Karl Mannheim, a Hungarian sociologist from the first half of the twentieth century, and his theory of generations, which suggests that people's basic values are formed in early adulthood and tend to persist thereafter.[76] Older generations were socialized in times when worries about money, health, and natural threats were much more of a concern, Inglehart suggests. Drawing as well on Maslow's theory of the hierarchy of needs (which I discussed in Chapter 2), Inglehart argues that economic improvements have allowed younger generations to focus instead on issues of aesthetics and self-actualization, such as freedom and quality of life.[77] Increased concern for the environment, argues Inglehart, is one product of this "postmaterial" socialization among younger generations. Inglehart's basic point is that we're better off now—pretty much all of us, not just the elites—and can afford to worry about the environment.

Inglehart and his collaborators have assembled an impressive array of survey evidence to back up their claim that a shift to postmaterial values is underway in countries across the globe. As older generations die off, Inglehart argues, the overall proportion of postmaterialists will increase, with far-reaching implications for politics and lifestyle choices. Inglehart has been conducting periodic surveys over the past 30 years to determine whether the proportion of postmaterialists is actually on the rise. (The World Values Survey that I discussed previously is led by Inglehart in part to test the postmaterialism hypothesis.)

In these surveys, Inglehart categorizes people as *postmaterialist, materialist,* or *mixed* by asking them which two of the following four choices should be their country's top goals:

1. Maintaining order in the nation

2. Giving people more say in important government decisions

3. Fighting rising prices

4. Protecting freedom of speech

Goals 1 and 3 are the materialist responses. Goals 2 and 4 are the postmaterialist ones. Any other combination is scored as mixed.[78]

Inglehart finds that a substantial shift is indeed underway. For example, in 1972, in the United States, 9 percent of those surveyed gave postmaterialist responses, and 35 percent registered as materialist. But by 1992, postmaterialists were up to 18 percent, and the materialists were down to 16 percent. The proportion of mixed had also gone up, from 55 to 65 percent, also indicating a shift in a postmaterialist direction.[79] Inglehart has documented similar changes in Britain, France, West Germany, Italy, the Netherlands, Denmark, and Ireland, as well as a slight postmaterialist shift in Belgium.[80]

Inglehart has also demonstrated a correlation in these surveys between postmaterialist values and support for environmentalism. For example, in one 1986 survey conducted in 12 European countries, 53 percent of postmaterialists strongly approved of the environmental movement, but only 37 percent of materialists did.[81] In that same survey, Inglehart notes, countries with more postmaterialists were also those that had taken more actions to protect the environment, indicating that postmaterialism also translates into environmental action. This is compelling evidence, and many researchers have applied this analysis to understanding the rise of the environmental movement in countries across North America and Europe.[82]

Questioning Postmaterialism

As compelling as this evidence for a change toward postmaterialist values may be, however, there are complications to Inglehart's interpretation of its link to rising environmental concern.

To begin with, postmaterialism cannot account for the environmentalism of the poor, as Guha, Steven Brechin, Willett Kempton, and other critics have pointed out.[83] The bulk of people in poor countries cannot be said to have postmaterialist values, and yet environmentalism is an increasingly central political issue in these countries—and in part for strong materialist reasons: Threats to the environment threaten human well-being. Inglehart himself has recently noted the existence of environmentalism in poor countries and has suggested that its origin must lie in materialist concerns, rather than postmaterialist ones.[84] But we also need to be careful not to presume that the poor are incapable of anything other than materialist concerns. For example, some of the broadest visions of the environment are the holistic traditions of many native peoples.

There are also some complications to a postmaterialist explanation for environmentalism in the rich countries. First, there is Inglehart's implication that greater wealth should correlate with declining concern for economic and security threats, and thus higher concern for the environment. And yet, as we have seen, the wealthy in the United States, at least, increasingly show lower levels of environmental concern than do people with lower incomes. Second, when you consider it carefully, it is evident that Inglehart's explanation rests on the idea that environmental issues are not real material concerns for the West. Environmentalism seems to represent for Inglehart the idyllic musings of a gilded society—children's passions for stuffed animals, for example, or Ansel Adams's photographs of nature's beauty. Many environmentalists, however, would no doubt respond that ecological threats are far from being issues of postmaterialism—of Maslovian aesthetic self-actualization. Ecological threats are material threats. Not that there isn't an important aesthetic (and moral, as I previously argued) dimension to environmentalism. But there are also important material considerations involved—considerations that concern both the rich and the poor. In other words, Inglehart may be falsely repeating in more sophisticated terms the old elitism charge against environmentalism.

Nonetheless, Inglehart's evidence of a link between environmental concern and people's response to his four choices for top goals cannot be discounted. The question is, how do we interpret this finding?—an ever-present question in all research, and one that I will return to at the end of the chapter.

Paradigm Shift

Other theorists agree that a broad, slow change is underway in the dominant outlook of the inhabitants of industrial countries, but reject the implicit charge of elitism in Inglehart. One such tradition is the *paradigm shift* thesis—an approach that the prominent environmental sociologist William Freudenburg recently called a "landmark" in the field.[85] Rather than seeing environmentalism as an affectation of the comfortable, this theory suggests that in response to discrepancies between evidence of environmental threats and ideologies that do not consider environmental implications, people are slowly but steadily adopting a more environmentally aware view of the world. People are becoming more aware of the real material effects that industrial life has on the environment, and their ideologies are beginning to change to match this new understanding. The central feature of the emerging new paradigm is that we no longer see humans as exempt from environmental implications. We are coming to see ourselves as connected to the environment, not separate from it. It's a deceptively simple theory, originally proposed by the sociologists Riley Dunlap, William Catton, and Kent Van Liere in the late 1970s and by Stephen Cotgrove and Lester Milbrath in the early 1980s.[86]

I say "deceptively simple" in part because measuring environmental awareness turns out to be quite difficult. Paradigm shift researchers have generally approached the question by distinguishing between two paradigms, one old and one new. Under the old paradigm, humans are exceptional creatures who are able to overcome environmental limits, and the basic goal of human society is technological mastery over nature for the purpose of wealth creation. Researchers have variously termed this view the "human exemptionalism paradigm," the "dominant social paradigm," and the "technological social paradigm." Under the new paradigm, humans are a part of nature and need to maintain a sense of balance and to live within limits in an interconnected world. Researchers have variously termed this latter view the "new environmental paradigm," the "alternative environmental paradigm," and the "ecological social paradigm."[87] The most common terms are the *human exemptionalism paradigm* and *new environmental paradigm*, usually referred to in shorthand as the HEP and the NEP. Based on this contrast, paradigm shift researchers have devised batteries of survey questions to assess how strongly a person adheres to one paradigm or the other.

The next measurement complexity is assessing why people's ideas might change from one paradigm to another. In a 1982 survey of the state of Washington, a research team from Washington State University approached this issue by distinguishing between *environmental beliefs* and *environmental values*—between how a person thinks the environment is and how she or he thinks the environment ought to be. For example, the researchers assessed environmental beliefs by asking people how much they agreed with a battery of statements such as, "The earth is like a spaceship, with limited room and resources." They appraised environmental values with statements like, "People should adapt to the environment whenever possible."

It might be expected that beliefs and values would correspond fairly closely, but the Washington State survey found considerably stronger support for environmental beliefs than environmental values. Although 78 percent of Washington residents expressed some degree of an "ecological social paradigm," 57 percent of the respondents were what the researchers called "strong believers," and only 25 percent were "strong valuers."[88] The Washington State researchers argue that what is at work is the slow process of people's values catching up with their beliefs about external material realities—such as rising levels of pollution, the disappearing wild, and technological failures like Chernobyl and Bhopal.

Similar to Thomas Kuhn's theory of scientific revolutions, the Washington State team's conclusions suggest that we tend initially to hold onto our paradigms even in the face of contradictory evidence. But eventually, we seek to bring the two—paradigms and evidence of the external world—into better correspondence, leading to paradigm change. As the Washington State team wrote, "Our analysis suggests that shifts in social paradigms are influenced primarily by external discrepancies."[89]

Questioning Paradigm Shift

But we must be cautious in interpreting these survey results, too. To begin with, there is the obvious problem of reducing such a complex matter as environmental ideology to only two categories. Such reduction may obscure ideological complexities more than it clarifies them.[90] A lot more may be going on. Dunlap and his colleagues later devised a new version of the NEP scale, however, which helps researchers tap into a wider range of facets of environmental concern.[91]

A closely related problem is deciding which ideology represents the environmental ideal. There are likely as many environmentalisms as there are people. Whose environmentalism, then, does one choose as the standard? Paradigm shift researchers have tried to draw this standard from the works of leading environmental writers.[92] But writers do not represent everyone (or there would be no need to conduct a survey), nor do they always agree (or they probably wouldn't bother to write). Inevitably, researchers have had to draw on their own understandings of environmental writers and environmentalism. There is thus a danger that paradigm shift researchers are in part assessing the degree to which the rest of the world agrees with them about what environmentalism is.

Survey-based research also doesn't give respondents a chance to explain why they answered the survey questions in the ways that they did. Survey researchers have to presume ahead of time the kind of phrases and questions that might reflect the way people see things. But there is no opportunity on a survey to determine if a person interprets a question differently from the way the researchers intend. For example, I might reject the notion that "the earth is like a spaceship, with limited room and resources" because I don't like the mechanical and techno-logical image of a "spaceship." I might see the Earth more as an organism. In other words, I might disagree with the statement even though I agree with the belief the statement is meant to assess.

Finally, there is the difficulty of assessing long-term ideological change with surveys of current public opinion. Paradigm shift researchers have on only a couple of occasions been able to resurvey the same population at a later date, and have found only modest and somewhat inconsistent shifts when they have done so.[93] The kind of ideological change that paradigm shift researchers hope to evaluate may be taking place over too long a time for surveys to document. The problem is, you can't go back and administer a question-naire to the people of the past. Thus, survey research needs to be balanced with historical research.

These critiques do not invalidate paradigm shift research, however. Nor do they invalidate Inglehart's postmaterialism research, to which all of these critiques mentioned equally apply, except for the last about historical evidence. (Inglehart has amassed some impressive historical information by separating out age cohorts.) Measuring and understanding public opinion is an inherently difficult task. All survey-based research faces these measurement problems. Without some simplifying assumptions, the question of ideological change probably could not be researched. And even a grainy image of the overall state of the public mind is useful and impor-tant to have.

Moreover, it seems hard to deny that material factors must have ideological consequence. If our patterns of thought, our mental reflexes, had no bearing on our material conditions, we would likely not last long. Or, to put it another way, if you keep stubbing your toe when you kick the environment, chances are you will eventually stop to reconsider why you were kicking it in the first place. And maybe that is what we are finally starting to do.

Ecological Modernization

Or so *ecological modernization* theorists would argue, but from a somewhat different stand-point. Based in part on the earlier work of Ulrich Beck (of whom we will hear much more in Chapter 9) and Joseph Huber of Germany, among others, the Dutch environmental soci-ologists Arthur Mol and Gert Spaargaren and the American environmental sociologists Maurie Cohen and David Sonnenfeld contend that the recognition of environmental prob-lems is starting to reshape the institutions and everyday social practices of modernity in fundamental ways. "The basic premise," writes Mol, "is the centripetal movement of eco-logical interests, ideas and considerations . . . which results in the constant ecological restructuring of modern societies."[94] Material conditions (environmental problems) shape ideas (those interests, ideas, and considerations), which in turn reshape material conditions (the constant ecological restructuring). But this shaping and reshaping is not just a matter of individual ideologies of environmental concern, such as the theories of postmaterialism and paradigm shift discuss. This shaping and reshaping occurs as well at the level of our institutions and the social practices we find ourselves engaged in, whatever we may think about them individually.

Central to the process of ecological modernization is what Mol terms the "emancipation" of "ecological rationality."[95] What he has in mind here is that we increasingly consider more than economic, technological, political, and social reasons in making decisions about how to organize our lives. Ecological rationality has now come to have an independent force in social debates. We're on the road—maybe just the beginning of it, but on it nonetheless—to the huge, even radical, changes that we need to make to maintain what ecological modernization theorists call our "sustenance base." But the problem we have faced was never modernization itself, say ecological modernization theorists. The problem was "simple modernization" driven by economy, technology, politics, and society that did not take into account ecological rationality. Put ecology in, in a serious and far-reaching way, and we can put our institutions and our daily lives on the ecological path. Put ecology in, and we can repair the "design fault," as Mol and Spaargaren term it, of an economic and technological order based on the presumption that ecology is a free service we need not pay much attention to.[96] Put ecology in, and we can overcome the long-standing conflict of business and the environment.

At first glance, ecological modernization can seem overly cheerful and hopeful. But look at the facts, environmental modernization scholars say. "Capitalism is changing constantly," write Mol and his colleague Martin Jänicke, "and one of the main current triggers in this change is . . . environmental concerns."[97] Many industries have made significant progress in retooling their businesses, embracing what has come to be called *industrial ecology,* and seeing environmental issues as opportunities and as indications of inefficiencies in their operations. Most wealthy nations now have developed environmental laws and regulations to encourage their economies in recognizing the services we gain from the environment and in becoming more eco-efficient, often with substantial savings to the economy. For example, in 2003, the Bush administration of the United States released a report showing vast economic savings from environmental regulation. The study, by the Office of Management and Budget, found that the $23 to $26 billion spent on retrofitting power plants to meet clean air standards saved the economy $120 to $193 billion in money that didn't need to be spent on health problems and lost workdays. For every dollar spent, $5 to $7 was saved.[98] This recognition—particularly by an administration with a reputation for being unfriendly to environmental concerns—is an indication of the welcoming of ecological rationality into our basic social institutions.

Ecological modernization is also reshaping our daily practices. New standards are changing the machinery of our lives. Refrigerators, air conditioners, and plumbing systems are far more efficient on the whole than they used to be, using much less energy and water. People are experimenting with new ways of living that reduce their personal consumption of the ecological services we were ignoring. Biking, car sharing, recycling, precycling, composting, more efficient housing urban design, and more are all on the rise. Consumption and lifestyle choices matter a lot, say ecological modernists, and many citizens are working hard on this, with some success.

One of the ways that ecological modernizationists believe we are moving in a green direction is, controversially, through globalization. I say "controversially" because many environmentalists (and indeed many environmental sociologists) are quite critical of globalization, as they fear it can weaken environmental protection by giving priority to trade above all else. But as Mol argues, many of our greatest environmental problems are global in scope, and many of our greatest environmental successes have been through global treaties that address them.[99] Take the Montreal Protocol, which has done so much to reduce the production of CFCs (chlorofluorocarbons) worldwide and to protect the upper-atmosphere ozone layer. Or take ISO 14000, the environmental management guidelines of the International Organization for Standardization, which many businesses are increasingly trying to follow, particularly in Europe. Meeting ISO

14000 standards gives businesses confidence in the supplies and products they buy from each other, and thus greater confidence that consumers will trust the environmental practices behind their products. As of 2010, some 200,000 public and private sector organizations in 155 countries had implemented ISO 14000 standards.[100]

Which doesn't mean that ecological modernization is a sure thing—that we can sit back and let it happen, now that ecological rationality has been emancipated and is starting to globalize. It would be a serious misinterpretation of ecological modernization theory to assume that it will happen on its own, through some kind of automatic process. It will take, and is taking already, a lot of hard work by social movement organizations, by politicians, by government agencies, and by imaginative businesses, building collaborations between them. It will require *political modernization*, as ecological modernizationists call it—that is, forms of government that help bring diverse interests and ideas together in a cooperative, and yet at times critical, way. And it will take the active engagement of citizens making choices in their consumption that push the government and the economy along in their embrace of ecological rationality. No, ecological modernization is not an autonomous force. It is something we have to do, ecological modernizationists argue, in both senses of the phrase.

Questioning Ecological Modernization

Scholars have taken a great interest in ecological modernization theory in recent years and have subjected it to rather vociferous criticism.[101] The volume of criticism, though, does not necessarily mean that a theory is a poor one. Rather, it can mean that a theory gets enough right that others pay attention to it, even if they disagree with much of it. This is how scholarship moves along. Also, Mol and Spaargaren have been unusually forthcoming in responding to the critics and in accepting many of their points, so as to improve ecological modernization theory.

Perhaps the main objection can be summed up in one word: *modernization*. The theory's embrace of that word and what it stands for—the value of science, technology, industry, capitalism, modern forms of government, and modern value systems—seems to some to ask that environmentalists marry their enemies. Aren't science and technology's arrogance, industry's obliviousness, capitalism's treadmill and growth mania, government's globalization dreams, and modernism's universalistic values the roots of the environmental litany? Ecological modernization theory, say these critics, is at best accommodationist and at worst a rhetorical ruse to allow the current power structures in society to have their merry way, perhaps with a few minor reforms. It's a kind of Wonder Bread theory, some critics feel, that claims it builds stronger ecological societies 12 ways, step by step, day by day. Still creating pollution at your factory? Just tell everyone you're doing your ecologically modernist best.

Ecological modernizationists agree that they seek the solution to problems of science, technology, industry, and capitalism in science, technology, industry, and capitalism. But these institutions won't look the same anymore. "Ecological modernization theory puts forth a radical reform programme," writes Mol. "The institutions of modern society, such as the market, the state and science and technology, will be radically transformed in coping with the environmental crisis, although not beyond recognition."[102] Ecological modernization theory may be reformism, in other words, but it is radical reformism.

Or is it? Radical reformism seems to some a tremendously optimistic claim. As one critic put it, ecological modernization is "hobbled by an unflappable sense of technological optimism"— not to mention market optimism and governmental optimism.[103] Ecological modernization theorists now accept this point, at least to a degree. Scholars like Spaargaren and Cohen describe the challenge of ecological modernization theory as learning how "to navigate between the dark green romantic dismissal of modernity and the naïve endorsement of market-driven, liberal eco-technotopias."[104] And they agree that early work on the theory was indeed hobbled by a bit of that naïve endorsement—by what the ecological modernization theorist Martin Hajer has called a "techno-corporatist" vision. Hajer advocates instead a "reflexive" vision—that is, one that emphasizes democratic institutions of discussion and debate that allow societies to "reflect" on where they are going, rather than being blindly led along through technological and corporate sleepwalking.[105] You could call the early vision "weak" ecological modernization and Hajer's "strong" ecological modernization, or perhaps "thin" and "thick."[106] By embracing the necessity of democratic debate, ecological modernization theorists agree that we can't just wait for this stuff to happen. Building the institutions that enhance democratic debate is what ecological modernizationists mean by "political modernization."

Besides, ecological modernizationists ask, have you got a better idea—that is to say, a *realistic* better idea? If we have to wait for capitalism to fall before we get anywhere, we may have to wait a very long time. By then, the ecological mess could be insurmountable, as it indeed may already be. At least ecological modernization theory works from within the general situation we're likely to have for some time to come. And it claims that, while by no means certain, ecological modernization is at least possible.

Really? Some critics also argue that ecological modernization applies with any great success to only a few countries in Europe, most especially Germany, the Netherlands, and the Scandinavian countries. It has little to say about the United States, for example, where environmentalists complain of virtually no progress since the great (and, at the time, revolutionary) acts of the 1960s and 1970s were passed: the Wilderness Act (1964), the National Environmental Policy Act (1969), the Clean Air Act (1970), the Endangered Species Act (1973), the Resource Conservation and Recovery Act (1976), and the Clean Water Act (1977). There has been little other than rear-guard action ever since, trying to hold onto some of the gains as mainly corporate interests systematically chip away at them, as the American environmental sociologist Fred Buttel has argued.[107]

Ecological modernizationists respond that yes, there's a long way to go, and yes, ecological modernization is currently most developed in Western Europe. Yet with appropriate forms of globalization, it could spread much further, much more rapidly. The current configuration of the World Trade Organization is not a help. Nor is the North American Free Trade Agreement, the Central American Free Trade Agreement, and other free trade agreements that put trade ahead of labor and environmental concerns. Nor is a growing trend toward unilateralism. But treaties like the Montreal Protocol, Agenda 21, and the Kyoto Accord can make a huge difference. As of 2009, there were 464 international treaties that have an environmental impact.[108] Moreover, many international development organizations like the World Bank are adopting greener policies.

Another criticism is that ecological modernization has little to say about issues of environmental justice. Nor does it have much to say about the rights and beauty of habitat. Its focus is almost entirely on issues of sustainability. Ecological modernization theorists accept these points, though, and have incorporated them into the theory, especially in writings on environmental governance and international development.[109]

Finally, some critics worry that ecological modernization is normative—that it both describes a direction it thinks it is possible we could go and says that is the way we should go, to which ecological modernizationists have a simple and direct answer: So? Don't you want to make the world a better place?

The Democratic Basis of Contemporary Environmental Concern

But why do people increasingly see that an important way to make the world a better place is through improving environmental conditions? If a mind is not already attuned to the implications of what paradigm shift theory calls "external discrepancies," and if information about the existence of these discrepancies is not heard by that mind, ideological change is likely to be slow—even in the face of substantial material threats to the environment. In the late nineteenth century, as the Industrial Revolution was still gathering steam, relatively few people thought much about the way the new industries were polluting, so strong was the faith in technological progress. But a little over 100 years later, people are both much more attuned to environmental concerns and much more likely to hear about discrepancies between their values and reality. Both this attuning and this hearing have been given ears by the spread of democracy, its sensibilities, and its institutions. In other words, the origin of environmental concern lies not only in the material concerns explored by postmaterialism, paradigm shift, and ecological modernization theories. It lies equally in the other side of the ecological dialogue: in our ideas.

Democratic Sensibilities

"It is impossible," the historian Keith Thomas suggests, "to disentangle what the people of the past thought about plants and animals from what they thought about themselves."[110] A less hierarchical ordering of human relations is historically associated with a less hierarchical ordering of human environmental relations. I don't want to overstate the point; there are plenty of exceptions. But the rise of democratic sensibilities in the social realm does seem to be generally associated with the rise of parallel sensibilities in the environmental realm.

Consider the rise of environmental concern in the eighteenth and nineteenth centuries, what is often termed the "first wave" of environmentalism.[111] This is the period of the first national parks and the first environmental pressure groups; Henry David Thoreau and Ralph Waldo Emerson, Gilbert White and William Wordsworth; and so many more important figures in environmental thought and important "firsts" in environmental protection. Strikingly, this period also saw the rise of new ideas about how human society should be arranged. In 1690, John Locke published his *Two Treatises of Government*. In 1754, Jean-Jacques Rousseau published his *Discourse on the Origins of Inequality*. Inspired in part by the ideas in such books, the United States declared its revolutionary independence from Britain in 1776, and France began its democratic revolution in 1789. Country after country in the nineteenth century went on to make its own democratic reforms.

These new sensibilities concerning the rights of each human were accompanied by, or soon led to, similar concerns about what constituted decent treatment of the nonhuman aspects of creation. To a democratic mind, environmental exploitation just didn't feel right. Of course, the democratic mind had a long way to go (and still does). Slavery, colonization, racism, sexism, classism, and the rest had hardly disappeared, nor had the "environmental movement" as we understand it today begun. (Nor is the environmental movement always democratic, as the next chapter discusses.) But beginning in the eighteenth century and continuing throughout the nineteenth century, a broadening conception of human rights was accompanied by a broadening conception of environmental rights and a greater valuation of both.[112] It is likely no accident that the great early writers for environmental reform were often important figures in social reform as well, such as John Ruskin, William Morris, and Henry Salt, not to mention Thoreau.

Steadily throughout this period, democratic concern grew not only for the rights (and beauty) of habitat, but also for the equitable distribution of environmental goods and bads. This was also, of course, the beginning of the Industrial Age. But the material threats that industrialism posed to the environment were not by themselves enough to evince a concern for environmental justice. Democratic concern played a key role.

A voice of democratic concern for environmental justice is heard in this 1845 description of conditions in Manchester, England, the center of the young Industrial Revolution:

> Looking down from Ducie Bridge, the passer-by sees several ruined walls and heaps of debrís with some newer houses. The view from this bridge, mercifully concealed from mortals of small stature by a parapet as high as a man, is characteristic of the whole district. At the bottom flows, or rather stagnates, the Irk, a narrow, coal-black, foul-smelling stream, full of debrís and refuse, which it deposits on the shallower right bank. In dry weather, a long string of the most disgusting, blackish-green, slime pools are left standing on this bank, from the depths of which bubbles of miasmatic gas constantly arise and give forth a stench unendurable even on the bridge forty or fifty feet above the surface of the stream. Above the bridge are tanneries, bonemills, and gasworks, from which all drains and refuse find their way into the Irk, which receives further the contents of all the neighbouring sewers and privies. . . . Here each house is packed close behind its neighbour. . . . The whole side of the Irk is built in this way, a planless, knotted chaos of houses, more or less on the verge of uninhabitableness, whose unclean interiors fully correspond with their filthy external surroundings. And how could the people be clean with no proper opportunity for satisfying the most natural and ordinary wants? Privies are so rare here that they are either filled up every day, or are too remote for most inhabitants to use. How can people wash when they have only the dirty Irk water at hand? . . . Everything which here arouses horror and indignation is of recent origin, [and] belongs to the industrial epoch.[113]

That voice of democratic concern was a young Friedrich Engels.

A short while later, George Perkins Marsh, a scholar and diplomat from Vermont, began questioning the long-term sustainability of industrialism's rapacious appetite for resources. In his 1864 book, *Man and Nature, or, Physical Geography Modified by Human Action*, Marsh argued that humans had unknowingly become a geological force in their own right—clearing forests, altering climate and the flow of rivers, shifting the pace of erosion—often to the detriment of human interests. Vast areas of Michigan's forest, for example, had disappeared in a few short decades of unrestrained cutting; Marsh's own Vermont had suffered the same fate. The result was soil erosion and more severe seasonal floods, since the forests were no longer there to absorb rainwater. Marsh's main goal was "to point out the dangers of imprudence and the necessity of caution in all operations which, on a large scale, interfere with the spontaneous arrangements of the organic or the inorganic world." We must now ask "the great question," concluded Marsh: "Whether man is of nature or above her."[114] This, perhaps, was the most difficult democratic question of all.

The same association between democratic and environmental sensibilities may be behind survey results showing an association between "postmaterialist" values and environmental concern. As I argued earlier, there is good reason to question this link with environmental concern, despite the association surveys have found. The reason for this association may be that Inglehart's four-item materialism/postmaterialism scale measures *democratization*, not a shift toward postmaterialism. In fact, Inglehart deliberately chose the two postmaterialist items on his scale—"giving people more say in important government decisions" and "protecting freedom of speech"—to reflect democratic attitudes, as he regarded such attitudes to be one feature of postmaterialism.[115] But can we become democratic only by overcoming material concerns? Furthermore, a low ranking of "maintaining order in the nation" (one of his two "material" items in the scale) may reflect democratic concern about authoritarianism—and not lack of material worries. And given that inflation is no longer a major economic issue in most wealthy countries, "fighting rising prices" (the other of his two "material" items) may not be an accurate gauge of material worries, either.

In other words, we are left with three items in Inglehart's scale that potentially tap democratic sensibilities and one item that is probably irrelevant. Thus, the correlation Inglehart found between what he called "postmaterialism" and environmental action may actually show instead that democratic sensibilities continue to influence environmental ones.

Democratic Institutions

The institutions of democracy have been no less important to environmental concern than have democratic sensibilities (see Figure 7.7). For instance, the free press has played an absolutely central

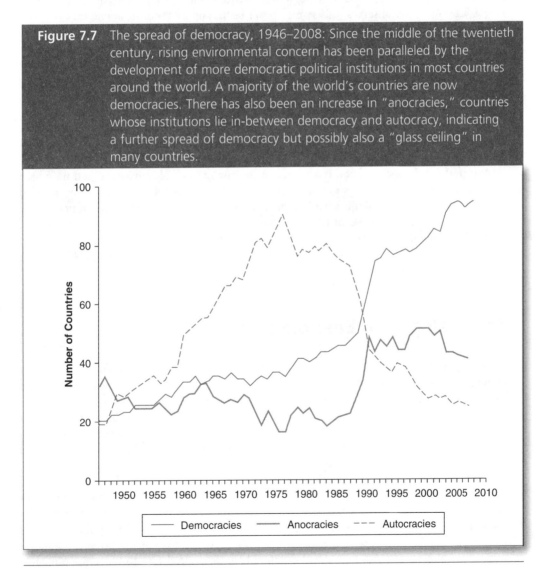

Figure 7.7 The spread of democracy, 1946–2008: Since the middle of the twentieth century, rising environmental concern has been paralleled by the development of more democratic political institutions in most countries around the world. A majority of the world's countries are now democracies. There has also been an increase in "anocracies," countries whose institutions lie in-between democracy and autocracy, indicating a further spread of democracy but possibly also a "glass ceiling" in many countries.

Source: Marshall, Monty G., & Benjamin R. Cole (2009). Copyright 2009 Center for Systematic Peace. Reprinted with permission.

role in alerting the public to environmental problems, in creating environmental "hearing." And without democratic rights to assemble, to protest, to vote, and to establish organizations and political parties that challenge the existing order, environmentalism would likely never have developed into a popular movement anywhere. No popular movement of any kind is likely without such abilities, as authoritarian governments have long recognized.

Consider how long it took for environmentalism to gain attention in the state socialist countries of the former Soviet Bloc. Environmental problems went so long unattended that these countries now face some of the world's worst local environmental conditions.[116] When Mikhail Gorbachev came to power in the Soviet Union in 1985, bearing his policy of glasnost, which allowed a degree of political expression unprecedented in the Soviet Bloc, its environmental movement suddenly came of age. Indeed, concern over environmental conditions was one of the major political forces behind the 1989 revolutions. The importance of this concern is reflected in the striking name of a Bulgarian environmental group that emerged in 1990: *Ecoglasnost*. As Guha observes, this name "bears testimony to the inseparable link between democracy and environmentalism."[117]

This "inseparable link" has also been an important factor in the development of the environmentalism of the poor countries of the Southern Hemisphere. Poor countries are often—indeed are usually—authoritarian ones. If environmental concern in poor countries has sometimes been hard to discern, in part this has been because the absence of democracy's freedoms has prevented the expression of such concern. Those with the courage of a Ken Saro-Wiwa are rare. As Guha has also observed, "It is no accident that one of the more robust green movements in the South is to be found in India, a democracy for all but two of its sixty years as an independent republic."[118]

And perhaps democratization is a source of the environmental movement, and not just a necessity for advancing it. The successes of the environmental movement may have as much to do with new sensibilities about what "we have to do" as it does the social recognition of the inefficiencies and hidden costs of taking ecological services for granted. The "design flaw" we are overcoming may be as much our social exemptionalism—and its concomitant constraints on public debate and on the accountability of governments and markets to the public—as it is our ecological exemptionalism.

Most likely, it is a good bit of both.

The Dialogue of Environmental Concern

Environmental concern is both new and old. Much of what makes this concern so constant are enduring moral issues, namely, social inequality and the challenges it poses to community. The desire to find a moral realm outside the reach of power and self-interest—the desire to find a natural conscience—represents another manifestation of the role social inequality plays in social–environmental relations, a role that is one of the principal themes of this book. Democracy, both in its sensibilities and its institutions, is another way that we try to deal with the challenges inequality poses to us.

Another enduring source of environmental concern is the constancy of material environmental threats. Industrialization and urbanization have brought with them an enormous range of benefits, of environmental goods. But the costs, the environmental bads, have not been inconsequential, and their distribution has been far from even-handed. Material threats have continually confronted us with the moral question of inequality—of the power of some to impose on others—and thus with the question of how we are to define and organize our communities, including that biggest community of all.

The material has not only shaped the moral, though. Our moral sensibilities concerning the importance (or lack of importance) of democracy and a natural conscience have shaped our ability to conceptualize environmental problems as problems—yet another dialogue between the material and the ideal. The democratic mind is an environmental mind, and vice versa.

The modern environmental movement can be described as a dialogue that connects the natural conscience, the condition of democracy, and the material state of the environment. Because of this dialogue, we have come to see the environment as morally attractive, morally compelling, and morally threatened. Raising our concerns in so many ways, the environment has become one of the leading issues of this or any day.

Postscript

My mother told me another story a few years ago. (She's a great storyteller.) This one beautifully illustrates the dialogic connection between the material and the ideal, and the way that the untroubled material practices of one day can seem quite problematic under the ideological sun of a different day. I have to retell it.

My family is fortunate enough to have a summer place in the Thousand Islands section of the St. Lawrence River, along the New York and Ontario border. I've been up there every summer of my life, and my mother has been up there every summer of hers. My grandmother was born there, and many of my great-grandmothers before that. (My grandmother's forebears were among the region's first White settlers.) The river is 4 to 5 miles wide in our area; more than 250 feet deep in places; sparklingly clear; and strewn with tree-green, granite islands. The "garden of the Great Spirit," the native people call it, and rightly so. It's one of the world's most beautiful places, at least to me.

But as much fun as the boating and swimming and hanging out with family and friends may be, island life is not without its hassles. One of the main ones is bringing groceries from the mainland and disposing of any garbage afterward. Everything has to be loaded from car, to boat, to dock, to cottage on the inbound trip, and from cottage, to dock, to boat, to car on the outbound trip. It gets a bit old. These days we haul off whatever we don't consume and can't compost, but in my mother's youth the approach was, shall we say, more environmentally casual. What wasn't consumed and couldn't be composted was taken out into the center of the river and sunk. A weekly task was loading up an old rowboat with empty bottles, cans, and other detritus, rowing out to the middle of the channel, and chucking it all in. Rather horrifying to think about today—especially considering that at the time most cottages drew their water directly from the river, instead of from the wells that most families have by now installed.

My mother reminded me of this old practice when, one morning, I found myself groaning over a garbage haul to the mainland. "Better than sinking it all out in the channel, like we used to do when I was a kid," she told me.

"Yeah, I guess so," I said, setting my last load into the boat. "Actually, definitely so," I continued as my mood improved with the loading done. A thought struck my mind: "But hey, Mom, didn't you ever think about the pollution you were causing, dumping straight into the river? I mean, the cottage water used to come straight from the river then." How could they not realize that they were sullying their own drinking water, I was wondering. Let alone the broader ecological effects of the dumping.

She thought a minute. "I do remember wondering about it," she replied. "Yes, in fact I remember asking my mother one time when she and I were out in the rowboat, sinking used cans and bottles. I must have been 10 or so. It must have been about the summer of 1940. And she said—I can remember this clear as a bell—'Don't worry about it. The river's bigger than all of us.'"

The river's bigger than all of us. I was astounded. "Do you think she really believed that," I asked, "or do you think she was making up a story to shut up an uncomfortable question from her own kid?"

"I think she really believed it. We all did back then. It was just part of the culture. We did lots of things you wouldn't dream of doing today. We just couldn't imagine that people could have such an effect on the environment. That things were so connected. That it just wasn't right. It was a different time."

Yes, it was. But not so different that a 10-year-old girl did not pause to question its ways.

CHAPTER 8

The Human Nature of Nature

When I hear that nature is a ruthless competitive struggle I remember the butterfly, and when I hear that it is a system of ultimate mutual advantage I remember the cyclone.

—Raymond Williams, 1972

My family's favorite brand of ice cream claims to contain something it calls "natural vanilla flavor." I never really thought about it much until my son Sam, who was then about 8 years old, asked me what was natural about it.

"Natural products are things people have done less to—you know, less manufacturing and processing," I told him. "Artificial vanilla flavor comes from a chemical made in a factory, and natural vanilla flavor comes from ground-up beans from vanilla plants," I said, pointing out the dark flecks in the ice cream. (I'm not even sure that's true, but that is what I said.)

Then I got to thinking that a heck of a lot of processing must go on in making "natural" vanilla flavor. Processors probably dry the beans in one machine, grind them up in another, and package them with yet a third. To grow the beans, farmers likely use a crop variety that scientists have specially developed for the purpose. No doubt they grow them in a field cleared from a tropical forest with machines, kept clear of weeds and insects with machines, and harvested with machines—not to mention the generous use of farm chemicals at several points along the way. Suddenly those beans didn't seem so "natural" anymore.

I tried to explain all this to Sam, which was probably a mistake. Soon I found myself in a major philosophical wrangle with my own 8-year-old kid. "Everything's natural," Sam said, defending the ice cream we were both licking up.

Given that anything that exists must be possible, and given that nature is what makes everything possible, Sam's defense was not unreasonable. But there are a number of difficulties with such an all-encompassing view, and I tried to describe them.

"Then maybe nothing's natural," he suggested before I got very far. That's not exactly what most environmental sociologists think either, but before I could explain why, a light bulb turned on somewhere in Sam's brain.

"Dad, I know. Everything that comes from the Earth is natural, and everything else isn't." He paused for a moment to work this thought through. "That means that solar power isn't natural because it comes from the sun and not the Earth."

Such are the hazards of having an environmental sociologist for a father.

But Sam (who, I must hasten to add, no longer holds these views) was only trying to deal with some central philosophical issues that have beset environmentalism from its earliest beginnings in the ancient world through to the current day. They are central sociological issues, too, for our philosophical resolutions have laid the moral foundations of environmental concern. In this chapter, we explore the sociological origins and implications of the resolutions we have sought to Sam's dilemma: What is nature?

Whatever else it may be, nature is without a doubt one of the most powerful of human concepts. Some of the most noble and selfless things that people have ever done have been in the name of nature. Consider the Greenpeace activists who have gone out on the open seas in small rubber boats to put themselves between dolphins and a careless tuna industry. Or consider the anti-roads protesters in England who barricaded themselves in hand-dug tunnels underneath proposed highways in order to prevent the movement of heavy construction machinery. People have often put their lives on the line—and sometimes over the line, with fatal results—in defense of nature. They have sought to defend not only nature in the environmental sense, but also nature as a democratic ideal. The notion of the "natural" rights of every person to a healthy, happy, fulfilling life in freedom has moved nations, governments, and their citizens to revolutionary acts of kindness and goodness.

But we also have to recognize that nature is one of the most dangerous of human concepts. The cultural history of nature is far from uniformly noble. For example, ideas of natural differences among humans—differences of sex, of race, of ethnicity, of talent and skill—have led to the enslavement of millions, the outright slaughter of further millions, and the oppression and mistreatment of millions more. We ignore this other side of nature at our peril.

Considering the difficult cultural history of nature, we would do well to exercise every critical caution in evaluating the uses to which we put the idea. To say that something is natural or to counsel that the appropriate course of action is to follow nature is to suggest something of immense significance for our moral values, for our identities, for our everyday lives, and thus for our politics. The previous chapter introduced the concept of the *natural conscience,* our common search for a realm that lies beyond the taint of social power and human interests and that might serve as an unbiased and external source of moral value and identity. But is such a realm possible? Can we truly escape the influence of our interests? This chapter takes a critical look at the relationship between ideas of nature and human interests and at the implications of this relationship for the moral and political claims we make.

These are difficult issues that challenge us all. But since these issues are central to how we constitute our human community and our ecological community, it is crucial that we accept the challenge.

The Contradictions of Nature

"Can and ought we to follow nature?" the environmental philosopher Holmes Rolston once asked.[1] Answering this question depends upon resolving a basic contradiction in the idea of nature, the contradiction between what can be termed *moral holism* and *moral separatism.*[2] The contradiction emerges because to ask whether we can and ought to follow nature immediately depends upon answering another question: Are people part of nature or separate from it?

Think of the contradiction this way: One of the truisms of the contemporary environmental movement is that people are part of nature and that environmental problems have emerged because we have tried to pretend otherwise. This is a holist moral point of view. But if we are a part of nature, there is no need to advise us to follow it, for we must already be doing so. How could we do anything but follow it? Under moral holism, "follow nature" is superfluous advice.

If people are separate from nature, the moral situation is no better. Since nature is different from us and is moved by different principles, it has no meaning for us. In fact, we probably can't follow nature, no matter how hard we try. Under moral separatism, then, "follow nature" is irrelevant advice.

As the sociologist Kai Erikson once observed, a moral value requires "a point of contrast which gives the norm some scope and dimension."[3] It has to give us a way to make moral distinctions. Otherwise, it will be a superfluous value. A moral value also requires a point of connection, however, if it is to be relevant to our actions. For nature to serve as a source of value that we can "follow," it has to be conceived in ways that provide points of both contrast and connection.

The contradiction between holism and separatism extends into the problem of power and interest, which is so central to the natural conscience. A natural conscience depends upon a person's subjective belief in a moral realm that lies beyond the reach of social power and the possibility of manipulation to suit the interests of some more than others. Such a realm requires separation from society if it is to be free from social influence, and it requires holist unity if it is to be relevant to social issues. But that unity immediately reintroduces the possibility that human interests may have been tinkering with what we conceive nature to be.

Any successful philosophy of nature must offer some kind of solution to these moral contradictions if it is to serve as a basis for a natural conscience.[4]

Ancient Problems, Ancient Solutions

The earliest philosophers of nature were well aware of the problems of holist superfluousness and separatist irrelevance. Aristotle, for example, worried about the "monism," as he termed it, in ideas like Plato's view that Goodness created the whole world and that, therefore, the whole world was good. Under such a philosophy, Aristotle wrote,

The being of the good and the being of the bad, of good and not good, will be the same, and the thesis under discussion will no longer be that all things are one, but that they are nothing at all.[5]

Aristotle suggested a way that we could distinguish the natural and the not-natural without falling into the problem of monist holism. That which comes into being "by nature," Aristotle argued, was that which had its "source of change" within itself.[6] Examples might be the movement of wind, the growth of plants and animals, and the dynamics of fire. Since humans have an internal source of change—our body's processes of growth and development—much that humans do is also according to nature. But something whose source of change is outside of itself is not due to nature, said Aristotle. So anything that humans make is not due to nature, although the materials humans use may well be. Aristotle gave the example of a wooden bed. Wood grows according to natural principles of growth internal to the tree, but the shape it takes as a bed is the result of an external source of change: humans. So wood is natural, but a bed is not.

This is a clever solution to the contradictions of nature. Humans are a part of nature because they have a source of change inside themselves. But when humans use that source of change to alter something outside themselves, they are not acting in accordance with nature. Aristotle thereby provided points of both moral connection and contrast.

Aristotle also argued that nature is ultimately the result of the workings of "the Good" (an idea he retained from Plato, as Chapter 7 describes). Appealing to this godlike ideal allowed Aristotle to claim that nature was interest-free—that nature had been created by something that was removed from human influence—thus allowing him to establish a natural conscience.

This was not a perfect solution, however. Maybe we are acting in accordance with our internal source of change when we do something like construct a bed or build a particular form of society or foster certain societal-environmental relations. Who is to say? We may do unnatural things for natural reasons. So are those actions truly unnatural? Aristotle's approach also could not distinguish beds made by humans from dams made by beavers and hives made by bees. A beaver dam must be as unnatural as anything humans make, for it equally has its source of change outside itself: in beavers. In fact, every action of every creature induces changes in the world around it and must therefore be unnatural, collapsing the point of moral contrast Aristotle hoped to establish.

The ancient Taoists offered a solution of their own. Recall from Chapter 7 the Taoist distinction between *wu wei*, acting without deliberate effort, and *yu wei*, acting deliberately. *Wu wei* guides one along the lines of Tao, and *yu wei* leads away from it, a clear moral contrast. And because Tao is something larger and prior to humans, it lies beyond the taint of human interests and desires, thereby serving as a basis for a natural conscience. Taoists found a point of moral connection by suggesting that Tao is the underlying first principle of everything, including humans. Tao gives everything a *te*, its essential qualities. People can ignore their *te* and act *yu wei*, or they can encourage *te* by acting *wu wei*—a separatism within a holism.

This solution, though, was not without problems, either, and later Taoist philosophy tried to grapple with some of these dilemmas. One problem was that acting without deliberate effort seemed a contradiction in terms. Taoism seemed to be counseling people to make a deliberate choice to act without deliberation, to desire to be without desire. Thus, in order to adopt the Taoist solution, you unavoidably introduce desire and interest, undermining its value as a source of a natural conscience.

Perhaps more problematic, though, is the issue of how one is to know when one is acting according to one's *te* and when one is not. Perhaps my *te* is to construct buildings and bridges. Perhaps my *te* is to be ruler of the world—not because I myself desire it, but because my *te* just happens to flow along such lines. In fact, it would be going against Tao not to exercise such a *te* to the fullest, suggested Kuo Hsiang, a Taoist from the first century CE. In his words,

Wu wei does not mean folding one's hands and remaining silent. It simply means allowing everything to follow what is natural to it, and then its nature will be satisfied. . . . Hence, let everyone perform his own proper function, so that high and low both have their proper places. This is the perfection of the principle of *wu wei*.[7]

In other words, Kuo Hsiang found that *wu wei* justified social hierarchy and keeping the high and the low in "their proper places." Given Kuo Hsiang's own position among the high—he was a scholar at a time when few had the luxury of literacy—there must be some suspicion of human interests in his formulation. Of course, practically any human action can be justified by such a use of *te* and *wu wei,* and thus the Taoist point of moral contrast also collapses.

Later thinkers offered solutions to these problems. For example, a standard Taoist response to the problem of deliberately choosing to be undeliberate is that it's a matter of getting things going, like setting afloat down a river on a raft. You do have to make the choice to get on the raft, to be sure, but after that, the river of Tao takes over.[8] Still, it is probably a lot easier to get on that raft if your sense of where it will take you is where you already wanted to go.

The Contradictions of Contemporary Environmentalism

These contradictions continue to confront all of us, and contemporary environmentalism is constantly rethinking its solutions to them.

Take the very idea, so pervasive in environmentalism, that nature is good and that it is best to do things the "natural" way—to use natural materials like cotton and wood and paper, instead of nylon and concrete and plastic, for example, or ground-up vanilla beans instead of "artificial" vanilla flavoring. The word *natural* sets up a clear moral contrast between good and bad. But what law of nature does nylon, concrete, plastic, or artificial vanilla flavor break? These substances are derived from the materials of nature and the Earth: petroleum for nylon and plastic; sand, water, limestone, and other crushed rock for concrete; and whatever it is they use to make artificial vanilla flavor.

Sometimes people say that the difference is that materials like nylon, concrete, and plastic have seen more processing and refining (the basis for distinguishing the natural that I initially suggested to my son). They are more the product of human actions than are cotton, wood, paper, and "natural" food. This is an empirically dubious claim, though, as anyone familiar with what goes on in the growing, harvesting, and processing of cotton, wood, paper, and "natural" food can attest to.

The moral claim being made is also dubious. Such an argument rests on a distinction similar to Aristotle's: Things that come about through their own internal source of change (or *more* through their own internal sources of change) are natural (or *more* natural) and not the product of human desires and interests. But are not humans a part of nature? Is that not also one of the great moral arguments of environmentalism? Anything humans do must therefore be natural, including making plastic—as well as cities, automobiles, toxic waste, and even atomic bombs. Indeed, what human has ever broken a law of nature? If it were a law of nature, how could a human have broken it? By this logic, nothing could be more—or less—natural than an atomic bomb.

Or take the idea of wilderness, which is often seen, particularly among North Americans, as the fullest expression of the true ends of nature and environmentalism. Thoreau sought in wilderness a restful release from the pollution of politics, and millions still do today. But what is wilderness? At least as the political process has defined it, wilderness is national parks and other areas of land set aside from deliberate human interference. Yet isn't the very setting aside of land a deliberate human action, thereby making wilderness the product of human intention rather than the result of its absence?

Perhaps the Taoist response applies here—that establishing a national park is like setting a raft afloat. But then a critic might ask, if humans are natural, why should we exclude houses, roads, open pit mines, and McDonald's from wilderness areas? And why should we see cities and farms as less wild and less in need of the attention of environmentalist concern?

The moral problem is that environmentalism needs to establish the relevance of nature to humans by arguing for our unity with it and, at the same time, to contrast humans and nature in order to set aside a realm that is free from the pollution of human interests. Upon this basic moral contradiction modern environmentalism uneasily rests.

Nature as a Social Construction

The point I've been leading up to is one that virtually all environmental sociologists recognize today: Whatever else nature might be, it is also a *social construction*. Nature is something we make as much as it makes us. How we see nature depends upon our perspective on social life. And as this perspective changes across time and place, history and culture, nature changes with it.

The British sociologist Raymond Williams, in a classic 1972 essay, put it this way:

> The idea of nature contains an extraordinary amount of human history. What is often being argued, it seems to me, in the idea of nature is the idea of man; and this not only generally, or in ultimate ways, but the idea of man in society, indeed the ideas of kinds of societies.[10]

(Williams, writing in an earlier day, used *man* to mean all people.)

Williams's point is that we look at nature through social categories formed by human interests. Our image of nature depends upon *social selection* and *social reflection*. We tend to select particular features of nature to focus upon, ignoring those that do not suit our interests and the worldview shaped by those interests. Moreover, the categories we use to comprehend nature closely reflect the categories we use to comprehend society; social life is so fundamental to our experience that all our categories reflect our social perspectives. Because of social selection and social reflection, "nature" is an inescapably social—and political—phenomenon.

I will give several extended examples in a moment. But first let me point out that such a perspective on our perspectives of nature is itself a perspective on nature—one perhaps best labeled *postmodernist*. Since the 1960s, an increasing number of social theorists have been arguing against the objectivist notion of a world in which there is only one truth about any one thing, the truth that best represents "the way things really are." Such a simplistic view is characteristic of the technological and scientific faith of modern society, say postmodern theorists. Truth is socially relative. What you see depends on who you are and not just on the character of the world "out there," postmodernists argue. And so, too, for an idea long associated with our conceptions of truth: nature.

Naturalizing Capitalism

We can see both social selection and social reflection at work in what is arguably the most important of all theories of nature: Charles Darwin's theory of natural selection—the idea that species change over time as the most successful individuals pass on their distinctive traits to following generations. (Let's not be confused by terms here. Darwin's theory of "natural selection" is not at all the same as the theory of "social selection" in nature, although, as we shall see, the former does exemplify the latter.)

When Darwin's book on natural selection, *On the Origin of Species,* appeared in 1859, it was an instant sensation, and intellectuals gave it close scrutiny and debated its implications. Among the earliest readers were, perhaps surprisingly, Karl Marx and Friedrich Engels. They immediately picked up on a correspondence in Darwin's theory that irritated them considerably: It very closely resembled the economic theories of free market capitalism that were so fundamentally altering the character of English society and, increasingly, world society. As Marx noted to Engels in a private letter in 1862, "It is remarkable how Darwin recognizes among beasts and plants his English society with its division of labour, competition, opening of new markets, 'inventions,' and the Malthusian 'struggle for existence.'"[11]

Consider some of the key mechanisms and results of natural selection that Darwin identified:

- *Life's incredible diversity:* To Marx, that sounded like the division of labor advocated by capitalist economists.
- *Competition for reproductive success:* Marx heard here the echoes of capitalist competition.
- *The link between new species and the discovery of new ecological niches:* This sounded suspiciously like the opening of new markets.
- *Selection of the best features from the range of variation that any species exhibits:* Marx saw here the idea of technological progress through new inventions.
- *The struggle for survival caused by the tendency of populations to increase unless checked by external constraints:* That sounded like Thomas Malthus's theory of population.

Marx's argument was that the categories Darwin used to understand evolution were originally social categories. Darwin's natural selection theory was a mirror of England—a social reflection. Engels later extended the point to social selection (that's selection, not reflection, or not reflection alone), noting that Darwin placed selective emphasis on competition in nature as opposed to cooperation, a more socialist vision of nature. Thus, the theory of natural selection was actually a theory of natural capitalism.

Darwin, in fact, agreed to some extent. In later writings, he described how he hit upon the theory of natural selection when, in 1838, he "happened to read for amusement Malthus on *Population,*" shortly after returning from his voyage around South America on board the HMS *Beagle.*[12] (During this famous voyage, Darwin made many of the observations, such as the diversity of Galapagos finches, that served as key evidence in *On the Origin of Species.*) And Darwin drew additional ideas, as well as the phrase "survival of the fittest," from the writings of Herbert Spencer, a mid–nineteenth-century social theorist.[13]

Metaphors and general patterns of understanding easily flit back and forth between our theories of the realms we label "society" and "nature." Chemists talk about chemical "bonds," and sociologists talk about social ones. Physical scientists talk about natural "forces," and social scientists talk about the social kind. Ecologists talk about community, and so do sociologists— a parallel I have repeatedly emphasized in this book. Thus, perhaps it should come as no surprise that the two scientists who first hit upon the theory of natural selection—Darwin and his lesser-known contemporary, Alfred Russel Wallace—were living in the midst of the world's first truly capitalist industrial society: 1840s and 1850s England.

Nevertheless, the flitting back and forth of concepts between science and social life deserves special scrutiny because of the way it sometimes allows science to be used as a source of political legitimization. It was this process of naturalization that most concerned Marx and Engels. As Engels put it,

The whole Darwinist teaching of the struggle for existence is simply a transference from society to nature. . . . When this *conjurer's trick* has been performed . . . the same theories are transferred back again from organic nature into history and it is now claimed that their validity as eternal laws of human society has been proved.[14]

There is no evidence that Darwin himself had a conscious mission of proving the value of capitalism through discovering it in nature. But witness the way we routinely talk about the economic "forces" of capitalism, such as innovation and competition, as if they were pseudo-natural processes, implying that any other arrangement would be somehow unnatural. Consider the way we often hear the marketplace described as a "jungle" in which you have to "struggle to survive." Naturalization can subtly give us the sense that the current form of our economy is inevitable and that it is foolishly idealistic to think otherwise.

We should be wary of this subtle power. Whether or not we agree with capitalism, or with its current manifestations, we should expect that the argument for any economic arrangement would be made on more explicit grounds.

Nature and Scientific Racism: Morton's Craniometry

What may be even more worrisome, though, is the tragic history of naturalizing arguments by scientists attempting to prove inherent differences in the capabilities of human "races." The controversial 1994 book *The Bell Curve* was only one in a long series of such attempts.[15] But what is distinctive (and hopeful, in a way) about that book is that unlike many earlier efforts in scientific racism, it was almost immediately widely discredited by the scientific community.[16]

By contrast, it was decades before scientists began to see the flaws in *craniometry*, a nineteenth-century scientific fad that compared the cranial capacities of different races. Samuel George Morton, a Philadelphia doctor of European descent, was the leading figure in craniometry. Morton spent years assembling a collection of more than 600 human skulls from all over the world—no easy feat in the 1830s and 1840s, when intercontinental travel was limited to sailing ships. He published three books on his detailed studies, which showed, lo and behold, that Europeans had the biggest brains on the planet. This work made Morton famous. When he died, his obituary in the *New York Tribune* read, "Probably no scientific man in America enjoyed a higher reputation among scholars throughout the world, than Dr. Morton." His obituary in a South Carolina medical journal read, "We of the South should consider him our true benefactor, for aiding most materially in giving to the negro his true position as an inferior race."[17]

Do Europeans in fact have the biggest brains? Let's inspect how Morton reached this conclusion. Morton's method was first to assign skulls to a particular race. Then, he would turn the skull upside down and fill it up with mustard seed poured in through the foramen magnum, the hole at the base of the skull where the spinal cord enters. He then poured the seed back out into a graduated cylinder to determine the volume. By this method, he determined that Whites had an average cranial capacity of 87 cubic inches, Native Americans 82 cubic inches, and Blacks 78 cubic inches.[18]

Morton later became discouraged with using mustard seed. He noticed that, because of their lightness, mustard seeds packed poorly and gave widely variable results when the measurements were repeated. So he switched to lead shot, which gave far more consistent results. He repeated his earlier measurements, and the average volumes for all the skulls went up. But the average for Black skulls went up 5.4 inches, for Native Americans 2.2, and for Whites 1.8. As biologist and historian of science Stephen Jay Gould pointed out,

> Plausible scenarios are easy to construct. Morton, measuring by seed, picks up a threateningly large black skull, fills it lightly and gives it a few desultory shakes. Next, he takes a distressingly small Caucasian skull, shakes hard, and pushes mightily at the foramen magnum with his thumb. It is easily done, without conscious motivation; expectation is a powerful guide to action.[19]

Morton still could record a comforting hierarchy of races using lead shot, though. Even with the new method, Whites were clearly on top. But the initial use of mustard seed was only one of many problems with Morton's work. Gould reanalyzed all of Morton's data and found them littered with statistical errors. (As a dedicated scientist, Morton was committed to empirical accuracy and published all his raw data, allowing Gould to reanalyze them a century and a half later.) Averages for Whites were rounded up and for Blacks rounded down; large skulls from non-White races in his collection were often excluded, without explanation, from final tabulations; and the sample itself was highly selective.[20]

Two problems with the selectivity of the sample stand out in particular. First, Morton did not consider the effect of stature on cranial capacity. Taller people tend to have bigger brains, just as they tend to have bigger feet. People from different regions of the world vary in their average height. For example, a quarter of Morton's sample of Native Americans came from the graves of Peruvian Incas, a small-bodied people with correspondingly small brain sizes, but only 2 percent (three skulls) came from the Iroquois, a tall people with correspondingly large brains. Morton did note in his raw data the height of the skeletons most of the skulls came from, but he did not understand (or did not wish to understand) the influence stature had on his results.

Second, Morton did not consider the effect of gender in his sample. Women tend to be smaller than men and thus to have smaller brain sizes. Not surprisingly, Morton's sample of English skulls, the group that recorded the largest average cranial capacity, came entirely from males. His sample of Black "Hottentot" skulls, which gave a very low figure, came entirely from females. Here again, Morton noted the gender of most of the skulls in his collection but did not understand (or did not wish to understand) the significance gender had for his sample.

Gould took these errors into account, recalculated Morton's averages, and found no significant differences between "races." But even if racially significant differences in cranial capacity remained, we have no basis for assuming this would indicate greater intelligence. As Gould points out, elephants have far larger brains than humans, and yet we do not consider them more intelligent than us.[21] Also, given that all Morton measured in the end was difference in stature, his results would equally well demonstrate that people with the biggest feet have the biggest brains as it would that people of a particular skin color have the biggest brains. But as foot size does not have anything like the cultural significance of skin color, Morton did not pause to consider such an interpretation.

Nature and Scientific Racism: Huntington's Environmental Determinism

Another infamous example of scientific racism was the study by Yale University geographer Ellsworth Huntington of the relationship between climate and the degree of civilization of a people, written up in his 1915 book, *Climate and Civilization.*

Huntington attempted to determine this relationship by the supposedly objective method of sending out a questionnaire to "about 50 geographers and other widely informed men in a dozen countries of America, Europe, and Asia" and asking them to rate the level of civilization in their locales.[22] Huntington compared these results with the level of what he called "human energy on the basis of climate," by which he meant the effect of the local climate on people's ability to get up and do the things that need to be done (to borrow a line from *A Prairie Home Companion*, the long-running weekly U.S. radio show). Climates that are too warm, Huntington reasoned, encourage laziness; those that are too cold encourage lethargy—and so on for factors such as rainfall, sunshine, and cloudiness. Huntington then drew up a global map of the level of human energy and another for the level of civilization, and compared them.

Lo and behold again, he found just what he was likely looking for—that the climates of Europe and the northeastern United States were associated with the highest levels of civilization. Europe and the northeastern United States were, of course, predominantly White regions of the world, and Huntington himself was White. (It is perhaps also worth noting that Yale University is in the northeastern U.S. city of New Haven, Connecticut.) And on this basis, Huntington concluded, "The climate of many countries seems to be one of the great reasons why idleness, dishonesty, immorality, stupidity, and weakness of will prevail"[23] (see Figures 8.1 and 8.2).

This study was intended to be serious science, and Huntington was a famous man. Today, his method seems ludicrously biased. Huntington's "geographers and other widely informed men" almost certainly rated the level of civilization of a place on the basis of what their own experience told them was "civilization." Most likely, they saw lives similar to their own as civilized and ones dissimilar as uncivilized and filled out their questionnaires accordingly.

Of particular interest here is Huntington's environmental argument. Although extremes of cold and heat definitely create difficult circumstances for humans—which is why few people live in the Sahara Desert or at the Arctic Circle—the influence heat and cold have on matters like "idleness," "stupidity," and "weakness of will" is far from certain. In fact, one could easily reverse Huntington's line of reasoning and argue that difficult climates might actually produce

Figure 8.1 Ellsworth Huntington's map of "civilization" around the world. Using a Eurocentric standard, Huntington argued in 1915 that the amount of "civilization" across the world could be related to the amount of "human energy" created by different climatic regions. See Figure 8.2 for his map of "human energy."

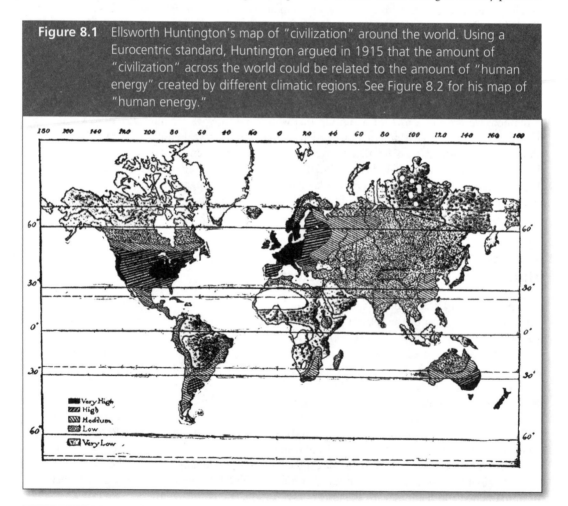

Source: From Huntington, Ellsworth (1915). *Civilization and Climate.* New Haven, CT: Yale University Press.

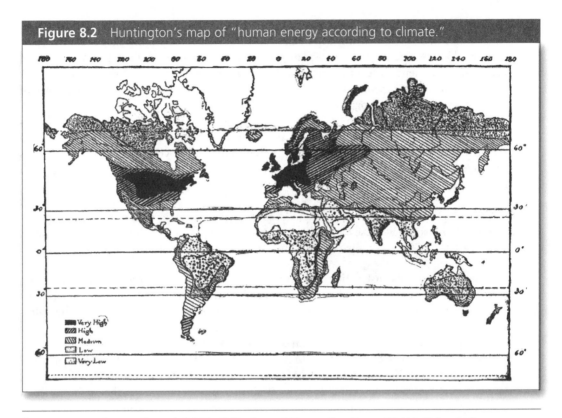

Figure 8.2 Huntington's map of "human energy according to climate."

Source: From Huntington, Ellsworth (1915). *Civilization and Climate*. New Haven, CT: Yale University Press.

peoples with the *greatest* strength of will, the least idleness, and highest intelligence, for otherwise they could not survive such extreme conditions. And indeed, the technological (not to mention the cultural) achievements of the Inuit in the Arctic tundra, the Tibetans in the high Himalayas, and the Bedouins in the deserts of the Middle East and North Africa rank, in my view, as some of the greatest in human history.

Why did Ellsworth Huntington and Samuel George Morton make these racist arguments, and why did they achieve high prominence and acceptability in their day? Probably because the socially powerful are not impervious to criticism about their dominant positions, and they seek some source of ideological comfort.[24] This is part of the power—and danger—of nature as a social idea: the way we so often attempt to use it to legitimate social inequality. If nature is a realm beyond social interest, then its truths must be beyond blame and responsibility. Anything nature may apparently demonstrate about human differences can be ascribed to something external to social power. Nature can then serve as an interest-free foundation for human interests, the unmoved mover of power relations. But is it ever really that?

Environment as a Social Construction

The same processes of social selection and social reflection in our understandings of nature can be seen in environmentalism itself, argue environmental sociologists influenced by postmodernism, such as Melanie Dupuis, John Hannigan, Steven Yearley, and others. Their research touched off a storm of controversy in the field, especially during the late 1990s, and it seemed to some to challenge the legitimacy of environmentalism. If power relations underlie environmentalism's claims about truth, what basis can they offer for assessing sustainability, environmental justice, and the rights and beauty of habitat? Social constructionists thus threatened the natural conscience of environmentalists. And given that most environmental sociologists are themselves environmentalists, social constructionism threatened the natural conscience of many environmental sociologists, too.

The result was the *realist–constructionist debate,* as it came to be called. The main counter-charge against the constructionists was that they were denying the reality of environmental problems. They were contributing to the immobilization of environmental politics, realists contended. As no less a figure in environmental sociology than Riley Dunlap wrote, along with his longtime colleague William Catton,

> Treating global environmental change . . . as a social construction discourages investigation of the societal causes, consequences and amelioration of global environmental problems. . . . [T]his seems particularly unwise in the case of global environmental change.[25]

In fact, most social constructionists in environmental sociology (at least, all that I know) are themselves also committed environmentalists, just like the realists. As John Hannigan has written, "I fully recognize the mess which we have created in the atmosphere, the soil and the waterways."[26] Rather than denying environmentalism, social constructionists typically envision their work as friendly critique—albeit sometimes rather tough-minded.

It is always best to be criticized by your friends before your enemies can do so. As well, social constructionists like Kate Burningham and Geoff Cooper "question the necessity of attempting to contribute directly to practical or political aims."[27] They don't mean that environmental sociology should never be so direct in its aims. But they also ask environmental sociology to step back from normative and political concerns, however good-hearted, so as to be open to results "that are not envisaged by the researchers."[28]

Most realists accepted this point, or came to. Still, they complained that the social constructionists were one-sided in the application of their critical eye. Where were the studies of the ideology of environmental skeptics? Realists such as Dunlap and his colleague Aaron McCright in response went out and conducted such studies, becoming constructionists.[29] (I'll describe one example of this work, the construction of climate change skepticism, below.) And social constructionists came to accept this point in turn, becoming realists. Most famously, Latour admitted that "dangerous extremists are using the very same argument of social construction to destroy hard-won evidence that could save our lives."[30]

The heat of the realist–constructionist debate has now waned considerably. But as irritated as scholars were with each other at the time, light came with the heat. We now understand realism and constructionism, with their roots in the age-old discussion of materialist versus idealist approaches, as the fundamental conceptual tension in environmental sociology. Top scholars in the field continue to write works that seek ways to make this tension a productive one. One new approach is the *critical realism* espoused by the environmental sociologists Michael Carolan (a contributor to this book) and Ray Murphy—by which they mean a realism that embraces the recognition that we can understand reality only by way of our ideas.[31] Dunlap also now accepts that constructionism has an important place in environmental sociology and hopes "to see greater efforts to merge the strength of the two approaches."[32]

With this productive tension in mind (a tension that I have tried to maintain throughout this book), let's take a closer look at the constructionist critique of environmentalism.

Constructing the Ozone Hole

From September 21 to September 30, 2006, atmospheric scientists at NASA (the U.S. National Aeronautics and Space Administration) and ESA (the European Space Agency) watched in horror as Antarctica experienced its worst recorded ozone event ever (see Figure 8.3). Nearly two decades after the signing of the Montreal Protocol to reduce ozone-destroying CFC (chlorofluorocarbon) production in 1987, and more than a decade after CFC production ceased worldwide in 1995, the whole world had hoped for better results. Scientists measure ozone concentration in the atmosphere in Dobson units. (Imagine taking all the ozone above some spot and squishing it down until it was all at the Earth's surface. If the resulting layer were 1 millionth of a meter thick, that would be 1 Dobson unit.) A standard figure is about 300 units—not very much, but very important to life on this planet. At the worst moment during this event, NASA's *Aura* satellite registered just 85 Dobson units. Paul Newman—not the actor, a NASA scientist—said "the average area of the ozone hole was the largest ever observed, at 10.6 million square miles." That's larger than North America. That's big.[33]

But although this Paul Newman is not an actor, he was telling a story. For there is no ozone hole.

Now, I am not suggesting that there is a government conspiracy at work to divert our attention from flying saucers, communist infiltration, terrorist plots, and corporate skullduggery. Nor am I suggesting that everything is just fine and dandy with atmospheric ozone in Antarctica and elsewhere. Not at all. We have a real problem, a terrible and dangerous problem with the upper-atmosphere ozone layer, so vital to protecting us from excess ultraviolet light from the sun. But the problem is not a "hole."

Rather, it would be more accurate to say that the problem is "ozone thinning," and more accurate yet to call it "ozone depletion." Indeed, "ozone depletion" is what scientists usually call it in academic publications. At the worst moment of the September 2006 Antarctic ozone event, there were still 85 Dobson units of ozone above the worst spot (which happened to be the East Antarctic ice sheet). That was nearly a quarter of what it should have been, and very scary. But there was still some ozone there, and it was spread out over many miles of air column over East Antarctica. It was there, and it was spread throughout a thick layer—albeit worrisomely depleted.

So why don't the NASA and ESA scientists call the problem "ozone depletion" when they speak to the public? Quite simply, "depletion" or "thinning" is nowhere near as scary an image as "hole." And the scientists want us to be scared. I do, too. It's a very serious problem.

But nonetheless, this imagery is an instance of social selection in environmental understanding, leading to a particular *representation* for a particular *narrative* purpose. We are always representing the world and telling stories about it, choosing some aspects to focus attention on and neglecting or obscuring others. It could never be any other way. The only perfect representation of the world is the world itself, and you can't put all of that in a sentence, on a page, or into a picture. You have to select. You have to represent. And even if you are a NASA scientist, you have to have some basis, some purpose, for selecting what to represent. You have to have a narrative.

NASA and ESA also tell a visual narrative of the ozone hole. Take another look at Figure 8.3. (This image comes from NASA, but ESA offers similar ones on its website.) Note first of all how NASA has chosen to represent the depleted areas with darker colors. Why? There is no scientific reason for the choice. But it sure makes the depleted regions look more like a hole. Then, also note how the colors go dark very fast around 220 Dobson units, heightening the visual feeling of a sharp edge to the hole. Now, 220 Dobson units is the accepted scientific definition of what constitutes the beginning of the hole. But note how these color choices make it difficult to pick out

Figure 8.3 The social construction of the "ozone hole" by NASA, for September 24, 2006, in Dobson units. Note how NASA has given the depleted regions both a darker color and more defined land masses, giving the visual feeling of a hole in the atmosphere. As well, NASA uses a sharp change in coloration at about 220 Dobson Units of ozone, emphasizing the feeling of an edge to the depleted region. (Original in color.)

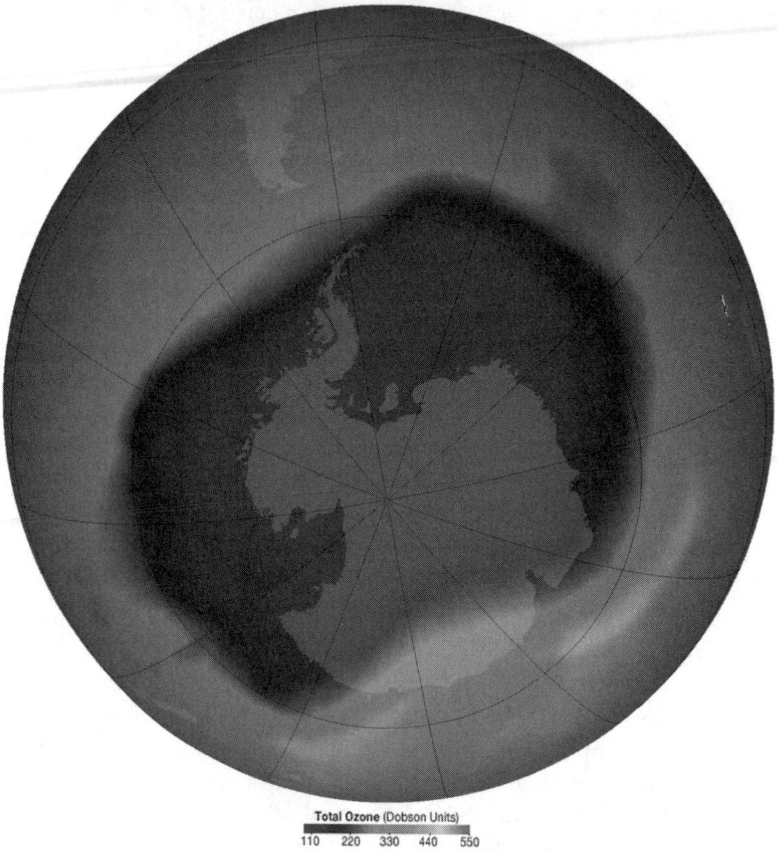

Total Ozone (Dobson Units)
110   220   330   440   550

Source: NASA (2007a).

variations in depletion inside the hole, including that 85 Dobson unit low over East Antarctica. All the visual emphasis is on the 220 Dobson unit edge. And finally, note NASA's narrative flourish (which the ESA images do not repeat) of making the Antarctic land mass look more distinct underneath the depleted regions, as if the viewer were looking through a hole in the sky.

I'm glad NASA does all this compelling storytelling about ozone. It is a selective view of the environment. It has an agenda. It seeks power in its effort to persuade, the power to change our environmentally destructive practices. But this is a power agenda I strongly support. I hope the readers of this book strongly support it, too. And I hope we can use the lessons of constructionist critique to learn how to make this agenda even more persuasive.

Constructing Wilderness

Now let's take a social constructionist look at wilderness protection. As I discussed earlier, wilderness is a contradictory idea. The highest exemplar of the natural, it is also based on the often forcible absence of, and regulation of, one widespread aspect of the natural world: people. Most wilderness areas are routinely patrolled by rangers (frequently armed) who monitor boundaries and police human activities within. Often our wilderness parks have been created through the removal of the people who were living there previously (see Figures 8.4 and 8.5).

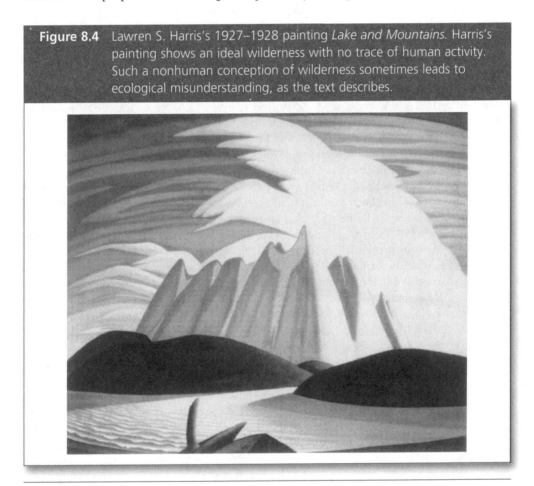

Figure 8.4 Lawren S. Harris's 1927–1928 painting *Lake and Mountains*. Harris's painting shows an ideal wilderness with no trace of human activity. Such a nonhuman conception of wilderness sometimes leads to ecological misunderstanding, as the text describes.

Source: Lawren S. Harris, *Lake and Mountains*, 1927–1928. Used with permission.

As Ramachandra Guha has pointed out, such an approach to environmental protection is based on a culturally specific understanding of what the environment is and could be, one specific to North America in the main.[34] The idea of the national park originated in the United States, and the initial model was one of a wilderness region with few people. Because of its low population density and vast areas of rugged terrain, the United States had large regions with few people during the great age of park establishment in the late nineteenth and early twentieth centuries. It was relatively easy to remove the few people who were there. The native peoples had already been removed, for the most part, and the settler populations in these rugged and remote zones were thin on the ground. Such is not the case in most of South and Southeast

Asia. There, people live virtually everywhere, and setting aside any large area usually means setting substantial numbers of people aside, too.

Also, people and wildlife in populated areas may coexist harmoniously, which is often why enough wildlife yet remains to attract interest in establishing a park. For example, when the Keoladeo Ghana bird sanctuary was established in Bharatpur, India, local villagers were told they would no longer be able to graze their livestock in the area. The villagers protested, and a battle broke out with the police in which several villagers were killed. After the grazing ban was enforced, scientists discovered that, in fact, bird populations declined. It turned out that grazing kept down the high grass, creating better habitat for the insects that many of the birds depended on.[35]

Although local villagers usually take the bulk of the blame for the loss of wildlife in their areas, the more important culprits often are poachers supplying primarily urban clients, earlier hunting by tourists and local elites, and the general decline in suitable habitat because of the spread of plantation forestry and plantation agriculture.

Also, establishing a wilderness park may actually increase the number of humans in an area by promoting tourism. In the case of Indian tiger reserves, Guha notes, the effect has been to replace local people with urban and overseas tourists and a full complement of upscale hotels, restaurants, and stores. As it takes capital and knowledge of middle-class life to set up tourist accommodations, the remaining local people often do not get much economic benefit from the establishment of a tiger reserve, aside from a few low-paying jobs.

Nancy Peluso has described a similar history in southern Kenya.[36] There, the Masai tribe of migratory pastoralists has been confined to increasingly smaller reserves and "sedentarized" to make way for game and wildlife reserves, particularly for rhinos and elephants. The process began in the early 1900s when, in response to the demands of big game hunters and conservation groups, the British colonial government established the Southern Game Reserve. At the time, however, officials did not see the Masai as a threat to wildlife, and the Southern Game Reserve was set up within the Masai reserve. But since the 1950s, a "wilderness" model of wildlife protection has taken hold. The old Southern Game Reserve has been replaced with four reserves that exclude the Masai completely and are surrounded by extensive buffer zones in which the Masai are required to adopt "group ranches" for their cattle, imposing yet another alteration of the Masai culture and economy.

There is little evidence, however, that the traditional Masai lifestyle was in conflict with the survival of rhinos and elephants. Peluso points out that rather than being a threat to wildlife, with whom they had coexisted for thousands of years, the Masai helped maintain the savanna environment through their migratory cattle grazing. As long as it is managed properly, cattle grazing can actually increase the availability of grass for wildlife by stimulating regrowth. Grass grows most rapidly and nutritiously when it is kept short (although not so short that it has trouble providing energy to its root system). Also, cattle fertilize as they graze, providing the regrowing grass with helpful nutrients. Rather than the low-intensity cattle grazing of the traditional Masai, the main sources of rhino and elephant decline have been big game hunters and the ivory trade.

Excluding the Masai from the new reserves, in fact, had just the opposite effect of what was intended. The Masai got angry and began killing some elephants and rhinos. Also, some Masai collaborated with the ivory trade in protest. Alienated from their traditional lands and traditional ways for the supposed benefit of wildlife, the Masai suddenly found themselves competing, rather than coexisting, with the rhino and the elephant. Corrupt park rangers and government officials, however, have been the main culprits in the continuing loss of wildlife to what Peluso terms the "ivory wars."[37]

The issue of constructing wilderness through forced removal of local people is not limited to developing nations. In the 1960s, the Tennessee Valley Authority (TVA) set out to create the

Land Between the Lakes National Recreation Area by displacing the roughly 1,000 families that lived there. The lakes themselves had been created by dams and earlier removals from a region that was originally known as the Land Between the Rivers. Lying at the Kentucky–Tennessee border, the Land Between the Rivers is a 50-mile long inland peninsula, originally bounded by the Tennessee River on the west and the Cumberland River on the east as they both flow north, never more than a few miles apart, and into the Ohio River. The Land Between the Rivers was a tight community of farmers and woods people. They also maintained an informal commons in the woods that they called "the Coalins," although no one now can remember why it had that name. The Land Between the Rivers people hunted in the Coalins, but managed it carefully so that there would always be hunting for the future. Oddly, this careful management led, in part, to their later undoing.[38]

In the early twentieth century, the state of Kentucky noticed that wild turkeys were extinct everywhere else in the state but in the Coalins. So in 1919, the state established the Hillman Game Reserve in the Coalins, naming it for an industrialist who once operated some iron furnaces in the area. The federal government took the land over in 1935 and in 1938 renamed it the Kentucky Woodlands Wildlife Refuge. Wildlife managers used the Land Between the Rivers flock to reintroduce turkeys throughout Kentucky, as well as deer, which had also become quite scarce. All of Kentucky's wild turkeys and deer today descend from the flock and herd the people of the Land Between the Rivers had maintained, except in places where managers have recently brought in turkeys and deer from elsewhere to broaden the genetic pool.[39]

Meanwhile, the TVA was busily damming rivers throughout the area of the Tennessee Valley. They dammed the Tennessee on the west of the Land Between the Rivers in 1945 and the Cumberland on the east in 1959, displacing hundreds of farms and thousands of people in the process. But the uplands area between the new lakes was still inhabited, with farms interspersed through the woods, although much of it, the former Coalins, was now the Kentucky Woodlands Wildlife Refuge. In 1965, the TVA declared its intention to put the lakes together with all the land in between them, including the refuge and all the remaining farms, creating the Land Between the Lakes National Recreation Area.

The Land Between the Rivers people fought back hard. But they lost. The removals came and were often brutal. Federal marshals. Handcuffs. Bulldozers. Burning. And the latter two sometimes were of houses still full of possessions. One resident who had resisted leaving was called away to a meeting with officials to work things out. Meanwhile, the bulldozers went about their business, shoving down the walls and everything inside into a pile and setting it all afire. A sick, elderly woman was carried out in her bed into her yard while her house was pulled down in front of her eyes. Sometimes the bulldozers came even though people were doing their best to comply with the removal orders but were having trouble getting a mover or a place to stay. Suddenly moving 1,000 families put a considerable strain on the local real estate market, and many people simply couldn't find a new place to live that they could afford. No matter for the marshals.

There are still plenty of people in the region today, probably as many as there ever were on a daily basis, and perhaps more. There is also a fair bit of farming, again perhaps more than ever before. But these people are the visitors and the employees of the Land Between the Lakes. And the farming is done by contractors that the Land Between the Lakes officials brought in when they realized that active farmland is a crucial part of the local habitat for turkey and deer, similar to when park managers realized that the Masai's cattle grazing improved conditions for wildlife.[40] The Land Between the Rivers people are gone now. Their homes are gone. Their communities are gone. Even their name, the Land Between the Rivers, is gone. All that is left are their cemeteries, one old church building that the TVA had overlooked, and bulbs from their former home gardens that still flower in the spring—plus their memories (see Figure 8.5). But imagine if they hadn't protected the turkeys in the Coalins years ago.

Figure 8.5 David Nickell, a member of the displaced Land Between the Rivers community, stands in front of the jonquils that still flower every spring at the site of his boyhood home, removed to make way for the Land Between the Lakes National Recreation Area at the Kentucky–Tennessee border. Wilderness preserves around the world have often been established by evicting local peoples.

Source: Author.

Wilderness is, in the end, a state of mind more than a state of nature. Habitat provision is certainly vital for protecting biodiversity. The question is, are habitat and humanity natural opposites, or merely conceptual ones? As the environmental historian William Cronon has written, "It is not the things we label as wilderness that are the problem—for nonhuman nature and large tracts of the natural world do deserve protection—but rather what we ourselves mean when we use that label."[41]

Guha's, Peluso's, and Cronon's point, then, is not that the United States should give Yellowstone Park to housing developers or that Canada should turn Banff into a ski resort. Theirs is a more practical suggestion: that we should not universally and imperialistically promote such a culturally specific notion as a "no people" vision of wilderness in places where it may lead to social injustice and may even undermine biodiversity.

For all its moral attractions as a realm free of politics, wilderness is nevertheless deeply political. There are unavoidable human consequences to the provision of habitat for wildlife—which does not mean that protecting wildlife habitat is wrong or unimportant. It does mean, though, that we need to think through the implications for both humans and nonhumans.

Tourism and the Social Construction of Landscape

Tourism, it is sometimes said, is the largest and fastest-growing industry in the world. Such a statement is hard to verify—the definition of the tourist economy can be as narrow as hotel bookings and plane fares or as broad as the cost of the rubber one wears off one's shoes walking in the neighborhood park. But tourism indisputably has become big business. And in a variety of ways, tourism's rise is closely associated with environmentalism's rise. After all, much of what tourists come to see are the parks created by the environmental movement. Plus, environmentalism has long stressed giving value to places far from one's own home.

Much of what tourists travel for is the look of a place and the occasion (or excuse) that look provides for what we must see as the *culture of leisure*. Although the local culture and environment may be part of the experience, most tourists spend the bulk of their time in the separate world of the hotel, the historical site, the beach, and the wilderness park—all places rarely frequented by local people, unless they happen to work there. In other words, tourists tend to objectify a landscape, using its visual qualities as a cultural opportunity for leisure.

To some extent, the tendency of tourists to concentrate on the look of a place is a matter of access. Local property boundaries limit physical access to those deemed to have rights to a particular area. But visual access is hard to limit in a similar way—people can't easily be confined to keeping their gaze trained only on those spots where they have been granted a specific legal right to look. The landscape can be veiled through the strategic placement of walls and roads, but these only create a different visual effect. A landscape would still exist for outsiders to "consume" through what the sociologist John Urry has aptly termed the "tourist gaze."[42]

Of course, local people experience landscape visually, as well. Since their lives and work are tied up in the local land, though, their visual experience has many more levels to it. They see the land differently; it means something different to them. As Thomas Greider and Lorraine Garcovich have written, "Every river is more than one river. Every rock is more than one rock. . . . Every landscape is a symbolic environment. These landscapes reflect our self-definitions that are grounded in culture."[43] We see landscape differently, and for different purposes, depending upon who we are.

The social significance of a landscape depends as well on some things that we can never see: the social associations we experience in it. A landscape can be alive with a kind of spirit, connections, and associations that we may have when we are in that place. These connections and associations may be general feelings, but they may also be of specific things, even specific people who somehow seem to be there though we know they are not. We commonly invest a landscape with, well, ghosts—a sense of the presence of that which is not, and those who are not, physically there. Through these ghosts of place, we build a feeling of attachment to that place through our attachment to the ghosts we sense there.[44]

A person local to an area, for example, may vividly recall events that friends and family participated in at a particular place in the landscape. He or she will likely be able to conjure up the mental images of friends and family involved in those events and to sense a connection to those places through those social attachments. The field where childhood friends used to play. The tree planted by a parent. His or her own ghost at the playground of the old school. The particularly deep sense of personal presence that generally comes with legal ownership of a place.

Tourists, of course, do not have these local associations and generally do not "see" these local ghosts—except for those conjured up, often simplistically, at historic sites, in heritage museums, and in tourism literature.[45] But tourists, too, need to "see" the landscape in some way to give it significance. The purely visual has no meaning. There is a close interrelationship between what we see and what we "see," for locals and tourists alike.

Tourism promoters play to the tourist's need to "see" by giving access to some readily accessible ghosts. As the environmental sociologist Clare Hinrichs describes, the state of Vermont in the northeastern United States has been particularly effective at giving tourists something to see in their imaginations. "To speak of Vermont is to conjure up a broader vision of a balanced, beneficent rurality," Hinrichs writes.[46] This conjured ghost of rural life can then be attached to the state's mountains, ski slopes, ice cream, cheese, maple syrup, and pancake mix—even to its mail-order Christmas wreaths. "Send a bit of the Vermont woods home for Christmas," Hinrichs reports one ad as reading, allowing one to tour Vermont, to "see" it, without ever leaving home. The landscape itself becomes an exportable commodity.

Local people are often ambivalent about the consequences. Turned into postcards and advertising slogans, a landscape can bring in the tourist's dollars. But often the benefits of tourism are not widely spread. The jobs are generally low paid and seasonal. Tourist dollars may also give local people the feeling that they have lost control of their community, and in the case of indigenous peoples, their culture as well. Tourism can also be an environmental disaster. Accommodations for tourists frequently burden local water supplies, eliminate sensitive habitat, and increase pollution, as well as intruding clumsily on the spirit (and spirits) of a place. The wilderness and rural charm that brought the tourists to begin with may soon become no more than a postcard and a slogan, even for those who live there.

Thus, the experience of a landscape—both what we see and "see"—is political. Should we therefore ban tourism and declare that tourists have no right to see or "see" the land in places where they do not live? That would be silly and impractical, and just as political as allowing tourism. Rather, it seems to me, we should treat the visual and imaginative qualities of a place as a kind of commons, something we all share. Landscape provides an important point of social connection in a world that is increasingly divided into no-go zones to which access is granted only to those with money. But to share does not necessarily mean to escape conflict. Indeed, it is often conflict that creates the incentive for sharing. And we should recognize that tourists and the tourism businesses they support are usually at an advantage in conflicts over landscape. Tourists tend to be wealthier than local people, and tourism businesses are often backed by corporate interests and by national and regional governments eager to generate foreign exchange and tax revenues. In such a situation of inequality, it is particularly challenging to manage a commons democratically.

The Social Construction of Environmental Exclusion

Now let us consider the use of environmental arguments for social exclusion.

This is an unfortunately common phenomenon. Suburban and exurban zoning controls, for example, are often instituted with the expressed intent of maintaining open space, wildlife habitat, and the "character" of an area, among other goals. Such controls, though, can have the not-so-hidden effect of raising the cost of acquiring a residence in these areas, excluding poorer people from moving in and sometimes forcing them to move out. Moreover, rural areas in most of Europe, Canada, and the United States, the principal exception being the American South, are almost entirely White, and most suburban areas remain primarily so. Restrictive zoning policies help perpetuate this historical segregation—a point not lost on at least some rural and suburban residents.

In a 1990s study of an exurban Canadian village on the far commuting fringe of Toronto, Stanley Barrett found that some residents had moved "out in the country" in part to escape the multiracial and multiethnic city. One White woman who had recently moved from the city spoke very directly on the issue: "That is one thing that I like about it here—none of those people," she said, referring to Blacks and Asians.[47]

Environmental arguments are also sometimes used to argue for restrictions on immigration. As a member of several national environmental organizations, I used to get a lot of environmental junk mail (something that ought to be considered an oxymoron). I don't get as much now, since I started using a number of junk-busting techniques. Still, environmental organizations I've never contacted or never even heard of will sometimes send me urgent appeals, action alerts, fake surveys of my opinion on various topics, and other grabs for my cash. A number of years ago, I must have gotten into some marketer's database of people concerned about population growth, something I do believe is an important environmental issue. At the time, the U.S. Congress was considering some important immigration bills, and organizations like Population–Environment Balance, Negative Population Growth, Zero Population Growth, and FAIR (the Federation for American Immigration Reform) used the occasion to flood my mail slot with appeals and surveys. Their literature's frequent use of xenophobic and other dubious arguments caught my eye (and raised my ire), and I saved some of it. Here's how the letter I received in 1990 from Population–Environment Balance, signed by their "Honorary Chairman," Garrett Hardin, began:[48]

Dear Friend:

Why do so many of America's environmental problems persist despite our efforts to solve them? We Americans work hard to improve our individual lives and our country as a whole. Yet, even though we continue to make these sacrifices, we continue to suffer from worsening problems such as environmental degradation, traffic jams, deteriorating infrastructures, and homelessness.

The pitch here is that "we Americans" are not to blame for environmental problems, traffic jams, infrastructure deterioration, and homelessness. We work hard and sacrifice every day. The letter continued:

THERE IS A SOLUTION! . . . But before we can discuss the solution, we must fully understand *the primary cause of these problems—overpopulation.* . . . You may ask, how did we become overpopulated? Natural increase—more births than deaths—is one factor. The reality is, however, that *illegal as well as legal immigration is a major cause of population growth in the already overcrowded United States.* . . . Unrestrained, the effects of this immigration-generated population growth on our environment and quality of life will continue to become more obvious and more serious.

In other words, the "SOLUTION" is getting rid of *them.* Although the letter points out in passing that "natural increase . . . is one factor" and elsewhere states Population–Environment Balance's support of birth control and sex education, nowhere does it specifically recommend smaller family size in the United States. Nor does it mention reducing the consumption levels of "we Americans" as a solution to environmental problems. Instead, the letter shifts the blame onto legal and illegal immigrants, who it implies are, among other things, clogging the streets with their cars.

The letter goes on to call for a limit of 200,000 legal immigrants a year, about a third of the then-current figure of 650,000, what it calls the "replacement level" that would balance annual emigration from the United States. The letter also links population growth to a host of other issues, including crime, stress, urban sprawl, taxes, living expenses, and unemployment (but without explaining what the links might be), pushing as many "hot buttons," as marketers say, as possible.

Garrett Hardin was a widely respected environmentalist who died in 2003, author of one of the most famous essays of all time on environmental issues, "The Tragedy of the Commons," which I discuss in Chapter 10. He was also known for promoting "lifeboat ethics"—the dubious idea that there is room for only so many and that rather than sink the whole boat, we will simply have to deny some people a place onboard.[49] Here in this letter, we learn who those entitled to a place in the boat are: the current citizens of the world's wealthiest and most powerful country. The basis for that decision is evidently the social power of those who already have a seat.

The Social Construction of Environmental Nonproblems

When environmentalists criticize the negative consequences of industrialism, and when scientists publish research that documents those consequences, industrial interests usually object, not surprisingly. One way industrialists object is by supporting politicians who will support them. This is the materialist route: Use money and other structures of power to move politics in your direction. The other way industrialists object is by supporting narratives that will support them. This is the idealist route: Tell a different story and spread it far and wide. In practice, as elsewhere in social life, each supports the other, the material and the ideal, in the construction of what we might term *environmental nonproblems*.

Let's consider the rise of climate change skepticism. Recent polls in some countries show decreasing public agreement that climate change is a real and serious problem.

First, though, we need to note that the overwhelming majority of the public in countries where we have polling data express considerable concern about climate change. A 2009 poll by the University of Maryland found that the public in 15 of 19 countries felt their governments needed to place a higher priority on climate change. In total, some 60 percent favored higher priority and only 12 percent wanted lower priority to be placed on climate change.[50] In a 2009 Pew Research poll that included 25 countries, a majority of the public in every single nation agreed that "global warming" was a serious problem, and most thought it was "very serious."[51] Yet another 2009 poll of 25 countries found that global public concern about climate change has never been higher.[52]

But in Britain, Japan, and the United States, concern is now substantially lower than it was. In Japan, the percentage of those who think global warming is a serious problem dropped from 78 percent in 2007 to 73 percent in 2008 to 65 percent in 2009.[53] In 2005 in Britain, 82 percent of the public described themselves as "very concerned" or "fairly concerned" about climate change; by 2010, that figure had fallen to 71 percent. As well, 91 percent of British adults in 2005 agreed that they "personally think the world's climate is changing"; in 2010, only 78 percent did.[54] The decline in the United States has been even sharper. In 2006, polls showed 79 percent agreed that there is "solid evidence the earth is warming," while in 2010, agreement was down to 59 percent. And when asked whether the earth was warming because of human activity, 50 percent of American respondents agreed in 2006 and just 34 percent agreed in 2010.[55]

Moreover, a wide partisan gap has opened up in the United States.[56] In 2010, polls indicated that 53 percent of Democrats agreed that the climate is warming because of human activity, while only 16 percent of Republicans agreed.[57] This gap is quite recent, as discussed in Chapter 7. Is this decline and gap because some people still don't know very much about global warming? Polls in the United States suggest no. The partisan gap in the United States is highest among those who say they understand global warming "well" or "very well."[58]

Rather, say scholars like Dunlap and McCright, industrial interests have been working hard, and effectively, to change the story. A 2007 George Mason University poll of a random sample of climate scientists in the United States found that 97 percent agreed that "global average temperatures have increased."[59] Two international polls of climate scientists, one in 2008 and one in 2009, found agreement just as strong.[60] Yet the percentage of the public who think most climate scientists agree that global warming and climate change are real was 65 percent in the United States in 2008 (54 percent among Republicans) and 56 percent in Britain in 2010.[61]

Industrial interests have used several techniques to undermine public agreement about the scientific status of climate change, as Dunlap and McCright document.[62] First, they have attacked the status of the science through "spinning" results, promoting the few contrarian scientists, and intimidating scientists through personal attacks in the media. Second, they have used political appointees to edit and suppress government reports, to interfere in the work of individual government scientists, and to prevent government scientists from speaking to the media. Third, they have used the existing rules of political process to slow down government decision making through holding frequent legislative hearings and to establish onerous rules and procedures for the release of findings by government agencies. And fourth, they have exploited the media's love of controversy and desire to appear neutral by always including an opposing voice in a news story, even if that voice has little scientific credibility, Dunlap and McCright note.

Plus, industrial interests have leapt on rhetorical opportunities, like the 2009 controversy over emails by climate scientists that an unknown hacker pulled off a University of East Anglia server in Britain. Several of these messages contained some rather unvarnished comments, as is not uncommon for email, as well as some phrases that, out of context, sounded like the scientists were deliberately fudging the data on climate change. The hacker released the email to a climate change skeptic blog, who soon released it to other bloggers, and the story went viral. "Climategate" is what the media delighted in calling the story.

Note the close interrelationship many of these techniques have with political control of government, showing the interaction of the material and the ideal. Industry cannot be expected to produce independent research. Consequently, democracies everywhere look to government scientists and university scientists supported by government grants, as well as foundations, to provide an independent voice. But because it depends so much upon government support, independent research is also vulnerable to political control. Politicians are often happy to oblige political interests offended by independent research—especially when doing so presents an opportunity to sort and polarize voters, thus solidifying their base of support.[63]

The Dialogue of Nature and Ideology

Compromised is not what we wanted nature to be. But for all the moral security we expect it to provide, nature is a contradictory and contested conceptual realm. Although we often see it as a source of conscience that is free of the pollution of social interests, nature is inescapably social. Nature is a social and political phenomenon as much as a physical one, as many environmental sociologists have recently argued. The French social theorist Bruno Latour even suggests that there is no such thing as nature, only "nature-culture."[64] The American environmental sociologist Bill Freudenburg and his colleagues make a similar point, arguing that we need "to resist the temptation to separate the social and the environmental, and to realize that the interpenetrating influences are often so extensive that the relevant factors can be considered 'socioenvironmental.'"[65]

Actors and Actants

It is important to recognize some limits here, though. What we have long thought of as nature is certainly a social construction. All human ideas are necessarily social constructions; we are social, and therefore so are our ideas. But is that all nature is? Is it only a social construction? Indeed, is social construction all *anything* is—including the very idea that ideas are social constructions?

Social constructionism is an important insight, yet it presents several theoretical dangers. To begin with, it can become a kind of universalism itself, as everything reduces to the social constructionist perspective. This is an internally contradictory result. The moral and theoretical foundation of social constructionism is the need to recognize the existence of "truths," not truth. Second, if one conceives social constructionism as advocating a subjective point of view, it runs the risk of becoming a kind of Sophist solipsism—helpful as a form of social critique about power and perspective but incapacitating for any effort to improve social and environmental conditions, lest we commit another social construction. This critique is one that realists have often raised. A social constructionist might come to regard any advocacy of social and environmental reform as hopelessly intertwined with power interests and therefore might cynically retreat from any social involvement. Finally, untempered social constructionism can wallow in a purely ideal realm, unwilling to engage the material side of the ecological dialogue.

There are several responses to these potential theoretical dangers. First, we need to recognize that there is nothing wrong with committing a social construction. The value and appropriateness of a social construction depends, of course, on what the social construction is. But we also need to recognize that there is more to human experience than ideas—and more to human experience than social life. It is not only ideas, and not only society, that fill our days. There is a material side to life and to social life, and that material side interacts with our ideas and our patterns of social organization.

The point is, we are unlikely to make up whatever vision of nature we want, at least not for long. There is a certain *social inconvenience of "nature"* that will, in time, guide the kinds of social visions we take toward it. King Canute, ruler of England 1,000 years ago, thought rather well of himself and supposedly once commanded the tide not to come in—with predictable results. He never tried it again.[66] Social understandings of the environment that don't work don't last long. Material consequences matter.

A perspective introduced in Chapter 3 can help us here, many environmental sociologists find: *actor network theory,* or ANT, as the acronym pleasantly works out. Chapter 3 described how ANT can help us see technology and humans as linked together in networks of common consequence, each acting on the other. The same thought applies to elements of what we often call "nature" and the "environment," argue Bruno Latour and his colleagues Michel Callon, John Law, Annemarie Mol, and others. By having mutual consequence for the other, each actor in a network shapes what the other can do, granting possibility and establishing constraint. And by shaping what an actor can do, in a way the network shapes what each actor is. For as pragmatist philosophy has long argued, something *is* what it *does.* Common consequence becomes mutual constitution.

The word *actor* has a distinctively human sound to it, though, so Latour suggests that we call all these network elements, human and otherwise, *actants,* lest we fall into our old oppositions again. What makes humans the actants we are is how the rest of the world interacts with us. Ideas matter a lot in this interaction. How do I know I am human? Because the world responds to me in the ways I conceptualize it should respond to humans—as opposed to, say, a bird or a fish or a rock or the wind that blows. Of course, I did the conceptualizing (with a lot of help from my society) to begin with, and I respond to the world's responses in ways that stem from that conceptualization. I know the actants I interact with through their *performances,* as Latour calls them, in combination with my conceptualization of their performances. And my conceptualizations of myself and the performances I give in return are what constitute me as an actant in the network, as well.

ANT is thus a very dialogic perspective. It helps us think about the "inconvenience" of the world, while at the same time recognizing our own constructions of our experiences of that inconvenience. Latour likes to say that his approach gets us past a lot of unhelpful oppositions, especially nature versus society and fact versus belief. He sees our understandings of the world as what he calls *factishes*—hybrids between facts and beliefs. Factishes help us in the essential act of what Latour calls *fabrication,* the simultaneous creation of knowledge and actor networks. (Latour likes puns. He wants to point out here the human role in "fabricating" our knowledge but also that there is a "fabric" of actants in a network who have an equal hand in fabricating knowledge.) Actor network theory thus helps us to see the interaction of—and the mutual constitution of—material and ideal factors in the dialogue of ecology, in "nature-culture."

The history of the theory of natural selection is a good example of the mutually constituting interaction of the material and the ideal. The correspondence between the rise of capitalism and the theory of natural selection may suggest that society was reflected in Darwin's vision of nature. But that doesn't mean Darwin just made it all up. Rather, it could be argued that before the rise of a form of social organization like capitalism, people did not have the conceptual resources—the forms of understanding—needed to envision natural selection. Before this form of social "seeing" developed, people were not able to see natural selection. A certain cast of mind is prepared to see things that other casts of mind cannot—but there has to be something there to "see." Indeed, the fossil record provides abundant material evidence—abundant performances, in Latour's terminology—that corresponds with Darwin's basic propositions.

But then, in turn, these correspondences help amplify the cast of mind that Darwin turned toward them. Darwin was thus changed by his interactions with fossils, Galapagos finches, and other actants in the network that was created by, and helped create, the theory of natural selection. Darwin was a different actant as a result, just as the theory of natural selection in time vastly changed human interactions with nonhumans, leading among other things to the discovery of genes, and thus changed the performances of those other actants. For just as an actor's performance is shaped by the audience's response to it, so are nonhuman actants' performances shaped by the response of humans to them, and vice versa. What something is, is what it does in the world. And what it does in the world is shaped by how the world responds to it.

Resonance

Nonetheless, however much we fabricate (in Latour's sense) the conditions of our lives through involvement in actor networks, a network of mainly human actants feels quite different to us from one with both human and nonhuman actants. We respond to such a network differently. We perform in it differently. And we think about it differently.

But not as differently as we often say is the case. Similarities in the categories we use to understand human networks and human/nonhuman networks are striking. Moreover, these categories experience parallel changes. Again, the theory of natural selection is a case in point.

In the twentieth century, there developed a more socialized version of capitalism than Darwin knew. Social services, worker protection, free education, and assistance to the poor became standard features of most democratic societies. At the same time, a more socialized understanding of evolution developed: ecology, which stresses holism and mutual interdependence in the nonhuman world. And as we have come to adopt a postmodern skepticism about "progress" in the human world, biologists have adopted similarly skeptical views about "progress" in evolution. In place of the every-day-we're-getting-better-in-every-way implications of Darwin's ideas about competitive selection of the fittest, biologists now stress the importance of random events like the meteorite crash that doomed the dinosaurs. They also question the extent to which animals are perfectly adapted to their environments, another Darwinist idea that seemed to imply progress in evolution.

The evidence is out there—the performances are out there among nonhuman actants—for these new views of evolution. Geologists are pretty sure that on the Yucatan Peninsula of Mexico, they have found the crater caused by that meteorite. Such a finding, though, has social effects. For example, it probably feeds our feeling that the direction we are going in is not certain to end well—that, to echo the title of an Elvis Costello song, "accidents can happen." It fits with, and therefore promotes, our current sense of anxiety. Thus, our ideas of the condition of our human networks reflect our ideas of our other networks, and vice versa. It's a two-way process, a search for a common language for understanding the conditions of our lives. It's another dialogue.

But the dynamic of these mutually supporting parallels is largely an unconscious process. I like to think of it as a kind of intellectual *resonance*. That is, we tend to favor patterns of understanding that work well across—that resonate with—the range of our experience. We feel an intuitive sense of ideological ease when a pattern we know from one conceptual realm seems to apply in another.[67] Darwin's theory of natural selection resonated with the Victorians' enthusiasm for the new capitalist way, just as the devastating crater that wiped out the dinosaurs resonates with our turn-of-the-twenty-first-century mood of doubt. It just somehow seems righter—the way a more resonant musical instrument just somehow seems to sound better—when we encounter such parallels.

The value of social constructionist arguments is that they alert us to the ideological implications of resonance and urge us to confront our intellectual ease. The results may be emotionally and politically challenging, but sometimes that's just what we need.

Maybe the best way to conclude is to go back to my philosophical wrangle with Sam. The answer to the question, "What is nature?" is that everything is and everything isn't. Both are true. Nature is an inescapably human conception, just as humans are inescapably a conception of nature. We construct nature, and nature constructs us. It's a paradox, but it makes ecological dialogue possible—and necessary.

CHAPTER 9

The Rationality of Risk

"Safety," the wife of Pablo said. "There is no such thing as safety. There are so many seeking safety here now that they make a great danger. In seeking safety now you lose all."

—Ernest Hemingway, 1941

Click. It's April 15, 1996. Click. The pixels start glowing. Click, click, click. *The Oprah Winfrey Show.* There's Oprah with three guests. Gary Weber of the National Cattlemen's Beef Association. Dr. Will Hueston of the U.S. Department of Agriculture. Howard Lyman of the Humane Society. It's the controversial episode about Mad Cow Disease or BSE (Bovine Spongiform Encephalopathy) and the human manifestation that comes from eating the infected cows, new variant Creutzfeldt-Jakob Disease (nvCJD), which is terminal and incurable. Oprah is introducing Lyman to the studio audience.[1]

"Howard Lyman is a former cattle rancher, turned vegetarian," says Oprah. "You hear me? Former cattle rancher turned *vegetarian*—we want to know why—and executive director of the Humane Society's Eating With Conscience Campaign. You said this disease could make AIDS look like the common cold?"

"Absolutely," replies Lyman.

"That's an extreme statement, you know?" Oprah challenges.

"Absolutely," says Lyman again, taking the challenge in stride. "And what we're looking at right now is we're following exactly the same path that they followed in England. Ten years of dealing with it as public relations, rather than doing something substantial about it. One hundred thousand cows per year in the United States are fine at night, dead in the morning. The majority of those cows are rounded up, ground up, fed back to other cows. If only one of them has Mad Cow Disease, [that] has the potential to affect thousands. Remember today, [in] the United States, 14 percent of all cows by volume are ground up, turned into feed, and fed back to other animals."

Oprah seems a bit shocked. "But cows are herbivores. They shouldn't be eating other cows."

"That's exactly right," says Lyman. "And what we should be doing is exactly what nature says. We should have them eating grass, not other cows. We've not only turned them into carnivores, we've turned them into cannibals."

"Now see, wait a minute, wait a minute." Oprah goes back to the challenge mode. "Let me just ask you this right now, Howard. How do you know the cows are ground up and fed back to the other cows?"

"Oh, I've seen it. These are USDA statistics. They're not something we're making up."

Oprah addresses the audience. "Now doesn't that concern you all a little bit, right here, hearing that?"

The audience intones a collective "Yeah!"

"It has just stopped me cold from eating another burger!" adds Oprah.

The audience breaks out in applause.

Click. Talk of risk is still in the air today, perhaps even more so than in 1996. We are fast entering an "age of risk," say some observers.[2] Along with cloning, globalization, virtual reality, and other defining terms of our time, the twenty-first century appears to be culturally and politically saturated with the language of "risk." Hazardous waste, lead, VOCs (volatile organic compounds). Radon, dioxin, smog. Cancer, heart disease, AIDS. Car accidents, plane crashes, train derailments. Global warming, the ozone hole, rising sea levels. Blizzards, hurricanes, cyclones. Everywhere we turn, it appears we are confronted with some new risk. A few years ago, it was Mad Cow Disease. Today it is pesticide residues. And tomorrow . . . ? In the words of one prominent environmental sociologist, Ulrich Beck, we have become a "risk society."[3]

But what is risk? And why are we suddenly all talking about it? Do we really live in such a perilous time that risk defines the spirit of our age? Are we all seeking an unattainable safety like the characters in Hemingway's novel, or is there something else going on? As of February 2010, only 3 cows in the United States and 18 in Canada had been identified with Mad Cow Disease.[4] And no person is yet known to have come down with nvCJD from eating U.S. or Canadian beef. As of January 2011, there have been worldwide only 219 documented cases of nvCJD—171 in the United Kingdom, and only 3 in the United States and 1 in Canada.[5] Three of the North American cases were people who picked up the disease while living in the United Kingdom, and the fourth picked it up while living in Saudi Arabia.[6] At the time of Oprah's show on the subject, the worldwide figure was about 15 cases—and none in North America.[7] Yet Oprah swore off hamburgers, and her fans roared their approval.[8]

This chapter brings the perspective of environmental sociology to these fears, doubts, and uncertainties. In so doing, I offer a distinction that I believe can help us better understand our worries: a distinction between *risk* and *risky.*[9] When I refer to *risk* in this chapter, I mean our sense of what we should worry about, and how much—the ideal side of worries and fears. When I refer to *risky* in this chapter, I mean the organizational, technological, economic, and biophysical potential of our circumstances for disrupting our goals and intentions—the material side of worries and desires. I try to show that this distinction between risk and risky leads, once again, to an appreciation of the dynamics of ecological dialogue.

Many things in this world can go wrong precisely because there are many things in this world to go wrong.[10] For example, you could get hit by a car walking to work. A rare and expensive book you purchased online could get damaged or lost in the mail. A truck carrying toxic chemicals could run off the road and release noxious gases into your neighborhood. A thunderstorm in the middle of the night could cause a power outage, which causes the alarm on your digital clock not to go off, which causes you to miss the bus, which causes you to miss your final exam in your environmental sociology class, which causes your A to slip to a C+ because your instructor does not allow rescheduling for missed exams, which causes . . .[11] These are all examples of the risky quality of life, of the material worries that daily confront us. But is this to say that there is more risk in life today? You are, after all, unlikely to experience any of these calamities, and you are highly likely to live a longer and healthier life than the people of the past. But you may worry less, or just as much, or even more, albeit about different things. You may, after reading this paragraph, rush out and buy a battery-operated backup alarm clock or install an alarm clock app on your cell phone for the next time you have an exam to take or a plane to catch. You may have one of these already, because of your sense of the risk in your life—your conceptual view of the worries involved in getting successfully through the day.

This chapter, like the previous three, enters the dialogue of ecology from an ideal "moment"— that is, from the moment of the ideal. And as in the previous three, the material side of things is never out of the room, excluded from the conversation. While the chapter focuses on the question of risk, the question of the risky is also continually in our minds. Unavoidably so, for the two—risk and the risky—help constitute and reconstitute each other. In this constitution and reconstitution, as we shall see, risk and the risky raise important questions about some of the most fundamental issues this book has traced: knowledge, democracy, and dialogue.

Rational Risk Assessment

Let's click this time to an *Oprah* episode that never happened. That button there. (You don't know about that one?) Click. Oprah is just introducing her fictional guest. Maynard Haskins, we'll call him.[12]

"Maynard Haskins is a former race car driver turned anti-car activist," Oprah is saying as the pixels come to life. "You hear me? Former race car driver turned *anti-car activist.* We want to know why. He's the executive director of AutoBAN, the Automobile Banning Activists Network, and the author of a new book by that title—*Auto Ban: How We Can Save Millions of Lives and Rebuild a Humane Society.*[13] You said that automobiles make Mad Cow Disease look like the common cold."

"Absolutely," replies Haskins.

"That's an extreme statement, you know?" Oprah challenges.

"Absolutely," says Haskins again, taking the challenge in stride. "But it's a far bigger epidemic than Mad Cow Disease, or even than AIDS. Almost everyone who has gotten Mad Cow Disease has died from it, but that's just 219 people worldwide and only 3 in the United States. By comparison, in 2009 in the United States, 33,808 people were killed in automobile accidents, and 2,217,000 were injured—and that was for just one year. The latest mortality figure for confirmed deaths from AIDS in the United States is for the year 2007, when 17,197 died from the disease.[14] That's a lot, but it's about half as many as died in automobile accidents."

Oprah seems a bit shocked. "But cars are supposed to be safer now. Seatbelts, airbags, better design, and better roads are supposed to be making driving less dangerous."

"That's exactly right," says Haskins. "And they have gotten a bit better. But we're driving so much more. And what we should be doing is exactly what nature says. We should be feeding our cars hydrogen or running them some other nonpolluting way. Because cars actually kill almost as many people with air pollution as they do in accidents—another 32,000 a year in the United States. Worldwide, 1.2 million people die in automobile accidents every year, and at least another 400,000 die from the air pollution cars cause, making 1.6 million deaths in all. Actually, to be strictly fair in our comparisons, we should note that worldwide more people die from AIDS each year—2.1 million in 2007—mostly in poor countries. But we haven't included in these automobile figures any of the millions that die each year from the effects of obesity—heart disease, diabetes, some cancers, and more. Lack of exercise from increased car use is a major part of the world obesity pandemic, although it is hard to calculate exactly how much. Taking that into account, cars almost certainly kill way more worldwide than AIDS does."

"Now see, wait a minute, wait a minute." Oprah goes back to the challenge mode. "Let me just ask you this right now, Maynard. How do you know all this?"

"Oh, I've seen it. These are statistics from reputable researchers from the United States government, the United Nations, the European Union, the World Health Organization, and elsewhere. They're not something we're making up."[15]

Oprah addresses the audience. "Now doesn't that concern you all a little bit, right here, hearing that?"

The audience intones a collective "Yeah!"

"It has just stopped me cold from driving another car!" adds Oprah.

The audience breaks out in applause.

Click. Possibly, a mainstream talk show on a topic like this might take place someday. But it is very unlikely that Oprah, or anybody else, would respond by swearing off cars—to the cheers of the studio audience. And yet, comparison of death rates shows cars to be far, far more risky than the likelihood of catching nvCJD, the human manifestation of Mad Cow Disease, in the United States or even in Britain, where most Mad Cow deaths have occurred. Still, Oprah was not alone in swearing off beef, at least for a time. Millions in Britain did, temporarily sending the price of beef to the cellar, and thousands of British farms with it. And some in the United States did it, too, after the renewed attention to Mad Cow Disease started by that sick cow found in a Washington State herd in December 2003. How many people have sworn off cars?

The issue at hand here is *rational risk assessment*—the idea that we should compare our best knowledge about the rates and probabilities of hazards and choose the least dangerous alternative. Anyone who doesn't is being irrational, right? Risks, according to this perspective, are calculable "facts" to be measured, recorded, and evaluated. In addition, the rational assessment perspective believes not only that risks *can* be evaluated independent of political, social, or cultural context, but that they *should* be.

It's a version of what is more generally called *rational choice theory,* which Chapter 2 discusses. The basic premise of the rational choice perspective is that the individual is a purposeful, calculating actor, seeking to maximize his or her interests. As Jon Elster explains, the rational choice perspective views action as beginning with "the logically most simple type of motivation: rational, selfish, outcome-oriented behavior."[16]

Rational risk assessment has had an extensive following among natural scientists, engineers, government officials, and corporate managers and is usually closely associated with risk-benefit analysis. And while rational risk assessment has been less attractive to social scientists, as we shall see, its influence can still be seen among them.[17] The rational perspective was the earliest theoretical framework for understanding risk. Consequently, it found a place for itself at the forefront of the field that it still retains in many quarters.[18]

Much of what makes this theory so attractive is that it provides a simple understanding of human behavior that can be easily translated into public policy.[19] Other theoretical approaches to risk that try to take account of culture, norms, social structures, and power are difficult to reduce to the simple line graphs, pie charts, and percentage tables that policy makers and the public have become used to seeing. In short, in a fast-food, fast-analysis world, the rational risk assessment provides an easily understood and homogenized formula.

Questioning the Rational Risk Assessment Perspective

Theoretical perspectives, however, are like a house. No matter how sound and well built they may appear on the surface, if their foundations are weak or if they are built upon unstable ground, they can eventually topple. The reign of the rational risk assessment perspective in the field of risk is threatened for this very reason, argue a growing number of social scientists—including me.

Recall the observation in Chapter 2 that people do not always act in their own interests—that people often deliberately act in ways that are other-serving, as much as in ways that are self-serving. People are moved by their sentiments as much as by their interests. That is to say, risk assessments are based as much on norms and values as on calculation. For example, cars are highly valued in modern culture, however foolishly from the perspective of our interests in staying alive.

Of course, the point of rational risk assessment is that we should not be so foolish. As Maynard Haskins does not share the widely held cultural value of cars, he would no doubt quite agree. But most people, for better or worse, do appear to share this cultural value, at least to some degree, and continue to drive cars, even though rational risk assessment advises against it. Culture, in this case at least, trumps rationality.

Well, no, not really, a rational risk assessment advocate might respond. Culture is itself a kind of interest and a kind of rationality. Our norms and values and other-serving ends are as much a part of our interests as our self-serving ends. We can add all this into the calculation of the threat a risk poses to us. No trouble. In fact, a number of rational choice theorists make precisely this more sophisticated argument.[20]

There are two problems with such a response. The first is a philosophical one—that turning everything into interest makes a tautology of rational choice and rational risk assessment. If everything we do is motivated by an interest, then whatever people are doing must already be rational, within the context of the constraints with which they must contend. There can be no irrationality, and therefore no need ever to make a rational argument for doing something differently.

The second problem is that of incommensurability. Let us say that we may wave aside the problem of tautology. Let's call that critique a kind of unhelpful philosophical logic trick. If so, then rational risk assessment and rational choice are still left with the problem of how to compare the utility and interests of, say, death by car accidents, pollution, and obesity with the cultural value of the car as a "dream machine." Cultural values simply are not in the same units of measurement as death rates. Comparing the importance of cultural values with death rates becomes a hugely uncertain endeavor, not easily amenable—and perhaps not amenable at all—to calculation. What is the value of a life? What is the value of a value? Sometimes people will kill to defend a cultural value. Sometimes people will sacrifice their own lives to defend a cultural value. However much others may regard such acts as abhorrent, we do need to recognize that some people sometimes regard their interests in this way. And who is to judge who is right? Which raises another problem of incommensurability: Not only may one person hold incommensurable values, but the values of two or more people may also be incommensurable.

Rational risk assessments often try to get around the problem of incommensurability through risk-benefit analysis. The idea is to put every aspect of a problem in the same unit of analysis so that everything can be directly compared. The usual unit is money. And the usual way to get something like cultural values, death rates, and the value of clean air into monetary units is to assess what people are willing to pay for them, through surveys—contingent valuation is the technical name for it—and other procedures. But contingent valuation presumes that people find it possible to put prices on matters like the value of a life or that they tacitly put prices on such things through their spending behavior. In other words, it presumes the problem of incommensurability away.

As well, risk-benefit analysis tends to be riddled with accounting problems in following the full consequences of a risk through an economy. Much is often left externalized, and critics' standard responses to any risk-benefit analysis is to argue that something left out needs to be put in, and vice versa. Consequently, there are usually contentious conceptual issues about what counts as being an economic consequence and what does not. For example, many might reject the notion that deaths from weight problems can be considered a risk associated with cars, as there are ready ways to make up any calorie imbalance, such as choosing to eat less and exercise more.

Which leads to another conceptual problem with rational risk assessment and rational choice: the degree to which it really is about choice. Because when you think about it, "rational" and "choice" are something of a contradiction in terms.[21] At least in its purest form, rational choice sees us all as carbon-based calculators, constantly assessing our environments in the never-ending quest to maximize utility. If all you are doing is calculating the highest utility, there is no choosing to be done. You are just a computer spitting out answers into your consciousness. To say that individuals have "choice" as rational and calculating actors is no different from suggesting that your computer has a "choice" when you have it add 2 plus 2.

Now, perhaps these criticisms of rationalist perspectives are a bit over the top. After all, most contemporary rational risk assessment and rational choice theorists do not have such a narrow view of rationality. Among other things, they recognize that people's processing ability in making "choices" is limited by time, information, and understanding. Some of them are also working on incorporating norms and values into the theory.[22] But once you begin to deviate from a narrow understanding of rationality—by including norms, values, culture, emotions, and whatever else you care to incorporate into it—where do you stop? The slope is a slippery one. If you include culture in your understanding of rationality, why not trust? And if trust, why not love, and so on. And soon you are left with an understanding of rationality that tells us nothing (well, almost nothing). All it says is that individuals have reasons for doing what they do, which doesn't really add to our understanding.

Plus, perhaps most importantly, rational choice tends to neglect the context in which people exercise rationality. Call it the problem of *contextuality*, which starts from this observation: We often do not choose the context in which we make choices. So is our choice, then, really a choice? The risks we bear in life are often imposed upon us. Often we may not recognize that imposition. For example, most people regard using cars as a personal choice of the most convenient way to travel, and not a choice they make within the context of the social organization of convenience, described in Chapter 3. Powerful social forces, most notably real estate developers and the automobile and oil and gas industries and the influence they have gained in the halls of government, have over time created much of the historical impetus for our car-dominated lives. Consequently, it is often hard to choose to travel a different way. Even if you think trains are safer than cars (as a rational risk assessment would quickly show to be the case), you can't take one if there isn't one. Even if you realize you could get more exercise if you walked or took your bike, you'll find it hard to take the time if development has followed the sprawl model. It may also be quite dangerous if there are no sidewalks or bike paths. Choice, then, is always socially constrained, and thus is not choice in the pure sense we usually associate with the word.

But people usually do not see their use of the car as an imposition. They typically see it truly as a choice, as something voluntary—which is much of the reason why few of us have sworn off cars. As I'll describe in more detail in a little bit, people are far more likely to accept a risk if they perceive it as voluntary. Furthermore, factors such as emotions, a sense of civic duty, and feelings of trust have also been shown to play a role in how we think about risks and what our responses should be toward them.[23] We are embedded within a social context. What we think and how we act are shaped by the social environment we find ourselves in.

Finally, rational risk assessment assumes that the "facts" relevant to an issue of risk can be known. However, the facts are not always so clear, and by extension, then, neither is risk. Risks are not something we can typically see "out there." You cannot see dioxin, radon, or radiation. But even these phenomena are not risks until we define them as such. Rational risk assessment is really rational assessment of the risky, not of risk. Or, better put, rational risk assessment assumes that risk and the risky are the same.

But they are not. What we understand to be risk is embedded within social relations. Our knowledge is influenced by relations of power and of trust.[24] Power relations will influence both the knowledge that comes to your attention and the knowledge that you seek out. Corporations, as I'll discuss later in the chapter, often manipulate information about how risky their products are, and many of us are not well placed to seek other sources. Trust relations will influence what you make of that knowledge when it comes your way. Someone can tell you until they are blue in the face that cellular telephones represent a health risk, whether that person is a Nobel Prize–winning scientist or your neighbor next door. Yet if you do not trust that person's knowledge on this subject as being the truth, or if you do not trust the source from which he or she heard this information, then you will likely not perceive cellular telephones as representing any particular risk.

I return to the relationships among risk, power, and trust at the end of the chapter. It is important to introduce them here, however, because they highlight the inherently social character of risk. And it is this social nature of risk that the rational choice perspective misses most.

The Culture of Risk

In response to the rational choice perspective's rather limited view of risk, centered on interest and utilities, other perspectives have been developed that give more consideration to values, worldviews, and culture. Led by anthropologists and cultural sociologists, a growing number of risk theorists suggest that perceptions of and responses to risk are fundamentally shaped by our cultural conditions. Enter the cultural perspective on risk.

Risk and Threats to the Group

Perhaps the most famous work to have come from this perspective is a book titled *Risk and Culture*, written in 1982 by anthropologist Mary Douglas and political scientist Aaron Wildavsky, which I discussed in Chapter 7 with regard to the "White male effect." According to Douglas and Wildavsky, people tend to accept risks that help to reinforce the social solidarity of their institutions and to reject those that do not. One example Douglas and Wildavsky give is a belief of the Hima, a Ugandan pastoral people, that would seem strange to most readers of this book. According to the Hima people, cattle can die as a result of contact with women, or if someone from within their social group eats food produced by another group while drinking cow's milk. These beliefs, suggest Douglas and Wildavsky, function to reinforce the traditional sexual division of labor among the Hima as well as their separate identity from neighboring farming people. These beliefs come from institutions and strengthen those institutions. Risks, therefore, are not something that exist "out there," distinct from our cultural surround, but instead are artifacts of that surround that dialogically tend to reinforce it as well.

The sociologist Kai Erikson earlier made a related argument in *Wayward Puritans,* his famous study of crime and deviance among the seventeenth-century Puritans of New England.[25] Erikson argued that at times of weakness and uncertainty in a group's social cohesion and sense of its boundaries, the perception of a threat can serve to draw the group together. Even more, Erikson argued that the experience of weak cohesion may lead people to perceive threats to begin with. What turned some old women in Salem, Massachusetts, into witches was not their practice of witchcraft. Rather, it was the people of Salem's practice of group cohesion in an uncertain time. As a new settlement in a new land, social order was very much in question. The discovery of witches of whatever kind "restates where the boundaries of the group are located," in Erikson's words, by clearly identifying a threat to those boundaries[26] (see Figure 9.1).

Perhaps as much could be said of the threat of terrorism today. Perhaps as much could be said of the threat of meat eating, or of cars, or of nuclear waste, or of any perception of risk that leads to the strengthening of group feeling or the constitution of a new social group. As groups form in protest of a perceived risk or threat, they give the adherents a feeling of membership and order—a comforting feeling that can be hard to find in an atomistic society whose rules of conduct seem so much up for grabs.

Let me be quick to emphasize, though, that the implications of the perception of risk for group feeling do not necessarily imply that all danger is witchcraft, pure figments of an over-wrought social imagination. People do die from Mad Cow Disease. People do die in car accidents. People do die from nuclear waste. The Twin Towers did fall. Life is often risky. But there is more to risk than the risky, and much of that more is the influence of group cohesion processes on our perception of the world and its worries.

Figure 9.1 The Salem witch trials.

Source: © Bettmann/Corbis.

Culture and Choice

There is also more to the perception of risk than the need for group cohesion. The work of Douglas and Wildavsky, and of Erikson (from whom we will later hear again), provides a helpful corrective to the individualism and materialist confidence of rational risk assessment. However, we need to be careful not to overbalance the accounts and place too much emphasis on how larger cultural patterns and structures shape perceptions. We need to be wary of the problem of *functionalism*—that of seeing the functions of a larger whole as the only thing that controls the parts. It is people that make up society just as much as larger social structures and patterns do.

Recent cultural accounts of risk have been working on an understanding of culture that is not so top-down. Here, culture is something that both constrains and enables perceptions of risk and the actions we take accordingly. Culture is both something that shapes us and something we shape within those constraints. Culture is something that influences our choices and gives us choices.

Let's consider, for example, the widely noted phenomenon that people are apparently more accepting of "voluntary risks" than "involuntary risks," which I mentioned a few pages back.[27] Recently, scholars have argued that this distinction is too simplistic, since it ignores how culture shapes the choices we have before we even make a decision.[28] Take smoking. We may think that smoking represents a clear example of a case where someone has accepted a voluntary risk—the major risk ultimately being premature death. But smokers do not smoke simply for the sake of smoking. Smokers may, for instance, smoke to suppress their appetites in order to achieve that "ideal" body figure. Or they may smoke to fit into a social group where smoking is considered "cool." Looking good and fitting in are culturally defined attributes of a person that are difficult to ignore in social life. There are, of course, other ways to look good and to fit in than through smoking. But for many people in many social contexts, those other ways may be less readily apparent to them or less available to them. Plus, there are the friends one may lose through smoking. Which risk to risk? When we consider the full context of a person's life, we discover him or her enduring risks as much as choosing them.

Nevertheless, people may not perceive smoking as an involuntary risk, depending on how they conceptualize the context in which they make their choices. Or they may see it in this way, at least in part. How people themselves conceptualize the voluntariness or involuntariness of a risk is key to their "choice" of what is most risky.

Closely linked to that perception is—going back to that phrase two paragraphs earlier—what is "readily apparent" to someone. Central to choice is knowledge. For example, tobacco companies knew the health consequences of smoking yet kept much of that information from the public for decades, and carefully shaped people's knowledge environment through advertising and pressure on government officials. The public therefore "voluntarily" chose to smoke, but did so without a full and wide-awake awareness of the material riskiness of smoking.

This brings us to what has been called *vertical knowledge gaps and horizontal knowledge gaps*.[29] Because of the complexity of modern social life, knowledge gaps emerge among experts and also between experts and nonexperts. We cannot possibly know everything there is to know, from how to rebuild a carburetor, to how to mend a hole in one's roof, to how subatomic particles react when they are exposed to high-intensity electromagnetic fields. The world has simply become too complex for that. It has become increasingly difficult to know all of the facts upon which to base one's actions. *Vertical knowledge gaps* are those instances where information is not communicated across the levels of society's hierarchies. This lack of communication might be part of a deliberate effort to keep options off the table, as in the suppression of evidence about the health consequences of smoking. The evidence was in hand but was not communicated from the "top" of society to the "bottom." *Horizontal knowledge gaps*, on the other hand, refer to those instances when knowledge exchange within a level in the social hierarchy breaks down, for example, between scientists. To refer again to the smoking example, the suppression of evidence by tobacco corporations meant that corporate scientists and university scientists did not share the same pool of knowledge. Often, knowledge gaps can be both horizontal and vertical at the same time, as there was a dimension of the vertical in those different knowledge pools of corporate and university scientists. As a result, even "experts" can be in the dark.

In short, if someone were to offer you the last piece of chocolate cake on the table without informing you that it was moldy, could it be said that you voluntarily chose to risk eating a piece of moldy cake?

Rational Risk Assessment as Cultural Practice

Let's go back to the title of Douglas's and Wildavsky's book, *Risk and Culture*. What I have covered so far in this section on culture is a kind of reining in of risk, pointing out the limits of rationalism because of the cultural conditions under which we make decisions. Now I try to take a cultural perspective deeper, arguing that risk is itself a kind of culture. We might term this perspective as moving from "risk and culture" to "risk *as* culture."

Recall the questions from the beginning of this chapter: What is risk? Why are we suddenly all talking about it? As I discussed earlier in the chapter, it is probably not because we live in more perilous times. The people of the past surely faced worries and hazards equal to or greater than those of the current day. Among other things, they didn't live as long as people do today, as the technological optimist Julian Simon (see Chapter 4) was fond of pointing out. The people of the past also experienced much uncertainty in their lives, and they sought ways to deal with the

Risk can be artificial.

unknown: religion, magic charms around the neck, and rituals performed under a full moon at midnight in a graveyard with a dead cat, such as Mark Twain so wonderfully described in *The Adventures of Tom Sawyer* (see Figure 9.2). Uncertainty is in part a matter of control. Through rituals with dead cats in the graveyard and sacrifices at the altar, the people of the past found ways to comprehend terror and the unknown and thus to gain a measure of power over it.

We are not so different today. We face much uncertainty—many worries, hazards, and scary possibilities that seem beyond our powers to control. Although lives in the modern, wealthy West are normally not so short, and perhaps not so brutish and nasty, there remains much that we cannot be sure about, many dark possibilities that we have to ponder. Living longer does not purge us of doubts and dangers. We still need ways to contend with them.

Figure 9.2 Huck Finn's cure for warts, from *The Adventures of Tom Sawyer.*

"'LEMME SEE HIM, HUCK'"

Source: From Twain, Mark. (1904). *The Adventures of Tom Sawyer*. New York: Harper & Brothers.

What is different about the worries of the present day is not the number of hazards we face or the degree of uncertainty we feel about our lives. Rather, it is the language we use to think and to talk about our worries, a language that, in keeping with the spirit of our time, is highly rationalistic. Consider the connotations of the very word we increasingly favor to talk about our troubles: *risk*. The term immediately conjures up numbers and calculations in a way that words like *hazard* and *concern* and *danger* do not. Risk is imbued with the image of science, of studies that have been done or could be done. Risk turns witchcraft into statistics. Risk turns subjective uncertainties into objective probabilities, sanctified by the iron laws of mathematical logic and scientific method. Along with the "iron cage" of rationalism, to use Max Weber's phrase (see Chapter 6), came the iron cage of risk.

This, then, is what *risk* is—a modern cultural means, a contemporary conceptual language, for confronting uncertainty. The people of the past did not think as much in this rationalist way. They had notions of numbers and chance and probability, of course. But the coming of a scientific and moneyed world order developed rationalism into a defining feature of daily life. Indeed, the word *risk* is a relatively recent arrival in the English language, dating from the mid–seventeenth century—a time of greatly accelerating development of capitalism and its calculating spirit.[30] As the famous sociologist Georg Simmel observed a century ago,

> Gauging values in terms of money has taught us to determine and specify values down to the last farthing. . . . The ideal of numerical calculability has been made possible in practical, and perhaps even in intellectual, life only through the money economy.[31]

Risk (and its equivalents in other Western tongues, such as *risque, riesgo, risco,* and *Risiko*) has come to be the word we most associate with the extension of rationalism into our outlook on danger and uncertainty.

Not that the people of the present have fully embraced the rationalization of risk. For example, a 2010 Gallup poll found that 95 percent of Americans believe that prayer is effective.[32] A 2003 Harris Poll found that 84 percent of American adults believe God can perform miracles.[33] We remain fascinated by stories of the occult and by movies that weave magic and graveyard rituals into the plot. The New Age movement has drawn heavily on Wicca and other traditions in an often vigorously nonrationalistic way.[34]

The language of risk, however, has become—and remains—the most politically legitimate way for us to discuss and debate life's dangers and uncertainties. Part of the reason for the political legitimacy of rationalism is its implicit democratic appeal. There are some close ideological connections at work between rationalism and democracy. Rationalism appears democratic because of its universalist and objectivist claim that it represents the perspective of the entire demos, via the sanctity of the methods of science. Rational knowledge is knowledge that is supposed to be everybody's knowledge, at least potentially. It is unnamed, unidentified with a particular standpoint or culture, applying to all and representing all, and thus apparently democratic. Governmental agencies, not coincidentally, typically defend their authority through such rationalist and implicitly democratic claims.

There may indeed be much democratic potential in rational risk assessment, given science's commitment to debate, as I come to at the end of the chapter. The point of a cultural perspective on risk is not to throw rationalism—and with it the material side of ecological dialogue—out the window. Rather, the point is to balance rationalism with the ideal side of the dialogue, showing the equal importance of how we socially construct our understanding of the material.

With that balance, we can recognize how rationalism can be a highly effective political tool for those who are able to wield its language to their advantage, often overriding its democratic potential. For rational risk assessment quickly leads to the rationalist claims of the state's regulatory authority. To speak of risk is to speak not only of control but also of power. It represents the extension of rationalism into our comprehension of the unknown, giving us a sense of control over uncertainty in a way that appeals to the modern mind. It as well represents the legal authority of expertise to order our lives in ways that may advantage the interests behind the experts more than those of us in front of the experts, down in the audience. Risk explains, and it also explains away. It gives control, and it takes control, and therefore we often feel culturally trapped in the iron cage of risk.

The Sociology of Disasters

But life is indeed risky sometimes. It has its material worries and uncertainties. And sometimes, unfortunately, those worries and uncertainties become realities, disastrous realities. These disasters not only affect individual people, but can also threaten the social fabric that links us one to another. Sometimes the experience of disaster can bring people together, as the flight attendant discussed at the end of Chapter 2 found, thereby allowing for both individual and collective recovery.[35] But other times disasters unfold in such a way that the fabric of human community is left in shreds, leaving its members flapping in the wind in shock, destitute, and alienated.[36] This social shredding is particularly true for disasters whose cause bears distinctly human fingerprints.

Take the common phrase "natural disaster." Scan a recent newspaper article describing a flood or a storm or a wildfire or an earthquake, and one is likely to find such phrases as "an act of God" and "Nature strikes again" strewn throughout the clipping. As long as we attribute the cause of a disaster to "natural" (or "supernatural") causes, it remains in the morally innocent realm of what earlier chapters termed the *natural conscience*.

But consider this: Much of the flooding that occurs today is the result of two centuries of human-induced landscape change: deforestation, the destruction of wetlands, the channeling of rivers, the spread of roads and other impermeable surfaces, and the like. Global warming also affects rain patterns and the frequency of extreme weather events like hurricanes and cyclones. Is flooding, then, a natural or "unnatural" act? The wildfires that rage in most years in the American West are in considerable measure the product of forest management practices, and the damage they cause must be attributed at least to some extent to the decisions of developers and governments to locate housing in vulnerable areas. We don't control earthquakes, of course. But the buildings that are most likely to fall are the shoddy ones built by unscrupulous contractors, sanctioned through payouts to equally unscrupulous government officials, as the residents of northwestern Turkey found on August 17, 1999, and the residents of Port-au-Prince in Haiti found on January 12, 2010. And while the Japanese authorities can't be blamed for the March 11, 2011, earthquake and tsunami, there are many now who doubt the wisdom of the planners and designers who put six nuclear reactors all together and so near the coast at Fukushima Daiichi, as well as running all the wiring for the backup cooling systems through the reactor basements, where flooding could take them all out at once.

For example, New Orleans has been questioning the naturalness of its horrific flooding ever since Katrina struck on August 29, 2005. Katrina was not the first big hurricane to score a bull's-eye on New Orleans, but it assuredly caused the worst damage, by far. Why? The widely reported poor condition of the levees had something to do with it. But the main culprit was a shipping canal begun in 1963 that connected New Orleans to the Gulf of Mexico in a straight shot, bypassing the Mississippi River's long and winding way. The Mississippi River Gulf Outlet canal is the official name; everyone calls it the "Mr. GO." But the fresh water of the Mississippi doesn't go down the Mr. GO. Rather, salt water from the Gulf comes up it. And ever since the Mr. GO was dug, this salt water has been killing the cypress swamps that, in addition to the natural levees of the Mississippi River, have long protected New Orleans. Plus, the Mr. GO poured the rising Gulf waters from Katrina straight into the city (see Figure 9.3). As the environmental sociologist William Freudenburg and his colleagues have written, the Mr. GO "created a hole in the wetlands, as well as in the life of a great city, that may never be filled."[37] In other words, this wasn't a case of a natural disaster, but rather a disaster for nature—and thus for us.

Figure 9.3 Were the Hurricane Katrina floods a natural disaster? The "Mr GO" Canal, or Mississippi Gulf Outlet Canal, punched a hole through New Orleans' protective barrier of cypress swamps when it was begun in 1963, funneling Gulf of Mexico water directly into the city and killing the swamps by allowing salt water intrusion. Both effects contributed greatly to the flooding associated with Hurricane Katrina in 2005.

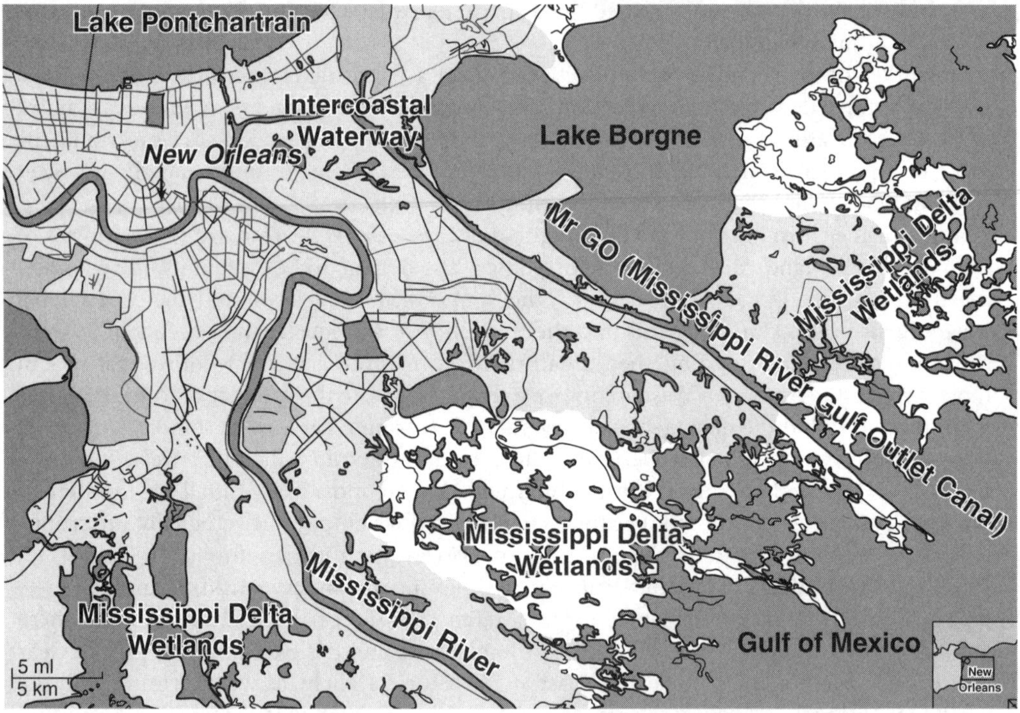

Source: Based on information from Freudenburg et al. (2007).

Realizations like these anger people, understandably. But more than that, they undermine our trust, not just in the security of the world, but also our trust in each other. This is the type of trauma that can be especially difficult to recover from.

A New Species of Trouble

The work of Kai Erikson has helped point sociologists' attention to this trauma. Erikson studied the devastating effects of a 1972 flood on the mining communities along Buffalo Creek in the West Virginia Mountains.[38] The flood that literally wiped out these small towns, and the lives of 125 people, was not solely the result of an unusually heavy rain. The local mining company, which employed someone from most households in the valley, had constructed an earthen dam at the top of the valley, so as to form a settling pond for the murky waters pumped out of the lower reaches of the mines. The dam was poorly made and poorly maintained, and one February morning it simply gave way, unleashing a coal dust–blackened tidal wave. In a matter of 3 hours, the water scoured the 17-mile valley of its 16 villages. As one witness of that terrible morning described,

> I cannot explain that water as being water. It looked like a black ocean where the ground had opened up and it was coming in big waves and it was coming in a rolling position. If you had thrown a milk carton out in the river—that's the way the homes went out, like they were nothing. The water seemed like the demon itself. It came, destroyed, and left.[39]

What Erikson discovered in the wake of that demon flood was not a "city of comrades," the term given by earlier disaster researchers to describe the renewed social solidarity that sometimes results after a disaster has subsided.[40] Instead, he found what has recently been termed a "corrosive community"—a place where community itself becomes soiled, ruined, and fractured and people flounder in a deep, unshakable malaise.[41] Central to that corrosion was the sense of betrayal the residents felt. The coal company, and some misguided relief efforts, were supposed to provide a network of social trust, the looking-out-for-one-another that helps make us all safe. Instead, the people of Buffalo Creek were left bereft of the moorings of social support that, until they are gone, we often do not recognize as essential to the stability of a self.

"A profound difference [exists]," notes Erikson, "between those disasters that can be understood as the work of nature and those that need to be understood as the work of humankind."[42] Erikson calls these human-induced disasters a "new species of trouble," and they have several characteristic and interrelated features:

- The crumbling of trust and the shredding of community ties;
- A chronic social trauma that victims only slowly recover from, if at all; and
- A pervasive sense of dread about what the future will bring.[43]

According to Erikson, these new troubles contaminate our minds as much as they may damage our bodies. Whereas "natural" disasters suggest a lack of control over processes beyond the reach of humans, technological disasters suggest a perceived loss of control over processes thought to be within the reach of humans and our commitment to care for one another. People find this loss deeply troubling.

The element of dread, Erikson notes, is particularly characteristic of disasters caused by a toxic release, such as happened in Bhopal, or in Grassy Narrows, or as many fear is ongoing with regard to pesticides in our food (see Chapter 5). Instead of leaving easily noticed visible damage, like collapsed bridges and devastated homes, toxics can be more insidious. They often lurk in a realm beyond the visible and can last for generations.[44] As has been said of the Chernobyl tragedy, most of its victims are yet to be born.[45] These new traumas do not take the form of broken windows and shattered bone. We should be so lucky. Rather, they manifest as a pervasive sense of threat from what may come: birth defects, lumps within the lungs, bleeding welts on the skin, premature death.

The dread associated with even the possibility of such harm numbs the spirit and clouds a person's sense of purpose and ability to navigate life. New species of trouble, then, do not exist merely in the here and now. Rather, they are something we carry with us into the future—the unwanted baggage required for riding on the airline of modernity.

Normal Accidents

While society cannot fully prevent "natural" disasters, there is a general belief that we can at least anticipate and prepare for such events, and we find a certain degree of social and psychological comfort in this.[46] Technological disasters, on the other hand, are viewed in just the opposite manner—they are understood as preventable. But there is no way to fully anticipate and prepare for some technological disasters. As the sociologist Charles Perrow notes, many of our technologies have become simply too complicated and involved.[47]

Perrow asks us to look at technologies as systems, and from two dimensions, so as to understand their potential for catastrophic failure. The first is the extent to which a system's interactions are linear or complex, and the second is the extent to which a system is *loosely coupled* or *tightly coupled.*

In a linear system, components are spread out, easy to segregate, substitutable, and have little feedback between each other. An example might be automobile transportation. A complex system is the reverse: Components are close together, hard to pull apart or substitute, and have a lot of feedback. The classic example would be a nuclear power plant.

Coupling refers to the degree of "slack or buffer or give between two items," writes Perrow.[48] A nuclear power plant is both complex and tightly coupled. If one component fails, it may create considerable stress on several others. A university is a complex system that is loosely coupled. If one student fails, others are unlikely to fail because of it. Mining might be another. If one shaft caves in, others are unlikely to cave in because of it. Automobile transport may be linear, but it is tightly coupled. If one driver makes an error, it is likely to have immediate consequences for several other drivers, but not the whole system of drivers. An elementary school would be a linear system that is loosely coupled. It is similar to a university in its degree of coupling, but far less complex (see Figure 9.4).

Perrow uses this simple classification to make a profound observation. In systems that have complex and tightly coupled interactions, we should expect the occasional occurrence of a serious accident. Complex interactions confuse and confound those who try to deal with a breakdown in a system, and tight coupling gives them very little margin for error. The operators of a nuclear power plant, say, have very little time to figure out what is going on when meters start giving unusual readings, and there is little charity in the system for a mistaken interpretation. Perrow calls such events "normal accidents." Yes, they are accidental. But they are also, if not predictable, certainly unsurprising.

The fact is, no engineering manual of procedures can cover every eventuality. Even if a manual that large could be written, it would still have to be comprehended by those trying to respond to, say, two valves inadvertently left closed by a maintenance crew. A repair tag on the control panel that obscures the indicator saying the valves are closed. A pressure relief valve that subsequently fails to close when it should have. A faulty indicator that tells the control room that this valve did close when it didn't. A few operators who don't quite follow procedures.

These contingencies are precisely what led to the infamous accident at Three Mile Island. The two valves left closed prevented sufficient coolant from reaching the reactor core during a test. The operators shut down the backup cooling system too soon. The pressure relief valve left open allowed 32,000 gallons of overheated coolant to steam out of the core and into the air above the plant, carrying substantial radioactivity with it.

The governor of Pennsylvania soon made an announcement advising roughly 3,500 pregnant women and preschool children within 5 miles of the plant to evacuate the area. So what would you do, even if you weren't pregnant or a preschooler? Probably what 150,000 people did—flee, an average of about 100 miles, apparently.[49] But that was either not far enough or not fast enough, for the region now has unusually high levels of several cancers[50] (see Figure 9.5).

Perrow's reasons for making this argument are several. Foremost, he wants to make a practical point: that complex, tightly coupled systems are risky. We now have a litany of examples to illustrate this point: Three Mile Island, Chernobyl, Bhopal, Toulouse, the *Challenger* accident,

Figure 9.4 Perrow's theory of "normal accidents": Systems that are more complex and more tightly coupled will be more prone to unpredictable accidents with catastrophic potential.

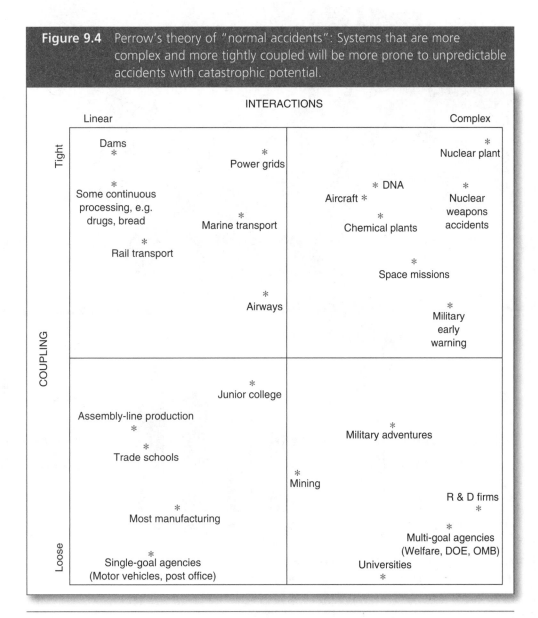

Source: Adapted from Perrow, Charles. *Normal Accidents*, Rev. ed. Copyright © 1999 by Princeton University Press. Reprinted by permission of Princeton University Press.

Figure 9.5 Fallout and post-accident lung cancer rates in the Three Mile Island region.

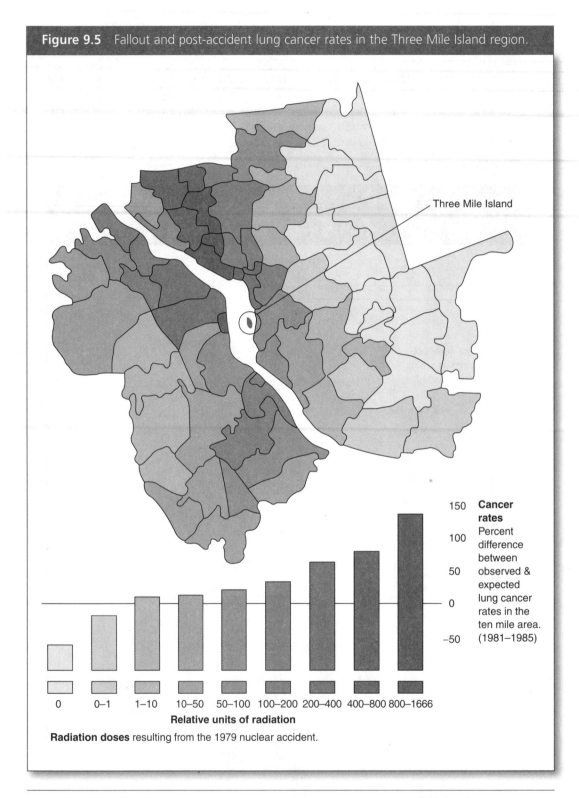

Source: From Wing et al. (1997).

the *Columbia* accident, Fukushima Daiichi, September 11. The last is a particularly troubling example from a systems theory point of view, because here we learned that two systems we thought of as separate—airplanes and skyscrapers—were not.

All of these disasters, and perhaps especially September 11, lead to a second observation Perrow wants us to recognize: We could use a good dose of humility with regard to our rationalist confidence about technology. No amount of rational calculation of risk probabilities could have predicted any of these disasters. Rather than confidently calculating odds of the 1,000,000 to 1 variety in making decisions about riskiness, we need a more qualitative approach that understands accidents (including deliberate accidents, like terrorism) as a normal property of systems. "Unruly complexity" is what the environmental sociologist Peter J. Taylor calls it. You just can't control something so complex, at least not completely.[51] Expect the unexpected, both Taylor and Perrow tell us.[52] And that should give us considerable pause with regard to systems with a high potential for catastrophic implications.

Finally, Perrow wants us to appreciate that technological systems are social systems. It is not human error on the part of operators that leads to technological disasters. Rather, it is forms of social organization—which is really what technology is, as Chapter 3 describes—that create situations where even highly trained, highly capable people are unlikely to be able to respond quickly enough, and correctly enough, to prevent a catastrophe. Thus, a technological disaster is really a *social* disaster in two senses—both in the organization of a dangerous situation and in its effects on us and our sense of confidence in each other.[53]

A Risk Society?

Like Dr. Frankenstein, we are in danger, as we seem increasingly to fear, of losing control of our own technological creations. We are as well losing the technological and rationalist confidence we once enjoyed, perhaps chastened by normal accidents, new species of trouble, risks that now seem less than voluntary, and other inadequacies of rational risk assessment. It is this Frankensteinian state of technological doubt and contention that the German environmental sociologist Ulrich Beck has termed a "risk society."[54]

Beck argues that Western societies are headed this way, and it is leading to a major reconfiguring of the basis of social conflict. Formerly, the central conflicts in society were class-based struggles over money and other resources. But in the risk society, conflict shifts to non–class-based struggles over pollution and other social and environmental bads.[55] In short, we in the West are moving, says Beck, from conflicts over the distribution of goods to conflicts over the distribution of bads.[56]

Significantly, many of the bads of modern society—pollution, technological hazards, ugly development—are hard to escape. They affect everyone. We are increasingly subject to a perverse new form of equality, an "equality of risk," as Beck calls it. In Beck's words, "The driving force in the class society can be summarized in the phrase, *I am hungry!* The movement set in motion by the risk society, on the other hand, is expressed in the statement, *I am afraid!*"[57]

Maslow's theory of the hierarchy of needs assumes that modern people have largely moved past issues of material wants and worries, and can now afford to focus on "higher" values, like aesthetics and social relations (see Chapter 2). But, says Beck, we are slipping on the ladder. In Beck's words, "security is displacing freedom and equality from the highest position on the scale of values."[58] Gone is our confidence in what Beck calls the *risk contract*, the system of rules and rights that governments have established to protect citizens from the hazards of modernity, or at least to compensate them.

Part of this fear is the public's pervasive sense that science is out of control and that scientists are, too. This sense, of course, increases the risk that people feel. It also leads people to question and doubt science. Our faith in the technological god has been shattered by pesticides in our food, by toxins in our water, by Chernobyls in our air, and by scientists in our news media who can't agree with each other. A risk society is not an optimistic society.

To some extent, the public's concerns are materially warranted, Beck agrees. Bad things do often happen even to the best of people in an industrial society. Our technologies and sciences do indeed sometimes result in a Pandora's box of horrors: accidents, poisons, diseases, disasters, and more. But Beck also stresses that risk society is about more than the *materially* risky. It is also about the *ideas* of risk we bring to bear on the risky. These ideas are embedded in stories we find variously compelling, and that social actors perform in front of us. Contemporary social life is a kind of political theater, Beck suggests, in which what he calls the *staging of risk* plays out.[59]

This shift from society's focus on "goods" to its focus on "bads" reflects a restructuring of social organization. In the words of Beck, in class society, "being determines consciousness," and in risk society, "consciousness (knowledge) determines being."[60] What he means is that in class society, your material position—your income, your employment, your place of residence and upbringing—dominated your sense of who you were. In risk society, it is your ideas and beliefs that matter most, including worries you may have about the trustworthiness of the social and technological world. We are thus moving from a risky life to a life of risk.

Yet here Beck sees potential signs of hope. If ideas and beliefs now matter most, then maybe we can regain control over where science and technology are taking us. Maybe we can move from the modernization that was associated with class society to a new modernization, a "second modernization" that Beck terms *reflexive modernization*—a form of modernization in which we think critically and engage in democratic debate about science and technology. In Beck's words, the goal of reflexive modernization is "to break the dictatorship of laboratory science . . . by giving the public a say in science and publicly raising questions."[61] By *reflexive*, then, Beck does not mean "reflexes" or a mirror's "reflection," but "self-confrontation" in which we collectively reflect on the meanings of modernity, science, and rationality.[62] And if we do not, we may plunge into a period of escalating fear and conflict over risk.

Indeed, risk is fast becoming global. "Risk society," says Beck, "means world risk society."[63] Yet it remains to be seen whether the global dimensions of risk result in new global coalitions or in new global conflict.

Beck's ideas have gained a lot of attention in recent years from environmental sociologists as well as political sociologists. His suggestion that risk is fundamentally reshaping the politics of our times has caused many scholars and many politicians (at least within Germany, where Beck is from) to reconsider the significance of environmentalism. For many years, political commentators have generally viewed environmentalism as just another issue, one more demand on our political attention. If Beck is right, environmentalism is, or is fast becoming, the defining political issue of our time.

Questioning the Risk Society

Although the idea of risk society has been getting a lot of attention, it has also received considerable criticism.[64] Much of the criticism has focused on the notion of the perverse equality of risk society compared with the class inequalities of industrial society. "Poverty is hierarchic, smog democratic," Beck has written.[65] To be sure, smog affects everyone, rich and poor. So do global warming, the ozone hole, species loss, landscape destruction, pesticide residues in food, and the like. But we cannot doubt that the rich are in a far better position to avoid the worst consequences of all these threats, as Chapter 5 describes. The distribution of environmental bads (as well as environmental goods) is far from equal. It is true that even the wealthy must contend with environmental bads, and in this fact lies some hope for building political coalitions that cross the lines of race and class and nation. Still, that doesn't mean the rich bear the risks equally with the poor. Wealth remains an important cleavage in environmental politics.

Beck now accepts this point in his recent writings, to some extent.[66] But nonetheless, Beck's framing of the issue of risk seems of little help in understanding the environmentalism of poor countries. One of the central driving forces of political conflict in poor countries originates very much from the feeling that "I am hungry." Yet poor countries nonetheless have vigorous environmental movements, as Chapter 7 discusses, just as the poor of rich countries have developed vigorous environmental movements. These movements are heavily based on class conflict—which is not to say that risk isn't involved. Disruptions in the environmental resources upon which you depend are particularly worrisome when you're poor.

Also, it may be that Beck's risk society theory fits his native Germany better than it fits most other wealthy countries.[67] The spectacular success of the German Green Party is, well, spectacular. No other country has seen the environmental agenda take such a central political role (see Figure 9.6). Meanwhile, environmental issues have a lower, in some cases far lower, political profile in most other Western nations. Beck is right, though, to point to the frequent centrality of risk issues when green concerns do receive attention, such as in the debate over Mad Cow Disease.

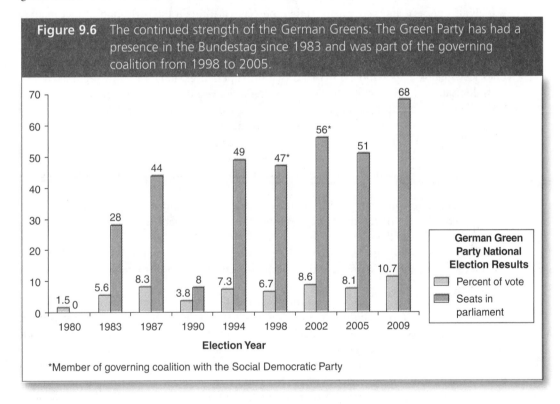

Figure 9.6 The continued strength of the German Greens: The Green Party has had a presence in the Bundestag since 1983 and was part of the governing coalition from 1998 to 2005.

*Member of governing coalition with the Social Democratic Party

But the main difficulty with the risk society theory is that it overstates the degree to which we have overcome economic anxiety and economic conflict. In this, Beck's ideas bear some similarity with those of Inglehart and his theory of postmaterialism, also discussed in Chapter 7. Wealth creation remains the central political goal of every country in the world, rich and poor. Meanwhile, recent years have seen a great increase in economic inequality both between and within nations, reversing the equalizing trend that had prevailed during the middle decades of the twentieth century. In this, Beck seemingly ignores a persistent feature of contemporary times: People all over the world, in both rich countries and poor countries, still find themselves going to bed every night feeling "I am hungry."

But like many of the other theories discussed in this book, the fact that risk society has attracted so much critical attention does not necessarily discredit it. Quite the reverse. Because Beck's work helps clarify so much, other scholars have worked out its implications with unusual care, sometimes uncovering problems. As I note later in the chapter, such is always the case with the best of science.

Risk and Democracy

One of the important implications of Beck's ideas is that for all its problems, rationality is not irredeemable. It won't look the same anymore if we do manage to redeem it. But there are aspects of the spirit of rationality that are not without merit. What we need is a subtler, wiser, less arrogant rationality. What we need is a rationality we can argue with, that respects the reasoning of others, and that takes that reasoning into consideration when it answers with reasons of its own. What we need is, well, a *dialogic rationality*.

Part of the problem is that we long misunderstood what science was about. We thought science was about universal and permanent truths, consecrated through the perfection of method. We thought the point of science was to provide us with final answers. But final answers are the purpose of dogma, not science. Science does give us answers, but it does not give us *final* answers.

In other words, science is about debate and the willingness to continually refine and change one's views in the light of more observations, more experiences, and more good arguments. The truly dedicated scientist is not the one who wishes to make a discovery that will stand up for all time. If our understandings about something never changed again, this would only indicate that the topic of the "discovery" was so irrelevant to the human project that no one ever bothered to pursue the matter further. No, the most dedicated scientist is the one who most truly and fervently wants his or her research to go out of date as quickly as possible. For if it does go out of date, this is a sure indication that the topics and insights of the work were important enough to consider further.[68]

There is always something more to know, and something more to say. When we look to science to resolve the uncertainty of risk, we get the role of science exactly backward. Science *is* uncertainty. Thus, in a way, science is risk itself. For science is a matter of answering every question and questioning every answer. It never stops. If this unfinalizable dialogue becomes what we mean by rationality, we can then perhaps recover the word from its high-handedness and exclusivity.

We have much work to do to attain that recovery, however. Currently, the all-too-common use of the language of science and rationality is to exclude others from the conversation. Science, we hear, is something that scientists do. If you are not a scientist, your views are not valid. Scientists use this rhetorical move to defend their authority, and government and business interests often call upon it, too. Yet science is not something that scientists do. It is something that people do. But it is only science when they commit to engaging the views of others, to taking those views into consideration, to offering reasoned responses to them, and to being open to change in response to the evidence and arguments offered by others.

There can be citizen scientists, citizens who are scientists—citizens who offer reasoned evidence and arguments on scientific issues and who commit to changing their views in the face of the evidence and arguments of others.[69] Citizens can analyze evidence and even gather their own evidence in ways that avoid the known mishaps of method, a practice sometimes called *popular epidemiology*.[70] There can be scientist citizens, scientists who are citizens—scientists who respect and respond to the reasoned views of the public and engage the concerns of the public in their work. And citizen scientists and scientist citizens can work together, and indeed increasingly do so through the practice of participatory research in which citizens and scientists form a common research team.[71]

It is here that the democratic potential of rationality lies. What might make risk assessment democratic is not the assumption of a universal perspective. Once again, it is the reverse. What might make risk assessment democratic is a willingness to engage a universe of *perspectives,* plural.

The Dialogue of Risk

But in order to get this dialogue of risk going, we're going to have to greatly improve our communication with one another. And unfortunately, there are powerful social interests who frequently and deliberately stand in the way. Let's look at a few examples from food and agriculture (but it increasingly goes on in all sectors of research and the economy).

Take the example of the famous *Oprah* episode on Mad Cow Disease, with which I began this chapter. Shortly after Oprah made her rather offhand remark about never eating another hamburger, she found herself served with a libel suit by the Texas Cattlemen's Association for defaming their product. The Cattlemen claimed that the bottom fell out of the cattle market immediately following Oprah's declaration about burgers, causing a 10 percent drop in cattle futures. Moreover, in 1995, Texas became one of an eventual 13 states to pass "food disparagement" statutes that provided restrictive libel standards for anyone stating in public that a particular food product was unsafe. Under such laws, it is not entirely clear that the first President Bush would have been clear of being libelous when he stated in 1990, "I'm President of the United States and I'm not going to eat any more broccoli."[72]

So the Cattlemen took Oprah to court, twice, and lost both times. Upon emerging from court victorious, Oprah declared, "Free speech not only lives, it rocks!"[73] But by the time it was all over in 2000, Oprah had spent more than $1 million defending herself.[74] The Cattlemen had made their point. If you're going to say anything bad about food in public, you had better have the funds at hand for a substantial legal battle. SLAPPs—strategic lawsuits against public participation—is what the environmental sociologist Penelope Canan and her colleague George Pring call this practice, and not without reason.[75] Such is increasingly the price of free speech.

Scientists researching environmental risk also sometimes have to face off with corporations in order to communicate their work to the public. Corporations can wield considerable influence in the government agencies that fund much scientific work, and they have used that influence to suppress the publication of findings and to control the work that gets funded to begin with. James Zahn was an award-winning swine researcher at the U.S. Department of Agriculture's Agricultural Research Service, and he worked in an Iowa lab on issues of pollution from hog confinements. Zahn's research found that air emissions from hog confinements contained antibiotic-resistant strains of bacteria that might threaten human health. Hog-farming industry groups soon found out about the research—Zahn wasn't trying to hide it—and put pressure on Zahn's superiors at the lab to put the findings in the deep freeze. They complied. They told Zahn he could not submit the work for publication in a scientific journal, nor could he speak in public about it. The results eventually did get around by word of mouth, and the story eventually hit the papers late in 2002. But by that time Zahn had quit the lab in frustration and had taken another job in another state. At this writing, the research still has not been published.[76]

During this controversy, it emerged that the U.S. Department of Agriculture (USDA) for some-time had been maintaining a list of "controversial" research topics that require permission from the department's Washington office before funding for them can be approved. (Zahn believes the department amended the list specially to include his work.) The *Des Moines*

Register, the leading daily paper in Iowa, requested to see this list and reported that it "appears to require special permission to study anything involving agricultural pollution of air, water or soil."[77] Such work can still be funded, but the delay of the special permission requirement makes it harder to meet grant application deadlines and discourages scientists from investing time in a research topic that might not be approved. Scientists need to publish research to keep their jobs, and practices like these encourage them to choose safe topics—which in this day and age increasingly means topics that do not offend corporate interests.

Zahn's case was not an isolated incident, as other examples from agricultural and food research show. James Russell, a USDA researcher at Cornell University, had to abandon some of his research after his department was pressured by industry interests and by other researchers associated with those interests. And it was the mildest of findings. Russell had found that if cattle are fed hay instead of grain in the days before they are slaughtered, E. coli contamination in the meat was reduced. Such a finding suggests that E. coli contamination can occur in meat (which is widely documented) and that confinement feeding operations, which rely on grain for the feed, are part of the problem.[78] So the pressure for suppression was soon on.

Then there's the case of North Carolina State University professor JoAnn Burkholder, who found that hog confinement operations in North Carolina had polluted a stream with 15,000 times the legal limit of bacteria. That's 15,000 times the limit! In 1997, she publicized her findings so parents would know to keep their children from playing in the stream. Her university was soon flooded with requests from agricultural interests for her dismissal. She even received death threats.[79]

It takes only a few stories like this to scare researchers and thus have a far broader effect. But perhaps the most controversial is the "Pusztai affair," the case of the British scientist Arpád Pusztai, who lost his job in 1998 after speaking in public—with permission from his superiors—about his rat-feeding studies of genetically modified potatoes. Using a specific kind of genetically modified (GMO) potato (which had been modified with a gene from snowdrop flowers), he found that rats fed on them developed shrunken internal organs, had their immune system depressed, and grew more slowly.[80] Pusztai's research was eventually published in the respected British medical journal *The Lancet,* but amid the storm of a parliamentary inquiry, a critical review of his work by the British Royal Society, and reports of a threatening phone call to the editor of *The Lancet* for having published the paper.[81] All this for some modest, tentative, preliminary findings that had negative implications for powerful corporate interests.

Dialogue and the Precautionary Principle

Censorship and intimidation are hardly conducive to dialogic rationality concerning risk. We need to encourage debate, but debate that engages each other's views, rather than ignoring them, silencing them, or shouting them down. This is particularly the case in situations of uncertainty and disagreement concerning potential harm.

The global debate over GMOs is a good example. Many people see the prospect of genetically modified crops with considerable alarm. Some think there is nothing to worry about at all. Many find themselves somewhere in between, with concerns about transgenetic crops—that is, crops that involve genes from different species—but tolerable comfort with genetic modification that stays within a plant's hereditary genome. Some find studies like Pusztai's disquieting. Some find comfort in studies with GMO-friendly results. Some are puzzled over the contradictory findings and don't know what to think. All told, a 2010 Eurobarometer survey of the 27 European Union member states found only 23 percent who supported the use of GMOs in food.[82] Support was higher in the United States at 45 percent in a 2005 Gallup poll, but not overwhelming.[83]

In response to situations like these, a number of scientists, government officials, environmental advocates, and even corporate executives have begun promoting the *precautionary principle*. In January 1998, a group of 32 representatives from Canada, Europe, and the United States gathered at the Wingspread Conference Center in Racine, Wisconsin, to draft the Wingspread Statement on the Precautionary Principle. The statement defines the precautionary principle as follows:

> When an activity raises threats of harm to human health or the environment, precautionary measures should be taken even if some cause-and-effect relationships are not fully established scientifically.[84]

The British mathematician Peter Saunders, an advocate of the principle, puts it more plainly:

> All it actually amounts to is this: if one is embarking on something new, one should think very carefully about whether it is safe or not, and should not go ahead until reasonably convinced it is. It is just common sense.[85]

Look before you leap. Think before you act. And, most important, talk things out with others.

We'll need some time to do that, of course, which is the controversial aspect of the precautionary principle—especially for corporations eager to start getting a return on their investment in research and development. Some people will never be satisfied, critics of precaution respond. It's just a delaying tactic. Such an attitude, though, betrays the social commitment to dialogic rationality—to genuinely engaging the points brought up by others—for it implies that the impatient party has already made up his or her mind about who is right: me.

With regard to GMOs, many prominent scientific, professional, and governmental bodies have suggested that precaution is just what is needed. Has the debate proceeded far enough already? The journal *The Lancet* in a 2002 editorial stated no, "Consumers are probably right to be skeptical at present."[86] In 1999, the British Medical Association advocated a moratorium on the introduction of genetically modified crops until more is known, although the group later considerably softened its position. The current view of the World Health Organization (WHO) is that because "GM organisms contain different genes inserted in different ways," their "safety should be assessed on a case-by-case basis."[87] In view of this continued uncertainty, as of September 2010, at least 169 regions and more than 4,700 municipalities in Europe had declared themselves GMO-free zones, including all of Poland, Greece, and Austria; the Australian state South Australia has also banned GMO crops.[88]

Of course, at some point we have to make a decision before everything is known and everything is said. That's because everything will never be known and everything will never be said. And to this role we rightfully look to the decision-making power of government. But let's stop and talk about it first, at length, honestly, freely, and with respect for what others have to say.

Trust and the Dialogue of Risk

How do we get the dialogue of risk going? One thing is for sure, comments like the following from the U.S. Environmental Protection Agency's *Handbook for Environmental Risk Decision Making* don't help: "The bottom line is that our society is like a bunch of spoiled brats who want an affluent lifestyle based on a throw-away society, supplied by synthetic chemistry and risk free at the same time."[89] Nor does it help when the CEO of a midsized company proclaims the importance of "managing fear" and of "finding a way to help the public manage its growing anguish and still accomplish our business objectives."[90] Belittling the public is not the way to engage others in open, democratic dialogue.

No, what we need to repair most of all is a sense of trust, if we are to build a dialogic rationality of risk.[91] For knowledge is a social process, as Chapter 10 will describe. Most of what we know we learned, at least in part, from others. It has to be that way. Who has time in this life to experience the whole world? Do you need to do the lab tests yourself to suspect that dioxin is carcinogenic? Do you need to have lived in the deposition zone from Chernobyl or Three Mile Island or Bhopal to consider it awful to live with the dread of a chemical time bomb ticking away inside you? No, you don't. But you need a basis for trusting the reports of others about all of these things.

The best way to build that trust is when people are open to questions that probe their reports of their experiences in the outdoors, in the lab, in their towns, in their families, in their own bodies—and when they are open to any considered reasons others might have for disagreeing. That doesn't mean we all have to agree. That doesn't mean that there cannot be uncertainty. For perfect agreement and certainty would lead us back to dogma, not science. And no one should trust dogma. Rather, trust should be about a willingness to disagree and debate and still retain a sense of social connection—a willingness to keep the conversation going, always open to further news, further ideas, further perspectives brought into our lives, and a confidence that others will have the same willingness.

Risk in the sense of uncertainty, then, can never be finally resolved. Nor should we try, for this is the paradox of risk: To seek its final resolution is to give up on the surest means we have for dealing with its consequences—the unfinalizability of a democracy of knowledge.

PART III

The Practical

CHAPTER 10

Mobilizing the Ecological Society

We have a purpose. We are many. For this purpose we will rise, and we will act.

—Al Gore, 2007

There is an old fable that I enjoyed telling to my children when they were young, "Androcles and the Lion." It is indeed old, 2,000 years at least. Some say it dates from the time of Caligula, who ruled the Roman Empire from 37 to 41 CE, and that it was written by Apion, a scholar from that day, and that it may be based on a real incident. Others say that it originally comes from Aesop, the Greek storyteller from the sixth century BCE. The earliest extant version of it is in the *Noctes Atticae*, or *Attic Nights*, of Aulus Gellius—20 volumes of random stuff that Gellius, a minor Roman official, scribbled down to pass the time while on a posting to Athens, which was in Attica.[1] Gellius himself, at least, says he got it from Apion, not Aesop, and that Apion claimed to have been an eyewitness to the story. In any event, it is plenty old.

The more I reflect on it, the more I see why people have continued to tell this story for so long. The more I reflect on it, the more I see it is not just a story for the young. The more I reflect on it, the more I see it has an important message about community and environment and how to bring the two together into that biggest community of all. Here's how it goes (or at least how I like to tell it).[2]

There once was a slave named Androcles who belonged to the Roman governor of Africa. The governor was a cruel master and used to beat Androcles mercilessly. One day, Androcles saw his chance to escape, and he ran away into the wilderness. After running for hours, he spied a cave where he thought he could rest, hide, and spend the night. But as he approached the cave, he heard a terrible roaring echoing from inside it, and a huge lion came out into the cave's mouth. Androcles was frightened, of course, but he noticed that the lion was favoring one of his feet. Androcles looked more closely and could see something sticking out of the paw of the hurt foot. It was an old nail. Forgetting his own safety, for he probably could have outrun the lame lion, Androcles cautiously approached the brute. He took the hurt paw into his hands and pulled out the nail, and then did his best to clean up the infected sore. It just seemed like the right thing to do. After all, Androcles knew what it was like to suffer.

The lion was ecstatic and gratefully licked Androcles's face. The two, man and lion, became fast friends. They lived together in the cave and learned to hunt together, using the lion's teeth and claws and Androcles's hands and wit. They became inseparable, despite their differences, and in many ways precisely because of those differences.

But they were a little careless one day. Some of the Emperor Caligula's soldiers were out hunting for a lion for a show in the Circus Maximus back in Rome and caught the beast in a net. One of the soldiers recognized Androcles as the governor's escaped slave, so they captured him, too. When the soldiers brought Androcles back to the governor, he flew into a rage about the poor slave. In those days before television, people enjoyed going to the circus to watch lions eat defenseless captives, and other gruesome sports. The governor condemned Androcles to the Circus Maximus to be used for this unhappy purpose.

On the day of the event, great excitement filled the air as the crowd filed into the arena. Even the Emperor Caligula was there. After all, it was good for an emperor's popularity to be seen putting on a satisfyingly bloody circus show, and Caligula was in political trouble because of his lavish spending on an expansion of his palace.[3] Besides, Caligula was a rather bloodthirsty fellow himself. Tension mounted as the preliminary acts—foot races, weight lifting, gladiator fights, a chariot race—were held. Finally, Androcles was thrown into the ring, naked and unarmed. The lion, who had been starved for days, was also released into the ring.

Snarling and roaring, the lion approached Androcles and prepared for a lethal pounce onto the modest frame of this gentle soul. But as he drew near to Androcles, the lion recognized who it was. The mighty cat lay down in front of Androcles, looked up at him, and began to mew softly.[4]

(Continued)

(Continued)

A few people in the stands began to jeer. They wanted blood. So some of Circus Maximus's animal handlers came out with long pikes to poke and anger the lion into action. But the lion rose up, shook his great mane, and roared fiercely at the handlers until they retreated. Then, the lion lay down once again at Androcles's feet, purring and swishing his tail.[5]

The circus crowd fell absolutely silent. Caligula, too, was astonished. He asked to have Androcles brought near to his viewing platform so he could question him. Androcles explained the strange history of his friendship with the lion, shouting up to the emperor high above on his portable throne. Caligula thought for a moment and then commanded that the story be written out on a tablet and passed through the crowd so all would know. After all, there were no loudspeakers in those days.

Once the tablet had made its way through the multitudes, with those who could read explaining the matter to those who could not, Caligula rose up. Everyone immediately fell silent again to hear. Caligula shouted out, "Should we release Androcles and the lion?" The crowd roared its approval. Caligula held up his hand to silence them again, and then pro-claimed, "The vote is clear. Let them both go free!" The crowd's roar after that could be heard clear across Rome.

For the next few months, Androcles and the lion walked through the city together, the lion on a light leash so as not to frighten anyone. People would give Androcles money and sprinkle the lion with flowers. And everyone who met them anywhere exclaimed, "This is the lion, a man's friend; this is the man, a lion's doctor."[6]

But eventually, Androcles and the lion grew tired of the fuss, even though they now had plenty of money to live on. They returned to the wilderness and lived out the rest of their days together, the closest of companions.

And so we learn that no act of kindness is ever wasted.[7]

How do we mobilize ourselves to build an ecological society, a community of care big enough to include humans, lions, and all other creatures—a community that recognizes, cele-brates, and maintains our mutual dependences and independences? The story of Androcles and the lion points out three foundations of ecological mobilization that environmental sociology also points out. I'll call them the *three cons* of *conceptions, connections,* and *contestations.*

The prefix *con-* is wonderfully multiple in its meanings. It can refer to "knowledge," as in the words *connoisseur, consideration* (which means knowing the stars), *con* (as in the phrase "con job"), and *cunning* (which transforms *con-* into *cun-*). *Con-* can also mean "together" and appears in this usage in a host of words, such as *consult, confide, convene, conversation, con-verge, conclave, concert, consent,* and (transforming *con-* into *com-*) *community* and *commonal-ity.* Plus, *con-* can mean "against," as in the pros and *cons* of something, and in its full form as *contra-* in the words *contradiction, control, contrast,* and *controversy.* An effective environmen-tal movement needs to be "pro" all of these basic meanings of *con-.* To mobilize the ecological society, we need ecological knowledge—conceptions. We need the solidarity of community ties—connections. And we need political strategy—contestations.[8]

Androcles had the ecological conceptions to begin with. Because of his own experiences as a slave, he could relate to what the lion was going through. The two then built a strong solidarity of mutual connections that carried them through the years. Finally, based on that solidarity, the lion contested the government's animal handlers, and Androcles made a compelling appeal, winning a remarkable political victory that brought freedom for human and lion alike. Whether the story is fact or fiction—and it has certainly had its share of reshaping over the years by various storytellers, including me—it is true in a deeper sense. As we shall see.

Mobilizing Ecological Conceptions

One of the oldest answers to the question of how to mobilize an ecological society is education. The environmental movement has put a huge amount of effort into environmental education. There are literally thousands of local environmental education centers and school programs across the world. There are dozens (it could even be into the hundreds) of professional associations of environmental educators at the regional, national, and international levels—for example, the North American Association for Environmental Education, founded in 1971; the Australian Environmental Education Association, founded in 1980; the Maine Environmental Education Association, founded in 1982; and the Japan Society of Environmental Education, founded in 1990. Many countries publish environmental education journals, including Australia, Canada, Hungary, South Africa, the United States, and more, and there are several international journals. Since 2003, there has been an annual World Environmental Education Congress. There can hardly be an environmental organization, either governmental or nongovernmental, that does not put significant effort into education and public outreach. And think of all the TV programs and popular magazines that have carried environmental stories, from the Nature Channel to *National Geographic*. In these many ways, the environmental movement has been working to put into practice the widely cited definition of environmental education from UNESCO's (United Nations Educational, Scientific and Cultural Organization) 1978 "Tbilisi Declaration," to whit,

> Environmental education is a learning process that increases people's knowledge and awareness about the environment and associated challenges, develops the necessary skills and expertise to address the challenges, and fosters attitudes, motivations, and commitments to make informed decisions and take responsible action.[9]

Sounds great. It is great. Much good has come of it. Nonetheless, after decades of environmental education, we still have massive environmental problems and significant issues with every goal mentioned in the Tbilisi Declaration's definition of environmental education.

The trouble is, knowing something doesn't mean you can do much about it. If people find their lives organized so that it is hard for them to put their ecological knowledge into practice, then they are unlikely to do so. Why? For exactly that reason: because it is hard—especially when one tries to act as an individual. The pattern of our economy, technology, built environment, and ideologies presents tremendous obstacles to getting something changed when we try to act on our own.

Plus, when people hear that they are doing something wrong that they feel they can't do much about, they will likely resist the implied sense of guilt by resisting knowledge. They may well accuse the bearer of environmental knowledge of playing a game of shame and blame to gain a position of moral superiority over the ecologically guilty. So they turn the page. They click the remote. They surf to another site.

Does this mean that environmental education doesn't accomplish anything? Hardly. I wouldn't be writing this book if I believed that. But it does mean we should be wary of a *behaviorist approach* to environmental problems—the idea that if we change an individual's attitudes, his or her behaviors will soon follow. Rather, we need to look at knowledge situationally, understanding the social contexts by which, and in which, people find themselves motivated to take action. Counseling individual action overwhelms and disappoints people. Plus, the "knowledge and awareness" the Tbilisi Declaration advocates is not something one can insert into someone's brain as he or she comes down an assembly line, a missing part that we slot into a skull as it goes by. These are social matters, not ones of individual mechanics.

The Cultivation of Knowledge

Think about what goes on in anyone's day. It is awash with information. For it is not just the environmental movement that is trying to grab people's attention. Every social movement, industry, and government agency is out there trying, as the Tbilisi Declaration describes, to increase "knowledge and awareness"; develop "skills and expertise"; and foster "attitudes, motivations, and commitments" that result in what each of those organizations regards as "informed decisions" and, we must hope, "responsible action." The Internet. The newspaper. Television. Radio. Mail campaigns. Viral marketing. Flyers passed out on the street. Advertisements on buses, billboards, and T-shirts. Everyone is trying it all. And no one can pay attention to it all. There is simply too much. So which sources will someone key into, and which will he or she ignore?

Plus, the sources often do not agree. (Indeed, if they did agree, they probably wouldn't feel a need to put out a message of their own.) Which is confusing and confounding (two more *con-* words), as none of us is an expert in everything, even within our own fields of endeavor. I think I know quite a bit about environmentalism and environmental issues after many years of studying these matters. But there is a lot I don't know. For example, I've had a good look through the *Fourth Assessment Report* of the United Nations Intergovernmental Panel on Climate Change (IPCC), the one that states that climate change is "unequivocal" and that there is "very high confidence" that this change is influenced by human actions.[10] I even have a degree in geology and another in forestry (and two joint degrees in environmental studies and sociology), so I have some technical background in some of the relevant natural sciences. But these are interdisciplinary matters, as I think pretty much everyone agrees now. And if something is recognized as being an interdisciplinary concern, that is another way of saying that no one person understands the whole thing. Moreover, who has read every one of the citations in the IPCC *Fourth Assessment Report*? There are thousands of them. Not that reading them is enough to be sure of their veracity. Maybe the experiments and measurements and models were done wrong. Maybe the sources aren't all reliable—as was indeed the case with a few of them, as the "Climategate" controversy showed.[11] Scientists make mistakes. Everyone does. Scientists sometimes misunderstand what they see. We all do. But I don't have the time or the resources or the expertise to do all those experiments for myself to see if they got it wrong.

Which means each of us has to trust someone else who knows more about some aspect of something than we can determine on our own—and not just about technical matters of environmental science. Throughout the day, we ask others about things that have worked for them or that they have heard worked for others. Do I need to sample every poisonous mushroom for myself to believe that they are poisonous? That would kill me. Do I need to read every book, newspaper, and website myself to decide which ones are worthwhile? I'll never live so long. So we each rely on others—others we trust—to help guide us successfully through the day.

But what if my friends are wrong? What if that supposedly poisonous mushroom was, in fact, safe and delicious? What if some bit of the knowledge in all those books, newspapers, and websites that my friends indicated, either explicitly or implicitly, that I shouldn't bother with was, in fact, exactly what I needed both to better myself and to better the world? I may never know.

The point is that education is not just about communicating facts. It never has been. It is also fundamentally about trust and the people by whom I gain a sense of what knowledge to pay attention to and what knowledge I can safely not pay attention to. Because of the centrality of trust, then, knowledge is not just knowledge. *Knowledge is a social relation.* And education is a social relation, too.

I like to think of it as a matter of the *cultivation of knowledge*.[12] By that I mean, what I take to be knowledge is a matter of my identity and a matter of the social relations of trust that shape my identity and come from my identity. It's an interactive matter. It's ongoing. And it's cultivated within culture and my resulting sense of lines of difference and lines of similarity with others.

What I know is who I am. (I'm an environmental sociologist.) Who I am is what I know. (I therefore know a lot about environmental sociology.) Who I am and what I know is whom I know and whom I trust. (I gained this identity and most of this knowledge from other sociologists, environmentalists, and environmental sociologists, who give me professional recognition and whose work and experience I use as a base for my own.)

Some deep commitments are at work here. A person's identity is who their friends and associates are and who their friends and associates are not. Given that your knowledge is linked to your identity and that both your knowledge and your identity are linked to others, a lot is at stake in the cultivation of knowledge. Your self. Your friends. Your associates. Your confidence. Your confidences. These are matters that are close to the bone of how we consider our location in the world. These are matters that are hard to change.

To cultivate knowledge is also to cultivate a sense of the *ignorable*. I don't mean ignorance. I don't mean stupidity about reality. Rather, I mean that which we can safely disregard—which is central to what we consider important. We gain knowledge by paying attention. But to pay attention to one thing is to not pay attention to something else—indeed, it is to not pay attention to far more than we pay attention to. To be where I am, in tune and attentive to the place and the people, is to be not everywhere else and not with all those other people. To decide what counts for knowledge—useful knowledge that is appropriate to my life, as I understand it—I must have some way to screen out far more that I will never know. I can try to read the *New York Times* every day (and I do). But can I also read *The Washington Post,* the *Los Angeles Times,* the *Wall Street Journal,* the *Times of London, Der Spiegel, Le Monde,* and *Al Jazeera*? Every day? Cover to cover? And how about *Sierra Magazine,* the *Ecologist, E/The Environmental Magazine, High Country News,* and *Environment Times*? When I am already reading the *National Review, The Spectator,* and *The Economist*? So how do I know that what I focus on is what I should focus on if I haven't looked at those things I am not looking at and will never look at? From the cues of culture, the culture with which I identify and find trust in my navigation of the world.

Cultivating Knowledge in the Fields of Iowa

How, then, does anyone ever change and begin to tune in to other cultivations of knowledge, other conceptions of self and reality such as, say, ecological ones?

I found myself asking this question some years ago when I lived in Iowa and encountered a marvelously successful organization: Practical Farmers of Iowa, or PFI for short. PFI is Iowa's largest sustainable agriculture organization, with over 2,000 members, roughly half of whom are farmers. That may not seem like a lot when one considers that Iowa, a major farm state, has over 90,000 farms. But in 1985, the year the group began, there were hardly any sustainable farmers in Iowa, which has as industrialized an agricultural landscape as one can find anywhere on the planet. Some 60 percent of the state is covered by just two species of plants: corn and soybeans. Each spring, Iowa's grain farmers gear up the machinery and chemistry to keep yanking this biomass out of the ground and into the mouths of hogs and cattle and, increasingly, into the mouths of ethanol and biodiesel factories. We're talking factory farming in the extreme (see Figure 10.1). This is what most Iowa farmers do. But they don't have to farm this way. In the years since 1985, PFI farmers and many other farmers have shown that sustainable agriculture works. It produces strong yields and solid incomes, while supporting families, communities, and ecologies. So why don't more farmers change, and, conversely, if so few do, why do any at all?

Figure 10.1   Industrial agriculture in Iowa. An articulated tractor discs and applies chemical fertilizer to a field, 2000.

Source: Photo by Helen D. Gunderson. Used with permission of Helen D. Gunderson.

The standard answer is that the *structure* of agriculture—markets, laws, subsidies, technologies—prevents them from changing to sustainable methods. After all, the government pours vast subsidies into corn and soybean production, which mightily maintain the existing pattern of markets, laws, and technologies, while giving a direct boost to farmers' bottom line. But that doesn't explain why some farmers change, especially as PFI farmers mostly come from the same situations as their conventionally farming neighbors. PFI farmers drive tractors. They wear "feed caps"—baseball-style caps that read "Dekalb" or "Cenex" or "Cargill" or "Pioneer Hybrid" or the name of some other company that sells supplies to farmers. They were mostly raised in the communities where they now live, and often on the same farms. They are not a bunch of old hippies gone to seed.

However, conventional farmers in Iowa have plenty of incentive to change. The structure of agriculture is not kind to most of them. That's why farm numbers have continued to plummet in Iowa long after the fabled farm crisis of the 1980s and, in fact, were plummeting well before the farm crisis, too. As farm numbers continue to fall, so do the number of local businesses that service the farm economy. The loss of those businesses boards up main streets, churches, schools, and even houses across rural Iowa. The result is an odd paradox: lush fields worked by expensive, modern equipment spreading out to the horizon and away from abandoned Victorian homes with sagging porches and glassless windows and small-town main streets lined by plywood instead of plate glass (see Figure 10.2). So the structure of agriculture is rough on farmers' communities. The stress of the work is often equally rough on their families. The danger of the work is hard on their own health. And the monocultures in the fields—pumped up and propped up by can't-miss chemistry and "Big Iron," as farmers say—literally send the land down the creek. In other words, the structure of agriculture is hard on the environment, too.

Yes, given that their farms, communities, land, and sometimes their families and health are all eroding away, there is plenty of reason for conventional farmers to change to sustainable practices. They rarely do change, though, because what they are really farming out in their fields is something far more important to them than their crops. They are farming their selves. They are farming their sense of themselves as men and women, knowledgeable about and competent in what they do, maintaining a respectable place for themselves in the fabric of

Figure 10.2 The smoke and dust rises as a 1907 Victorian farmhouse in Pocahontas County, Iowa, is razed to the ground. Industrialization of agriculture has increased farm sizes so much that many farmhouses are abandoned and fall into ruin, and rural communities with them.

Source: Photo by Helen D. Gunderson. Used with permission of Helen D. Gunderson.

their communities, in which they have lived most of their lives, and sometimes all their lives. They do this farming with a stock of knowledge, built up over many years, often hard won, that serves as a continual investment in their identities, the more they draw on and add to this knowledge. It's a lot to give up.

Now, conventional farmers could be going to the series of field days that PFI puts on at members' farms across the state, demonstrating sustainable practices. Sometimes they do. They could be accessing the information on sustainable practices offered by Iowa State University's College of Agriculture and agricultural extension staff. Again, sometimes they do. They could be attending the winter meetings that the state's organic and sustainable farmers put on that discuss the strong markets for sustainably raised farm products and that bring farmers together with eaters interested in helping promote agricultural change. At times, indeed, they do go. And when conventional farmers tune in to these events, they are often surprised to find out that PFI farmers have most of the same passions for farming life that they do—that there is a network of friends here that one can trust to help navigate through the poison mushrooms and the flood of potential knowledge each day brings. That discovery brings them back, again and again.

One of the main leaders of PFI, a man I'll call Earl, explained this discovery to me one hot summer afternoon at a field day at a PFI member's farm.[13] About 25 people were huddled under the shade of an old oak, and a few of us, including me, in the shadow of a huge John Deere tractor. Most of the crowd were PFI members. But there were several new faces, too. The farmer (not Earl) was explaining the importance of not putting too much fertilizer on a corn crop.

"If your corn is looking dark green right through to the end of the year, you're throwing your money away," he told the crowd—meaning that if your corn is dark green that long you've spent more on fertilizer than you'll get back in yield, however good the crop may look. Plus, you've just polluted the groundwater because you've probably used more fertilizer than the crop can absorb. In farming, as in other endeavors, wasting money often means wasting the environment.

I knew Earl, whose own farm was just up the road, from earlier PFI events. He must have been finding it hard to concentrate in the heat, too, and he wandered back from the group by the oak to where I was standing. "All this fertility stuff must be a bit boring for you," he said to me, in a kind of classroom whisper between naughty pupils. Earl knew by then that I was a sociologist and that I didn't have a farm background.

"Well," I replied, a bit uncertainly, not knowing where this was going, "there's a lot of it I can't follow."

"Besides," said Earl, "it's all social. That's where the real change has to come from. All this, this is just technical."[14]

But how do these social relations of knowledge get going? The 9 years of research I conducted on PFI with my colleagues Sue Jarnagin, Greg Peter, and Donna Bauer led us to this conclusion: It most commonly starts with what we came to call a *phenomenological rupture* in people's existing cultivation of knowledge, a wrenching experience that causes them suddenly to doubt the bases of trust upon which they had long committed themselves to this cultivation. That rupture is not something PFI creates. The strains of industrial agriculture are what rip the fabric of trust, often through a financial crisis or a health crisis, according to the farmers we interviewed who had switched from conventional farming to sustainable practices.

Dick and Sharon Thompson (their real names) are perhaps the best-known PFI farmers, and they have won several major environmental awards for their work (see Figure 10.3). They were among the first to switch to sustainable practices, back in the 1960s, in fact, and went on to help

Figure 10.3 Dick and Sharon Thompson, winners of many environmental awards, at their farm in Boone, Iowa, 1999.

Source: Photo by Helen D. Gunderson. Used with permission of Helen D. Gunderson.

found PFI in 1985. Here's how Dick described the rupture, and his spiritual experience of it, in response to a crisis in the health of his livestock and his own overworked body:

> In January of 1968, while chopping stalks in field number six, going north, I was—I'd had it. All the work. The pigs were sick. My cattle were sick. I hollered "help." That's about the only way I know how to explain this. But some things started to happen. And a lot of things that happened seem to happen early in the morning. That thoughts come into my mind that I know that are not mine. So I want to share this. The creator wants to put a receiver, a still small voice, way down deep inside each one of us, for communication. It's our choice. It's not forced on us. If you want it, you can have it.[15]

Other farmers also described the unsettling period of rejecting a valued knowledge cultivation as a spiritual crisis. For example, one experienced it as a calling from God, speaking down from the sky one day when he was up on a ladder, painting his barn, and closer to the heavens.[16] Another called it a "planned event" orchestrated by a higher being.[17]

Dale, another PFI member, explained the disorientation of the rupture well. He had been one of those market-oriented farmers, spending every free moment in front of a computer, watching the movement of prices on the Chicago Board of Trade. He made some good choices about when to sell and on futures contracts he signed, but also some bad choices. He had dug himself into a big hole with a string of bad guesses, but then thought he had the situation pegged. He had accumulated a good stock of grain to sell, even though it hadn't rained in Iowa for a while. Because of the lack of rain, prices were good, and getting better day by day. Monday, he thought, I'll sell. But that was 2 days too long.

"You know, I had the boat loaded. But it rained over the weekend, so that was the end of that deal." Dale laughed, wistfully. "The boat sunk in the harbor!"[18]

The financial crisis that ensued nearly drove him out of farming, like so many others before. And he began to doubt everything that he thought he knew about how to be a successful farmer. "I kind of lost confidence in myself, because some of the things I was doing before failed me. Naturally after your ideas and things had failed, well, then, soon you get kind of gun-shy. It's taken me probably a couple years to gain my confidence back, so to speak."[19]

How did he get it back? He happened on a notice for a PFI field day. He'd seen such notices before. He knew a couple of other farmers in his area were members. But this time he decided he would go. Two years later, he had his confidence again. He had trust again, in himself and in others. He had a cultivation of knowledge back.

Today he is farming differently, too. No longer a fence-row-to-fence-row grain farmer, he has a diversified operation based on integrating grass, livestock, and some grain crops through a 5- to 7-year rotation that breaks up pest cycles and fertilizes the ground without chemicals and with very little erosion. He is getting a lot more money per acre, too, because he is spending less and getting better prices by not emphasizing low-value commodities like grain. So he is no longer in the game of forever plotting to get his neighbor's land to make a low-margin, high-volume income. In so doing, he is making space for more farmers and a stronger local community. And he also likes himself better.

"But you know, going back, all these chains of events that have happened in my life, I'd have to say that I've been a better person because of it. And I have no remorse for the money I've lost or whatever. Because, the thing is, to me, it made me a better person."[20]

New knowledge, new self. New self, new knowledge. New self, new knowledge, new friends. New self, new knowledge, new friends, new farm—a farm that sustains the land and its people.

Cultivating Dialogic Consciousness

Central to knowledge cultivation is not only the existence of social relations of knowledge, but also the character of those relations. As I'll come to, PFI emphasizes a horizontal, dialogic process of engaging others and linking knowledge and identity. There are top-down, monologic ways, too. These monologic ways are by no means uncommon or unsuccessful, at least in the short term. Think of the "you" ads discussed in Chapter 2. Think of promotional campaigns fronted by a good-looking and famous person that many people would like to consider as a personal friend. Think of the religious leader who pitches a particular value by connecting it to traditions. Think of the manipulation of nationalist sympathies by unscrupulous politicians. These are all common practices.

The theory of *frame analysis* describes top-down methods of knowledge cultivation well. Monologic ways of cultivation depend on framing issues so that people will respond to them as the framer desires. This is the skill of the "good communicator"—the orator who takes the symbols of the day and uses them to his or her advantage, or reframes them to the same effect. The key process is what David Snow and Robert Benford call *frame alignment,* by which an individual's frames become congruent or complementary with a knowledge cultivation.[21] Snow and Benford identify four basic tactics: frame bridging, which links frames; frame amplification, which invigorates the values behind a frame; frame extension, which widens a frame's boundaries; and frame transformation, which reconfigures a frame's meaning. Throughout, the effort is to link knowledge to social relations, often subtly (and, to those on the outside, often not-so-subtly) reshaping them.

Probably any social movement, or indeed any social encounter, engages in at least a bit of framing. After all, rhetoric is inescapable. We are always representing, always engaging in social construction. There are many ways to describe any situation. As Chapter 8 discusses, you can never describe absolutely everything about anything. We all have to choose our words, and we do so with our audience in mind. It has long been this way. Two millennia ago, Aristotle wrote out a whole book on the matter. In his *Art of Rhetoric,* he distinguished among three forms of persuasion, noting that we generally use all three at once:

- *Ethos,* convincing by getting the audience to feel that the speaker is ethical and trustworthy;
- *Pathos,* convincing by getting the audience to emotionally experience the speaker's point of view; and
- *Logos,* convincing by getting the audience's agreement on the logic of the argument (yes, even social constructionists agree this is important).[22]

Frame analysis restates (and reframes) this ancient insight.

Many environmental sociologists have applied frame analysis to understanding the success of the environmental movement and the success of countermovements to environmentalism. Barbara Grey has used it to understand conflicts over Voyageurs National Park in Minnesota.[23] Dorceta Taylor has used it to understand the success of the environmental justice movement.[24] Andrew Rhys Jones has used it to understand how the media portrays global warming.[25] Brian Walton and Connor Bailey have used it to understand the recent success of the wilderness preservation movement in Alabama, after years of little headway.[26] At least a bit of monologue is part of every social situation and every social movement.

But do we want an environmental movement that stresses monologue as its main technique of knowledge cultivation? Paulo Freire, the great Brazilian philosopher of pedagogy, would say no. If we want a humane, liberatory education that is truly based on trust, and not the appearance of trust, the main approach should be what Freire called *conscientization*—the building not of head-nodding agreement, but of critical consciousness in dialogue with the world.[27] Head nodding leads to nodding off. Critical consciousness leads to a wide-awake creativity.

Thus, dialogic education, and a dialogic knowledge cultivation, does not emphasize persuasion. Rather, it develops people's critical capacities and welcomes their differences and disagreements as ways we grow and grow together.

PFI wonderfully cultivates this critical friendship in which friends stimulate each other to new knowledge for everyone, as I came to appreciate. PFI farmers, I discovered, love to argue with each other—not with negative, alienating disagreement, but through fostering mutual checks and balances with the experiences and insights of others, developing the ties of cultivation as they develop knowledge.

Central to PFI's critical approach are the group's participatory research trials, conducted on-farm and often with the help of university researchers, in which farmers become scientists and scientists become farmers. Many of the techniques of sustainable agriculture are new or have been little studied by the professional research community. Plus, being a better scientist is a great way to be a better farmer. And as Chapter 8 discussed, to do science is to be open about one's reasoning and to be open to the reasoning of others. On-farm, participatory research trials exemplify both of these opennesses by using methods that are accessible and explainable and by presenting the results to other farmer-scientists and scientist-farmers for their critical feedback.

One tool of mutual critique that PFI uses is randomized, replicated plots, where you lay out an experiment on your farm in a way that allows you to take account of the variability of the soil and microclimate, so you know your result is due to your treatment and not due to an unrecognized environmental difference. Knowing what your result is due to also means you can explain your result to someone else. Plus, you write down what you do. That, too, makes it a lot easier to explain your experiment (see Figure 10.4). The university scientists use the same logic of openness. Glen, a PFI farmer who recently switched his farm over to organics, explained how PFI's dialogic approach helped him make the change:

"I probably wouldn't have taken that step . . . if I hadn't had the knowledge to be able to document what I was doing, and some people there to hold my hand," Glen described. "Just sitting down with people and arguing about how do we structure our costs and things."

Figure 10.4 Cultivating knowledge in a machine shed. PFI farmers discuss the results of the group's research trials, along with a university researcher, Iowa State University professor Kathleen Delate.

Source: Photo by Helen D. Gunderson. Used with permission of Helen D. Gunderson.

I was really struck by his use of the word "arguing," something that Midwesterners are generally known for trying to avoid.

"It has been very valuable. In my organic operation now, I have weed control issues using the flame weeding and rotary hoeing and things." Flame weeding is when farmers pass a gas flame along the rows of a crop when the weeds first start to come up, not to burn them but to superheat their sap so it bursts plant capillaries and wilts the weed; you have to do it when the crop is strong enough to take the heat, but before the weeds are. Rotary hoeing is another way to control weeds without chemicals. You use an array of barbed disks to crumble the surface of the ground, breaking up the contact of weeds with the soil; again, you have to do it when the weeds are small. Organic farmers in PFI, and elsewhere, have been trying to figure out when to use one versus the other.

"Actually, that was my PFI trial this year," Glen explained. PFI farmers get together in the winter to coordinate their research trials for the group and then debate the results everyone gets on their individual farms. "I had a statistically significant difference in weeds and yield. So I was pretty pleased with the trial."

"Using the flamer versus the rotary hoe?" I asked.

Glen nodded. "In my mind, the flamer is the last tool in the toolbox for early season weed control. . . . I choose to use the flamer last because I think it is the most severe to the crop. There are people that argue that point with me, and that's next year's trial."[28]

So the dialogue of critical consciousness continues, ever developing the knowledge cultivation of PFI. Which is exactly what the word *consciousness* means: *con-*, for "together," and *sci-*, another root meaning "to know"—yielding "knowing together." Indeed, it's the only way we ever really know.

Mobilizing Ecological Connections

How do we get that togetherness together? According to Garrett Hardin, it won't be easy.

The Tragedy of the Commons

In a famous 1968 article, "The Tragedy of the Commons," Hardin described the problem in stark terms. Imagine you are a shepherd grazing your sheep on your village's common pastureland, back in the hills above the village. As a member of the village, you have the right to graze your sheep there, just as every other village member does. You've got only 10 sheep, though, and after a while you think, "Well, I'd be a bit better off if I added a few more to my flock." Meanwhile, your fellow villagers are thinking the same thing about their own flocks. Pretty soon, as everyone adds a few more animals, there are a lot more sheep in the common pastureland.

The pastureland is only so big, though. Eventually, overgrazing occurs. The grass cover gets thin, and the land starts to erode. Everybody's sheep start to die. You wind up with fewer sheep than you began with, and the eroded common land is no longer capable of supporting as many sheep as it originally could: economic and environmental disaster.

Here's how Hardin, rather melodramatically, described the situation:

> The inherent logic of the commons remorselessly generates tragedy. . . . The rational herdsman concludes that the only sensible course for him to pursue is to add another animal to his herd. And another; and another . . . But this is the conclusion reached by each and every rational herdsman sharing a commons. Therein is the tragedy. Each man is locked into a system that compels him to increase his herd without limit—in a world that is limited. Ruin is the destination toward which all men rush, each pursuing his own best interest in a society that believes in the freedom of the commons.[29]

Hardin intended this parable as a master allegory for all environmental problems. Three examples he mentioned in the article are traffic, pollution, and overfishing. Think of streets as a kind of commons, something we all collectively own—which, in fact, they usually are. As a member of the community, I am free to drive on my city's streets as much as I want. But what if everybody decides to get about this way? The result is traffic jams, smog, and the loss of alternatives as mass transit shuts down.

Or think of the lake where your summer cabin sits as a kind of commons. It's expensive to put in a good septic system, and it wouldn't hurt you much to flush into a shallow leaching field close to the water's edge, where, as it happens, it would be the cheapest and easiest place to put the field. The lake is pretty big, and it can handle a little bit of pollution. Besides, it would be hard for anyone to determine that you're the one with the shallow leaching field close to the shoreline. Lots of cabins ring the lake. But what if everybody on the lake did what you're doing?

The oceans are a commons, too. If I fish for a living, I might as well cast as big a net as I can. What I do myself won't have that much effect on overall fish stocks. Anyway, the other fishermen and -women are probably going to do the same, right? And soon the fish are gone.

Hardin's analysis is far from perfect, as I'll come to in a moment. But it is hard to ignore the fact that traffic jams, for example, are on the rise. In the car-dependent United States, traffic congestion cost Americans a collective 4.8 billion hours and 3.9 billion gallons of fuel in 2009. That's close to an hour a week per commuter in the United States. And it's getting much worse, year by year. By comparison, traffic congestion wasted a collective 0.99 billion hours in 1982.[30] As for pollution, another of Hardin's examples, many recreational lakes have been badly polluted by their users, and total annual air pollution deaths worldwide are huge: 2.6 million per year, as Chapter 9 discussed. As for fishing, stocks are in terrible shape in many parts of the oceans and have simply collapsed in the Grand Banks, leaving hundreds of fishing communities from Newfoundland to New England economically devastated.

These are all examples of a more general class of circumstances, what social scientists call the *problem of collective action*: In a world of self-interested actors, how can we get people to cooperate for their own benefit? Individual actors pursuing their rational self-interest often lead us to irrational collective outcomes that, in fact, undermine the interests of those who enact them. The result is a striking paradox of social life: We often do not act in our own interests when we act in our own interests. Or, to put it another way, when we all do what we want, it often leads to outcomes nobody wants.

Why It Really Isn't as Bad as All That

Hardin's account of the "tragedy of the commons" remains one of the most discussed theories in environmental sociology, even 40 years after it was written. The phrase "tragedy of the commons" is familiar to many in the general populace. Academics regularly employ it in analyses. In a quick search of a database at my university library, I found dozens of academic articles just from the year 2010 that discussed the concept. In as specialized a realm as academia, this is a lot. Several of these articles extend the allegory of the commons far beyond environmental concerns, applying it to analyses of narcissism, altruism toward strangers, online forums, and partner selection.[31]

Much of the reason for the continuing attention, though, is to point out how spectacularly oversimplified and overstated Hardin's allegory is and how it diverts attention from some fundamental social processes at work in environmental problems.[32]

To begin with, Hardin seemed to blame common ownership of resources for the tragedy. But, in fact, we can find countless examples of highly successful use of commons for resource management. Grazing lands all across Africa, Asia, and South America; traditional systems of fisheries management in India and Brazil; even the private homes of modern families, which are a kind of commons in miniature and remain a highly popular form of social arrangement—these are just a few of the many examples of generally successful commons management.[33] Indeed, common ownership is the primary way that people have managed their affairs for centuries. And it has, at least until recent years, largely worked.[34]

Rather than the tragedy of the commons, Hardin's allegory is better characterized as the *tragedy of individualism.* For what breaks down Hardin's commons is not collective ownership itself, but rather the inability (and perhaps unwillingness) of the herders to take a view wider than their own narrowly conceived self-interests.

Herders, in fact, are unlikely to conceive of their interests so narrowly, at least in traditional commons. For one thing, Hardin assumes that no one will notice the overgrazing until it is too late. But herders out there in the pasture every day with their sheep are likely very quickly to note the deteriorating condition of the grass. For another thing, Hardin assumes that the herders do not communicate with one another. More likely, as soon as the herders notice the beginnings of overgrazing, they will walk over to each other's houses in the village and have a few words about the situation. They will likely convene a gathering of some sort to try to work out an arrangement that restores the grass, while following local norms about the number of sheep each herder is fairly entitled to graze.

More significant, however, is the reliance of Hardin's allegory on a rational choice view of human motivation. People are, simply put, more complex—and thankfully so. We are moved by more than our own narrowly conceived self-interests, as Chapters 2 and 9 described. Equally important are the sentiments—the norms, the feelings of affection (and lack of affection) for others—we have in social life. These sentiments are a crucial aspect not only of our humanity but, as we shall see, of our interests as well.

The Dialogue of Solidarities

Let's return to the story of "Androcles and the Lion" and the home truths it recounts, despite being a fable. It is not the usual sort of evidence that sociologists draw upon, but hear me out.

To begin with, why did the lion spare Androcles in the ring, as he came up to him, all snarl and roar? At that moment, the lion could have had no idea that refusing to eat his former partner would result in freedom. Indeed, the Circus Maximus animal handlers might have decided to kill this apparently hopeless lion for failing to put on a good show. (The Circus Maximus was like that.)

And why did Androcles initially pull the nail from the lion's paw? At that moment, Androcles could have had no idea that pulling the nail would result in his gaining a friend and hunting partner. (Hunting partnerships between humans and lions are, after all, rather unusual.) And neither could he have known that they would eventually be able to return to the forest to live out their days together.

The reason was, according to Apion, that the lion and Androcles were moved by more than narrow calculations of their own pure self-interests. They were moved as well by their sentiments: Androcles for a lion in pain, and the lion for a friend and former companion; Androcles for reasons of commitment to certain norms of behavior, and the lion for reasons of friendship, of affective commitment. These sentimental commitments, in turn, led to the promotion of their interests *although they could not have known that at the time.* This is a crucial point of criticism of the rational actor model described in the tragedy of the commons. What it means is that sentiments may promote interests but do not reduce to them.

At the same time, interests promote sentiments. A large part of the reason Androcles and the lion liked each other is that, beginning with Androcles's act of pulling the nail and extending through the lion's refusal to eat Androcles, they had learned to rely on each other to promote each other's needs. Because they helped each other out, they liked each other and shared a sense of commitment to common norms of social behavior. And because they liked each other and shared a commitment to common norms, they helped each other out. The story is thus another example of a dialogue, this time what I like to call the *dialogue of solidarities.*

I use the plural because this dialogue is based on the interaction between two mutually supporting bases for social commitment: a *solidarity of interests* and a *solidarity of sentiments.* The interests of both Androcles and the lion were served through their relationship. But as well, they sensed the existence of sentimental ties—affection and common norms—between them. And the one constantly shaped and maintained the other.

All this emphasis on sentiment may sound a little idealistic, the kind of rare altruism we sometimes hear about in stories or, as in this case, in an ancient fable. But sentiment is actually quite common—and quite necessary—in social relationships, at least those that endure across time and space.

Consider, for example, a domestic union of some kind, two recent college graduates perhaps. They each have interests, such as careers. They support each other through graduate school. They make their job choices with the other partner's interests in mind. They manage their home in ways that allow each to succeed at work. And thus they maintain a solidarity of interests.

These domestic partners may not each be getting the same interest satisfied through their domestic union, however. Indeed, likely not, for everyone is different, as we know, which means everyone's interests are at least a bit different. Maybe only one of these domestic partners actually wants to go to graduate school. Or maybe one is a musician and the other a school teacher, leading to quite different rhythms of time demands and resource needs. As long as they can work out a way to coordinate these different interests, that's fine. The important thing is not that their interests are the same, but that they are *complementary.*[35]

However, there are always time delays involved in complementary and cooperative action. How does one partner know that the other will come through when it is the other partner's turn to make a career sacrifice? There are also always issues of space in complementary and cooperative action. The two domestic partners cannot keep each other under constant surveillance. How does each know that the other can be relied upon to coordinate shopping, to maintain monogamy (if the union is based on that understanding), to cover for each other when situations require it?

The answer is, again, *trust.* This trust can exist because each believes the relationship to be based upon more than the narrow calculation of self-interests. Because each has affection for the other or because each has a sense of common commitment to common norms of interaction—or both—they can trust that the other will come through across the isolating reaches of time and space. Without this sense of trust that a solidarity of sentiments gives, no solidarity of interests can last long.

The process works the other way, too. The persistence of a solidarity of interests is one of the principal ways that each partner comes to sense real affection and common normative commitment on the part of the other. If one partner violates that trust by not looking out for the other's interests, chances are, frankly, that pretty soon they won't like each other anymore, nor have faith that they share some crucial norms. Trust is the essential glue of both a solidarity of interests and a solidarity of sentiments.

So, to return finally to the tragedy of the commons, one of the main reasons why herders on a commons have usually managed to keep from overgrazing the pastures is that *they trust each other.* These are their neighbors, after all, and likely their kinfolk, too. These are the people they relax with, dance with, worship with, and marry. Of course, villages sometimes fall into considerable internal conflict, and when they do, those sentimental ties may go. If so, the grass on the pastures will likely go, too.[36]

The dialogue of solidarities is a kind of ecologic dialogue, a constant and mutually constituting interaction between the realm of the material (a solidarity of interests) and the realm of the ideal (a solidarity of sentiments). From this dialogue emerge solidarities of solidarities, if you will, within families, organizations, businesses, neighborhoods, villages, towns, cities, counties, provinces, states, nations, species, ecosystems, and all other kinds of commons. What I mean is, from this dialogue emerges *community*.

Not only is there dialogue in the philosophical sense at work. A dialogue of solidarities depends upon dialogue in the everyday sense of the term: dialogue as communication. And communication means the mobilization of the social relations of knowledge. In other words, right in the center of a dialogue of solidarities is the cultivation of knowledge, all mutually building and depending upon trust and upon each other. For from the cultivation of knowledge, we gain identification with norms and commitments—with sentiments. From the cultivation of knowledge, we also gain identification with where our interests lie. Neither interests nor sentiments are given in this life. Rather, they are created, and re-created, throughout our lives in interaction with each other and the world.

If the paradox of collective action is that people often do not act in their own interests when they act in their own interests, the solution is clear: Also act on your sentiments. But consider those sentiments and those interests broadly and openly. That is, consider them dialogically.

A Tale of Two Villages

At least this is the everyday wisdom my colleague Peggy Petrzelka found among the Imazighen people of the Atlas Mountains of Morocco, widely known as the Berber.[37] (*Imazighen* is the name they prefer.) Along the Imdrhas River Valley lie two villages, some 13 kilometers apart: Tilmi and M'semrir (see Figure 10.5). It's not great cropland, and the Imazighen in the area have traditionally relied on grazing sheep and goats for income and sustenance. It's not great grazing land either, however. The land is steep and the climate is dry. So local villages use what they call the *agdal* system of collective management of the grazing lands, which have traditionally been held almost entirely in common.

Under *agdal*, grazing schedules and any disputes are worked out through a local representative council of herders, known as the *jemaa*. The head of the *jemaa* is called the *Amghrar,* and he (it is traditionally a he) is elected by the local villagers. If signs of overgrazing start showing up or if there's been a particularly dry spell, the *jemaa* will close certain areas of the commons to allow regeneration. The *nuadar,* two men from each village, are selected annually to keep watch on the commons to make sure that the guidelines of the *jemaa* are being followed. If someone violates the guidelines, they may be forced to pay an *izma*, a penalty. When fence repair, harvesting, or other work needs to be done, the villagers organize *touiza*—communal work teams. It's a system that has worked for centuries.

Has worked. Peggy, who speaks Arabic, got a chance to live for most of a year in the area, during the course of a fellowship. She soon noted what many in the area now frequently complain about: that in M'semrir, the *agdal* system is breaking down. The grass looks bad. Stocking rates are double what they should be. Violators are getting away without paying *izmas*. Much of the land has been privatized. Some people seem to be getting quite a bit richer, and satellite TV dishes have sprouted from a number of rooftops. *Touiza* is disappearing. People are scared of the *Amghrar*. The *jemaa* is increasingly an in-group who distribute grazing rights to each other and their friends. People are angry with each other.

But in Tilmi, the grass still looks good. Stocking rates are just what they should be. Very little of the traditional commons land has been privatized. The *jemaa* distributes grazing rights in ways that everyone Peggy spoke with found generally equitable. *Touiza* is still going strong. There are very few satellite dishes. When they disagree with him, Tilmi residents tell the *Amghrar* to his face. That's because they like him and are confident that he likes them, even when there are disagreements. Which there aren't very often, because people in Tilmi still like each other.

Figure 10.5   The dialogue of solidarities in two Imazighen (Berber) villages in Morocco. In Tilmi, the village up the Imdras River valley and off the main road, solidarity is stronger, and the pasture lands are in better shape. In M'semrir, capitalist individualism has undermined the dialogue of solidarities, and the pasture lands are suffering.

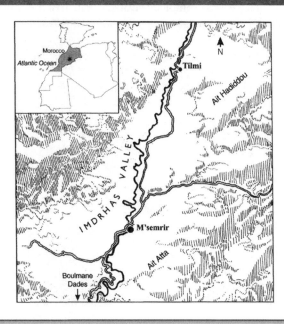

Source: Petrzelka, Peggy, & Bell, Michael M. 2000. "Rationality and Solidarity: The Social Organization of Common Property Resources in the Imdrhas Valley of Morocco." *Human Organization* 59(3):343–352. Reprinted by permission of the Society for Applied Anthropology.

In fact, the people in Tilmi like each other so much that they dance together. A lot. It may sound romantic, but most evenings when the weather is fine and the work is done, a group of villagers get together to sing and dance in the village center. When there are family celebrations— a wedding, a circumcision—virtually the entire village attends, and the dancing can go on for days, until 2 or 3 in the morning. And they sing when they practice *touiza*, helping each other harvest their personal garden plots, or as they repair the road or clear snow. All this astounded Peggy. Yes, it may sound romantic, because it is romantic. But it is also what they really do.

In M'semrir, however, people don't dance much anymore. There may be a bit at family celebrations, but the whole village is no longer invited. Just close family and friends. In Tilmi, weddings are usually held together during the same season of the year, and the brides walk through the village together amid the throwing of dates, almonds, and figs from the roofs of the grooms' houses to the crowds below. But in M'semrir, weddings are individual and scattered throughout the year, and the rich and festive foods are thrown only to the guests.

Peggy went for a walk one day with Amina, a woman from M'semrir, up into the hills above the village. They paused for a rest on a high rock overlooking M'semrir and the Imdrhas Valley below. They got to talking about changing traditions in M'semrir.

"We used to gather everyone and have one big party—now everyone has their own tradition," Amina remarked.

She pointed out what used to be the communal property, now divided into small private plots.

"Nizha," she said, using the Arabic name locals informally conferred on Peggy, "the words of today are not like the words of yesterday, and that which we did early is not that which we do today."

Why, then, this difference between the two villages? The Moroccan government has been working hard to "develop" the local economy, trying to increase the nation's productivity and also people's personal incomes. So they've developed regional market centers and have begun promoting tourism. They have also promoted privatizing much of the communal land, figuring that production would go up. But in the rugged terrain of the Atlas Mountains, it's harder to bring "development" to the more remote villages. M'semrir is lower down the Imdrhas Valley, more accessible to the Jeeps of government officials and the delivery vans of the central Moroccan economy. Tilmi may be only 13 kilometers from M'semrir, but those 13 kilometers are up a twisty, rutted, dirt road, and the officials, tourists, and other bearers of "development" just don't make it up there so often.

People in Tilmi have heard of privatization, though. They aren't that isolated. After all, they often go to M'semrir for its bigger, more vibrant marketplace. And they've toyed with some of the practices that the people of M'semrir have taken to. But thus far, they've only toyed with them. Thus far, they are still singing and dancing together. Thus far, they still have a dialogue of solidarities. Thus far, the grass is still green.

How Big Is Your Solidarity?

But can we create a dialogue of solidarities with aspects of our world that do not speak? Can we create community with nonhumans?

Apion said yes. Androcles and the lion managed it, even though the lion did not know how to speak Latin and Androcles, as far as we know, did not mew and growl. The philosopher Bruno Latour says yes, too. One of the most basic points of actor network theory is that networks of actors (what Latour prefers to call *actants,* as Chapter 8 describes) extend well beyond the human, forming a broader coalition that he sometimes calls a "collective." Now, this collective, this solidarity, is not necessarily symmetrical. Yes, we can imagine that lions have interests and sentiments and can form relations of trust, as anyone intimate with nonhuman mammals and other vertebrates can attest. They have ways of communicating their interests, sentiments, and trust. But butterflies and clams, rocks and soil, grass and trees? Here the interests and sentiments and trust will lie in our part of the collective, not theirs.

The environmental sociologist Ray Murphy has suggested that, as we do not communicate with the non-articulate actants in our collectives, an easier terminology is to think of these interactions as dancers who *prompt* each other. "The concept of 'prompt' captures the influence of nature's dynamics on conceptions, discourse, and practices, without claiming the latter are determined by those dynamics," Murphy writes. He goes on to say,

> Human agents dance with the moves of nature's actants to form hybrid constructions, with both influencing the other and both having some autonomy. The dance can be adroitly or ineptly performed. An approach that analyzes movements between human agents and nature's actants, like partners in a dance influenced by the other's creative movements, can bridge the nature/culture divide in sociology and transcend the limitations of a one-sided approach that focuses solely on nature's determinisms or human social constructions. The metaphor of dance captures the autonomous movements of nature's dynamics without implying intentionality by the non-human partner, only movement.[38]

In other words, an actant does not have to be able to speak to be part of an environmental movement. What an actant needs is enough connections to be invited to the dance.

Mobilizing Ecological Contestations

So, you've got the conceptions your group needs, having developed a cultivation of knowledge, one that I hope emphasizes dialogue and conscientization, not monologue and PR. And you've got connections going, bringing interests and sentiments into solidarity and dialogue through trust and through your cultivation of knowledge, and I hope also bringing nonhuman actants into the collective. Great. Your grassroots environmental movement is well on the way.

But what if government or corporations or the broader society doesn't agree with what your movement is trying to do? What if their interests and sentiments lead them in other directions? What if they'd really prefer it if you and your connected community of environmental conceptions just went away?

Indeed, if your group felt motivated to get something going and make a change, the chances are that not everyone will welcome what you want to do. Otherwise, it probably would have been done already. So let's explore the environmental sociology of contestation: how environmental social movements successfully confront resistance and, in the end, often broaden their solidarities.

Double Politics and the Political Opportunity Structure

Saul Alinsky was a tough old bird, but an inspiring one. Starting with the people of the "Back of the Yards" neighborhood of Chicago in the 1930s, Alinsky more or less invented the notion of grassroots community organizing. Back of the Yards is next to where Chicago's stockyards used to be—the district that Upton Sinclair made infamous in *The Jungle*. In the 1930s, Back of the Yards had appalling health conditions, poor housing, and the disorganized social life one often encounters among the disenfranchised and downtrodden. What Alinsky catalyzed, with notable success, was perhaps the first environmental justice movement, although no one called it that back then. Through the Back of the Yards Neighborhood Council, set up with Alinsky's help, local people organized a cleanup of the stockyards, built homes, developed local businesses, and were instrumental in the founding of the National School Lunch Program. The Council is still going strong. Alinsky's passions are also in the thousands of grassroots groups that have taken inspiration from what the Back of the Yards neighborhood has accomplished. Here is perhaps Alinsky's most famous quotation, from his *Rules for Radicals:*

> Change means movement. Movement means friction. Only in the frictionless vacuum of a nonexistent abstract world can movement or change occur without that abrasive friction of conflict.[39]

In other words, to think a grassroots movement can avoid friction is fiction. If there is something worth doing that hasn't been done, it is probably because some powerful interests out there stand in the way. And if there is something worth doing that hasn't been done, it probably won't be easy. Indeed, most anything worth doing isn't easy, or, again, it would already be done. So be ready to embrace conflict. Be ready for the sit-in, the march, the confrontation, the rough treatment. That was Alinsky's message and method (see Figure 10.6).

Other organizers, though, worry that this conflict-based approach can be off-putting to potential allies and that it works more by fighting fire with fire instead of with water. It resists the material expression of power as embodied in laws, regulations, police, locks, and fences with the material power of people out there on the street, blocking traffic and ready to fill the holding cells. But maybe a better strategy is to look to changing the minds of those who make the laws, write the rules, instruct the police, smith the locks, and build the fences. Rather than a conflict model of contestation, organizers in the tradition of Michael Eichler argue for a consensus-based approach. "Instead of taking power from those who have it," Eichler has written,

Figure 10.6 Town sign for Defiance, Iowa.

Source: Photo by Helen D. Gunderson. Used with permission of Helen D. Gunderson.

"consensus organizers build relationships in which power is shared for mutual benefit."[40] Rather than a materialist approach, Eichler works on the side of ideas.

So who is right? Which way is in fact the most successful?

It's yet another dialogue. The community sociologist Randy Stoecker suggests there is both a "vinegar" and a "honey" side to successful organizing—that both conflict and consensus have their place, and that most organizing involves a good bit of both.[41] I like to call it the *double politics* of contestation.[42]

Stoecker suggests that a grassroots movement begin this double politics by first analyzing the *political opportunity structure* that it will have to contend with to gain its goals. Political opportunity theory originally comes from Peter Eisinger, who defined political opportunity as "the degree to which groups are likely to be able to gain access to power and to manipulate the political system"[43]—which sounds very instrumental, with its bald use of the term "manipulate." But that's exactly the point. We're talking about strategy here. Many environmental sociologists have used this form of analysis to understand why some environmental movements, like the anti–nuclear power movement, have been so spectacularly successful, while others, like the anti–global warming movement in the United States, have gained so little traction.[44] Plus, many authors have elaborated and refined political opportunity theory over the years.[45] Stoecker usefully synthesizes these insights and adds a few twists of his own, pointing to four factors a grassroots group should think out in planning a successful strategy and in deciding whether to use more honey or more vinegar:

- The *openness* of decision makers to hearing grassroots concerns
- The *implementation power* decision makers have to do something about those concerns
- The *structure of alliances* that shape how decision makers will feel compelled to act
- The *stability* of all of the above

So there you are, having a meeting one evening with your grassroots environmental group, sitting in a local church basement on some old folding chairs, drinking watery coffee and munching brownies while someone stands at the flip chart. Here's what the group's first line of questions should be, advises Stoecker: Who are the relevant decision makers we have to contend with, and are they likely to be open to what we have to say? Not likely to be very open? A definite minus. Here's your second line of questions: But do those decision makers actually have the resources to implement what we'd like them to do—plenty of budget, staff, and legislative leeway? Yes? Well, that's a plus for sure. The third line of questions is this: How about the interest groups that shape these decision makers' sense of the politically possible and politically necessary. Are they for us or against us? A bit mixed? Could be worse. And the fourth line of questions, which concerns the stability of the answers to the other three is this one: Do the alliances have some internal divisiveness? Nice. Does that divisiveness leave the decision makers a bit uncertain as to what to do? Double nice, at least for us. And the implementation power of the decision makers seems pretty secure? If we can get things turned our way, that will be great. Sounds like a mixed strategy of judicious applications of honey and vinegar both is the way to go. Double politics to the rescue.

Most times, suggests Stoecker, the appropriate strategy will be like the above: some kind of mix of conflict making and consensus building. The specifics will vary, of course, and it will be important to pay close attention to those. Maybe you'll be able to identify a decision maker from the start who is with you. Maybe the trouble will be that the decision maker doesn't have the budget and staff to do much about your concerns and has to deal with alliances that don't agree with you, with mixed stabilities for these. The honey and the vinegar will have to be doled out differently, then, but you'll still want to use both.

Sometimes, though, the honey and vinegar scorecard will come up pretty much all vinegar. Decision makers aren't welcoming. Their implementation power is weak. The alliances are against you. And the situation seems quite stable. So do you get out your battered copy of *Rules for Radicals* and take to the streets, perhaps in some situations risking tear gas, rubber bullets, and worse? That is not an easy decision for any group, on both practical and ethical grounds. Is it practically worthwhile to attempt the conflict approach in such an unreceptive context? Is it ethical to put group members at such risk? Best check how strong your dialogue of solidarities is to begin with and consider how wide and significant the gaps are between your own group's interests and sentiments and those of the context you are trying to change.

But here's some better news. As Stoecker points out, contestation unfolds historically. The conflict-making approach of today, if reasonably successful, often lays the foundation for the consensus-building approach of tomorrow. Double politics is about gaining a *face* in the dialogue of decision making. Dialogue is not a "frictionless vacuum," a space free of power. One gains face to speak, and to be responded to, through power. Dialogue requires face, and face requires power. Conflict can thus in time lead to consensus.[46]

Conflict can also destroy face, however. It's a tricky matter, so don't try it alone. Conflict is dangerous, of course. But also, the old approach of "good cop/bad cop" is one of the most successful strategies of double politics. The people doing the conflict making don't have to be the same people doing the consensus building. In fact, they usually are not.

The Double Politics of Practical Farmers of Iowa

The history of PFI is a good example of double politics.

The 1980s was a difficult time in rural America. The "farm crisis" came about after a period of massive industrialization in American agriculture, funded by constantly rising farmland values in the 1970s. The Soviet Union was buying lots of grain, keeping prices high. The new machinery, chemicals, and hybrid crop varieties were yielding strong production. Banks were telling farmers to borrow and buy, borrow and buy, and borrow and buy. Farms got bigger. Tractors got bigger. The costs of both skyrocketed, but there was plenty of money around and nobody seemed worried. Then, President Carter tried to punish the Soviet Union for their December 27, 1979, invasion of Afghanistan by ordering a grain embargo. U.S. grain prices immediately plummeted. Farmland values went south soon afterward. Plus, inflation pushed interest rates to shocking levels. In January

1981, the U.S. prime rate hit 20 percent. Money got tight and so did the banks. No more easy farm loans. The spending spree was over, and there was little way to pay for the binge.

It was awful for rural life. Suicide rates tripled in rural states like Kansas and Nebraska. A farmer in Iowa shot his banker, and a farmer and son in Minnesota shot two of theirs. A Farmers Home Administration official in South Dakota, depressed from foreclosing farms, shot his family, his dog, and finally himself.[47] New farm advocacy groups—the American Agriculture Movement, the National Family Farm Coalition, and Prairie Fire Rural Action, to name a few— organized protests. In March 1986, hundreds of farmers blockaded a Farmers Home Administration building in Chillicothe, Missouri, with their tractors, and stayed until the end of the summer.[48] The Reverend Jesse Jackson showed up at that one, wearing bib overalls.[49] Farmers also organized local protests to disrupt farm auctions and foreclosures, including one in March 1985 in Plattsburg, Missouri, that brought 1,500 protestors and turned ugly, with injuries and arrests.[50] Rural people don't usually do this kind of thing.

As far as the officials at the College of Agriculture at Iowa State University in Ames, Iowa, were concerned, there were plenty of bad cops around. Farmers were angry with the university, too, for having led them down a rosy path of happy industrialism. Farm advocates were writing blistering critiques of their research priorities.[51] Even some academics were among the bad cops, publishing their own critiques.[52] So when a new farm group, Practical Farmers of Iowa, formed in 1985 in the next county over, the university was suspicious and standoffish. The university's face was under considerable challenge, with its structure of alliances increasingly in tatters, and the university mainly responded by distancing itself from the threat.

But in 1987, it all changed when the associate dean of extension at Iowa State at the time, Professor Jerry DeWitt, showed up in a three-piece suit to a PFI field day at Dick and Sharon Thompson's farm.[53] It began earlier that year when the Plant Pathology Department at Iowa State courageously invited Dick Thompson, the first president of PFI, to give a talk at a seminar series it was hosting on the farm crisis. Jerry DeWitt described the moment to me in his cluttered campus office, soda can in hand:

> It was the most formal I have ever seen Dick Thompson. I do remember that Dick Thompson was extremely nervous. And that was probably equal to the anxiety in the room of what was this guy going to say and what was he going to do. So it was sort of a moment of like two dogs looking at each other and not quite knowing what the other was gonna do, or who was gonna move first. It was sort of a stalemate. He was nervous, we were nervous. It was a quietly electric moment in Iowa State's history.

But Dick didn't yell. He didn't run down the university. He didn't blockade the auditorium door with his tractor. He gave a measured talk in which he asked the university to do some research to help farmers interested in something other than the Big Iron, Big Chemical way. He also asked the university to have a look at the scientific research trials that PFI farmers had already started doing on their farms, set up with the help of a visionary, in my view, Iowa State graduate student in agronomy, Rick Exner. In other words, Dick talked like the good cop of science.

The university didn't trust how PFI was doing science, though. PFI farmers were doing randomized, replicated plots, but they were doing them on a large scale that they could farm. When you are trying to control for unrecognized environmental effects in an experiment, a big dilemma in agronomic research is controlling the problem of *field variability,* as agronomists call it. Fields vary in slope, soil quality, and other environmental factors, often in significant ways over, say, 100 feet. One field, or even one side of a field, might give very different results from another. So agronomists typically do their research in small plots on the order of a few tens of feet square, or even smaller, and use the fact that there will still be many plants in even a small plot to do statistical comparisons. But you can't profitably farm little plots like that. Dick Thompson had started out doing research more on the half-field scale of things, so he could still farm it. The trouble was, the results of his trials might just have been measuring environmental differences across one of his fields, and not differences in his farming methods.

Dick had recognized that this was a problem. And he had gotten friendly with Rick Exner, the graduate student who later helped PFI set up its research trials. As a graduate student, Rick was not so tightly a part of the structure of alliances that kept the university moving along in its inertial way. He was one of those *nodal people*—someone who was positioned at the intersection of social networks—so valuable to any social movement.[54] Rick was someone Dick could approach without anyone losing face. They discussed the problem of field variability, and Rick presented it to Chuck Francis, an agronomist at the University of Nebraska, safely distant from Iowa State but still part of the network of university scientists. (In fact, Chuck has gone on to become a very well-known agronomist and advocate of sustainable practices. But this was early on in the acceptability of sustainable agriculture in colleges of agriculture.) Chuck suggested that PFI try what agronomists call *paired comparisons*. Rick brought that back to Dick, and Dick came up with the idea of doing the pairs in field-length strips, one pass of the tractor each, so the field could still be farmed. A farmer could do one practice up the field, get off the tractor and make whatever changes to the equipment the experiment required, and do the comparison back down the field on the next pass, randomizing which strips got which treatment. By doing the trials one tractor pass wide, meaning just a few rows per strip, field variability could be assumed to vary equally for each strip. Then any difference in the comparison would not be a result of differences across the field.

It had never been done at the university, though. A few weeks after Dick's talk at the university, Rick talked Jerry DeWitt and another professor at Iowa State into sitting down with him and Dick. They met at a McDonald's, the most nonthreatening location Rick could think of—off campus and anonymous.

"That was a very tense meeting," Jerry recounted to me:

> We went in there not knowing if we were going to argue or not. They said, "We've got something we think you ought to recognize." And I represented, in a sense of body, [the view] that probably was saying, "Well, what you're doing is not valid." Across that little table there was a lot of tension.

But Jerry took away with him their little diagram of randomized, replicated, field-length paired comparisons. A few days later, he brought it to yet another university professor, Reggie Voss.

"'What do you think?'" Jerry recalled asking him.

> Reggie looked it over and said, "Yep, that'll work. That's valid." Poof! When I heard Reggie Voss say, "Yes, that's valid, that'll work, that's fine," it was like, now wait a minute. For how many years have we been discounting what they're doing as not workable? And they have been thinking, we will never recognize their work. It took one meeting, an hour meeting, a piece of paper, and a why-don't-you-look-at-it. It took me 10 minutes to give it to somebody and Dr. Voss to simply look at it, and poof! All of that tension was over with.

Jerry had heard there was a field day that day at Dick and Sharon's farm, just a few miles from the university. He was so excited that he got into his university car and drove out there, in the three-piece suit of a dean, to shake Dick's hand. It was quite a scene when he arrived. A big crowd of farmers in jeans and feed caps. Dick standing 10 feet above them in the bucket of a front-end loader so he could be heard. Jerry in his suit. Rick Exner remembered the moment this way:

"So Dick got out of the bucket, talked to Jerry for about a minute, came back. He was quite pleased. DeWitt had gotten back in his car and left, and Dick said, 'Sounds like we're going to work together.'" Rick chuckled, happily thinking it over in his mind.

"So this was at Jerry's initiative?" I asked him.

"Jerry really gets the credit for this, yes."

Over the next year or so, Jerry used his implementation power as associate dean to broker an unusual collaboration between a university and a farmer's group. He found an office for PFI on campus and found some money to steer toward the group so they could hire a staff person for the office. PFI and the university would have equal say in hiring that person, and that person would be considered a university employee, with the university's health care and retirement benefits. And who became that employee? Rick Exner. Over the years, PFI has grown to the 13 staff it had as of 2010. By now, dozens of university faculty and students have conducted hundreds of research trials with PFI farmers, using the field-length paired comparisons and other techniques and always bringing farmers and researchers together as partners through participatory research. And by now, PFI has helped thousands of farmers implement sustainable practices through the group's cultivation of knowledge and dialogic sense of solidarity.

As for Jerry DeWitt, he later went on to direct the Leopold Center for Sustainable Agriculture at Iowa State—a different kind of administrator at a different kind of college of agriculture—until he retired in 2010.

"PFI has meant a lot to me personally," Jerry explained at the end of our interview, conducted before he retired.

I would not be who I am today—I would not be where I am today or doing what I'm doing today—if it weren't for PFI. It's been that important. Sometimes you look back and you can see events that were real turning points in your life that you might not have realized were at the time. Well, PFI has been that for me.

In other words, Jerry assigns the credit for this successful double politics just the other way around, to PFI and not to himself—a sure sign of a solidarity that is both wide and truly in dialogue.

The Pros of the Three Cons

As the success of PFI shows, cons can be pros—at least when we are talking about bringing together the three cons of conceptions, connections, and contestations. But the main thing is togetherness itself. In fact, if the reader will forgive me, the three words that I choose to represent the three cons did not derive from each etymological form of *con-*. Rather, they all derive from the "together" form. *Conception* means to put together. *Connection* means to tie together. And *contestation* means to bear witness together. Grassroots environmental movements succeed through this togetherness of togethernesses.

The trick is how to get a togetherness of togethernesses together. Environmental grassroots movements are not always as successful as a group like PFI has been, of course. If they were, environmental sociologists would not devote so much research to understanding how such achievements come about, and environmental activists would not be so interested in hearing the findings.

Environmental sociology has no precise recipes to offer, however. The origins of movement success are often best understood after the fact, so dependent are they upon the happenstance of, say, a phenomenological rupture or having a willing and visionary nodal person at hand. These are matters of what I like to call *dialogic providence*—not luck, exactly, but situational opportunities that provide scope for agency and change.[55] The reason why this providence is not just luck is that we can help create situations that invite these opportunities, even if we cannot predict what they will look like, when they will appear, or how they will turn out. The skill of a grassroots movement is in creating the situations as much as it is in acting upon the opportunities they may occasion. Like inviting people to a meeting at, of all places, a McDonald's restaurant.

No, there can be no recipe for success, just conditions that welcome success. Otherwise, there would be no success—only ineluctable outcomes in which agency is merely the working out of some socio-environmental mechanism, and is thus not agency at all.[56] And ineluctable outcomes are exactly what grassroots environmental groups are mobilizing to prevent.

CHAPTER 11

Governing the Ecological Society

The Earth is one but the world is not.

—Bruntland Commission, 1987

own a car. I admit it. And I often use it. I live in an 1,850-square-foot house with four bed-rooms, plus two bathrooms. It is not a solar house. We also own a share in a vacation house on the St. Lawrence River. My family owns a washing machine, a stereo system, several land-line phones, a cell phone, three clock radios, three laptop computers, and an MP3 player. I own nine pairs of footwear of various sorts; nevertheless, I just bought another pair. Our closet is filled with clothes we rarely wear, and yet we buy more. We have loads of books. I'm a semi-pro musician on the side, so we have a small orchestra of musical instruments including a piano, plus we have a sound system and some recording equipment. We also have a range of shop and yard tools, bicycles for everyone in the family, a kitchen full of dishes and cookware, and a house worth of furniture. We moved to a new state a few years ago, and the bill of lading from the moving company came to about 12,000 pounds. Plus, at the vacation home, there is another houseful of furniture, another collection of tools, and another collection of dishes and cookware—not to mention seven boats of various sorts, a pile of recreational gear, and a boathouse to fit them all during the summer and an old barn to store them all during the winter. That's a lot of stuff, an awful lot of stuff. And did I mention that I eat meat? Plus, we have two children.

So how could I fancy to call myself an environmentalist—let alone an environmental sociologist?

There are a few things I could point to in my defense. My wife and I own only one car, a 2010 Prius. This is our first new car; we kept our previous car until it was 15 years old, and the one before that until it was 18 years old. (Reuse is as environmentally important as recycling and reduction.) As I write, we've had the Prius for 17 months and have put 12,120 miles on it. That's about 8,600 miles per year, or 4,300 per driver—well below the U.S. average of roughly 14,500 annual miles per driver in recent years.[1] We bicycle to work (it's about 2 miles each way for both of us), and we bike for most of our shopping, even in winter (which gets pretty forbidding in Wisconsin, where we now live). We made sure to buy a house in a location that was relatively convenient for bicycling, and also for getting around by bus when the weather is bad.

Let's see . . . what else? We have no microwave oven, no television, no VCR, no camcorder, no DVD player. Although we do now own a cell phone, it's just a cheap pay-as-you-go model that we pretty much only use when we travel; most weeks it never gets switched on. We own a washing machine, but we use a solar-powered clothes drier in the summer—a clothesline, that is—and a line strung across the basement in the winter. (We don't have a mechanical clothes drier.) There was a dishwasher already installed in our new house, but it's broken now. We haven't missed it. When it was still working, we hardly ever used it. There was also central air-conditioning already installed, as in almost every home in our neighborhood. But we use it only once or twice a year, in part just to keep the thing in working order, should we ever decide to sell our house. We keep the house comfortable during a heat wave by shutting all the win-dows during the day and opening them all at night, when the temperature drops. Plus, we have a well-placed tree for shading the house. We keep no cupboard or closet full of household poisons. We grow some of our own vegetables during the summer, all organically, and we mow our small lawn, which is also organic, with a manual reel mower. We compost all our leaves, garden waste, and kitchen waste. The previous owners of the house installed a "rain garden" to promote infiltration and groundwater recharge, thus lessening the storm water and pollution load on the beleaguered lake in the park near our house. (Many of the sewers for our neighbor-hood dump into that poor little lake.) We've long used energy-efficient lightbulbs almost everywhere in the house, and we're very good about turning them off when a room's not in use. Our furnace is a high-efficiency model, and during the winter we keep the thermostat at 62 degrees Fahrenheit for the day and 56 degrees Fahrenheit when we go to bed and when we're out, although we sometimes bump it up a degree or two in the evenings or during the day when someone's at home.

All told, we use about 285 kilowatt hours of electricity a month for a household of three (not counting when our 21-year-old son is home from college), or about 95 kilowatt hours per person per month. That's almost a third of the 256 kilowatt hours per person per month average for all households in Madison, Wisconsin, our city.[2] And we pay an extra fee to our local utility to support the small wind farm they've bought into. Our gas use is quite low; our local utility puts us in their "excellent" category in terms of BTUs of gas use per square foot. Although we do have a lot of books, as well as a few CDs, we've pretty much stopped buying them now. We use the library instead. (We pop down to our local branch several times a week.) We're avid recyclers, too. Plus, we use almost no paper goods in the kitchen; we use cloth towels and napkins, which are reusable and usually better for the task. We cook almost everything from scratch—we like to cook—so there is no sodden mass of food-stained packaging to pitch out at the end of the day, aside from the occasional can or bottle. We produce about a paper shopping bag of nonrecyclable garbage a week, which isn't bad for an American family.

Then there's our diet. Most of the meat one can buy in the supermarket packs a huge environmental wallop. The grain that conventional farming shovels through livestock erodes the land, sucks the water, burns the tractor fuel, and relies on chemical pesticides and fertilizers. The feed efficiency of that grain—the pounds of grain it takes to produce a pound of meat on your table—is around 2 to 1 for chickens, 4 to 1 for pigs, and 7 to 1 for cattle, greatly decreasing world grain stores.[3] Plus, the animals live so close-packed that farmers include daily pharmaceuticals with the grain, not just when the animals get sick. Vegetarians often point to these stark facts, and fair enough—but only if one is talking about grain-fed livestock. We buy only grass-fed meat, which leaves the land covered year-round so there is little erosion and relies on rainwater, not irrigation. Pastures normally require few chemical pesticides and fertilizers, and none at all if the farmer has got the hang of organic techniques. The animals fertilize the fields as they go, and they do the harvesting for the farmer, too, so there's no need to burn much tractor fuel. Furthermore, the animals convert something humans can't eat—grass—into something we are able to eat, and 71 percent of agricultural land around the world is pasture anyway.[4] So grass-fed meat is ecologically very efficient. Plus, rotating grass through cropland every so often is a great way to break up pest cycles without chemical pesticides. The animals live well, too, getting exercise and fresh air, and don't require constant dosing with antibiotics and other medicines. We usually buy our grass-fed meat directly from farmers, whom we've gotten to know well enough to be sure that their practices are kind to the land and to the animals. Moreover, we only eat meat a few times a week.

And we aren't having any more kids.

I'm not a complete sinner, I think, but I'm certainly no environmental saint either. According to an online "ecological footprint" analysis I worked through once, my footprint is 16 biologically productive acres, in comparison with the U.S. average of 24. Not bad, huh? And my footprint is actually probably quite a bit smaller than that, as I took a 4-acre hit just for my meat eating. The online site didn't distinguish between grain-fed and grass-fed meat or the amount of meat one eats. I probably should take some hit for my meat eating, though, maybe an acre I'll guess, giving me a total footprint of 13 biologically productive acres, almost half the American average. But according to footprint analysts, there are only 4.5 biologically productive acres per person on the planet. If everyone lived like I do, we'd need 3.6 planets assuming my footprint is 16 acres, and 2.9 planets assuming my footprint is 13 acres.[5] Either way, that's pretty dire. So, am I just another environmental hypocrite, big on the guilt trip and fairly small on action, mostly talk and little walk?

From a certain political perspective, yes. As that is a perspective I share—the perspective of the committed environmental moralist—my environmental inadequacies often pain me deeply. Yet from a sociological perspective, my situation does not necessarily indicate some deep personal moral failing. In fact, to the extent that my situation is typical of others, it represents some important opportunities for social and environmental change. It suggests the possibility of collective action toward making a society that more closely resembles what we say we want it to be. And indeed, the overwhelming majority of the public in both rich nations and poor are concerned about the environment, even though very few could be said to have yet put that concern into full action.

The previous chapter took up how to move into environmental action from the grassroots up, through the three cons of conceptions, connections, and contestations. But do we only want an environmentalism that we have to move along ourselves? How about an environmentalism that also moves us? This concluding chapter considers this latter potential, focusing on the place of governance in an ecological society.

The A–B Split

But before we get to governance, let's take a closer look at my footprint problem and, by extension, that of many, many others.

Maybe I flatter myself, but I don't think my problem comes from a lack of environmentally committed attitudes. I am a classic example of the limits to what Chapter 10 called a "behaviorist approach" to environmentalism—the idea that environmentally committed attitudes lead to environmentally committed behaviors. Social psychologists have long noted that there is often a sharp disjunction between what people profess to value and believe and how they really act. I like to call it the *A–B split*, standing for "attitude–behavior split," and it is a characteristic that probably all of us share, at least to some degree.[6]

Sometimes we consciously recognize some of these inconsistencies in our lives (perhaps with a little help from family, friends, and others in our circle of personal critics). As behaviorism suggests, this recognition can at times impel us to adjust our behaviors to fit our attitudes. But other times, social psychologists find, we work to adjust our attitudes to fit our behaviors. This second direction of adjustment typically goes on unconsciously. Consider the way student radicals often become more conservative when they start raising families and enter the world of paid employment. Without deliberate intent, they find themselves taking on the very attitudes—and enacting the very behaviors—that they had protested against only a few years earlier.

But the point of the A–B split is not to suggest that today's radicals are tomorrow's hypocrites. Nor necessarily are all the rest of us—the rest of us who have adjusted what we believe to what we do or have gone on doing things that do not fit what we believe, which probably includes just about everyone on the planet. It is very hard to maintain a conscious sense of an A–B split. Such inconsistency strikes at the very core of our identities, our sense of who we are. (Consequently, we do not always respond gently to that circle of personal critics.) Understandably, people tend to avoid conscious recognition of an ideological mismatch if they can. But often they can't, and they try to adjust their lives and their thinking accordingly—which is also hard. In other words, an A–B split is a source of internal struggle and conflict. Contrary to the image of the complacent hypocrite, such a split is hardly something about which most people feel comfortable.

The sociological point here is that one of the main reasons people find their attitudes at odds with their behaviors (and often find themselves adjusting those attitudes to fit or putting the conflict out of their mind as much as possible) is the social constitution of daily life, a concept introduced in Chapter 1.[7] We do not have complete choice. We face material constraints that limit what we can do and influence what we are likely to do, to paraphrase Fred Cottrel once again.[8] I own a car and use a car because the automobile-based planning of the past 50 years has led to the scattering of businesses, shopping, schools, parks, and homes. Our city has, for the United States, a pretty good system of bike paths and an okay bus system. But they can't make up for the structured inconvenience of sprawl, and there is no passenger train.[9] So I often feel strongly pressured to use my car, and sometimes I do.

These material constraints, in turn, influence what I am likely to think. I'm not sure they limit what we can think—the mind seems boundlessly imaginative—but they sure mightily influence it. Okay, one might come up with an infinity of possible social constructions about one's life experience. But a person is unlikely to hold on long to those that simply do not fit his or her circumstances.

Unless the circumstances can be changed. The A–B split not only shows the constraints we face, but also presents us with opportunities—the opportunities of ecological dialogue. When we as a community consider our collective attitudes and our collective behaviors—when we consider the ideal and the material implications of the current arrangement of our social and ecological lives—we have an opportunity to reconsider them as well. The social constitution of our communities may be a large part of our problems, but the *social reconstitution* of our communities can be a large part of the solutions. Together we can create new social structures, new constraining influences that shape and guide our lives.

Social structures are not necessarily bad things. It depends on what they guide us into doing. Social structures do not necessarily create the A–B split (or what is really an ideal–material split). Properly rearranged, properly reconsidered, social structures can help heal the splits in our communities—including that biggest community of all, the environment of which we are (thankfully, I say) an inescapable part.

Virtual Environmentalism

But people are busy, terribly busy, caught as we are on the treadmills of production and consumption.[10] Although we are surrounded by modernity's supposed inducements of choice and leisure, modernity equally induces us into a treadmill-driven rush from home to work to the point that work becomes home. And when we come home—late, probably—we're still in such a rush that what we do at home becomes work, as the sociologist Arlie Hochschild observes.[11] The supper has to be cooked, the dishes done, the children put to bed, the toys picked up, the floor swept, the clothes washed, and the bills paid.

When work becomes home and home becomes work, daily decisions have to be made fast. This isn't going to change soon. If being environmental means a lot of extra thought about the consequences of each act of consumption, if it means switching to everyday practices that take a lot of time to do and to learn about, then daily decisions are unlikely to be made with the environment in mind. Environmentalism on these terms is unlikely to become a significant part of everyday life in a modern world. Our daily experience is too full already. And if being environmental costs a lot extra, one doesn't have to be a social scientist to recognize that will put a significant damper on its feasibility, too. People have mostly already figured out what they are going to do with the money they have.

Let's cut to the quick of it: If it is hard for people to be environmental, well, then being environmental will be hard.

What we need, then, is what might be termed *virtual environmentalism*—environmentalism you don't have to worry about because you just find yourself doing it anyway. Virtual environmentalism is environmentalism that lies behind and beneath our daily lives. Like environmentalism in the usual sense, virtual environmentalism is walking or taking your bicycle to work; buying food produced with sustainable production methods; replacing old appliances with energy-efficient ones; and using less heating, cooling, construction materials, and water. But virtual environmentalism means doing these things not because you've made a conscious decision to be environmentally good today, but because these were the cheapest, most convenient, and most enjoyable things to do. Virtual environmentalism is being environmentally good without having to be environmentally good.

I think it is safe to say that virtual environmentalism is a lot more likely to be popular with the general public than environmentalism by guilt, cajoling, shaming, issuing court summons, imposing fines, and locking offenders in jail. But it will become popular only if we change the structures of the cheap, convenient, and enjoyable by reorganizing the social organization of production and consumption.

Take walking or bicycling to work. In the Netherlands, nearly half of all trips are by bike or by foot, compared with 16 percent of all trips in Britain and just 7 percent of all trips in the United States; Germany is somewhere in between, with a third of all trips by foot or bike.[12] And no wonder. Germany and the Netherlands have made vast investments in infrastructure to accommodate people-powered transport. Bike lanes. Raised crosswalks. Pedestrian lights. Pedestrianized shopping districts. Snug and coordinated urban planning that puts shopping, schools, and workplaces close to where people live. As a result of this infrastructure, foot and bike travel is also pleasant and safe. Both per trip and per mile traveled, American pedestrians and bicyclists are killed at 3 times the rate of German pedestrians and cyclists and 6 times the rate of Dutch pedestrians and cyclists.[13] Not 25 percent more or even, say, 75 percent more—300 percent and 600 percent more (see Figures 11.1 and 11.2).

Figure 11.1 The bike parking lot at the train station at Ede, a small Dutch town. Scenes like this are common throughout the Netherlands, where half of all trips take place by bike or by foot. The Dutch also often combine bike transport with train transport, and may even keep a bike at each end of their commute, biking from home to train and then train to work or school.

Source: Author.

Figure 11.2 A three-level street in Wageningen, a Dutch university town. Vast government investments in infrastructure have made biking and walking safe and convenient in the Netherlands. Commonly, Dutch cities install raised bike lanes on either side of the car lanes, but lower than the sidewalks, helping separate all three modes of transport: car, bike, and foot. As well, raised crosswalks make pedestrians more visible and slow cars down exactly where a pedestrian might be, as this picture shows. Furthermore, the bike lanes narrow the space for cars, forcing them to slow down, especially when passing an oncoming vehicle.

Source: Author.

There is no law that requires a third of Germans and half of the Dutch to leave their cars behind when they head out the door. There is no vast supplier of halos of environmental virtue that these folks don on their heads when they step outside. They are just going about the business of the day in ways that are economical, expedient, and agreeable.

In short, virtual environmentalism means making environmentalism easy. The trouble is, it is often hard to make things easy. Social reconstitution usually requires a terrific effort. But when you do reconstitute society, you've really done something, something lasting and important—precisely *because* it is so hard to do. If social reconstitution were easy, it probably wouldn't be social reconstitution at all. It can be done. And it is done, all the time. But you have to do it. We can become environmental without trying—but only if we try.

Which means that social reconstitution depends as much upon our social constructions as it does upon our material constraints and possibilities. Maybe we'll try if we must. But think how much easier it will all work out if we try to reconstitute our lives because we also want to.

Seen in this way, the A–B split is not a problem that stands in the way of virtual environmentalism. Rather, it is what leads to getting virtual environmentalism going, when we think and act collectively. And when we do get that social reconstitution into place, we may well find that we're not completely satisfied with it—that the behaviors it encourages are not fully in line with our attitudes, maybe because we didn't get the result we wanted or because our attitudes changed once we did get it. That's Okay. We're learning. We're always learning, for it's another dialogue, an ecological dialogue that virtual environmentalism depends upon. Or, put another way, virtual environmentalism means turning the A–B split into the "A–B dialogue," where the difference between our attitudes and our behaviors is not a sign of our hypocrisy but a sign of our growing collective wisdom about what it is we'd like to do and how best to make it possible.

The Bottom and Top of Change

When we think and act collectively, that is. Chapter 10 discussed how collective action can emerge from the grassroots, as often it must, and overcome the "tragedy of the commons" and other challenges to social mobilization. Through knowledge cultivation, the dialogue of solidarities, and double politics, a grassroots group can gain the pros of the three *cons*.

But there are limits to what the grassroots can do on its own, especially if we want to make environmentalism something people do without trying to. There is both a "bottom" and a "top" to effective change in the social and environmental circumstances of our lives—an interactiveness of the grassroots with the shovels and rakes of government, economy, technology, and other social structures that govern daily life. It's another (you guessed it) manifestation of ecological dialogue. Let's call it the *dialogue of bottom and top*. Local communities and civil society represent more the ideal side of the dialogue, contributing beliefs and values, and the structures of governance represent more the material side.

Sometimes we hear advocates argue for a purely "bottom-up" approach, though.[14] There is much wisdom in this suggestion, especially when we consider the legacy of the "top-down" way. Who likes rules and regulations, especially when they are of the one-size-fits-nobody variety? Who likes being told no? Who likes living with bad planning decisions, made mainly on behalf of some well-positioned political interest? Who likes having one's choices dictated by corporate power? Who likes being coerced or even forced, sometimes by the barrel of the police officer's or soldier's gun? Who likes losing freedom?

But a purely bottom-up approach would face many challenges of its own. Not least of these is that we have many local communities in the world, not just one. The grassroots are plural. Each is at least somewhat differently situated in the world and will have at least somewhat different interests in, and sentiments for, the world. If nothing else, this pluralism complicates coordination, should the issue at hand be significant for more than one group or locality. As it probably is. But it also raises the issue of how we draw the boundary of a locality or group and who are the relevant stakeholders—and who are not. A bottom-up community may have a strong dialogue of solidarities going, and thus a strong sense of the mutualism of interests and sentimental ties. But what of the interests of those defined as outside the community? Can we be sure that there will be sentimental commitment to the interests of those others? In addition, there is no guarantee, just because an effort starts from the grassroots, that it is truly inclusive of even those in the local community. Local areas have their political moves, too. So, although a bottom-up approach sounds inherently democratic, it may wind up denying voice to at least as many as it gives voice.

Moreover, working bottom-up presents many resource needs for time, money, and people. A local community organization may have a fierce commitment to saving a nearby forest, say, or to ending the fumes coming from the vicinity's factory. But the company that owns the forestland or the one that owns the factory likely has people working 9/5, and maybe even 24/7, representing its interests. Local groups may do well to get a meeting together once or twice a week, and maybe to hire a part-time staff person—even if there are thousands of affected residents. It's hard to raise money, especially for a long effort, and to find committed people to serve on volunteer boards. In these cases, a purely bottom-up approach would be denying voice to itself.

Plus, bottom-up groups often lack the relevant expertise. Take the factory example again. The smokestack may have a constant brown plume coming out of it that can be smelled miles away. Nevertheless, in order to make the case that the pollution harms the local area, there will probably have to be a scientifically valid sampling regime and analysis of what the stack spews out and how far into the neighborhood the spew travels. Likely, too, the locality's argument with the factory will require some way, say, to connect the excess cases of childhood asthma the area is infamous for, or the unusual rate of lung cancer, with the kinds of chemicals the plume contains. These were precisely the issues faced by the people of Diamond, Louisiana, a small African American community living in the shadow of a Shell facility in Louisiana's "cancer alley" of chemical plants along the Mississippi River.[15] Yet documenting such harm is a difficult technical matter, likely requiring several different sorts of graduate-level training. And people with graduate-level training probably had the financial resources to avoid living in a community like Diamond to begin with.

And then there is political influence. The double politics of contestation may help a group gain the face needed to navigate the political opportunity structure. But wouldn't it be better if the top actively facilitated giving face to a grassroots group and voice to its demands?

The top potentially has a lot to offer the bottom with regard to these challenges of coordination, democracy, resources, expertise, and influence. It can provide communication for coordination, rules of fair dealing for ensuring local democracy, money and people for getting work done, knowledge and technology for documenting issues of concern, and balanced access to its structuring ability. And it can implement the changes that turn environmental virtue into virtual environmentalism.

The good news is that sometimes it does. Take the overfishing of cod in the North Atlantic. The situation is still dire in the Grand Banks and Georges Bank, but in the North Sea and the Gulf of Maine, government regulation and coordination have led to a dramatic recovery of cod stocks. The U.S. National Marine Fisheries Service no longer lists Gulf of Maine cod as "overfished," as of 2010.[16] North Sea cod stocks were up 52 percent from 4 years ago in 2010, and the European Union was even able to allow a slight increase in fishing. As one North Sea fisherman said, "We have not seen so much cod in the grounds for 20 years." And then he added, in recognition of the need for the top to ensure a continued recovery, "I don't think there's a fisherman who thinks we can keep fishing and fishing."[17]

Plus, the top often finds that it needs the bottom as much as the bottom needs the top. Changes that come only from on high encourage foot dragging on the part of those down below. The top can resist the bottom, as we well know, but the bottom can also resist the top, often covertly through the "arts of resistance," as the political scientist James Scott has called the technique of keeping your head down while quietly obstructing the aims of the powerful.[18] Consequently, it is far easier to lead the willing than the unwilling—particularly if the top has to operate within legal restrictions on its use of coercive force, as fortunately is now generally the case. In addition, surveillance and policing are quite expensive even for government. It is always one thing to put a law or regulation on the books. It is quite another to implement it. The rulebook of a state is filled with stuff the police and regulatory agencies don't bother to enforce (sometimes blessedly so). Moreover, one-size-fits-nobody is a problem for the top, too, although the top may not always clearly recognize it. The top seeks particular outcomes through the application of laws and regulations and has to contend with how lack of fit for particular localities may compromise those outcomes. If people on the bottom have a say in how laws and regulations are implemented and agree with the implementation to begin with, the grain of attainment of the top's desired outcomes is likely to be far finer.

Plus, the top faces the constant potential that its legitimacy may crumble—that it may experience what the sociologist Jürgen Habermas called a "legitimation crisis."[19] People are rarely unaware that hierarchy exists in society, and the top is in continual need of justifying its authority so people will continue to put up with it. Consequently, even if the top really couldn't care a fig for the needs and concerns of the bottom, it generally behooves it to give the impression that it does. And giving that impression will likely require at least a degree of actually attending to those needs and concerns. The skeptical are always eyeing the spectacle.

What I'm talking about here, of course, is power. For the bottom to have power, it needs the top. For the top to have power, it needs the bottom. And for effective social reconstitution to occur from either perspective, they both need to be in dialogue. That is, they both need, as Anthony Giddens has termed it, "dialogic democracy"—a democracy in which all, including the environment, are taken into account.[20]

But I am also talking about community. The dialogue of bottom and top depends upon a dialogue of solidarities—a dialogue of our interests and our sentiments and the species of mutual power that come from each. Otherwise it will fall apart. And the dialogue of solidarities likely requires a dialogue of bottom and top to coordinate the complementaries of people's interests and sentiments. The time gaps and space gaps of solidarity can be mightily, if warily, helped by a bit of bottom and top. A bit, and be careful. But the inequality of the bottom and the top does not necessarily imply injustice, although it is indeed all too frequently associated with it. Indeed, that inequality may be just what we need to ensure justice, as the political philosopher John Rawls argued (see Chapter 4).

Moreover, we don't necessarily all have to agree to get something good to happen. Recall those German and Dutch bicyclists and pedestrians, happily foregoing their cars as they save money; get to where they are going without stewing in traffic jams; and enjoy the sights, sounds, smells, and good health that come from self-powered transport. Global warming probably hardly ever crosses their minds. And maybe when it does, it is accompanied by a mental snort of skepticism. At least in Germany, a 2009 poll found that 40 percent of Germans do not regard global warming as a very serious problem, and 43 percent are unwilling to pay higher prices to alleviate it.[21] Now, possibly all these German global warming skeptics get about only by car. But I doubt it—although it wouldn't surprise me if the skeptics tend to drive more than others, given the social psychological struggle of the A–B split. They often find themselves being virtual environmentalists nonetheless.

Does this mean that these German and Dutch virtual environmentalists, with all their mental snorts, are doing the right thing for the wrong reasons? Maybe. But should that matter, as long as they are doing the right thing? Besides, economy, expedience, and sheer fun seem like good reasons to do what one does. If we can craft environmental practices so they also have all these other appeals, it won't make a difference if someone doesn't have much environmental concern. She or he will have plenty of other reasons to do the right thing.

In short, we're all different, and that's why we need community—including that biggest community of all.

Participatory Governance

The feeling that we have to find ways to bring bottom and top together has prompted a huge range of new programs and projects, as well as a vast literature that studies them. Command-and-control environmentalism is no route to virtual environmentalism, as many years of poached endangered species, midnight dumping, continued deforestation, and many other troubles now show us. There is widespread agreement that governing the ecological society will entail more than the work of government, in the narrow sense of the word. It will require the work of citizens, too. It will require a shift from government to what is often called *participatory governance*.

Consequently, a new language of management has spread across the world, with phrases like the following: participatory management, co-management, collaborative management, community-based management, deliberative environmentalism, community forestry, participatory rural appraisal, participatory development, participatory planning, participatory research, and participatory action research. Behind these phrases are still more: stakeholder analysis, civic engagement, civil society, social capital, deliberative democracy, dialogic democracy (which I mentioned earlier), focus groups, town meetings, citizen juries, citizen advisory committees, local knowledge, and more. These are all ways that environmental managers and environmental sociologists have articulated the importance of the dialogue of bottom and top.[22]

Why so many phrases for the same basic idea? For a good reason. Most observers concur that the "participatory turn"—yet another phrase—is a wonderfully fertile idea, with equally wonderful material potential. And some of the efforts to put it into practice have been tremendously successful, both for environmental quality and for the quality of our societies. Chapter 10 already discussed one such success, the participatory research of the group Practical Farmers of Iowa.

Participatory governance doesn't always work out, though, and I'll come to some of the troubles shortly. But first, let's consider a few more examples of its many successes.

Supplying Water in a Costa Rican Village

For years, international agencies have been drilling wells, planting trees, providing new crop varieties, building dams, and promoting tourism in "less developed" communities across the world, hoping to spur economic development. Sometimes this form of international aid has worked, but very often it has not. Local people have often looked on with pleasant smiles while the dams were put up and have shaken hands in apparent thanks when given trees to plant, only to fail to maintain the dams and the trees later. Eventually—after the development reports were filed away back at the international aid agency's headquarters—the dams crumbled and the trees died.

Astonishing as it seems in retrospect, supporters of this 1970s-style approach to development rarely bothered to ask a crucial question of local people: What do you want? Such a top-down style of development assistance not only alienated the people it was supposed to help, but because of the development officials' lack of knowledge of local conditions, top-down approaches often resulted in increased social inequality and environmental damage.

In the early 1990s, though, development agencies began to see both the practical and the democratic value of what has come to be called *participatory development*.[23] Involving local people as equal partners and leaders in development projects ensures a sense of ownership—of sentimental commitment—to a project. It also ensures that the project is more likely to do what people want, making the project fit their interests as well. This approach is so totally obvious in retrospect that it may seem incredible that development efforts ever took another course. But early development thinking often had little respect for the views of local people, seeing them as backward and incapable of understanding all the advantages of the modern techniques that were being offered to them, while assuming (rather contradictorily) that the modern way was what everyone wanted.

Some years ago, I was fortunate enough to see firsthand the results of a more participatory approach to local development. An old friend lives in Platanillo, Costa Rica, a farming village of around 500 people about 3,500 feet up into the Talamanca Mountains. We had lost touch since meeting in the 1970s, but I happened to be in the country on university business. The village had no phone at the time—this was before cell phones became widespread—and I wasn't even sure he was living there anymore. So I quite literally looked him up. I took the Platanillo bus up the dirt road into the mountains on a Saturday afternoon, got some directions from the barkeeper in the local tavern, and surprised my friend as he was returning from his fields for the day. He recognized me almost immediately, even after 17 years, and excitedly led me around his farm and the village.

One of the places he brought me to, with considerable pride, was the new water supply dam that he and some other villagers had installed earlier that year. The dam made a small impoundment on a stream up in the mountains above the village—not big enough to cause much damage should it give way someday, but large enough to supply all the houses on that side of the valley with running water. Before the dam was built, everyone was drawing water by hand from household wells, often dug dangerously close to outdoor toilets. Now, everyone in the neighborhood had safe running water piped into their houses.

The people in Platanillo had some outside help in building the water system. My friend mentioned that several development agencies were involved, although he didn't mention which ones. That didn't seem important. Instead, he talked about the neighbors with whom he had worked on the project, about the way the sluice gate worked, about the way they arranged for the land where the dam sat, about the village committee that is maintaining the dam, and other local details. This clearly was the villagers' own water supply.

What really struck me, as my friend described the new system, was how much he knew about it—far, far more than I know about the water supply system in my own community. After all, my friend had helped build and design the one in his community. Should those pipes or that dam or the watershed up above or down below ever develop any problems, he and his neighbors would know what to do and would feel a sense of investment and responsibility for carrying out any repairs. Which was a good thing, I thought. In such a remote place, if the local people didn't take care of a problem, it would be a long time before anyone else would.

As I took the bus back down the valley that evening, I passed a building in the next village down the road from Platanillo with a sign on it that said, "U.S. Peace Corps." I don't know if Peace Corps volunteers were involved in Platanillo's dam—my friend never said. But if they were, I thought, they sure understood the value of participation, and also of a helping hand from the top.

Growing Local Knowledge in Honduras

Jeff Bentley is not your typical social scientist. I knew that as soon as I laid eyes on him, when he came to give a seminar in my department. The title of his talk was suitably academic sounding—something like "Farmer-Scientists and Integrated Pest Management in Honduras," as I recall. But rarely, even in this informal age, is a seminar delivered by someone wearing old jeans whose bottom hems are frayed from continually catching beneath the wearer's construction-style boots. He did wear a sport coat, a tweed one, but it only made his ragged jeans and uncombed hair seem that much more incongruous in a university seminar room.

And yet Bentley held the packed room (including several conservatively dressed scientists from the entomology department) absolutely spellbound. Bentley at the time was employed by the Department of Crop Protection at the Escuela Agricola Panamericana in Zamorano, Honduras, trying out a radical new way of doing research on the farm problems of Honduras, working with the country's poor peasant farmers. In collaboration with Werner Melara and others at Zamorano, Bentley had been going into Honduran villages and conducting entomology seminars with local farmers. "We don't tell them what to do to solve their pest problems," Bentley said. "We try to give them the intellectual tools for solving the problems themselves."[24]

Ever since the "Green Revolution," the typical approach of agricultural scientists working on the problems of tropical agriculture has been to encourage peasant farmers to adopt hybrid crop varieties developed by the scientists themselves. Such varieties generally yield more but also have fewer defenses against pests. The scientists have developed an answer for that problem, too, though: pesticides. (It's a package deal.) But farmers have to buy the hybrid varieties and pesticides, rather than relying on seed saved from the previous crop and on lower-cost pest control practices. And if you're a poor Honduran farmer, money is something you don't have a lot of. Capital-intensive agriculture also promotes international economic inequality by draining scarce cash from the Honduran countryside. Plus, a high degree of literacy is required to read the label warnings on the safe and appropriate use of the pesticides. Thousands of people have been poisoned.[25]

Bentley's view is that any solutions farmers devise for themselves are far more likely to be relevant to their ecological, economic, cultural, and agricultural circumstances. Also, Bentley stresses the importance and validity of farmers' own knowledge about local conditions and local farming practices—their local knowledge.[26] Honduran peasant farmers are poor, not stupid, and they know a lot of relevant things that the scientists don't. After all, the peasant farmers live there.

University scientists do have a lot to offer local people, though, particularly concerning phenomena that are not easily observed. In Bentley's rural seminars, he helps the farmers see inconspicuous connections that the university scientists have figured out. Most local farmers don't understand insect life cycles, so he puts larvae in glass jars for several days so that people can watch caterpillars and grubs develop into adult insects. Local farmers almost never go out into their fields at night, so Bentley takes them out to watch insect activities by flashlight. And then he steps back and lets them apply the knowledge.

In one village, the local farmers had been spending quite a bit of money on pesticides to eradicate the fire ants that were infesting their fields, although they had no evidence that the ants were harming their yields. When Bentley took them out at night, though, they watched as the ants crawled up their corn plants and ate some other insects that were harming the crop. A local woman was very impressed with this observation and wondered how to encourage the ants. She recalled that ants were often attracted to the sugar in her kitchen, and she came up with the idea of making a dilute solution of sugar water and spraying it on infested plants to attract the ants.

This idea, suggested Bentley, has several advantages typical of local innovations. First, it's cheap, as sugar is relatively inexpensive. Second, it relies on easily accessible local materials—sugar and water. Third, it is something that the local people understand completely, which should allow them to refine the idea, generating further innovations. Fourth, it is safe, both for the environment and for the farmers. And fifth, as it is their own idea, local farmers feel a sense of ownership and are far more likely to be committed to making the idea work.

But does this idea from the bottom actually help control insect pests? Here's where the top—the scientists—can step in again, performing experiments and helping local people design their own experiments to assess the validity of the idea. With the Zamorano approach, scientists are still very important, but, as Bentley and Melara explain, "We depend on farmers to help tell us what to study and to work with us in actually carrying out experiments in their fields, fine-tuning the technologies to their conditions."[27]

The point of participatory development, in other words, is not that local people always know best. Rather, the point is to get a dialogue going between local people and scientists, between local knowledge and expert knowledge. Such a dialogue encourages the respect and concern of each party for the other and perhaps even genuine friendships, as each comes to know the other better: solidarities of interests and sentiments. Participatory development is thus *dialogic development.*

Clearing the Air in Three British Cities

Like Jeff Bentley, Steven Yearley often wears jeans with his sport coat. And he doesn't comb his hair very carefully, either. After all, Yearley is also an academic. And he is just as radical in his experimentation with participatory techniques of environmental management. But his jeans are black, instead of blue like Bentley's. His favorite coat is leather, not tweed. Plus, most significantly, he works in rich countries, usually Britain, rather than poor ones. (Rich countries need participatory governance, too.) I remember first meeting Steve at a conference in Britain and going on a long walk with him in the countryside in which we debated postmodernism and democratic theory.

People drive too much in Britain, as in many other rich countries. The resulting air pollution load is now widely recognized as a major health hazard, as I have described earlier in the book. Now, what to do? In Britain, the approach has been to have cities declare Air Quality Management Areas (AQMAs) where the air pollution exceeds government standards, and then to put into place an action plan to reduce the pollution back below the standards. That's straightforward enough. Scatter some pollution-monitoring stations around the city, do a study of traffic trends, include any notable stationary air pollution sources, factor in the effects of climate, put it all onto a map, and bang: Local officials have an air pollution model they can use to establish the AQMA and devise the action plan.

Still, there's a trouble: A model is a model. There are inevitable limits in the number of monitoring stations a city can establish and inevitable questions about their appropriate placement. But ordinarily, city officials don't worry all that much about these issues. The AQMA process is pretty much a top-down approach. Officials have lives to live, too—homes to return to at the end of the day and retirement accounts to make deposits into. The central government in London says that it is to be done a certain way . . . so let's make them happy, and then we'll be happy. (There are tops within the top.)

Hang on, said Yearley and his colleagues.[28] Local people have local knowledge that would be useful to get into the models. After all, they are out there every day. Each one of them is a monitoring station, a monitoring station that even talks. Plus, they will have to live with the results of the action plan—which will likely mean finding ways to encourage reduced automobile use, as well as impacts on property values. The whole business will go down better if they are involved in the process.

So Yearley and his colleagues held a series of focus groups in three cities that were implementing the AQMA process: Bristol, Sheffield, and York. They contacted bicycling advocates and traffic-calming campaigners, as well as advertising the events through newspapers and posters. At the focus groups, they put up a map of the city and drew people's impressions of air quality right onto it. Then they showed people the resulting map in a second series of meetings and through individual contacts, to check it and refine it. They digitized the result, put it at the same scale as the officials' map, and brought it to the city government offices.

In all three cases, local people's results largely matched the official maps, but with some significant points of difference. Local people usually had good explanations for the places where their assessments differed. The officials' models, they felt, understated the pollution from factories, as the officials assumed that the factories never exceeded the permitted levels of pollution releases. But local people knew from their own experience and from talking to factory employees that the factories often exceeded the permitted releases. Local people also felt the official models' assumption of average levels of pollution from motor vehicles did not take into consideration that in poorer areas the vehicles are likely older and in worse repair, and thus pollute more. Plus, cyclists contended that the scatter of monitoring stations did not accurately reflect how certain streets act as canyons that trap poor-quality air close to the ground.

Yearley and his colleagues call the process "participatory modeling." In all three cities, Bristol, Sheffield, and York, the officials took it seriously. Unfortunately, the participatory modeling in Bristol wasn't completed until after the AQMA for Bristol was designated, and in Sheffield the AQMA process had been subcontracted and officials had less control of the procedures. But in York, officials revised their model in light of citizen input—not only their model of air pollution but also their model of government and how it should relate to the citizenry that, after all, it is supposed to represent. As Yearley notes, "Public engagement can assist in bridging knowledge and policy."[29] For better participation means better science. Better science means better government. And better government, in turn, means better participation.

Governing Participation

So we know participatory governance can work to help alleviate environmental problems. We know because it has been done, hundreds if not thousands of times by now.

But, as I say, it has not always worked out well.[30] It can be hard to get people to show up at the meetings. Some get frustrated by all the talk when they do. The group process sometimes simply hasn't gelled, and misunderstandings have developed, in one documented case leaving the participants more at odds than when the process began.[31] There are often instances of those who felt they should have participated but did not feel invited to, leading to feelings of exclusion.

In response, some researchers and practitioners have devised guidelines for "best practices" for ensuring successful participatory processes.[32] Their goal generally is to ensure that all voices are heard and are welcome, that communication is open and available to all, and that the agenda is always open to change. These best practices typically advise a set of formal procedures—a kind of Robert's Rules of Order for participation—that practitioners can use. Participation can be a bit unruly, and they suggest, in effect, how to govern it.

But as Caroline Lee has argued, formal procedures imply a particular form of political culture that not everyone necessarily shares: a culture of the formal.[33] The nonelites that formal methods are intended to give voice to are often turned off by the fussy proceduralism and may be culturally ill-equipped to navigate it. Moreover, given the popularity of participatory methods, elites can gain some kudos from their colleagues and associates by agreeing to take part in the formal process—a process in which they are likely to perform well anyway, as it was likely designed by someone of a similar cultural background. The emphasis on proper procedures for participation can also encourage suspiciousness of those who, perhaps inadvertently, violate the protocols and engage in informal communication outside the approved venues.

Think of it as a thinning of solidarity. Formal procedures for participation run the risk, if overemphasized, of turning the participatory encounter into a mere exchange of interests. Lost is the chance to build the sentimental ties of friendship, which are rarely a formal matter, ungluing trust. As Francesca Polleta has observed, "Friends are unlikely to suspect each other of cutting corners or cutting deals, and their affection for each other makes the deliberative process tolerable, even pleasurable."[34] Formal procedures may also stifle the creativity that can come from free-flowing dialogue.[35] But we also should not forget the troubles that come when informal communication reverts to an "old boys" or "old girls" network. It's a tough balance to maintain, and many efforts at participatory governance have not hit it right.

Another trouble builds on the kudos factor for elites of using participatory approaches. Given the way the top needs the bottom for legitimacy, there is often a strong temptation for government or corporations to call in the participation consultants—there are loads of these now—and have them prepare the process so that the public gets the "right" answer. For example, in early 2008 in Britain, a flap emerged over the government's plan to start building nuclear power plants again.

The government officials backed up their plan by pointing to a public consultancy process they had hired in. When presented with the facts chosen by the consultant, a plurality of participants said bringing nuclear power back was the best option. Critics cried foul as the consultant—eager to serve its paying client, the government—hardly seemed disinterested in the outcome.[36]

And if the public gets the "wrong" answer in a participatory process, the skeptics doubt they would be listened to, anyway. As one such skeptic from California put it, "San Diego's history is littered with the skulls of bureaucratic brain-picking sessions that invited people from the neighborhoods to contribute, then discarded their ideas."[37] Call it "participation-washing."

As a result, phrases like "participation fatigue," "participation-itis," and even the "tyranny of participation" are starting to show up.[38] There is much to be learned from these critiques. Let me suggest a few aphorisms that capture some of that learning:

- Participatory practices are not *above* power; rather, they are all *about* power.
- One-size-fits-all formal procedures of participation are no way to solve the problem of one-size-fits-all government.
- You can't govern trust, but you need trust to govern.

Do we want to go back to letting the top call all the shots? No, nor are we likely to. The thing about dialogue is that it is so satisfying when we get it right that it ever encourages us to get it even better. This is one of the most important of all dialogues: dialogue about dialogue itself.

The Reconstitution of Daily Life

The dialogue of bottom and top is not only about participatory governance, however. It is about all the ways of getting the bottom and top to work together to bring about virtual environmentalism and other features of an ecological society. This will mean changing how we socially organize our lives, and it will mean changing how we socially construct our lives. It is a matter of both the material and the ideal factors, and bottom and top factors, that shape the ways our daily lives are socially constituted.

It is also a matter of recognizing that we are each, in a way, both bottom and top. The existential philosopher Jean-Paul Sartre, a noted misanthrope, used to like to trot out this little definition of hell: other people. What this means is that all actions are, in fact, interactions. What any one of us does has consequences of some sort for the world, and thus for others in the world. In this way, whether deliberately or not, we constrain what others can do and thus govern them to some degree. We also enable them, too, at least for some actions, to be more optimistic (and I think more accurately balanced) observers, as the sociologist Anthony Giddens points out.[39] My virtual environmentalism can ease yours, and vice versa, as the bottom extends to the top and the top extends to the bottom, and we reconstitute each other's daily lives. Other people may be hell, but they can also be heaven.

Greening Capitalism

Perhaps the most important feature of our daily lives that we need to reconstitute, argue many, is our economy.[40] As the great British sociologist Raymond Williams wrote, it is vital that we overcome "the intellectual separation between economics and ecology. It will be a sign that we are beginning to think in some necessary ways when we can conceive of these becoming, as they ought to become, a single discipline."[41] Indeed, each has the same etymological root in the Greek word *ecos,* or "home," that deepest center of daily life.

The first step is to beware of the easy populism of the-market-knows-best arguments. I mean this kind of common claim: Got a problem? Just let the wonderfully adjusting and democratic market take care of it. Every dollar, euro, and pound another vote. As for government? Just get it out of the way and let the popular voice come through in people's buying decisions. If people want ecology, well, then they'll ask for it in how they shop.

Not so fast. As Chapter 3 discussed, sociologists have pointed out that the state is inseparable from the market. Without governance in the form of trade regulation, a money supply, police enforcement of property rights, and so much more, we would have the Wild West, not markets. The notion of a "free" market is always a question of freer for whom, and it is generally used to gain more freedom for the speaker and less for someone else, via the state. For the governing power of the state is no less called upon when a regulation is removed or refused than when it is put on the books, what Chapter 3 called the difference between "negative regulation" and "positive regulation."

So the bottom should not be squeamish about calling on the top to reconstitute the pattern of our economy so that it becomes a more sympathetic feature of our ecological home. Not only is the relationship of the market and the state a central feature of the dialogue of bottom and top, it is an unavoidable one. And these are some of the changes that are most crucial to bringing about virtual environmentalism. Yes, let the market work its marvelously adjusting magic (at least, given that capitalism seems here to stay for a good while yet). But structure that magic so that the freedoms people seek in the market support positive externalities for ecology. Like all those Dutch and Germans finding that the easiest, cheapest, and nicest way to get to work is by foot or on their bikes. Or like the new law in the United States that will phase out most of the current generation of energy-sucking incandescent bulbs by 2014, or the new law in the European Union that phases out all incandescent bulbs by 2012, creating the market incentive to make energy-sipping compact fluorescents and other technologies far cheaper.

One route to the positive externalities of virtual environmentalism is through structuring real estate development so that what gets built uses land efficiently, promotes community, builds tax base, and is beautiful enough that people will want to live there. *Smart growth* is what planners call it, and cities across the industrialized world are discovering its ecologic, economic, and social good sense. The basic idea of smart growth is to reject the standard polarization between anti-growth naysayers and pro-growth yeasayers, familiar to development controversies across the United States, for example. Smart growth says, yes, there are serious problems with how development usually goes on in the United States. But we can use the power of development forces to "grow out of" sprawl. Pressures for growth provide the capital to reshape what we have done and give us the opportunity to rethink what we might do. Besides, there are good economic reasons for reshaping what we've done and might do— let alone the environmental reasons. It's expensive to construct and maintain the necessary roads, sewer lines, and power lines and to provide police, fire, and emergency services to spread-out developments. Although sprawl is often defended for adding a tax base to a community, the cost of providing for it can easily be more than the added government revenue. Plus, developers are figuring out that sprawl raises their own costs and, as distances increase beyond tolerable commuting, limits their market.

Smart growth is often coupled with an architectural style and approach to planning called *new urbanism.*[42] The basic idea of new urbanism is to model new developments on the kind of traditional neighborhoods that cities routinely turn into historic districts. If we think such areas are nice enough to make special efforts to preserve them and to visit them as tourists, new urbanists ask, why not design all our neighborhoods that way? New urbanism is thus in many ways an old urbanism, the urbanism of a time when cities were built for people rather than cars. And if we build with people first in mind instead of cars, the result will be not only pleasing to the eye but pleasing to the balance sheets of local governments and local developers, because of new urbanism's efficient land use. That's the "smart growth" part. But also, new urbanism advocates argue that such an approach helps reduce the impact of development on community in the ecological sense and helps promote more interactiveness in the community in the social sense (see Figure 11.3).

New urbanism designers typically recommend the following guidelines for people-friendly development: Build houses up, not out, so lots can be narrower and land-use efficiency can go up. Bring back the front porch, the sidewalk, and the alleyway, all zones of interaction between neighbors. Make most streets through streets so all the traffic doesn't get channeled onto a few trunk roads, causing traffic jams even in suburbs. Bring back the corner shop. Provide a diversity of housing types within a neighborhood so that people with all kinds of household situations can live there, from singles, to families with children, to the elderly. Locate stores and schools and workplaces near homes—and without the traffic and oversized parking lots that make most commercial life so unappealing and environmentally unsound today. Don't mix stores and housing types higgledy-piggledy, but instead institute far more detailed zoning plans than the current big-blob style of zoning with huge areas devoted to a single type of use. Increase density, so walking and public transit are more realistic options (see Figure 11.4).

Another route to virtual environmentalism is internalizing costs. That is the goal of *green taxes,* sometimes called *Pigouvian taxes,* after Nicholas Pigou, the English economist who proposed the idea in the early 1900s. Green taxes are an attempt to make the price of goods and services reflect their true costs and to shift the burden of government revenue generation away from regressive taxation schemes like sales taxes and value-added taxes. Finland, for example,

Figure 11.3 A street in Kentlands, Maryland, the best-known "new urbanism" development. Note the space-saving, close-together houses with small front yards and the community-building presence of porches. Although this view shows single-family homes, Kentlands has a wide variety of housing—as is characteristic of "smart growth" initiatives.

Source: Author.

Figure 11.4 A street in Providence, Rhode Island, developed in the 1890s. New urbanism takes as its model "old urbanism" developments like this one. Note here, too, the close-together houses, the small yards, the front porches. Also note the mixing of single-family homes with the duplex in the foreground.

Source: Author.

now has a carbon tax aimed at internalizing the costs of global warming and other pollution issues associated with fossil fuel use. Britain has a landfill tax. The Netherlands and several Scandinavian countries now have energy taxes. London, Milan, Singapore, Stockholm, and Valletta (the capital of Malta) have recently instituted "congestion pricing" that taxes drivers for bringing their cars into already congested areas. New York City seriously considered it a few years ago, and public opinion supported it by a 2-to-1 margin, although it was eventually scuttled by the New York State legislature.[43]

Green taxes offer a lot of possibilities, but like any taxation scheme, they have to be handled with great care. Taxes are perhaps the most hotly contested of any issue these days. If they are not supported by public sentiment and if they harm public interests, perhaps by being instituted in regressive ways, green taxes will be a political disaster. The dialogue of bottom and top is crucial. Also, powerful interests often get the upper hand in taxation debates, as when Belgium instituted a pesticide tax that exempted farmers and when the early versions of energy taxes in Scandinavia exempted some energy-intensive industries. Business interests also appeared to play a significant hand in the dropping of New York City's proposed congestion tax. But we are learning the lessons of these early experiments and becoming wiser in how to use green taxes to help build an economy that reflects what things really cost.

There is also increasing excitement these days among business leaders about *industrial ecology,* as Chapter 7 discussed—about treating industry as a part of ecologic systems as opposed to a means of dominating ecologic systems.[44] The key principle of industrial ecology is regarding pollution as a sign of inefficiency in an industry. Waste products should be regarded as wasted opportunities, not leftovers to be gotten rid of in the cheapest and fastest and least conspicuous way possible. "Closing the loop" is the way advocates of industrial ecology often describe the greener approach. Environmental standards such as ISO 14000 alert industry to places where the loop is perhaps not yet closed and opportunities are being wasted. Industrial ecology thus advises businesses to see environmental standards and environmental regulation as business opportunities rather than obstructions to be fought or dodged.

A related idea is *dematerialization*—finding ways to accomplish our economic goals with a lot less material use of ecology's productive capacity. Think of it as a loop made from a much finer strand. "Factor 4" advocates argue that the technology is pretty much here already to reduce material use by a factor of 4—25 percent of what we now use—without materially affecting quality of life.[45] "Factor 10" advocates push us to get it to 10 percent.[46] Businesses that manage these reductions in material use will also be saving themselves that much in the cost of materials to produce their products. Here again, good ecology is good economics.

How is a customer to know that environmental standards have been followed or that dematerialization practices are in place at a company? Recent years have seen a great growth in *green labeling* schemes that emphasize transparency and traceability, verified through third-party certification. Many of these labels are backed by governmental bodies, such as the organic labels now so widespread in countries across the world. But many are operated by nonprofits or on behalf of industries. For example, the Rainforest Alliance, a worldwide environmental nonprofit that works to protect biodiversity, offers the Rainforest Alliance Certified label. An industry-organized example is GlobalG.A.P., for "Global Good Agricultural Practices," standards that are now widespread among European food retailers and are increasingly found elsewhere. *Fair trade* labels have also been growing; they verify that the workers who make the products have received a living wage for their labor. A problem with fair trade is that it is very focused on producers and doesn't may much attention to what happens after that. Fair trade coffee and tea are only the best-known of these certification schemes. Now fair trade practices have been set up for chocolate, fruit, sugar, wine, cotton, handicrafts, and more, involving, as of the end of 2007, some 1.4 million small producers around the world, and worth $2.4 billion annually[47] (see Figure 11.5).

Economists often talk about the importance of "value chains" in the economy, a sequence of economic activities that continually "add value" to a product. An example might be all the steps along the way to turn cotton plants into the shirt you may be wearing. But what green and fair trade labels do might be called establishing *values chains*—adding values in the plural, not merely market value alone, as Thomas Lyson, George Stevenson, and Rick Welsh point out.[48] Values chains operate in two directions, allowing producers and consumers to communicate their commitment to a wider sense of what is valuable, across this increasingly global world.

Sometimes green standards and fair trade standards have not gone hand in hand, however. Some criticize GlobalG.A.P. and organic labels for, in fact, discriminating against small farmers, especially in poorer countries, both through requiring low prices and requiring procedures and record-keeping practices that small farmers do not have the equipment or training for. Small farmers often can't pay the costs required to participate in these labels, giving an advantage to outside investors who then employ local people at lower wages, giving rise to what Hugh Campbell has called a new form of colonialism.[49] Fortunately, efforts are growing to link the fair and the green, and many products are now reaching consumers certified with fair trade and green labels together.

Another concern is that the increasing growth of nongovernmental third-party certification is transferring environmental governance away from government. Because this is governance outside of government, citizens have little say in how the certifying is done, stretching the dialogic limits of trust and chains of values. This is particularly a concern for green standards controlled by an industry trade association such as GlobalG.A.P., some scholars argue.[50] As Lawrence Busch and Carmen Bain have written,

Figure 11.5 A collection of green and fair trade labels from around the world. Backed by third party certification arrangements, such labels have become an important means for producers and consumers to communicate their values to each other. See Figure 5.4 for a collection of organic certification labels.

Source: Author.

While it may be true that the private sector has acted where governments have failed to do so, the growing shift in standards setting and enforcement from the public to the private arena is problematic. Standards are pervasive; therefore, debates surrounding their content and their organization have import for our lives not just as consumers but, more importantly, as citizens.[51]

In other words, let's be careful here.

But the real heart of capitalism, with its problem of the original capitalist and resulting treadmills of production and consumption, is property relations, as Karl Marx recognized long ago. Peter Barnes, a capitalist with a heart and a founder of the ethical investment fund Working Assets, argues that "capitalism as we know it over-rewards people who own private property. It's a *system* flaw, not a personal flaw."[52] The solution? Barnes suggests that much can be done to green capitalism with a property structure that is perfectly legal and already on the books: trusts. Unlike a conventional corporation, the property of a trust is managed by trustees who are legally required to act on behalf of the beneficiaries of the trust. They are not allowed to act in their own interests. Barnes proposes that we establish trusts whose beneficiaries are the common good of all citizens and future generations. The effect would be to reestablish common property as a major sector of the economy.

This is no pipe dream. It's already happening. Land trusts have become an important tool of conservation and are now major holders of land in many regions, particularly in the United States, where there are now some 1,600 of them.[53] The Vermont Land Trust is one of the most successful and has protected about 4.3 percent of the land area of the state, either through direct ownership or through easements—nearly half a million acres.[54] This is not private land, as it ordinarily cannot be sold for profit, nor is it public land, as it is not held by government. It is, in effect, a form of commons—regulated commons, without the infamous tragedy that Garret Hardin worried about.

A related form of ownership, typically found in urban areas, is a "community land trust," in which land is held in common but houses on the trust's land are owned privately. The effect is to reduce the cost of housing by separating land value from the market value of a home, making it more affordable. Plus, houses in the trust cannot be sold speculatively; the resale price can be only a certain percentage over what the owner originally paid for it. But typically, ecological goals have not been part of community land trusts. A group of graduate students of mine have suggested a new form of trust, a hybrid of a conservation land trust and a community land trust they call an "eco-community land trust." With an eco-community land trust, as they envision it, the focus would be on both affordable housing and land conservation, linking social justice and environmental goals in the best spirit of environmental justice.[55]

Trusts aren't the only legal alternative to growth-maximizing capitalism out there, however. There are nonprofits. There are cooperatives. There are also family-owned businesses, which often operate with quite a different set of priorities from shareholder-owned businesses. And there is also government.

Not only are all these other forms of economic organization perfectly legal under capitalism, but they are also widespread, and in some cases, growing rapidly. Let's continue with the United States, that supposed paragon of capitalism, as an example. According to a recent report by the University of Wisconsin Center for Cooperatives, there are now nearly 30,000 cooperatives in the United States, they employ nearly 1 million people, they have assets of over 3 trillion dollars, and they have annual revenues of half a trillion dollars.[56] To put this in scale, the size of the U.S. economy is about $14 trillion, meaning that 1 in 28 dollars of output comes from a co-op. As for nonprofits in the U.S. economy, as of 2005 there were 1.4 million of them, and they had $1.6 trillion in revenue.[57] That means nonprofits are 11 percent of the U.S. economy. Government spending in the United States is, in most years, about 19 or 20 percent of the economy.[58] Combined, this is about 35 percent of the United States' capitalist economy, all perfectly legal.

There are even more radical (and also legal) proposals. An idea that is getting some serious attention these days, especially in Europe, is *degrowth economics*, or what is also called *sustainable decrease economics*—or, in French, *décroissance soutenable*. There is even a degrowth political party in France now, the Parti Pour La Décroissance. Serge Latouche, an emeritus professor of law and economics at the University of Paris-Sud, is perhaps the central figure behind this idea, which resonates strongly with the older steady-state economics of the American economist Herman Daly, and with Richard Douthwaite's writings on what he calls the "ecology of money."[59] The central idea is to reject what Latouche calls the "faith system" that economies must always grow, but not to abandon the notion of markets. Rather, he argues that we need to not just dematerialize the economy; we need to make it nonmaterial. In Latouche's words,

> But it does not necessarily mean ceasing to create value through non-material products. In part, these could keep their market forms. Though the market and profit can still be incentives, the system must no longer revolve around them.[60]

The economic sociologist Juliet Schor has another idea along these lines. We pretty much all work too hard under capitalism, one foot on each treadmill—our two feet getting further and further apart as both treadmills speed up. At the same time, many of us are working too little, although not by choice, because of unemployment. Schor suggests that we could solve both problems through what she calls the *80 percent solution*. Have the employed work 20 percent fewer hours so they have time to dematerialize their needs and break out of the environmentally destructive cycle of work-and-spend. Use those 20 percent fewer hours to put the unemployed to work, while encouraging a culture of *plenitude*—of satisfaction with abundance rather than constant lust for more, ever more.[61]

Are ideas like degrowth and the 80 percent solution possible? One of the great benefits of capitalism, we are often told, is its openness to creativity. If so, then let's use this creativity. Let's use it for environmentally appropriate ends. Let's try out a few different forms of economic relations that are kind to people and the environment and see if we can figure out how they might work. Let's recognize as well that the bottom-up freedom to create is always contextual, always regulated, always structured, always governed—that the top will always be there, too. And let's recognize that this is not necessarily a bad thing.

The Local and the Global

Another manifestation of bottom and top is the relationship of the local to the global. Here, too, we need to encourage a dialogue. The famous phrase, "Think global, act local" helps us along in seeing these connections. But the relationship should be just as much the other way around. We should think local when we act global, considering the local effects of global decisions.

As the theory of mobilities and environmental flows reminds us (see Chapter 5), air, water, food, wood, fossil fuels, other minerals, wildlife, people, and ideas are all constantly in motion. When a tanker load of oil makes its way through the Suez or Panama canal, it is not just a ship on the move; it is an ecological act. Part of the world is moving, flowing from one region to another, through another. It is no different when we bring a load of groceries home or carry out the trash, or even when we breathe in and out. The local is moving the global, and the global is moving the local.

In other words, our daily lives, as local as they may feel to us in the moment, are just as much manifestations of the global. This recognition, obvious enough when you think about it, suggests some fundamental redirecting of how we think about environmental issues. There should be nothing remote feeling about an international environmental treaty. We cannot govern our daily lives by considering only the local, or even a national, level. And conversely, when we act globally, we are acting locally. We are setting in motion flows that cross localities and help constitute what those localities now are, or we are stemming or redirecting such flows.

This does not mean that place no longer matters. If anything, it matters more, for the concept of environmental flows helps us to see that to do something in one place is probably also to do something in another. The world is not as bounded as we once thought it was, but it takes place in places now as ever.

One of the kinds of places that still matter is the state. As a number of observers have argued, the World Trade Organization (WTO), the World Bank, the United Nations, the European Union, and the recent great flowering of international treaties on environmental and other matters have not spelled the death of the state.[62] National boundaries remain edged with walls—some metaphorical and some literal—that impinge on environmental flows. Try bringing more than 3 ounces of fluids in your carry-on bags. Try getting work in another country. Try shipping across a national border something that is clearly benign, like a box of books, without filling out a customs form. Try inviting a group of landless peasants from Mexico to fill up the backseat of your car as you head back home to El Paso. Try setting up your own pipeline to send oil from the rainforests of Bolivia up to the United States, or from the tar sands of Canada down to the United States. People in uniforms will be getting up close and personal with you very soon.

No, ours is an international world, not a non-national one. But it is moving more than it used to, way more. All those ships, planes, trains, cars, trucks, pipelines, transmission lines, wireless towers, and satellites, as well as the soil, dust, and chemical pollution in the water and the air, together mean that more sheer mass of atoms and electrons moves from place to place on the surface of the earth today than was the case, say, 100 years ago. But our forms of governance have historically been based on thinking mainly about what goes on in one place at a time.[63] This nation. This town. This forest. This piece of property. The concept of environmental flows does not ask us to abandon our boundaries, but rather to learn to flicker across them, seeing difference and seeing connections at the same time, governing with here, there, and everywhere in mind.[64]

For if we are to make connections in a global world, there must still be difference. Otherwise, as I have earlier stressed, there will be nothing to connect.

Reconstituting Ourselves

In all these ways, we can achieve virtual environmentalism—the virtue of being environmental without being virtuous. And, I believe, we not only can. We must. For in the end, there is nothing virtual about virtual environmentalism. It's the real thing.

We are, however, unlikely to work to change our relationships in the biggest community of all unless we have personally committed to change. We will need a new collective sense of who we are and what we want to do. It's going to take some virtue to become virtual environmentalists.

It's important to recognize the interaction, the ecological dialogue, between the material organization of our lives and the ideas with which we construct our lives. We are more likely to regard the environment in environmentally appropriate ways when our community lives are constituted to encourage such regard. But we can't simply wait around for that social reconstitution to happen. We need to make it happen. Individuals are the agents of community change as much as communities are the agents of individual change.

In other words, our personal values and actions matter. We can be what Juliet Schor and Margaret Willis call *conscious consumers*.[65] And millions are trying, even in a country like the United States, with its relatively low level of environmental concern. A 2008 Harris Poll found that 53 percent of Americans say they have made a change in their lifestyle to promote environmental sustainability, and only 25 percent say they have not.[66] Some 91 percent said they now recycle, at least occasionally. Some 49 percent said they are now buying more locally. Some 23 percent said they are now composting. Some 16 percent said they are now carpooling. A 2010 Harris poll asked Americans how often in their daily lives they do a variety of environmental practices.[67] Here are some of their answers. Always or often turn off unneeded lights when you leave the room: 81 percent. Always or often buy food in bulk: 32 percent. Always or often purchase organic products: 15 percent (see Figure 11.6).

This stuff adds up, not only materially but socially and politically, providing the grassroots support at the bottom for the substantial reconstitution we will need from the top. There is a crucial ethical dimension to our ecological dialogues. Our virtues, at least, need to be more than virtual.

Which brings us back to community, for ethical ideas are always ideas about community relationships. Aldo Leopold put it well in "The Land Ethic," probably the twentieth century's most influential essay on environmental ethics: "All ethics so far evolved rest upon a single premise: that the individual is a member of a community of interdependent parts."[68] But how we draw the boundary of community membership shapes (and is shaped by) our sense of interdependence, and thus what and whom we feel a sense of moral concern for. (Moral concern and interdependent parts—the interplay of sentiments and interests again.) Our fellowship with others implies that they are entitled to our moral concern, just as we are entitled to their moral concern.

Figure 11.6 Frequency of various green behaviors in the United States.

"How often do you do each of the following in your daily life?"
Summary of those saying "Always" or "Often"

Base: All adults

	2009 Total	2010 Total
	%	%
Keep unneeded lights off or turn lights off when leaving a room	83	81
Recycle	68	68
Reuse things that I have instead of throwing them away or buying new items	65	63
Make an effort to use less water	60	57
Unplug electrical appliances when I am not using them	40	39
Purchase locally grown produce	39	33
Buy food in bulk	33	32
Purchase locally manufactured products	26	23
Purchase used items rather than new	25	24
Purchase all-natural products	18	16
Purchase organic products	17	15
Compost food and organic waste	17	15
Carpool or take public transportation	16	16
Walk or ride a bike instead of driving or using public transportation	15	15

Source: Harris Poll (2010 and 2011).

My point is that in the idea of community is the idea of justice. There is a constant tension between commitment to those included within the community's boundaries, and lack of commitment, and consequent inattention, to the troubles of those excluded from the community. Ideas of community and the boundaries of moral concern are thus closely intertwined.

Each of the three central issues of environmentalism—sustainability, environmental justice, and the beauty of ecology—challenges a different dimension of these boundaries of concern. Sustainability considers how we draw boundaries of concern between present and future generations. Environmental justice considers how we draw boundaries of concern between human groups. The beauty of ecology considers how we draw boundaries of concern between humans and the rest of creation.

This last boundary is perhaps the most difficult. How can we form a sense of community with the ecosystem, something we're not even sure is an intentional actor? How can we form a solidarity of interests and sentiments with something we're not even sure has interests and sentiments? Would not such an "ethical extension," as Leopold termed it, be mere anthropomorphism—treating the inherently nonhuman as the human—and therefore highly unstable?

The notions of the dialogue of solidarities and the dialogue of bottom and top can help us here, I think. We must begin by recognizing that all communities are imagined.[69] This is why trust is so important. We cannot get into the mind of the other, so we are always guessing, trusting, and closely watching for the signs of solidarity.

The same may be the case for human–environmental interactions. We need to imagine this form of community, too. And this imagination is what the environmental movement has long promoted, at least as I interpret the two sides of what has long been the main debate in environmental ethics: anthropocentric environmentalism versus ecocentric environmentalism.

Anthropocentric environmentalism suggests that we consider our own interests first in our interactions with the environment—interests in sustainability and environmental justice—and also that we consider the environment's interests in order to gain our own. (That last clause, I should point out, is what distinguishes anthropocentric environmentalism from mere anthropocentrism.) In other words, anthropocentric environmentalism says, treat the environment well and it will treat us well in return: hence, a solidarity of interests.

Ecocentric environmentalism, on the other hand, suggests that we consider the environment as a moral entity in its own right and with its own beauty and that we see ourselves as a part of that moral entity. It argues that we need to go beyond questions of calculated human interest and recognize the importance of what the environmental philosopher Paul Taylor, for example, termed "respect for nature."[70] But we are part of that beautiful entity for which respect is due: hence, a solidarity of sentiments.

As with purely human communities, the solidarity-of-sentiments side of environmental ethics is the harder argument to make. This difficulty stems, we cannot doubt, from the individual and instrumental thinking so characteristic of our time and place. The challenge of imagination is particularly hard here because the environment does not speak, at least not directly. Which may in part be why anthropomorphism, as in the story of Androcles and the lion in Chapter 10, is such a popular way to think about the environment: It helps us imagine the voice of the other in the ecological dialogue.

But I believe the environmental movement is right to try to make the case for a solidarity of environmental sentiments. Even though it is a hard case to make, the evidence suggests to me that it is a vital one. Moments that threaten to bust the glue of trust are too frequent. Also, we can't always wait to figure out what part of the ecosystem is crucial to our interests before we act, and sentiments may add some efficiency here. If for no other reason than they are good for our interests, we need to have sentimental bonds with the ecosystem as well. Correspondingly, the bonds of sentiment are unlikely to last unless they are also good for our interests—which, I believe, they are.

Thus, the wise anthropocentrist is also an ecocentrist, and vice versa, not one or the other. And that means that the issues of democracy and governance, of bottom and top, and even of virtual environmentalism that this chapter has considered, pertain to more than just us. We are not only governing humans. Therefore, fairness requires that not only humans do the governing.

So how do we bring the nonhuman world into the top of governance? The same way we do with humans: through dialogue, ecological dialogue. Maintaining this dialogue is the basic work of the democratic community, from the smallest to that biggest community of all. It is also the sustainable, just, and beautiful thing to do.

References

Abram, David. 1996. *The Spell of the Sensuous: Perception and Language in a More-Than-Human World.* New York and London: Vintage.

AAE (Asociación Empresarial Eólica [Spanish Wind Energy Association]). N.d. "Wind Energy Has Consolidated as the Third Technology of the Power System." Retrieved February 16, 2011, from http://www.aeeolica.es

Abell, Annette, Ernst, Erik, & Bonde, Jens Peter. 1994. "High Sperm Density Among Members of Organic Farmers' Associations." *The Lancet* 343:498.

Abramson, Paul R., & Inglehart, Ronald. 1995. *Value Change in Global Perspective.* Ann Arbor: University of Michigan Press.

"Acid Rain Problem Getting Worse." 2000. *China Times.* Retrieved August 22, 2003, from http://www.taiwanheadlines .gov.tw/20001128/20001128s2.html

Agarwal, B. 2001. "Participatory Exclusions, Community Forestry, and Gender: An Analysis for South Asia and a Conceptual Framework." *World Development* 29(10):1623–1648.

Ajzen, Icek. 1985. "From Intentions to Actions: A Theory of Planned Behavior." Pp. 11–39 in *Action-Control: From Cognition to Behavior,* edited by J. Kuhl & J. Beckman. Heidelberg, Germany: Springer.

Ajzen, Icek, & Fishbein, Martin. 1980. *Understanding Attitudes and Predicting Social Behavior.* Englewood Cliffs, NJ: Prentice Hall.

Ajzen, Icek, & Fishbein, Martin. 2005. "The Influence of Attitudes on Behavior." Pp. 173–221 in *The Handbook of Attitudes,* edited by D. Albarracín, B. T. Johnson, & M. P. Zanna. Mahwah, NJ: Erlbaum.

Alavanja, Michael C. R., Samanic, Claudine, Dosemeci, Mustafa, Lubin, Jay, Tarone, Robert, & Lynch, Charles F. 2003. "Use of Agricultural Pesticides and Prostate Cancer Risk in the Agricultural Health Study Cohort." *American Journal of Epidemiology* 157:800–814.

Alejandro, Oliva, Spira, Alfred, & Multigner, Luc. 2001. "Contribution of Environmental Factors to the Risk of Male Infertility." *Human Reproduction* 16(8):1768–1776.

Alexander, Jeffery C. 1996. "Reflexive Modernization: Politics, Tradition and Aesthetics in the Modern Social Order." *Theory, Culture and Society* 13(4):133–138.

Alinsky, Saul. 1971. Rules for Radicals: A Practical Primer for Realistic Radicals. New York: Random House.

Alvord, Katie. 2000. *Divorce Your Car! Ending the Love Affair With the Automobile.* Gabriola Island, Canada: New Society Press.

American Farmland Trust. 2002. "Farming on the Edge: Sprawling Development Threatens America's Best Farmland." Washington, DC: Author. Retrieved August 24, 2003, from http://www.farmland.org

Andelson, Robert V. 2004. "Hardin's Putative Critique." *American Journal of Economics and Sociology* 63(2):441–450.

Anderegg, William R. L., Prall, James W., Harold, Jacob, & Schneider, Stephen H. 2010. "Expert Credibility in Climate Change." *PNAS* 107(27):12107–12109.

Anderson, Benedict. [1983] 1991. Imagined Communities: Reflections on the Origin and Spread of Nationalism. Rev. ed. London and New York: Verso.

Anderson, Chris, Hartley, Jacki, & Robinson, Matt. 2007. *Eco-Community Land Trusts as Community-Based Land Reform: Contesting the Growth Machine for People and Their Environment.* Unpublished manuscript, University of Wisconsin–Madison, Department of Rural Sociology and Agroecology Program.

Anderson, Ian. 1995. "Australia's Growing Disaster." *New Scientist* 147:12–13.

Angenent, Largus T., Mau, Margit, George, Usha, Zahn, James A., & Raskin, Lutgarde. 2008. "Effect of the Presence of the Antimicrobial Tylosin in Swine Waste on Anaerobic Treatment." *Water Research* 42:2377–2384.

Animashaun, Kishi N. 2006. "Racialized Spaces: Exploring Space as an Explanatory Variable in Environmental Justice Analysis." *Dissertation Abstracts International, A: The Humanities and Social Sciences* 66(9):3472-A.

Anthanasiou, Tom. 1996. *Divided Planet: The Ecology of Rich and Poor.* Boston: Little, Brown.

Arcury, T. A., & Christianson, E. H. 1990. "Environmental Worldview in Response to Environmental Problems." *Environment and Behavior* 22:387–407.

Arens, Marianne, & Thull, François. 2001. "Chemical Explosion in Toulouse, France, Leaves at Least 29 Dead." *World Socialist.* Retrieved October 23, 2003, from http://www.wsws.org/articles/2001/sep2001/toul-s25.shtml

Argyle, Michael. 1987. *The Psychology of Happiness.* New York and London: Methuen.

Argyle, Michael. 1991. *Cooperation: The Basis of Sociability.* London: Routledge.

Aristotle. 1987. *A New Aristotle Reader,* edited by J. L. Ackrill. Princeton, NJ: Princeton University.

Associated Press. 2009. "Worst Industrial Disaster Still Haunts India: 25 Years After Lethal Bhopal Gas Leak, Injuries and Birth Defects Linger." *Associated Press,* December 2. Retrieved February 9, 2011, from www.msnbc.msn.com

Ayres, Robert U., & Ayres, Leslie W. 1996. *Industrial Ecology: Towards Closing the Materials Cycle.* Cheltenham, UK: Edward Elgar.

Bakhtin, Mikhail. [1965] 1984. *Rabelais and His World.* Bloomington: Indiana University Press.

Bakhtin, Mikhail. 1981. *The Dialogic Imagination: Four Essays.* Austin: University of Texas Press.

Bakhtin, Mikhail. 1986. *Speech Genres and Other Late Essays,* translated by Vern W. McGee. Minneapolis: University of Minnesota Press.

Balmforth, Richard. 2011. "Factbox: Key Facts on Chernobyl Nuclear Accident." *Reuters News Service.* Retrieved March 15, 2011, from http://www.reuters.com

Banerjee, Damayanti. 2006. "Between the Rivers: Reconstructing Social and Environmental Histories of Displacement." Doctoral dissertation, University of Wisconsin–Madison.

Banerjee, Damayanti, & Bell, Michael M. 2007. "Ecogender: Locating Gender in Environmental Social Science." *Society and Natural Resources.* 20(1):3–19.

Barclay, Eliza. 2007. "Clearing the Smog: Fighting Air Pollution in Mexico City, Mexico, and São Paulo, Brazil." Disease Control Priorities Project, World Bank. Retrieved December 7, 2007, from http://www.dcp2.org/features/47

Barnes, Peter. 2006. *Capitalism 3.0: A Guide to Reclaiming the Commons.* San Francisco, CA: Berrett-Koehler.

Barraclough, Geoffrey, ed. 1982. *The Times Concise Atlas of World History.* London: Times Books.

Barrett, Stanley R. 1994. *Paradise: Class, Commuters, and Ethnicity in Rural Ontario.* Toronto, Ont., Canada: University of Toronto Press.

Bartlett, John. [1855] 1980. *Familiar Quotations.* edited by Emily Morison Beck. Boston, MA: Little, Brown.

Barton, A. 1969. Communities in Disaster: A Sociological Analysis of Collective Stress Situations. Garden City, NY: Doubleday.

Beavis, Simon, & Brown, Paul. 1996. "Shell Oil Has Human Rights Rethink." *Guardian,* November 8, p. 1.

Beck, Ulrich. 1992. Risk Society: Toward a New Modernity. London: Sage.

Beck, Ulrich. 1994. "The Reinvention of Politics: Towards a Theory of Reflexive Modernization." Pp. 1–55 in *Reflexive Modernization Politics, Tradition and Aesthetics in the Modern Social Order,* edited by U. Beck, A. Giddens, & S. Lash. Cambridge, UK: Polity Press.

Beck, Ulrich. 1995. *Ecological Politics in an Age of Risk.* Cambridge, UK: Polity Press.

Beck, Ulrich. 1996a. "Risk Society and the Provident State." Pp. 27–43 in *Risk, Environment, and Modernity: Towards a New Ecology,* edited by S. Lash, B. Szerszynski, & B. Wynne. London: Sage.

Beck, Ulrich. 1996b. "World Risk Society as Cosmopolitan Society? Ecological Questions in a Framework of Manufactured Uncertainties." *Theory, Culture, and Society* 13(4):1–32.

Beck, Ulrich. 1997. The Reinvention of Politics: Rethinking Modernity in the Global Social Order. Cambridge, UK: Polity Press.

Beck, Ulrich. 1999. *World Risk Society.* Cambridge, UK: Polity Press.

Beck, Ulrich. 2009. *World at Risk.* London: Polity.

Becker, Elizabeth. 2002. "Big Farms Making a Mess of U.S. Waters, Cities Say," *New York Times,* February 10, p. 13. Retrieved January 13, 2002, from http://www.nytimes.com/2002/02/10/politics/10FARM.html

Becker, Gary S. 1986. "The Prophets of Doom Have a Dismal Record." *Business Week,* January 27, p. 22.

Beeman, Perry. 2002. "Ag Scientists Feel the Heat." *Des Moines Register,* December 1. Retrieved October 30, 2003, from http://desmoinesregister.com/business/stories/c4789013/19874144.html

Begossi, Alpina. 1995. "Fishing Spots and Sea Tenure: Incipient Forms of Local Management in Atlantic Forest Coastal Communities." *Human Ecology* 23:387–406.

Bell, David. 2006. "Cowboy Love: Male Homosexuality in Rural America." Pp. 163–182 in *Country Boys: Masculinity & Rural Life,* edited by M. Bell & H. Campbell. University Park: Penn State University Press.

Bell, Michael M. 1985. *The Face of Connecticut: People, Geology, and the Land.* Hartford: State Geological and Natural History Survey of Connecticut.

Bell, Michael M. 1994a. *Childerley: Nature and Morality in a Country Village.* Chicago, IL: University of Chicago Press.

Bell, Michael M. 1994b. "Deep Fecology: Mikhail Bakhtin and the Call of Nature." *Capitalism, Nature, Socialism* 5(4):65–84.

Bell, Michael M. 1995. "The Dialectic of Technology: Commentary on Warner and England." *Rural Sociology* 60(4):623–632.

Bell, Michael M. 1996. "Stone Age New England: A Geology of Morals." Pp. 29–64 in *Creating the Countryside: The Politics of Rural and Environmental Discourse,* edited by Melanie Dupuis & Peter Vandergeest. Philadelphia, PA: Temple University Press.

Bell, Michael M. 1997. "The Ghosts of Place." *Theory and Society* 26:813–836.

Bell, Michael M. 1998a. "Culture as Dialog." Pp. 49–62 in *Bakhtin and the Human Sciences: No Last Words,* edited by Michael M. Bell & Michael Gardiner. London: Sage.

Bell, Michael M. 1998b. "The Dialogue of Solidarities, or Why the Lion Spared Androcles." *Sociological Focus* 31(2):181–199.

Bell, Michael M. 1999. "Natural Conscience: Environmental Morality and the Constructionism–Realism Debate." In *Sociological Theory and the Environment: Proceedings of the Second Woudschoten Conference,* Vol. 2, edited by Auguus Gijswijt, Frederick Buttel, Peter Dickens, Riley Dunlap, Authur Mol, & Gert Spaargaren. Amsterdam, the Netherlands: Research Committee 24 (Environment and Society) of the International Sociological Association and the University of Amsterdam.

Bell, Michael M. 2001a. "Can the World Develop and Sustain Its Environment?" Pp. 440–459 in *Sociology for a New Century,* edited by York Bradshaw, Joseph Healey, & Rebecca Smith. Thousand Oaks, CA: Pine Forge Press.

Bell, Michael M. 2001b. "Dialogue and Isodemocracy: An Essay on the Social Conditions of Good Talk." *Revue Internationale de Sociologie (International Review of Sociology)* 11(3):281–297.

Bell, Michael M. (with Bauer, Donna, Jarnagin, Sue, & Peter, Greg). 2004. *Farming for Us All: Practical Agriculture and the Cultivation of Sustainability.* Rural Studies Series of the Rural Sociological Society. College Station: Penn State University Press.

Bell, Michael M. 2007. "In the River: A Sociohistorical Account of Dialogue and Diaspora." *Humanity and Society* 31(2–3):210–234.

Bell, Michael M. 2009. "The Problem of the Original Capitalist." *Environment and Planning A* 41(6): 1276–1282.

Bell, Michael M. (with Abbott, Andrew, Blau, Judith, Crane, Diana, Jones, Stacy Holman, Kahn, Shamus, Leschziner, Vanina, Martin, John Levi, McRae, Christopher, Steinberg, Marc, and Stowe, John Chappell). 2011. *The Strange Music of Social Life: A Dialogue on Dialogic Sociology,* edited by Ann Goetting. Philadelphia: Temple University Press.

Bell, Michael M., & Gardiner, Michael, eds. 1998. *Bakhtin and the Human Sciences: No Last Words.* London: Sage.

Bell, Michael, M., & Laine, Edward. 1985. "Erosion of the Laurentide Region of North America by Glacial and Glacio-Fluvial Processes." *Quaternary Research* 23:154–174.

Bell, Michael M., & Lowe, Philip. 2000. "Regulated Freedoms: The Market and the State, Agriculture and the Environment." *Journal of Rural Studies* 16:285–294.

Bell, Michael M., & Mayerfeld, Diane B. 1998. "The Rationalization of Risk." Paper presented at the Twelfth Congress of the International Sociological Association, Montreal, Canada.

Bell, Michael M., & van Koppen, Kris. 1998. "Coming to Our Senses: (In Search of) the Body in Environmental Social Theory." Paper presented at the Twelfth Congress of the International Sociological Association, Montreal, Canada.

Bellah, Robert. 1969. Tokugawa Religion: The Values of Preindustrial Japan. New York: Free Press.

Bentham, Jeremy. [1779] 1996. *Introduction to the Principles of Morals and Legislation.* Oxford, UK: Clarendon Press.

Bentley, Jeffery W. 1994. "Facts, Fantasies, and Failures of Farmer Participatory Research." *Agriculture and Human Values* 11(2/3):140–150.

Bentley, Jeffery W. 2006. "Folk Experiments." *Agriculture and Human Values* 23:451–461.

Bentley, Jeffery W., & Andrews, Keith L. 1991. "Pests, Peasants, and Publications: Anthropological and Entomological Views of an Integrated Pest Management Program for Small-Scale Honduran Farmers." *Human Organization* 50:113–24.

Bentley, Jeffery W., & Melara, Werner. 1991. "Experimenting With Honduran Farmer-Experimenters." *Overseas Development Institute Newsletter* 24:31–48.

Bentley, Jeffery W., Rodriguez, G., & Gonzalez, A. 1994. "Science and People: Honduran Compesinos and Natural Pest Control Inventions." *Agriculture and Human Values* 11(2/3):178–182.

Benton, Ted. 1994. "Biology and Social Theory in the Environmental Debate." Pp. 28–50 in *Social Theory and the Global Environment*, edited by Michael Redclift & Ted Benton. London: Routledge.

Benton, Ted. 2001a. "Environmental Sociology: Controversy and Continuity." *Sosiologisk Tidsskrift* 9(1–2):5–48.

Benton, Ted. 2001b. "Theory and Metatheory in Environmental Sociology. A Reply to Lars Mjoset." *Sosiologisk Tidsskrift* 9(1–2):198–207.

Berlin, Isaiah. [1958] 1969. "Two Concepts of Liberty." Pp. 166–218 in *Four Essays on Liberty*, edited by Henry Hardy. London and New York: Oxford University Press.

Berry, Wendell. 1977. *The Unsettling of America: Culture and Agriculture.* San Francisco, CA: Sierra Club Books.

Berry, Wendell. 2003. *Citizenship Papers.* Berkeley, CA: Counterpoint.

Beus, Curtis E., & Dunlap, Riley E. 1991. "Measuring Adherence to Alternative vs. Conventional Agricultural Paradigms: A Proposed Scale." *Rural Sociology* 56(3):432–460.

Bhattacharya, Shaoni. 2003. "European Heatwave Caused 35,000 Deaths." *New Scientist*, October 10. Retrieved November 1, 2010, from http://www.newscientist.com

Biehl, Janet. 1991. *Rethinking Ecofeminist Politics.* Boston, MA: South End Press.

Black, Fiona. 1991. *Aesop's Fables.* Kansas City, MO: Andrews McMeel.

Blackwood, Amy, Wing, Kennard T., & Pollak, Thomas H. 2008. *Facts and Figures From the Nonprofit Almanac 2008: Public Charities, Giving, and Volunteering.* Washington, DC: The Urban Institute.

Bland, William L., & Bell, Michael M. 2007. "A Holon Approach to Agroecology." *International Journal of Agricultural Sustainablity* 5(4):280–294.

Blowers, Andrew. 1997. "Ecological Modernization or the Risk Society?" *Urban Studies* 34:845–871.

Blühdorn, Ingolfur. 2000. "Ecological Modernization and Post Ecological Politics." Pp. 209–228 in *Environment and Global Modernity*, edited by Gert Spaargaren, Arthur P. J. Mol, & Fredrick H. Buttel. London: Sage Studies in International Sociology.

Boerner, Christopher, & Lambert, Thomas. 1995. "Environmental Injustice." *Public Interest* 95(118):61–82.

Bolin, Bob, Grineski, Sara, & Collins, Timothy. 2005. "The Geography of Despair: Environmental Racism and the Making of South Phoenix, Arizona, USA." *Human Ecology Review* 12(2):156–168.

Boserup, Ester. 1965. The Conditions of Agricultural Growth: The Economics of Agrarian Change Under Population Pressure. London: Allen & Unwin.

Boserup, Ester. 1981. *Population and Technology.* Oxford, UK: Blackwell.

Boserup, Ester. [1970] 1989. *Woman's Role in Economic Development.* London: Earthscan.

Bosma, M. P. J., van Boxtel, R. W., Ponds, H. M., Houx, P. J., & Jolles, J. 2000. "Pesticide Exposure and Risk of Mild Cognitive Dysfunction." *The Lancet* 356:912–913.

Brasher, Philip. 2003. "Study Finds No Atrazine–Cancer Link." *Des Moines Register*, July 18. Retrieved October 24, 2003, from http://desmoinesregister.com/business/stories/c4789013/21775812.html

Bray, Dennis, & von Storch, Hans. 2008. "CliSci2008: A Survey of the Perspectives of Climate Scientists Concerning Climate Science and Climate Change." *Institute for Coastal Research.* Retrieved January 27, 2011, from http://www.hzg.de/institute/coastal_research/

Brecher, Jeremy. 1994. Global Village or Global Pillage: Economic Reconstruction from the Bottom Up. Boston, MA: South End Press.

Brechin, Stephen, & Kempton, Willet. 1994. "Global Environmentalism: A Challenge to the Postmaterialism Thesis?" *Social Science Quarterly* 75(2): 245–269.

Brickman, Philip, & Campbell, Donald. 1971. "Hedonic Relativism and Planning the Good Society." Pp. 287–302 in *Adaptation Level Theory: A Symposium*, edited by M. H. Apley. New York: Academic Press.

Bridges, Olga, & Bridges, Jim. 1996. *Losing Hope: The Environment and Health in Russia.* Aldershot, UK: Avebury.

Brohman, John. 1996. Popular Development: Rethinking the Theory and Practice of Development. Oxford, UK, and Cambridge, MA: Blackwell.

Bromley, Daniel W., ed. 1992. *Making the Commons Work: Theory, Practice, and Policy.* San Francisco, CA: ICS Press.

Brown, Lester R. 2006. Plan B 2.0: Rescuing a Planet Under Stress and a Civilization in Trouble. New York: Norton.

Brown, Lester R. 2011. World on the Edge: How to Prevent Environmental and Economic Collapse. New York: Norton.

Brown, Lester R., Flavin, Christopher, & Kane, Hale. 1996. *Vital Signs, 1996: The Trends That Are Shaping Our Future.* New York and London: Norton.

Brown, Lester R., & Kane, Hale. 1995. Full House: Reassessing the Earth's Population Carrying Capacity. New York: Norton.

Brown, Lester R., Kane, Hale, & Roodman, David Malin. 1994. *Vital Signs, 1994: The Trends That Are Shaping Our Future.* New York and London: Norton.

Brown, Lester R., Lenssen, Nicholas, & Kane, Hale. 1995. *Vital Signs, 1995: The Trends That Are Shaping Our Future.* New York and London: Norton.

Brown, Lester R., Renner, Michael, & Flavin, Christopher. 1997. *Vital Signs 1997: The Environmental Trends That Are Shaping Our Future.* New York and London: Norton.

Brown, Lester R., Renner, Michael, & Halweil, Brian. 2000. *Vital Signs 2000: The Environmental Trends That Are Shaping Our Future.* New York and London: Norton.

Brown, Philip. 1990. "Popular Epidemiology: Community Response to Toxic Waste-Induced Disease." Pp. 77–85 in *The Sociology of Health and Illness in Critical Perspective,* edited by P. Conrad & R. Kern. New York: St. Martin's Press.

Bruno, Michael, & Squire, Lyn. 1996. "The Less Equal the Asset Distribution, the Slower the Growth." *International Herald Tribune,* September 30, p. 12.

Bruntland Commission. 1987. *Our Common Future.* Washington, DC: United Nations Environment Programme, World Commission on Environment and Development.

Brush, Stephen B., & Stabinsky, Doreen. 1996. Valuing Local Knowledge: Indigenous People and Intellectual Property Rights. Washington, DC: Island Press.

Bryan, Todd A. 2004. "Tragedy Averted: The Promise of Collaboration." *Society and Natural Resources* 17:881–896.

Buege, Douglas J. 1994. "Rethinking Again: A Defense of Ecofeminist Philosophy." Pp. 42–63 in *Ecological Feminism,* edited by Karen Warren. London and New York: Routledge.

Buie, Elizabeth. 1995. "Global Warming Question Has Scientists Under High Pressure." *Herald* (Glasgow, Scotland), August 2, p. 6.

Bullard, Charles. 1997. "Study Seeks Cut in Carbon Dioxide." *Des Moines Register,* April 28, pp. 1M, 5M.

Bullard, Robert D. 1993. Confronting Environmental Racism: Voices From the Grassroots. Boston, MA: South End Press.

Bullard, Robert D. 1994a. *Dumping in Dixie: Race, Class, and Environmental Quality,* 2nd ed. Boulder, CO: Westview Press.

Bullard, Robert D., ed. 1994b. *Unequal Protection: Environmental Justice and Communities of Color.* San Francisco, CA: Sierra Club Books.

Bultena, Gordon, Hoiberg, Eric, & Bell, Michael M. (1995). Unpublished research, Department of Sociology, Iowa State University.

Bureau of Economic Analysis. 2009. Gross Domestic Product: Third Quarter 2009 (Advance Estimate). *Report BEA 09–47.*

Bureau of Transportation Statistics. 2003. *2001 National Household Travel Survey.* Washington, DC: U.S. Department of Transportation.

Bureau of Transportation Statistics. N.d. *National Transportation Statistics 2010.* U.S. Department of Transportation. Retrieved February 16, 2011, from http://www.bts.gov

Burenhult, Goran, ed. 1994. *Old World Civilizations: The Rise of Cities and States.* San Francisco, CA: HarperCollins.

Burningham, Kate, & Cooper, Goeff. 1999. "Being Constructive: Social Constructionism and the Environment." *Sociology* 33(2):297–316.

Burton, Brian K., & Dunn, Craig P. 1996. "Collaborative Control and the Commons: Safeguarding Employee Rights." *Business Ethics Quarterly* 6:277–288.

Busch, Lawrence, & Bain, Carmen. 2004. "New! Improved? The Transformation of the Global Agrifood System." *Rural Sociology* 69(3):321–346.

Busch, L., & Lacy, W. B. 1983. *Science, Agriculture, and the Politics of Research.* Rural Studies Series of the Rural Sociological Society. Boulder, CO: Westview Press.

Busch, L., & Lacy, W. B., eds. 1986. *The Agricultural Scientific Enterprise: A System in Transition.* Rural Studies Series of the Rural Sociological Society. Boulder, CO: Westview Press.

Buttel, Frederick H. 1992. "Environmentalization: Origins, Processes, and Implications for Rural Social Change." *Rural Sociology* 57:1–27.

Buttel, Frederick H. 1996. "Environmental and Resource Sociology: Theoretical Issues and Opportunities for Synthesis." *Rural Sociology* 61:56–76.

Buttel, Frederick H. 1997. "Classical and Contemporary Theoretical Perspectives and the Environment." Paper presented at the Social Theory and the Environment Conference of the International Sociological Association, Zeist, the Netherlands.

Buttel, Frederick H. 2003. "The Political Economy of Environmental Flows: Globalization, Unipolarity, and the Need to Reinvent the National-State, Some More Than Others." *Governing Environmental Flows: Conference Proceedings* [CD-ROM]. Research Committee 24 (Environment and Society) of the International Sociological Association, Wageningen, the Netherlands, June 13–14.

Buttel, Frederick H. 2006. "Globalization, Environmental Reform, and U.S. Hegemony." Pp. 157–186 in *Governing Environmental Flows: Global Challenges to Social Theory,* edited by Gert Spaargaren, Arthur P. J. Mol, & Frederick H. Buttel. Cambridge, MA, and London: MIT Press.

Campbell, Hugh. 2005. "The Rise and Rise of EurepGAP: European (Re)Invention of Colonial Food Relations?" *International Journal of Sociology of Food and Agriculture,* 13(2):1–19.

Campbell, Hugh, Bell, Michael M., & Finney, Margaret, eds. 2006. *Country Boys: Masculinity and Rural Life.* Rural Studies Series of the Rural Sociological Society. College Station: Penn State University Press.

Campbell, Marie, & Manicom, Ann. 1995. *Knowledge, Experience, and Ruling Relations: Studies in the Social Organization of Knowledge.* Toronto, Ont., Canada, and Buffalo, NY: University of Toronto Press.

Cantrill, James G., & Oravec, Christine L., eds. 1996. *The Symbolic Earth: Discourse and Our Creation of the Environment.* Lexington: University Press of Kentucky.

Carolan, Michael S. 2002. Trust and Sustainable Agriculture: The Construction and Application of an Integrative Theory. Unpublished doctoral thesis, Iowa State University.

Carolan, Michael S. 2005. "Society, Biology and Ecology: Bringing Nature Back Into Sociology's Disciplinary Narrative Through Critical Realism." *Organization and Environment* 18:393–421.

Carolan, Michael S., & Bell, Michael M. 2003. "In Truth We Trust: Discourse, Phenomenology, and the Social Relations of Knowledge in an Environmental Dispute." *Environmental Values* 12(2):225–245.

Carolan, Michael S., & Bell, Michael M. 2004. "No Fence Can Stop It: Debating Dioxin Drift From a Small U.S. Town to Arctic Canada." Pp. 385–442 in *Science and Politics in the International Environment,* edited by Neil Harrison & Gary Bryner. Boulder, CO: Rowman & Littlefield.

Caron-Sheppard, Judi Anne, & Johnson, Billy. 2003. "Race and Waste: Environmental Justice in the Location of a Landfill." *Virginia Social Science Journal* 38:52–69.

Carson, Rachel. 1962. *Silent Spring.* Greenwich, CT: Fawcett Crest.

Cassidy, Elizabeth A., Judge, Rebecca P., & Sommers, Paul M. 2000. "The Distribution of Environmental Justice: A Comment." *Social Science Quarterly* 81(3):877–878.

Castells, Manuel. 1996. The Rise of the Network Society. Vol. 1, The Information Age: Economy, Society, and Culture. Oxford, UK: Blackwell.

Catton, William. 1982. *Overshoot: The Ecological Basis of Revolutionary Change.* Urbana: University of Illinois Press.

Catton, William. 2008. "A Retrospective View of My Development as an Environmental Sociologist." Organization and Environment 21:471–478.

Catton, William R., & Dunlap, Riley E. 1980. "A New Ecological Paradigm for Post-Exuberant Sociology." *American Behavioral Scientist* 24:15–47.

Centers for Disease Control. 2007a. "BSE (Bovine Spongiform Encephalopathy, or Mad Cow Disease)." Retrieved November 30, 2007, from http://www.cdc.gov/ncidod/dvrd/bse/

Centers for Disease Control. 2007b. "Fact Sheet: Variant Creutzfeldt-Jakob Disease." Retrieved November 30, 2007, from http://www.cdc.gov/ncidod/dvrd/vcjd/fact sheet_nvcjd.htm

Centers for Disease Control. 2007c. "HIV/AIDS Surveillance Report." Atlanta, GA: Department of Health and Human Services.

Centers for Disease Control. 2007d. "National Mortality Data, 2004." Hyattsville, MD: National Center for Health Statistics.

Centers for Disease Control. 2008. "Diagnoses of HIV infection and AIDS in the United States and Dependent Areas, 2008." *HIV Surveillance Report,* Vol. 20. Retrieved February 3, 2011, from http://www.cdc.gov

Central Intelligence Agency. N.d. *The World Fact Book.* Retrieved January 22, 2011, from http://www.cia.gov

Charlier, C., Albert, A., Herman, P., Hamoir, E., Gaspard, U., Meurisse, M., & Plomteux, G. 2003. "Breast Cancer and Serum Organochlorine Residues." *Occupational and Environmental Medicine* 60:348–351.

Chauhan, P. S. 1996. *Bhopal Tragedy: Socio-Legal Implications.* Jaipur and New Delhi, India: Rawat.

Cheney, Jim. 1994. "Nature/Theory/Difference: Ecofeminism and the Reconstruction of Environmental Ethics." Pp. 158–178 in *Ecological Feminism,* edited by Karen Warren. London and New York: Routledge.

Christoff, P. 1996. "Ecological Modernisation, Ecological Modernities." *Environmental Politics* 5(3):476–500.

Chuang Tzu. 1968. *The Complete Works of Chuang Tzu,* translated by Burton Watson. New York: Columbia University.

Clare, John D., ed. 1993. *Classical Rome.* San Diego, CA, and New York: Harcourt Brace.

Climatic Research Unit, University of East Anglia. 2007. "Global Temperature Record." Retrieved October 15, 2007, from http://www.cru.uea.ac.uk/cru/info/warming

Cochrane, W. 1958. *Farm Prices: Myth and Reality.* Minneapolis: University of Minnesota Press.

Cohen, Aaron J. H., Anderson, Ross, Ostra Bart, Pandey, Kiran Dev, Krzyzanowski, Michal, Künzli, Nino, et al. 2005. "The Global Burden of Disease Due to Outdoor Air Pollution." *Journal of Toxicology and Environmental Health, Part 1 A,* 68:1–7.

Cohen, Joseph, & Centano, Miguel. 2006. "Neoliberalism and Patterns of Economic Performance: 1980–2000." *Annals of the American Academy of Political and Social Science* 606(1):32–67.

Coleman, James C. 1990. *Foundations of Social Theory.* Cambridge: MIT Press.

Collingwood, R. [1945] 1960. *The Idea of Nature.* London: Oxford University Press.

Columbia Water Center. (2010). "Groundwater Depletion in Gujarat." Online video report. Retrieved January 11, 2011, from http://www.earth.columbia.edu/videos/watch/145

Commission of the European Communities. 2007. *Tackling the Pay Gap Between Women and Men.* Communication from the Commission to the Council, the European Parliament, the European Economic and Social Committee and the Committee of the Regions. Retrieved January 28, 2008, from http://ec.europa.eu/employment_social/gender_equality/gender_mainstreaming/equalpay/equal_pay_en.html

Commission on Population and Quality of Life. 1996. *Caring for the Future: Making the Next Decades Provide a Life Worth Living.* Oxford, UK, and New York: Oxford University Press.

Cone, Marla. 2010. "President's Cancer Panel: Environmentally Caused Cancers Are 'Grossly Underestimated' and 'Needlessly Devastate American Lives.'" *Environmental Health News.* Retrieved January 11, 2011, from http://www.environmentalhealthnews.org

Cooke, Bill, & Kothari, Uma, eds. 2001. *Participation: The New Tyranny?* New York: Zed Books.

Cooper, William. 1996. "Values and Value Judgments in Ecological Health Assessments." Pp. 3–10 in *Handbook for Environmental Risk Decision Making: Values, Perceptions and Ethics,* edited by C. Richard Cothern. Boca Raton, FL: Lewis Publishers.

Cotgrove, Stephen F. 1982. Catastrophe or Cornucopia: The Environment, Politics, and the Future. Chichester, UK, and New York: Wiley.

Cottrel, Fred. 1955. Energy and Society: The Relation Between Energy, Social Change, and Economic Development. New York: McGraw-Hill.

Couch, Stephen R., Kroll-Smith, Steve, & Kindler, Jeffery. 2000. "Discovering and Inventing Hazardous Environments: Sociological Knowledge and Publics at Risk." Pp. 173–195 in *Risk in the Modern Age: Social Theory, Science, and Environmental Decision-Making,* edited by Maurie J. Cohen. New York: St. Martin's Press.

Courtenay, Will H. 2006. "Rural men's health: Situating men's risk in the negotiation of masculinity." Pp. 139–158 in *Country Boys: Masculinity and Rural Life,* edited by Hugh Campbell, Michael M. Bell, and Margaret Finney. University Park: Penn State University Press.

Cresswell, Tim. 1996. *In Place/Out of Place: Geography, Ideology, and Transgression.* Minneapolis and London: University of Minnesota Press.

Cronon, William, ed. 1995a. "The Trouble With Wilderness, or, Getting Back to the Wrong Nature." Pp. 379–408 in *Uncommon Ground: Toward Reinventing Nature.* New York: Norton.

Cronon, William, ed. 1995b. *Uncommon Ground: Toward Reinventing Nature.* New York: Norton.

Csikzentmihalyi, Mihaly, & Rochberg-Halton, Eugene. 1981. *The Meaning of Things: Domestic Symbols and the Self.* Cambridge, UK: Cambridge University Press.

Dahlberg, Kenneth A., ed. 1986. New Directions for Agriculture and Agricultural Research: Neglected Dimensions and Emerging Alternatives. Totowa, NJ: Rowman & Allanheld.

Daly, Herman E. 1991. *Steady-State Economics,* 2nd ed. Washington, DC: Island Press.

Daniels, Glynis, & Friedman, Samantha. 1999. "Spatial Inequality and the Distribution of Industrial Toxic Releases: Evidence From the 1990 TRI." *Social Science Quarterly* 80(2):244–262.

Darwin, Charles. 1958. *The Autobiography of Charles Darwin,* edited by Frances Darwin. New York: Dover.

Davidson, Pamela, & Anderton, Douglas L. 2000. "Demographics of Dumping II: A National Environmental Equity Survey and the Distribution of Hazardous Materials Handlers." *Demography* 37(4):461–466.

Davies, James B., Sandström, Susanna, Shorrocks, Anthony, & Wolff, Edward N. 2008. "The World Distribution of Household Wealth." United Nations University World Institute for Development Economics Research (Discussion Paper No. 2008/03).

"Death Toll From Heat Wave in France Likely to Reach 10,000, Minister Says." 2003. *Canadian Press.* Retrieved August 21, 2003, from http://www.canada.com

"Death Toll Rises to 37 After Last Miners Found Dead in Central China Mine." 2010. *BNO News,* October 19. Retrieved October 25, 2010, from http://wireupdate.com

Deaton, Angus. 2008. "Worldwide, Residents of Richer Nations More Satisfied." Retrieved January 13, 2011 from http://www.gallup.com

Delhi Science Forum. 1984. *Bhopal Gas Tragedy.* New Delhi, India: Author.

Deller, Steven, Hoyt, Ann, Hueth, Brent, & Sundaram-Stukel, Reka. 2009. *Research on the Economic Impact of Cooperatives.* Madison: University of Wisconsin Center for Cooperatives.

Dennis, Leslie K., Lynch, Charles F., Sandler, Dale P., & Alavanja, Michael C. R. 2010. "Pesticide Use and Cutaneous Melanoma in Pesticide Applicators in the Agricultural Heath Study." *Environmental Health Perspectives* 118(6):812–818.

Denq, Furjen, Constance, Douglas H., & Joung, Su-Show. 2000. "The Role of Class, Status, and Power in the Distribution of Toxic Superfund Sites in Texas and Louisiana." *Journal of Poverty* 4(4):81–100.

Derezinski, Daniel D., Lacy, Michael G., & Stretesky, Paul B. 2003. "Chemical Accidents in the United States, 1990–1996." *Social Science Quarterly* 84(1):122–143.

Devereux, Stephen. 1993. *Theories of Famine.* New York and London: Harvester/Wheatsheaf.

Diamond, Irene, & Orenstein, Gloria E., eds. 1990. *Reweaving the World: The Emergence of Ecofeminism.* San Francisco, CA: Sierra Club Books.

Dick, F. D., De Palma, G., Ahmadi, A., Scott, N. W., Prescott, G. J., Bennett, J., et al. 2007. "Environmental Risk Factors for Parkinson's Disease and Parkinsonism: The Geoparkinson Study." *Occupational and Environmental Medicine* 64:666–672.

Dickens, Peter. 1996. Reconstructing Nature: Alienation, Emancipation and the Division of Labour. London: Routledge.

Dorfman, Paul, ed. 2008. "Nuclear Consultation: Public Trust in Government." *Nuclearconsult.* Retrieved January 14, 2008, from http://www.nuclearconsult.com

Douglas, Mary, & Isherwood, Baron. [1979] 1996. *The World of Goods: Towards an Anthropology of Consumption.* New York: Routledge.

Douglas, Mary, & Wildavsky, Aaron. 1982. Risk and Culture: An Essay on the Selection of Technological and Environmental Dangers. Berkeley: University of California Press.

Douthwaite, Richard. 1992. The Growth Illusion: How Economic Growth Has Enriched the Few, Impoverished the Many, and Endangered the Planet. Devon, UK: Green Books.

Douthwaite, Richard. 2000. *The Ecology of Money.* Devon, UK: Green Books.

Downey, Liam. 1998. "Environmental Injustice: Is Race or Income a Better Predictor?" *Social Science Quarterly* 79(4):766–778.

Downs, Anthony. 1972. "Up and Down With Ecology: The Issue-Attention Cycle." *Public Interest* 28:38–50.

Duany, Andres, Plater-Zyberk, Elizabeth, & Speck, Jeff. 2001. *Suburban Nation: The Rise of Sprawl and the Decline of the American Dream.* New York: North Point Press.

Dubey, A. K. 2010. "Bhopal Gas Tragedy: 92% Injuries Termed 'Minor.'" *Webcite,* June 21. Retrieved February 9, 2011, from www.webcitation.org

Duden, Barbara. 1992. "Population." Pp. 146–157 in *The Development Dictionary: A Guide to Knowledge as Power,* edited by Wolfgang Sachs. London and Atlantic Highlands, NJ: Zed Books.

Dunlap, Riley E. 1980. "Paradigmatic Change in Social Science: From Human Exemptionalism to an Ecological Paradigm." *American Behavioral Scientist* 24:5–14.

Dunlap, Riley E. 1992. "Trends in Public Opinion Toward Environmental Issues: 1965–1990." Pp. 89–116 in *American Environmentalism: The U.S. Environmental Movement, 1970–1990,* edited by Riley E. Dunlap & Angela G. Mertig. Philadelphia, PA: Taylor & Francis.

Dunlap, Riley E. 2002. "An Enduring Concern: Light Stays Green for Environmental Protection." *Public Perspective,* September/October, pp. 10–14.

Dunlap, Riley E. 2008. "Partisan Gap on Global Warming Grows." *Gallup Poll*. Retrieved January 26, 2011, from http://www.gallup.com

Dunlap, Riley E. 2010a. "At 40, Environmental Movement Endures, With Less Consensus." *Gallup News*, April 22. Retrieved January 25, 2011, from http://www.gallup.com

Dunlap, Riley E. 2010b. "The Maturation and Diversification of Environmental Sociology: From Constructivism and Realism to Agnosticism and Pragmatism." Pp. 15–32 in *International Handbook of Environmental Sociology*, 2nd Ed., edited by M. Redclift & G. Woodgate. Cheltenham, UK: Edward Elgar.

Dunlap, Riley E., & Catton, William R. 1994. "Struggling With Human Exemptionalism: The Rise, Decline, and Revitalization of Environmental Sociology." *American Sociologist* 25:113–135.

Dunlap, Riley E., & McCright, Aaron M. 2008. "A Widening Gap: Republican and Democratic Views on Climate Change." *Environment* 50 (September/October):26–35.

Dunlap, Riley E., & Van Liere, Kent D. 1978. "The New Environmental Paradigm: A Proposed Measuring Instrument and Preliminary Results." *Journal of Environmental Education* 9:10–19.

Dunlap, Riley E., & Van Liere, Kent D. 1984. "Commitment to the Dominant Social Paradigm and Concern for Environmental Quality: An Empirical Examination." *Social Science Quarterly* 65:1013–1028.

Dunlap, Riley E., Van Liere, Kent, Mertig, Angela, & Jones, Robert Emmet. 2000. "Measuring Endorsement of the New Ecological Paradigm." *Journal of Social Issues* 56(3):425–442.

Dupuis, Melanie. 2000. "Not in My Body: rBGH and the Rise of Organic Milk." *Agriculture and Human Values* 17:285–295.

Dupuis, Melanie, & Vandergeest, Peter, eds. 1996. *Creating the Countryside: The Politics of Rural and Environmental Discourse*. Philadelphia, PA: Temple University Press.

Durning, Alan T. 1992. How Much Is Enough? The Consumer Society and the Future of the Earth. New York: Norton.

Dworkin, Ronald. 2000. *Sovereign Virtue: The Theory and Practice of Equality*. Cambridge, MA, and London: Harvard University Press.

Earth Policy Institute. (n.d.). "Food and Agriculture: Grain Production per Person in Sub-Saharan Africa, 1960–2004." Retrieved January 22, 2011, from http://www.earth-policy.org

Earth System Research Laboratory. 2011. "Trends in Atmospheric Carbon Dioxide." Retrieved January 11, 2011, from http://www.esrl.noaa.gov

Easterlin, Richard A. 1973. "Does Money Buy Happiness?" *Public Interest* 30:3–10.

Easterlin, Richard A., McVey, L. A., Switek, M., Sawngfa, O., & Zweig, J. S. 2010. "The Happiness–Income Paradox Revisited." *PNAS* 107(52):22463–22468.

Edelstein, Michael R. 1988. *Contaminated Communities: The Social and Psychological Impacts of Residential Toxic Exposure*. Boulder, CO: Westview Press.

Eder, Klaus. 1996. *The Social Construction of Nature: A Sociology of Ecological Enlightenment*. London and Thousand Oaks, CA: Sage.

Edgar, Bill, & Meert, Henk. 2006. *Fifth Review of Statistics on Homelessness in Europe*. Brussels, Belgium: European Federation of National Associations Working With the Homeless.

Edwards, Bob, & Ladd, Anthony E. 2000. "Environmental Justice, Swine Production and Farm Loss in North Carolina." *Sociological Spectrum* 20(3):263–290.

Ehrlich, Paul R. 1968. *The Population Bomb*. New York: Ballantine Books.

Ehrlich, Paul R., & Ehrlich, Anne H. 1990. *The Population Explosion*. London: Hutchinson.

Eichler, Michael. 1998. "Organizing's Past, Present and Future: Look to the Future, Learn from the Past." *Shelterforce Online* 101, September/October. Retrieved December 12, 2007, from http://www.nhi.org/online/issues/101/eichler.html

Eisinger, Peter. 1973. "The Conditions of Protest Behavior in American Cities." *American Political Science Review* 67:11–28.

Eklavya. 1984. Bhopal: A People's View of Death, Their Right to Know and Live. Bhopal, India: Author.

Ellen, Roy, & Fuhui, Katsuyoshi. 1996. *Redefining Nature: Ecology, Culture, and Domestication*. Oxford, UK, and Washington, DC: Berg.

Elster, Jon. 1989a. *Nuts and Bolts for the Social Sciences*. Cambridge, UK: Cambridge University Press.

Elster, Jon. 1989b. *Solomonic Judgements*. Cambridge, UK: Cambridge University Press.

Energy Information Administration. 2007. *Emissions of Greenhouse Gases in the United States 2006*. Report number DOE/EIA-0573 (2006). Washington, DC: U.S. Department of Energy.

Engels, Friedrich. [1844] 1987. "Outlines of a Critique of Political Economy." Pp. 104–105 in *Perspectives on Population: An Introduction to Concepts and Issues,* edited by Scott W. Menard & Elizabeth W. Moen. New York and Oxford: Oxford University Press.

Engels, Friedrich. [1845] 1973. *The Condition of the Working Class in England, from Personal Observations and Authentic Sources.* Moscow: Progress Publishers.

Engels, Friedrich. [1898] 1940. *The Dialectics of Nature,* translated by Clemens Dutt. New York: International Publishers.

Environment Canada. 2003. "Acid Rain and Water." Retrieved August 22, 2003, from http://www.ec.gc.ca/acidrain/acidwater.html

Environmental Protection Agency (EPA). 2003a. "Atrazine Background." *Pesticides: Topical and Chemical Fact Sheets.* Retrieved January 27, 2004, from http://www.epa.gov/pesticides/factsheets/atrazine_background.htm

Environmental Protection Agency. 2003b. "Pesticides and Food: Why Children May Be Especially Sensitive to Pesticides." Retrieved October 22, 2003, from http://www.epa.gov/pesticides/food/pest.htm

Environmental Protection Agency. 2007. "Achievements in Stratospheric Ozone Protection." Retrieved June 6, 2007, from http://www.epa.gov/ozone/2007stratozoneprogressreport.html

Environmental Protection Agency. N.d. "2000–2001 Pesticide Market Estimates: Usage." Retrieved February 11, 2011, from http://www.epa.gov

Erikson, Kai T. 1966. Wayward Puritans: A Study in the Sociology of Deviance. New York: Wiley.

Erikson, Kai T. 1976. Everything in Its Path: Destruction of Community in the Buffalo Creek Flood. New York: Simon & Schuster.

Erikson, Kai. T. 1994. A New Species of Trouble: Explorations in Disaster, Trauma, and Community. New York: Norton.

Essential Action and Global Exchange. 2000. *Oil for Nothing: Multinational Corporations, Environmental Destruction, Death, and Impunity in the Niger Delta.* Retrieved September 1, 2003, from http://www.essentialaction.org/shell/report/

Esteva, Gustavo. 1992. "Development." Pp. 6–25 in *The Development Dictionary: A Guide to Knowledge as Power,* edited by Wolfgang Sachs. London and Atlantic Highlands, NJ: Zed Books.

European Commission. 2000. *2000 Report on the Forest Condition in Europe.* Press release. Retrieved August 22, 2003, from http://www.dk2002.dk/euidag/rapid/19/

European Nuclear Society. 2011. "Number of Nuclear Plants, World-Wide." Retrieved January 10, 2011, from http://www.euronuclear.org

European Space Agency. 2009. "Earth From Space: Declining Aral Sea." Retrieved January 11, 2011, from http://www.esa.int

Evans, Gary W., & Kantrowitz, Elyse. 2002. "Socioeconomic Status and Health: The Potential Role of Environmental Risk Exposure." *Annual Review of Public Health* 23:303–331.

Evernden, Neil. 1992. *The Social Creation of Nature.* Baltimore, MD: Johns Hopkins University Press.

Ewen, Stanley W. B., & Pusztai, Arpád. 1999. "Effect of Diets Containing Genetically Modified Potatoes." *The Lancet* 354(9187):1353–1354.

ExxonMobile. (n.d.). "Save the Tiger Fund." Retrieved January 15, 2011, from www.exxonmobil.com

Faber, Daniel, & Krieg, Eric. 2005. *Unequal Exposure to Ecological Hazards 2005: Environmental Injustices in the Commonwealth of Massachusetts.* Boston, MA: Philanthropy and Environmental Justice Research Project, Northeastern University.

Fair Trade Labeling Organizations International. 2007. "Figures." Retrieved February 29, 2008, from http://www.fairtrade.net/figures.html

Feder, Barnaby J., & Revkin, Andrew C. 2000. "Vast Effort to Fix Computers Defended (and It's Not Over)." *New York Times,* January 1. Retrieved February 1, 2004, from http://www.greenspun.com

Federal Highway Administration. N.d. "Our Nation's Highways: 2008." U.S. Department of Transportation. Retrieved January 21, 2011, from http://www.fhwa.dot.gov

Federal Ministry for the Environment, Nature Conservation, and Nuclear Safety. 2010. "Development of Renewable Energy Sources in Germany, 2009." Retrieved January 10, 2011, from http://www.bmu.de.

Feeny, David, Hanna, Susan, & McEvoy, Arthur E. 1996. "Questioning the Assumptions of the 'Tragedy of the Commons' Model of Fisheries." *Land Economics* 72:187–205.

Festinger, Leon. [1957] 1962. A *Theory of Cognitive Dissonance.* Stanford, CA: Stanford University Press.

Fickling, David. 2003. "Toothfish 'Poachers' Arrested After 7000km Antarctic Chase." *Guardian Weekly,* September 4–10, p. 3.

Finley, Moses I. 1963. *The Ancient Greeks.* New York: Viking Press.

Fischlowitz-Roberts, Bernie. 2002. "Air Pollution Fatalities Now Exceed Traffic Fatalities by 3 to 1." Earth Policy Institute. Retrieved October 8, 2003, from http://www.earth-policy.org/Updates/Update17.htm

Fishbein, Martin, & Ajzen, Icek. 1975. *Belief, Attitude, Intention, and Behavior: An Introduction to Theory and Research.* Reading, MA: Addison-Wesley.

Fisher, Dana, & Freudenburg, William. 2004. "Post-Industrialism and Environmental Quality: An Empirical Assessment of the Environmental State." *Social Forces* 83(1):157–188.

Fisher, Kimberly, & Layte, Richard. 2004. "Measuring Work–Life Balance Using Time Diary Data." *International Journal of Time Use Research* 1(1):1–13.

Flaccus, Gillian. 2007. "1,500 Homes Lost; $1B Loss in San Diego Area." *Seattle Times*, October 24. Retrieved January 23, 2008, from http://www.seattletimes.nwsource.com/html/nationworld/2003971082_wildfires24.html

Fletcher, Brian M. 2004. "Environmental Equity or Environmental Discrimination: An Assessment of the New York State Inactive Hazardous Waste Disposal Site Program." *Dissertation Abstracts International, A: The Humanities and Social Sciences* 64(11):4217-A-4218-A.

Fontanella-Khan, James. 2011. "India Car Sales to Slow Up in 2011." *Financial Times,* January 11. Retrieved January 21, 2011, from http://www.ft.com

Food and Agriculture Organization. 2005. *The State of Food Insecurity in the World 2005.* Rome, Italy: Author.

Food and Agriculture Organization. 2006a. "Food and Agriculture Statistics Global Outlook." Retrieved January 14, 2008, from http://faostat.fao.org/Portals/_Faostat/documents/pdf/world.pdf

Food and Agriculture Organization. 2006b. *The State of Food Insecurity in the World 2006.* Rome, Italy: Author.

Food and Agriculture Organization. 2007. *The State of the World Fisheries and Aquaculture 2006.* Rome, Italy: Author.

Food and Agriculture Organization. 2010a. "Crop Prospects and Food Situation." Retrieved January 11, 2011, from http://www.fao.org/giews/english/cpfs/index.htm

Food and Agriculture Organization. 2010b. The State of Food Insecurity in the World: Addressing Food Insecurity in Protracted Crises. Rome, Italy: Author.

Fortmann, L. 1996. "Gendered Knowledge: Rights and Space in Two Zimbabwe Villages." Pp. 211–223 in *Feminist Political Ecology: Global Issues and Local Experiences,* edited by Dianne Rocheleau, Barbara Thomas-Slayter, & Esther Wangari. New York: Routledge.

Foster, John Bellamy. 1999. "Marx's Theory of Metabolic Rift: Classical Foundations for Environmental Sociology." *American Journal of Sociology* 105(2):366-405.

Foster, John Bellamy. 2005. "The Treadmill of Accumulation: Schnaiberg's Environment and Marxian Political Economy." *Organization and Environment* 18:7–18.

Foster, John Bellamy, Brett Clark, and Richard York. 2009. *The Ecological Rift: Capitalism's War on the Earth.* New York: Monthly Review Press

Fotopoulos, Takis. 2007. "Is Degrowth Compatible With a Market Economy?" *The International Journal of Inclusive Democracy* 3(1). Retrieved February 28, 2008, from http://www.inclusivedemocracy.org/journal/v013/v013_n01_Takis_degrowth.htm

Foucault, Michel. 1969. The Archaeology of Knowledge and the Discourse of Language. New York: Harper Colophon.

"France Ups Heat Toll." 2003. *CBS News.* Retrieved January 8, 2004, from http://www.cbsnews.com/stories/2003/08/29/world/main570810.shtml

Frank, Andre Gundar. 1969. "The Development of Underdevelopment." In *Latin America: Underdevelopment or Revolution.* New York and London: Monthly Review Press.

Fraser, Steven, ed. 1995. *The Bell Curve Wars: Race, Intelligence, and the Future of America.* New York: Basic Books.

Freemantle, Michael. 1995. "The Acid Test for Europe." *Chemical and Engineering News* 73(18):10–17.

Freire, Paulo. [1970] 1993. *Pedagogy of the Oppressed.* New York: Continuum.

"French Minister Predicts More Heat Wave Deaths." 2003. *Associated Press, CTV,* August 31. Retrieved September 1, 2003, from http://www.ctv.ca

Fresco, Louise. 2003. "Fertilizer and the Future." *Agriculture 21.* Food and Agriculture Organization of the United Nations. Retrieved August 24, 2003, from http://www.fao.org/ag/magazine/0306sp1.htm

Freudenburg, William R. 2000. "Social Constructions and Social Constrictions: Toward Analyzing the Social Construction of 'The Naturalized' as Well as 'The Natural.'" Pp. 103–119 in *Environment and Global Modernity,* edited by Gert Spaargaren, Arthur P. J. Mol, & Frederick Buttel. London: Sage.

Freudenburg, William R., 2005. "Privileged Access, Privileged Accounts: Toward a Socially Structured Theory of Resources and Discourses." *Social Forces* 84(1):89–114.

Freudenburg, William R. 2008. "Thirty Years of Scholarship and Science on Environment–Society Relationships." *Organization Environment* 21:449–460.

Freudenburg, William R., Frickel, Scott, & Gramling, Robert. 1995. "Beyond the Nature/Society Divide: Learning to Think About a Mountain." *Sociological Forum* 10:361–392.

Freudenburg, William R., Gramling, Robert, Laska, Shirley, & Erikson, Kai T. 2007. "Katrina: Unlearned Lessons." *Worldwatch,* September/October:14–19.

Freudenburg, William R., & Jones, Timothy R. 1991. "Attitudes and Stress in the Presence of Technological Risk: A Test of the Supreme Court Hypothesis." *Social Forces* 69: 1143–1168.

Freudenburg, William R., & Nowak, Peter. 2000. "Disproportionality and Disciplinary Blinders: Understanding the Tail That Wags the Dog." Paper presented at the International Symposium on Society and Resource Management, Bellingham, WA, June.

Fung, Yu-Lan. [1931] 1989. *Chuang-Tzu: A New Selected Translation With an Exposition of the Philosophy of Kuo Hsiang.* Beijing, China: Foreign Languages Press.

Fung, Yu-Lan. [1934] 1953. *A History of Chinese Philosophy,* Vol. II, translated by Derk Bodde. Princeton, NJ: Princeton University Press.

Fung, Yu-Lan. [1948] 1966. *A Short History of Chinese Philosophy,* edited by Derk Bodde. New York: Free Press.

Funk, John. 1995. "Summer Chills Out: Season Had It All: Hot, Wet, Dry—But It's Fall Now." *Plain Dealer,* September 23, p. 1a.

Gadgil, Madhav, & Guha, Ramachandra. 1995. *This Fissured Land: An Ecological History of India.* Delhi, India: Oxford University Press.

Galbraith, John Kenneth. 1958. *The Affluent Society.* New York: New American Library.

Gallup International. N.d. "World Opinion: Governments Care Too Little about the Environment." *Recent International Surveys.* Retrieved March 21, 2003, from http://www.gallup-international.com/survey11.htm

Gallup Poll. 2007. "Nutrition and Food." *Poll Analyses.* Retrieved December 9, 2007, from http://www.gallup.com/poll/6424/Nutrition-Food.aspx

Gardiner, Michael. 1992. *The Dialogics of Critique: M. M. Bakhtin and the Theory of Ideology.* London: Routledge.

Gardiner, Michael. 1993. "Ecology and Carnival: Traces of a 'Green' Social Theory in the Writings of M. M. Bakhtin." *Theory and Society* 22(6):765–812.

Gauderman, W. James, Vora, Hita, McConnell, Rob, Berhane, Kiros, Gilliland, Frank, Thomas, Duncan, et al. 2007. "Effect of Exposure to Traffic on Lung Development From 10 to 18 Years of Age: A Cohort Study." *The Lancet* 369(9561):571–577.

Geertz, Clifford. 1983. *Local Knowledge: Further Essays in Interpretive Anthropology.* New York: Basic Books.

Gellius, Aulus. [c. 150 CE] 1927. *The Attic Nights,* translated by John Rolfe. Loeb Classics ed. Retrieved December 2, 2007, from http://www.brainfly.net/html/gellius.htm

Gellius, Aulus. [c. 150 CE] 2006. *Noctes Atticae [Attic Nights],* edited by Bill Thayer. Retrieved December 3, 2007, from http://penelope.uchicago.edu/Thayer/E/Roman/Texts/Gellius/5*.html

General Accounting Office. 2000. *Acid Rain: Emission Trends and Effects in the Eastern United States.* GAO/RCED-00-47. Washington, DC: U.S. Government Printing Office.

George, Susan, & Sabelli, Fabrizio. 1994. *Faith and Credit: The World Bank's Secular Empire.* Harmondsworth, UK: Penguin.

Ghai, Dharam, & Vivian, Jessica M., eds. 1992. *Grassroots Environmental Action: People's Participation in Sustainable Development.* London and New York: Routledge.

Giddens, Anthony. 1984. *The Constitution of Society: Outline of the Theory of Structuration.* Cambridge, UK: Polity Press.

Giddens, Anthony. 1990. *The Consequences of Modernity.* Cambridge, UK: Polity Press.

Giddens, Anthony. 1994. *Beyond Left and Right: The Future of Radical Politics.* Cambridge, UK: Polity Press.

Global Footprint Network. 2007. "Living Beyond Our Means: Global Ecological Footprint, 1961 to 2003." Retrieved November 7, 2007, from http://www.footprintnetwork.org/gfn_sub.php?content=global_footprint

Global Scan. 2009. "Climate Concerns Continue to Increase: Global Poll." *BBC World News Service.* Retrieved January 25, 2011, from http://www.globescan.com

Global Strategy Group. 2007. "Memorandum to Yale Center for Environmental Law and Policy." Retrieved November 10, 2007, from http://www.loe.org/images/070316/yalepole.doc

GMO-Free Europe. 2010. "List of GMO-Free Regions." Retrieved February 7, 2011, from http://www.gmo-free-regions.org

Goddard Institute for Space Studies. 2011. "GLOBAL Temperature Anomalies in 0.01 degrees Celsius." Retrieved January 10, 2011, from http://data.giss.nasa.gov.

Goldman, Benjamin A. 1996. "What Is the Future of Environmental Justice?" *Antipode* 28(2):122–142.

Goldman, Benjamin, & Fitton, L. J. 1994. *Toxic Wastes and Race Revisited.* Washington, DC: United Church of Christ Commission for Racial Justice.

Goldman, Michael, & Schurman, Rachel A. 2000. "Closing the 'Great Divide': New Social Theory on Society and Nature." *Annual Review of Sociology* 26:563–584.

Gott, Phil. 2008. "Is Mobility as We Know It Sustainable?" *Global Insight.* Retrieved January 21, 2011, from http://www.challengebibendum.com

Gould, Stephen Jay. [1981] 1996. *The Mismeasure of Man,* 2nd ed. New York: Norton.

Graedel, T. E., & Allenby, B. R. 1995. *Industrial Ecology.* Englewood Cliffs, NJ: Prentice Hall.

Grant, Don, Trautner, Mary Nell, Downey, Liam, & Thiebaud, Lisa. 2010. "Bringing the Polluters Back in: Environmental Inequality and the Organization of Chemical Production." *American Sociological Review* 75(4):479–504.

Gray, Richard, & Leach, Ben. 2010. "New Errors in IPCC Climate Change Report." *Daily Telegraph,* February 6. Retrieved February 6, 2011, from http://www.telegraph.co.uk

Greenlee, A. R., Arbuckle, T. E., & Chyou, P. H. 2003. "Risk Factors for Female Infertility in an Agricultural Region." *Epidemiology* 14:429–436.

Greenpeace. 2003. "Greenfreeze: A Revolution in Domestic Refrigeration." Retrieved August 22, 2003, from http://www.archive.greenpeace.org/ozone/greenfreeze/

Greider, Thomas, & Garkovich, Lorraine. 1994. "Landscapes: The Social Construction of Nature and the Environment." *Rural Sociology* 59(1):1–24.

Grey, Barbara. 2004. "Strong Opposition: Frame-Based Resistance to Collaboration." *Journal of Community and Applied Social Psychology* 14:166–176.

Guadalupe, Patricia. 2007. "Russians Far Less Satisfied With Environment Than Other G8 Residents." *Gallup News Service,* June 6. Retrieved January 27, 2011, from http://www.gallup.com

Guha, Ramachandra. 1989. "Radical American Environmentalism and Wilderness Preservation: A Third-World Critique." *Environmental Ethics* 11:71–83.

Guha, Ramachandra. 1995. "Mahatma Gandhi and the Environmental Movement in India." *Capitalism, Nature, Socialism* 6(3):47–61.

Guha, Ramachandra. 1997. "The Authoritarian Biologist and the Arrogance of Anti-Humanism: Wildlife Conservation in the Third World." *Ecologist* 27(1):14–19.

Guha, Ramachandra. 1999. *Environmentalism: A Global History.* Boston, MA: Addison-Wesley.

Gusinde, Martin. [1937] 1961. *The Yamana: The Life and Thought of the Water Nomads of Cape Horn,* 5 vols. New Haven, CT: Human Relations Area Files.

Gutro, Rob. 2006. "NASA and NOAA Announce Ozone Hole Is a Double Record Breaker." *News and Features, NASA.* Retrieved December 10, 2007, from http://www.nasa.gov/vision/earth/lookingatearth/ozone_record.html

Habermas, Jürgen. 1975. *Legitimation Crisis.* Boston, MA: Beacon Press.

Habermas, Jürgen. 1984. *The Theory of Communicative Action,* 2 vols. Boston, MA: Beacon Press.

Habermas, Jürgen. 1989. *Jürgen Habermas on Society and Politics: A Reader,* edited by Steven Seidman. Boston, MA: Beacon Press.

Haines, Andy, & Patz, Jonathan A. 2004. "Health Effects of Climate Change." *Journal of the American Medical Association* 291(1):99.

Hajer, Martin. 1995. *The Politics of Environmental Discourse.* New York: Oxford University Press.

Hall, Thomas D. 1996. "The World-System Perspective: A Small Sample from a Large Universe." *Sociological Inquiry* 66(4):440–454.

Handwerk, Brian. 2006. "Chilean Sea Bass: Back in Stores But Still in Trouble." *National Geographic News,* November 28. Retrieved June 6, 2007, from http://news.nationalgeographic.com/news/2006/11/061128-sea-bass.html

Handwerk, Brian. 2011. "Gulf Oil Spill Surprise: Methane Almost Gone." *National Geographic News,* January 6. Retrieved January 25, 2011, from http://news.nationalgeographic.com

Hannigan, John A. 1995. *Environmental Sociology: A Social Constructionist Perspective.* London and New York: Routledge.

Hansen, J., Ruedy, R., Sato, M., & Lo, K. (2010). "Global Surface Temperature Change." *Reviews of Geophysics, 48*(4):4004–4033.

Harari, Raul, Julvez, Jordi, Murata, Katsuyuki, & Barr, Dana. 2010. "Neurobehavioral Deficits and Increased Blood Pressure in School-Age Children Prenatally Exposed to Pesticides." *Environmental Health Perspectives* 118(6):890–897.

Haraway, Donna Jeanne. 1991. *Simians, Cyborgs, and Women: The Reinvention of Nature.* New York: Routledge.

Hardin, Garrett. 1968. "The Tragedy of the Commons." *Science* 162:1243–1248.

Hardin, Garrett. 1977. *The Limits of Altruism: An Ecologist's View of Survival.* Bloomington: Indiana University Press.

Hardin, Russel. 1991. "Trusting Persons, Trusting Institutions." Pp. 185–209 in *Strategy and Choice,* edited by R. Zeckhauser. Cambridge: MIT Press.

Hardin, Russel. 1993. "The Street-Level Epistemology of Trust." *Politics and Society* 21: 505–529.

Harris Interactive. 2008. "The Environment . . . Are We Doing All We Can?" *Harris Poll,* June 19, #63.

Harris Interactive. 2011. "Fewer Americans 'Going Green': New Poll Shows Decrease in 'Green' Behaviors Since 2009." *Harris Poll.* Retrieved February 11, 2011, from http://www.harrisinteractive.com

Harris, Leila. 2006. "Irrigation, Gender, and Social Geographies of the Changing Waterscapes of Southeastern Anatolia." *Environment and Planning D: Society and Space* 24:187–213.

Hart, Stanley I., & Spivak, Alvin L. 1993. *The Elephant in the Bedroom: Automobile Dependence and Denial: Impacts on the Economy and Environment.* Pasadena, CA: Hope Publishing.

Hartmann, Betsy. 1987. *Reproductive Rights and Wrongs: The Global Politics of Population Control and Contraceptive Choice.* New York: Harper.

Hassanein, Neva, & Kloppenburg, Jack R., Jr. 1995. "Where the Grass Grows Again: Knowledge Exchange in the Sustainable Agriculture Movement." *Rural Sociology* 60:721–740.

Hatanaka, Maki, Bain, Carmen, & Busch, Lawrence. 2005. "Third Party Certification in the Global Agrifood System." *Food Policy* 30:354–369.

Heiman, Michael K. 1996. "Race, Waste, and Class: New Perspectives on Environmental Justice." *Antipode* 28(2):111–121.

Hemingway, Ernest M. [1941]. 1976. *For Whom the Bell Tolls.* London: Grafton.

Herbert, Bob. 1997. "Nike's Boot Camps." *New York Times,* March 31, p. A15.

Herrnstein, Richard, & Murray, Charles. 1994. *The Bell Curve: Intelligence and Class Structure in American Life.* New York: Free Press.

Hickey, Sam, & Mohan, Giles, eds. 2004. *Participation: From Tyranny to Transformation? Exploring New Approaches to Participation.* London: Zed Books.

Hickman, Martin. 2010. "Living Proof That Conservation Works: Scientists Say the Fish Threatened With Extinction Is Back on the Menu Again." *The Independent,* May 15. Retrieved January 20, 2011, from http://www.independent.co.uk

Hightower, Jim. 1973. *Hard Tomatoes, Hard Times.* Cambridge, MA: Schenkman.

Hileman, Bette. 1995. "Scientists Warn That Disease Threats Increase as Earth Warms Up." *Chemical and Engineering News* 73(40):19–20.

Hilz, Christoph. 1992. *The International Toxic Waste Trade.* New York: Van Nostrand Reinhold.

Hinde, Robert A., & Groebel, Jo, eds. 1992. *Cooperation and Prosocial Behavior.* Cambridge, UK: Cambridge University Press.

Hines, Revathi I. 2001. "African Americans' Struggle for Environmental Justice and the Case of the Shintech Plant: Lessons Learned From a War Waged." *Journal of Black Studies* 31(6):777–789.

Hinrichs, C. Clare. 1996. "Consuming Images: Making and Marketing Vermont as Distinctive Rural Place." Pp. 259–278 in *Creating the Countryside: The Politics of Rural and Environmental Discourse,* edited by Melanie DuPuis & Peter Vandergeest. Philadelphia, PA: Temple University Press.

Hipp, John R., & Lakon, Cynthia M. 2010. "Social Disparities in Health: Disproportionate Toxicity Proximity in Minority Communities Over a Decade." *Health & Place* 16(4):674–683.

Hirsch, Fred. 1977. *Social Limits to Growth.* London: Routledge.

Hobley, Mary. 1996. *Participatory Forestry: The Process of Change in India and Nepal.* London: Overseas Development Institute.

Hochschild, Arlie. 1989. *The Second Shift.* New York: Avon.

Hochschild, Arlie. 1997. *The Time Bind: When Work Becomes Home and Home Becomes Work.* New York: Metropolitan Books.

Hofrichter, Richard, ed. 1993. *Toxic Struggles: The Theory and Practice of Environmental Justice.* Philadelphia, PA: New Society.

Hopkins, Terence K., & Wallerstein, Immanuel. 1982. *World Systems Analysis: Theory and Methodology.* Beverly Hills, CA: Sage.

Horace. [c. 20 BCE] 1983. *The Essential Horace,* translated by Burton Raffel. San Francisco, CA: North Point Press.

Houlihan, Jane. 2003. "Body Burden: The Pollution in People." Environmental Working Group report. Retrieved March 19, 2011, from http://www.rebprotocol.net/bodybur3.pdf

"How Safe Is GM Food?" 2002. *Lancet* 360(9342):1261.

Hubbard, Ruth. 1982. "Have Only Men Evolved?" Pp. 17–42 in *Biological Woman: The Convenient Myth,* edited by Ruth Hubbard, Mary Sue Henifin, & Barbara Fried. Cambridge, MA: Schenkman.

Huberman, Michael, & Minns, Chris. 2005. "Hours of Work in Old and New Worlds: The Long View, 1870–2000." Retrieved June 15, 2007, from http://www.econ.ubc.ca/ine/papers/wp027.pdf

Hudson, Robyn, Arriola, Aline, Martınez-Gomez, Margarita, & Distel, Hans. 2006. "Effect of Air Pollution on Olfactory Function in Residents of Mexico City." *Chemical Senses* 31: 79–85.

Huntington, Ellsworth. 1915. *Civilization and Climate.* New Haven, CT: Yale University Press.

Huntley, Steve, & Creighton, Linda L. 1986. "Winter of Despair Hits the Farm Belt." *U.S. News and World Report,* January 20, pp. 21–23. Retrieved December 14, 2007, from http://www.lexisnexis.com

Hurley, Patrick T., & Walker, Peter A. 2004. "Collaboration Derailed: The Politics of 'Community-Based' Resource Management in Nevada County." *Society and Natural Resources* 17:735–751.

Imbrie, John, & Imbrie, Katherine Palmer. 1979. *Ice Ages: Solving the Mystery.* Short Hills, NJ: Enslow.

Independent Commission on Population and Quality of Life. 1996. *Caring for the Future: Making the Next Decades Provide a Life Worth Living.* Oxford, UK, and New York: Oxford University Press.

Inglehart, Ronald. 1977. *The Silent Revolution: Changing Values and Political Styles Among Western Publics.* Princeton, NJ: Princeton University Press.

Inglehart, Ronald. 1990. *Culture Shift in Advanced Industrial Society.* Princeton, NJ: Princeton University Press.

Inglehart, Ronald. 1995. "Public Support for Environmental Protection: Objective Problems and Subjective Values in 43 Societies." *PS: Political Science and Politics* 28(1):57–72.

Intergovernmental Panel on Climate Change. 2002a. *Climate Change 2001: Synthesis Report: Third Assessment Report of the Intergovernmental Panel on Climate Change.* Cambridge and New York: Cambridge University Press.

Intergovernmental Panel on Climate Change. 2002b. *Climate Change 2001: Working Group II: Impacts, Adaptation, and Vulnerability.* Retrieved August 22, 2003, from http://www.unep.no/climate/ipcc_tar/wg2/481.htm

Intergovernmental Panel on Climate Change. 2007a. *Fourth Assessment Report: Climate Change Impacts, Adaptation and Vulnerability—Summary for Policymakers.* Working Group II. Geneva, Switzerland: Intergovernmental Panel on Climate Change Secretariat.

Intergovernmental Panel on Climate Change. 2007b. *Fourth Assessment Report: Technical Summary.* Working Group I. Geneva, Switzerland: Intergovernmental Panel on Climate Change Secretariat.

International Organization for Standardization (ISO). 2007. *The ISO Survey of Certifications, 2006.* Geneva, Switzerland: ISO Central Secretariat.

International Organization for Standardization. (2010). "ISO 14000: Environmental Management." Retrieved January 27, 2011, from http://www.iso.org

International Road Traffic and Accident Database. 2003. "Selected Risk Values for the Year 2001." Retrieved October 8, 2003, from http://www.bast.de/htdocs/fachthemen/irtad/english/we2.html

International Union for the Conservation of Nature (IUCN). 2007. "IUCN Red List." Retrieved November 7, 2007, from http://www.iucnredlist.org/info/2007RL_Stats_ Table%201.pdf

International Union for the Conservation of Nature. 2010. *IUCN Red List of Threatened Species. Version 2010.4.* Retrieved January 10, 2011, from http://www.iucnredlist.org

IPSOS. 2010. "Cardiff University—Energy Futures and Climate Change Survey 2010." Retrieved January 25, 2011, from http://www.cf.ac.uk

IRRI (International Rice Research Institute). 2010. "IRRI World Rice Statistics (WRS)." Retrieved January 22, 2011, from www.beta.irri.org

Irwin, Alan. 1995. *Citizen Science: A Study of People, Expertise, and Sustainable Development.* London: Routledge.

Jacobs, Michael. 1991. *The Green Economy: Environment, Sustainable Development, and the Politics of the Future.* London and Concord, MA: Pluto Press.

Jänicke, Martin. 2006. "The Environmental State and Environmental Flows: The Need to Reinvent the Nation-State." Pp. 83–105 in *Governing Environmental Flows: Global Challenges to Social Theory,* edited by Gert Spaargaren, Arthur P. J. Mol, & Frederick H. Buttel. Cambridge, MA, and London: MIT Press.

Jarvie, Jenny. 2007. "Georgia Governor Leads Prayer for Rain." *Los Angeles Times,* November 14. Retrieved December 7, 2007, from http://www.latimes.com

Jevons, William Stanley. [1865] 2001. "On the Economy of Fuel." *Organization and Environment* 14(1):99–104.

Jianhua, Feng. 2003. "Parking Strife Frustrates China's Auto Ambitions." *China Today.* Retrieved January 21, 2004, from http://www.chinatoday.com.cn/English/p20.htm

Joekes, Susan. 1987. *Women in the World Economy: An INSTRAW Study.* New York: Oxford University Press.

Johansen, Bruce E. 2003. "Nigeria: The Ogoni: Oil, Blood, and the Death of a Homeland." In *Indigenous Peoples and Environmental Issues: An Encyclopedia.* Westport, CT: Greenwood Press.

Johnson, H. Thomas. 1992. *Relevance Regained: From Top-Down Control to Bottom-Up Empowerment.* New York: Free Press.

Jones, Andrew Rhys. 2007. "How the Media Frame Global Warming: A Harbinger of Human Extinction or Endless Summer Fun?" *Dissertation Abstracts International, A: Humanities and the Social Sciences* 67(10):3988.

Jones, Jeffrey M. 2010. "Few Americans Oppose National Day of Prayer." *Gallup News,* May 5. Retrieved February 4, 2011, from http://www.gallup.com

Jones, Nicola. 2003. "South Aral Sea Gone in Fifteen Years." *New Scientist,* July 21. Retrieved August 22, 2003, from http://www.ecology.com/ecology-news-links/2003/arti cles/7–2003/7–21–03/south-aral-sea.htm

Jones, Robert Emmet, & Dunlap, Riley E. 1992. "The Social Bases of Environmental Concern: Have They Changed Over Time?" *Rural Sociology* 57(1):28–47.

Julien's Auctions. 2009. *The Collections of the King of Pop: Michael Jackson.* Online catalog. Retrieved January 11, 2011, from http://www.juliensauctions.com

Kahan, Dan M., Braman, Donald, Gastil, John, Slovic, Paul, & Mertz, C. K. 2007. "Culture and Identity-Protective Cognition: Explaining the White-Male Effect in Risk Perception." *Journal of Empirical Legal Studies* 4(3):465–505.

Kalof, Linda, Dietz, Thomas, & Guagnano, Gregory. 2002. "Race, Gender and Environmentalism: The Atypical Values and Beliefs of White Men." *Race, Class, Gender* 9(2):1–19.

Karmel, Philip E., FitzGibbon, Thomas M., & Cave, Bryan. 2002. "PM2.5: Federal and California Regulation of Fine Particulate Air Pollution." *California Environmental Law Reporter* 8:226–237. Retrieved August 22, 2003, from http://www.bryancave.com

Katz, Peter. 1994. *The New Urbanism: Toward an Architecture of Community.* New York: McGraw-Hill.

Kay, Jane Holtz. 1998. *Asphalt Nation: How the Automobile Took over America, and How We Can Take It Back,* Reprint ed. Berkeley: University of California Press.

Kellert, Stephen R., & Berry, Joyce K. 1982. *Knowledge, Affection, and Basic Attitudes Toward Animals in American Society.* Washington, DC: U.S. Government Printing Office.

Kiely, Timothy, Donaldson, David, & Grube, Arthur. 2004. *Pesticide Industry Sales and Usage: 2000 and 2001 Market Estimates.* Washington, DC: U.S. Environmental Protection Agency.

Kim, Lucian, & Levitov, Maria. 2010. "Russia Heat Wave May Kill 15,000, Shave $15 Billion of GDP." *Bloomberg.* Retrieved January 10, 2011, from http://www.bloomberg.com.

Kinealy, Christine. 1996. "How Politics Fed the Famine." *Natural History* 105(1):33–35.

Kirby, Alex. 1999. "Sci/Tech Parliament Ponders GM Potatoes." British Broadcasting Corporation. Retrieved October 30, 2003, from http://news.bbc.co.uk/2/hi/science/nature/291105.stm

Kitschelt, H. 1986. "Political Opportunity Structures and Political Protest: Anti-Nuclear Movements in Four Democracies." *British Journal of Political Science* 16:57–85.

Kleim R., & Lubin, I. 1997. *Reducing Project Risk.* Aldershot, UK: Gower.

Kleinhesselink, R. R., & Rosa, Eugene. 1994. "Nuclear Trees in a Forest of Hazards: A Comparison of Risk Perceptions Between American and Japanese Students." Pp. 101–119 in *Nuclear Power at the Crossroads: Challenges and Prospects for the Twenty-First Century,* edited by T. C. Lowinger & G. W. Hinman. Boulder, CO: International Research Center for Energy and Economic Development.

Klinenberg, Eric. 2002. *Heat Wave: A Social Autopsy of Disaster in Chicago.* Chicago, IL: University of Chicago Press.

Knight, Barry, & Stokes, Peter. 1996. *The Deficit in Civil Society in the United Kingdom.* Birmingham, UK: Foundation for Civil Society.

Koch, Wendy. 2010. "BP Oil Spill Hits Record as Gulf's Worst." *USA Today,* July 1. Retrieved January 25, 2011. from http://content.usatoday.com

Koestler, Arthur. 1967. *The Ghost in the Machine.* New York: Macmillan.

Kofman, Jeffrey. 2010. "BP Oil Spill: Clean-Up Crews Can't Find Crude in the Gulf." *ABC News,* July 26. Retrieved January25, 2011, from http://abcnews.go.com

Konisky, David M., & Schario, Tyler S. 2010. "Examining Environmental Justice in Facility-Level Regulatory Enforcement." *Social Science Quarterly* 91(3):835-855.

Korten, David. 1995. *When Corporations Rule the World.* West Hartford, CT: Kumarian.

Kraus, Stephen J. 1995. "Attitudes and the Prediction of Behavior: A Meta-Analysis of the Empirical Literature." *Personality and Social Psychology Bulletin* 21:58–75.

Krieg, Eric L. 1995. "A Socio-Historical Interpretation of Toxic Waste Sites: The Case of Greater Boston." *American Journal of Economics and Sociology* 54(1):1–14.

Krieg, Eric J. 1998. "The Two Faces of Toxic Waste: Trends in the Spread of Environmental Hazards." *Sociological Forum* 13(1):3–20.

Krieg, Eric J. 2005. "Race and Environmental Justice in Buffalo, NY: A ZIP Code and Historical Analysis of Ecological Hazards." *Society and Natural Resources* 18(3):199–213.

Kriesi, H., Koopmans, Ruud, Wilhem, Duyvendak, Jan, & Giugni, Marco G. 1992. "New Social Movements and Political Opportunities in Western Europe." *European Journal of Political Research* 22(2):219–244.

Kunreuther, Howard. 1992. "A Conceptual Framework for Managing Low-Probability Events." Pp. 301–320 in *Social Theories of Risk,* edited by Sheldon Krimsky & Dominic Golding. Westport, CT: Praeger.

Kunstler, James Howard. 1993. *The Geography of Nowhere: The Rise and Decline of America's Man-Made Landscape.* New York: Simon & Schuster.

Kunstler, James Howard. 1996. *Home From Nowhere: Remaking Our Everyday World for the Twenty-First Century.* New York: Simon & Schuster.

Künzli, N., Kaiser, R., Medina, S., Studnicka, M., Chanel, O., Filliger P., et al. 2002. "Public Health Impact of Outdoor and Traffic-Related Air Pollution: A European Assessment." *The Lancet* 356:795–801.

Kurzman, Dan. 1987. A Killing Wind: Inside Union Carbide and the Bhopal Catastrophe. New York: McGraw-Hill.

Kwok, R., & Rothrock, D. A. (2009). "Decline in Arctic sea Ice Thickness From Submarine and ICES at records: 1958–2008." *Geophysical Research Letters, 36*(15):L15501.

Lack, M., Short, K., & Willock, A. 2003. *Managing Risk and Uncertainty in Deep-Sea Fisheries: Lessons From Orange Roughy.* New South Wales, Australia: TRAFFIC Oceanta and WWF Australia.

LaMotte, Greg, & Davis, Patty. 1998. "Oprah: Free Speech Rocks: Texas Cattlemen Lose Defamation Suit." *CNN Interactive.* Retrieved January 31, 2004, from http://www.cnn.com/US/9802/26/oprah.verdict/

Land Trust Alliance. 2007. "Working to Save America's Land Heritage." *About Land Trusts.* Retrieved January 14, 2008, from http://www.lta.org/aboutlt/index.html

Landman, Anne. 2010. "BP's 'Beyond Petroleum' Campaign Losing Its Sheen." *PR Watch.* Retrieved January 15, 2011, from http://www.prwatch.org

Lane, Marcus B. 2001. "Affirming New Directions in Planning Theory: Comanagement of Protected Areas." *Society and Natural Resources* 14:657–671.

Langdon, Phillip. 1994. *A Better Place to Live: Reshaping the American Suburb.* Amherst: University of Massachusetts Press.

Lao Tzu. 1963. *Lao Tzu: Tao Te Ching,* translated by D. C. Lau. Harmondsworth, UK: Penguin.

Lappé, Frances Moore. 1977. *Food First: The Myth of Food Scarcity.* London: Souvenir Press.

Lappé, Frances Moore, & Schurman, Rachel. 1988. *Taking Population Seriously.* London: Earthscan.

Lash, Jonathan. 2005. "Turning the Corner in Mexico City." *WRI Stories.* Retrieved December 7, 2007, from http://www.wri.org/stories/2005/07/turning-corner-mexico-city

Latouche, Serge. 2003. "Will the West Actually Be Happier With Less? The World Downscaled." *Le Monde Diplomatique,* December. Retrieved February 29, 2008, from http://www.hartford-hwp.com/archives/27/081.html

Latouche, Serge. 2004. "Why Less Should Be So Much More: Degrowth Economics." *Le Monde Diplomatique,* November. Retrieved February 29, 2008, from http://mondediplo.com/2004/11/14latouche

Latouche, Serge. 2006. "How Do We Learn to Want Less: The Globe Downshifted." *Le Monde Diplomatique,* January. Retrieved February 29, 2008, from http://mondediplo.com/2006/01/13degrowth

Latour, Bruno. 1987. *Science in Action.* Cambridge, MA: Harvard University Press.

Latour, Bruno. 1993. *We Have Never Been Modern.* London: Harvester/Wheatsheaf.

Latour, Bruno. 1999. *Pandora's Hope: Essays on the Reality of Science Studies.* Cambridge, MA: Harvard University Press.

Latour, Bruno. 2004. *Politics of Nature: How to Bring the Sciences Into Democracy,* translated by Catherine Porter. Cambridge, MA: Harvard University Press.

Latour, Bruno. 2007. *Reassembling the Social: An Introduction to Actor-Network-Theory.* Oxford and New York: Oxford University Press.

Leakey, Richard, & Lewin, Roger. 1996. The Sixth Extinction: Patterns of Life and the Future of Humankind. New York: Anchor.

Lee, Caroline. 2007. "Is There a Place for Private Conversation in Public Dialogue? Comparing Stakeholder Assessments of Informal Communication in Collaborative Regional Planning." American Journal of Sociology 113(1):41–96.

Lee, Duk-Hee, Steffes, Michael W., Sjödin, Andreas, Jones, Richard S., et al. 2010. "Low Dose of Some Persistent Organic Pollutants Predicts Type 2 Diabetes: A Nested Case-Control Study." Environmental Health Perspectives 118(9):1235–1243.

Legot, Cristina, London, Bruce, & Shandra, John. 2010. "Environmental Ascription: High-Volume Polluters, Schools, and Human Capital." Organization & Environment 23(3):271–290.

Leopold, Aldo. [1949] 1961. "The Land Ethic." Pp. 237–264 in A Sand County Almanac. San Francisco, CA: Sierra Club Books.

Leopold, Aldo. [1949] 1966. A Sand County Almanac, With Essays on Conservation From Round River. New York: Sierra Club/Ballantine.

Lerner, Steve. 2005. Diamond: A Struggle for Environmental Justice in Louisiana's Chemical Corridor. Cambridge, MA: MIT Press.

Leroy, P., & van Tatenhove, J. 2000. "New Policy Arrangements in Environmental Politics: The Relevance of Political and Ecological Modernization." Pp. 187–209 in Environment, Sociology, and Global Modernity, edited by Gert Spaargaren, Arthur P. J. Mol, & Fred Buttel. London: Sage.

Lewis, Michael J. 2005. "Pleasure Domes for Millionaires and Other Lost Boys." New York Times, June 19. Retrieved January 11, 2011, from http://www.nytimes.com

Lichter, Robert S. 2008. "Climate Scientists Agree on Warming, Disagree on Dangers, and Don't Trust the Media's Coverage of Climate Change." STATS. Retrieved January 30, 2011, from http://stats.org

Lidskog, Rolf. 2001. "The Re-Naturalization of Society? Environmental Challenges for Sociology." Current Sociology/ Sociologie Contemporaine 49(1):113–136.

Lieh Tzu. 1960. The Book of Lieh-Tzu, translated by A. C. Graham. London: John Murray.

Lipset, Seymour M. 1959. "Some Social Requisites of Democracy: Economic Development and Political Legitimacy." American Political Science Review 53, March, pp. 69–105.

Lipset, Seymour M. [1960] 1981. Political Man: The Social Bases of Politics. Baltimore, MD: Johns Hopkins University Press.

Logan, John, & Molotch, Harvey. 1987. Urban Fortunes: The Political Economy of Place. Berkeley: University of California Press.

Lohbeck, Wolfgang. 2004. "Greenfreeze: From a Snowball to an Industrial Avalanche." Retrieved January 23, 2008, from http://www.greenpeace.org/international/press/reports/greenfreeze-from-snowball-to

Long, C. S., Zhou, S., Miller, A. J., Gelman, M. E., Flynn, L. E., Hoffman, D. J., et al. 2005. "A Review of the 2005 Antarctic Ozone Hole." Paper presented at the fall 2005 American Geophysical Union, San Francisco, CA. Retrieved June 5, 2007, from http://adsabs.harvard.edu/abs/2005AGUFM.A13D0986L

Long, Steven, & Ort, Donald. 2002. SoyFACE: A Changing Environment for Agriculture. Retrieved August 22, 2003, from http://www.soyface.uiuc.edu/news.htm

Lovejoy, Arthur O. 1936. The Great Chain of Being. Cambridge, MA: Harvard University Press.

Lovejoy, Arthur O., & Boas, George. 1935. Primitivism and Related Ideas in Antiquity. Baltimore, MD: Johns Hopkins Press.

Lowenthal, David. 1985. The Past Is a Foreign Country. Cambridge, UK, and New York: Cambridge University Press.

Lowenthal, David. 1997. The Heritage Crusade and the Spoils of History. London: Penguin.

Lowenthal, David. 2000. "A Historical Perspective on Risk." Pp. 251–257 in Risk in the Modern Age: Social Theory, Science, and Environmental Decision-Making, edited by Maurie J. Cohen. New York: St. Martin's Press.

Luhmann, Niklas. 1994. Risk: A Sociological Theory. New York: Aldine de Gruyter.

Lukaszewski, James E. 1992. "Managing Fear: Taking the Risk Out of Risk Communication." Vital Speeches of the Day 58:238–241.

Lyon, Alexandra, Bell, Michael M., Croll, Nora Swan, Jackson, Randall, & Gratton, Claudio. 2010. "Maculate Conceptions: Power, Process, and Creativity in Participatory Research." Rural Sociology 75(4):538–559.

Lyson, Thomas, Stevenson, George W., & Welsh, Rick, eds. 2008. Food and the Mid-Level Farm: Renewing an Agriculture of the Middle. Cambridge: MIT Press.

MacKensie, Debora. 1995. "Deadly Face of Summer in the City." New Scientist 147, September 9, p. 4.

MacKenzie, James J., Dower, Roger C., & Chen, Donald D. T. 1992. The Going Rate: What It Really Costs to Drive. Washington, DC: World Resources Institute.

Mairie de Toulouse. 2003. "Sinistre AZF: Le Point Sur la Reconstruction des Quartier Sinistrés." Retrieved October 22, 2003, from http://www.mairietoulouse.fr/Actualite/Dossiers_Actualites/AZF/AZF_reconstruction.htm

Malin, Stephanie A., & Petrzelka, Peggy. 2010. "Left in the Dust: Uranium's Legacy and Victims of Mill Tailings Exposure in Monticello, Utah." *Society and Natural Resources* 23(12)1187–1200.

Malthus, Thomas. [1798] 1993. *An Essay on the Principle of Population,* edited by Geoffrey Gilbert. Oxford and London: Oxford University Press.

Mamdani, Mahmood. 1972. *The Myth of Population Control: Family, Caste, and Class in an Indian Village.* New York: Monthly Review Press.

Manning, Jason. N.d. "The Midwest Farm Crisis of the 1980s." *The Eighties Club: The Politics and Pop Culture of the 1980s.* Retrieved December 14, 2007, from http://eightiesclub.tripod.com/id395.htm

Marsh, George Perkins. [1864] 1965. *Man and Nature.* Cambridge, MA: Belknap Press of Harvard University Press.

Marshall, Monty G., & Cole, Benjamin R. 2009. "Global Report 2009: Conflict, Governance, and State Fragility." Center for Systemic Peace.

Marx, Karl. [1844] 1972. "Economic and Philosophic Manuscripts of 1844: Selections." Pp. 52–103 in *The Marx-Engels Reader,* edited by Robert C. Tucker. New York: Norton.

Marx, Karl. [1859] 1972. "Preface to a Contribution to the Critique of Political Economy." Pp. 3–6 in *The Marx-Engels Reader,* edited by Robert C. Tucker. New York: Norton.

Maslow, Abraham. [1943]. 1970. "A Theory of Human Motivation." Pp. 80–106 in *Motivation and Personality,* 2nd ed. New York: Harper & Row.

Mauss, Marcel. [1950] 1990. *The Gift: The Form and Reason for Exchange in Archaic Societies,* translated by W. D. Hall. New York and London: Norton.

Mauzerall, D. L., & Tong, Q. 2006. "A Preliminary Estimate of the Total Impact of Ozone and PM2.5 Air Pollution on Premature Mortalities in the United States." In *Air Pollution Modeling and Its Application XVII,* eds. Carlos Borrego & Ann-Lise Norman, pp. 102–108. New York: Springer.

McCright, Aaron M., & Dunlap, Riley E. 2000. "Challenging Global Warming as a Social Problem: An Analysis of the Conservative Movement's Counter-Claims." *Social Problems* 47:499–522.

McCright, Aaron M., & Dunlap, Riley E. 2003. "Defeating Kyoto: The Conservative Movement's Impact on U.S. Climate Change Policy." *Social Problems* 50(3):348–373.

McCright, Aaron M. and Riley E. Dunlap. 2010. "Anti-Reflexivity: The American Conservative Movement's Success in Undermining Climate Science and Policy." *Theory, Culture and Society* 26:100–133.

McEvedy, Colin. 1995. *The Penguin Atlas of African History.* Harmondsworth, UK, and New York: Penguin Books.

McKibben, Bill. 1998. *Maybe One: A Personal and Environmental Argument for Single Child Families.* New York: Simon & Schuster.

Mead, Chris. 1997. "Birds and Roads: Wilderness and Wildlife at Risk." Lecture to the British Association for the Advancement of Science. Retrieved March 27, 2011, from http://www.birdcare.com/bin/shownews/85

Meadows, Donella H., Meadows, Dennis L., & Randers, Jorgen. 1992. *Beyond the Limits: Confronting Global Collapse, Envisioning a Sustainable Future.* Post Mills, VT: Chelsea Green.

Meadows, Donella H., Meadows, Dennis L., & Randers, Jorgen. 2005. *Limits to Growth: The 30-Year Update,* 3rd ed. Post Mills, VT: Chelsea Green.

Mearns, Robin. 1996. "Community, Collective Action, and Common Grazing: The Case of Post-Socialist Mongolia." *Journal of Development Studies* 32:297–339.

Meek, Ronald L., ed. 1971. *Marx and Engels on the Population Bomb.* Berkeley, CA: Ramparts.

Menard, Louis. 2001. *The Metaphysical Club: A Story of Ideas in America.* New York: Farrar, Straus & Giroux.

Mennis, Jeremy. 2002. "Using Geographic Information Systems to Create and Analyze Statistical Surfaces of Population and Risk for Environmental Justice Analysis." *Social Science Quarterly* 83(1):281–297.

Merleau-Ponty, Maurice. 1970. *Themes From the Lectures at the College de France 1952–1960.* Evanston, IL: Northwestern University Press.

Mertig, Angela G., & Dunlap, Riley E. 2001. "Environmentalism, New Social Movements, and the New Class: A Cross-National Investigation." *Rural Sociology* 66:113–136.

Merton, Robert K. 1948. "The Self-Fulfilling Prophecy." *Antioch Review* 8:193–210.

Merton, Robert K. [1968] 1973. "The Matthew Effect in Science." Chap. 20 in The *Sociology of Science: Theoretical and Empirical Investigations,* edited by Norman W. Storer. Chicago, IL: University of Chicago Press.

Merton, Thomas. 1965. *The Way of Chuang Tzu.* New York: New Directions.

Metzner, Andreas. 1997. "Constructivism and Realism (Re)Considered." Paper presented at the Social Theory and the Environment Conference of the International Sociological Association, Zeist, the Netherlands.

Milbrath, Lester W. 1984. *Environmentalists: Vanguard for a New Society.* Albany: State University of New York Press.

Mill, John Stuart. [1874] 1961. "Nature." Pp. 445–488 in *The Philosophy of John Stuart Mill,* edited by Marshall Cohen. New York: Modern.

Miller, G. Tyler, Jr. 1994. *Living in the Environment,* 8th ed. Belmont, CA: Wadsworth.

Miller, G. Tyler. 2005. *Living in the Environment,* 14th ed. Pacific Grove, CA: Brooks/Cole-Thomson Learning.

Mitchell, Jerry T., Thomas, Deborah S. K., & Cutter, Susan L. 1999. "Dumping in Dixie Revisited: The Evolution of Environmental Injustices in South Carolina." *Social Science Quarterly* 80(2):229–243.

Mitchell, Ross E. 2001. "Thorstein Velben: Pioneer in Environmental Sociology." *Organization Environment* 14(4):389–408.

Mjoset, Lars. 2001. "Realisms, Constructivisms, and Environmental Sociology: A Comment on Ted Benton's 'Environmental Sociology: Controversy and Continuity.'" *Sosiologisk Tidsskrift* 9(1–2):180–197.

Moghadam, Valentine M., ed. 1996. *Patriarchy and Development: Women's Positions at the End of the Twentieth Century.* Oxford, UK: Clarendon Press.

Mohai, Paul, & Bryant, Bunyan. 1992. "Environmental Racism: Reviewing the Evidence." Pp. 163–176 in *Race and the Incidence of Environmental Hazards: A Time for Discourse,* edited by Bunyan Bryant & Paul Mohai. Boulder, CO: Westview Press.

Mol, Arthur P. J. 1995. *The Refinement of Production: Ecological Modernization Theory and the Chemical Industry.* Utrecht, the Netherlands: Van Arkel.

Mol, Arthur P. J. 1996. "Ecological Modernisation and Institutional Reflexivity: Environmental Reform in the Late Modern Age." *Environmental Politics* 5:302–323.

Mol, Arthur P. J. 2001. *Globalization and Environmental Reform: The Ecological Modernization of the Global Economy.* Cambridge: MIT Press.

Mol, Arthur P. J., & Jänicke, Martin. 2009. "The Origins and Theoretical Foundations of Ecological Modernisation Theory." Pp. 17–28 in *The Ecological Modernization Reader: Environmental Reform in Theory and Practice,* edited by Arthur P. J. Mol, David A. Sonnenfeld, & Gert Spaargaren. London and New York: Routledge.

Mol, Arthur P. J., Sonnenfeld, David A., & Spaargaren, Gert, eds. *The Ecological Modernization Reader: Environmental Reform in Theory and Practice.* London and New York: Routledge.

Mol, Arthur P. J., & Spaargaren, Gert. 2000. "Ecological Modernization Theory in Debate: A Review." Pp. 17–49 in *Ecological Modernization Around the World,* edited by Arthur P. J. Mol & D. A. Sonnenfeld. London: Cass.

Mol, Arthur P. J., & Spaargaren, Gert. 2005. "From Additions and Withdrawals to Environmental Flows: Reframing Debates in the Environmental Social Sciences." *Organization and Environment* 18:91–107.

Morehouse, Ward, & Subramaniam, M. Arun. 1986. *The Bhopal Tragedy: What Really Happened and What It Means for American Workers and Communities at Risk.* A Report of the Citizens Commission on Bhopal. New York: Council on International and Public Affairs.

Morello-Frosch, Rachel, Pastor, Manuel, Jr., & Sadd, James. 2001. "Environmental Justice and Southern California's 'Riskscape': The Distribution of Air Toxics Exposures and Health Risks Among Diverse Communities." *Urban Affairs Review* 36(4):551–578.

Morgan, G. et al. 1998, "Air Pollution and Daily Mortality in Sydney, Australia, 1989 to 1993." *American Journal of Public Health* 88(5).

Morin, Richard, & Berry, John M. 1996. "As for the Economy Public Sees Thorns: Survey Finds Americans Gloomy." *International Herald Tribune,* October 14, pp. 1, 6.

Moritz, Mark, Soma, Eric, Scholte, Paul, Xiao, Ningchuan, Taylor, Leah, Juran, Todd & Kari, Saidou. 2010. "An Integrated Approach to Modeling Grazing Pressure in Pastoral Systems: The Case of the Logone Floodplain (Cameroon)." *Human Ecology* 38(6):775–789.

Movement for the Survival of the Ogoni People. 2007. "Update and Timeline of Recent Developments in Ogoni." Retrieved January 23, 2008, from http://www.mosop.net

Muir, Hazel. 2002. "Suffer the Children: The Effects of Radiation Don't Stop With the People Exposed to It." *New Scientist* 2342, May 11, p. 5.

Mumford, Lewis. 1934. *Technics and Civilization.* London: Routledge & Kegan Paul.

Murphy, Raymond. 1994a. *Rationality and Nature.* Boulder, CO: Westview Press.

Murphy, Raymond. 1994b. "The Sociological Construction of Science without Nature." *Sociology* 28(4):957–974.

Murphy, Raymond. 2004. "Disaster or Sustainability: The Dance of Human Agents With Nature's Actants." *Canadian Review of Sociology and Anthropology* 41(3):249–266.

Nash, Roderick. 1989. *The Rights of Nature: A History of Environmental Ethics.* Madison: University of Wisconsin Press.

National Aeronautics and Space Administration. 2003. "Historical C02 Record From the Siple Station Ice Core (1734–1983)." *CDIAC, Online Trends.* Retrieved December 18, 2007, from http://gcmd.nasa.gov/records/GCMD_CDIAC_C02_SIPLE_ICECORE.html

National Aeronautics and Space Administration. 2007a. "Ozone Hole Watch." Retrieved September 15, 2007, from http://ozonewatch.gsfc.nasa.gov/

National Aeronautics and Space Administration. 2007b. "2006 Was Earth's 5th Warmest Year." Retrieved June 26, 2007, from http://www.nasa.gov/centers/goddard/news/ topstory/2006/2006_warm.html

National Atmospheric Deposition Program. N.d. "Sulphate Deposition, 1985–2005." *NADP/NTN Animated Trend Maps.* Retrieved December 18, 2007, from http://nadp.sws.uiuc.edu/amaps2/s04dep/amaps.html

National Coalition for the Homeless. 2002. "Who Is Homeless? Fact Sheet #3." Washington, DC: National Coalition for the Homeless. Retrieved August 28, 2003, from http://www.nationalhomeless.org/facts.html

National Coalition for the Homeless (2009). "How Many People Experience Homelessness?" Retrieved January 11, 2011, from http://www.nationalhomeless.org

National Creutzfeldt-Jakob Disease Surveillance Unit (NCJDSU). 2007. "Variant Creutzfeldt-Jakob Disease Current Data (November 2007)." Retrieved November 30, 2007, from http://www.cjd.ed.ac.uk

National Health Service. 2006. "Statistics on Obesity, Physical Activity, and Diet: England, 2006." Leeds, UK: Information Centre of the Government Statistical Service.

National Health Service. 2010. *Statistics on Obesity, Physical Activity and Diet: England, 2010.* Retrieved January 11, 2011, from http://www.ic.nhs.uk/pubs/opad10

National Highway Traffic Safety Administration (NHTSA). 2007. "Motor Vehicle Traffic Crash Fatality Counts and Estimates of People Injured for 2006." National Center for Statistics and Analysis. Retrieved January 24, 2008, from http://www.nhtsa.dot.gov

National Highway Traffic Safety Administration. 2010. *Traffic Safety Facts 2009.* Early ed. Washington, DC: U.S. Department of Transportation. Retrieved January 21, 2011, from http://www.nhtsa.gov

National Oceanic and Atmospheric Administration. 2010. "Atlantic Cod." In *Fishwatch: U.S. Seafood Facts.* Retrieved January 20, 2011, from http://www.nmfs.noaa.gov

National Snow and Ice Data Center. 2002. "Larsen B Ice Shelf Collapses in Antarctica." Retrieved August 22, 2003, from http://nsidc.org/iceshelves/larsenb2002/

National Snow and Ice Data Center. 2007. "Press Release: Models Underestimate Loss of Arctic Sea Ice." Retrieved June 5, 2007, from http://nsidc.org/news/press/20070430_Stro eveGRL.html

National Snow and Ice Data Center. 2010. "Arctic Sea Ice Falls to Third-Lowest Extent; Downward Trend Persists." Retrieved January 10, 2011, from http://nsidc.org

National Weather Service. 2003. "Stratosphere: Southern Hemisphere Ozone Hole Size." Retrieved August 22, 2003, from http://www.cpc.ncep.noaa.gov/products/stratosphere/sbuv2to/ozone_hole.html

Natural Resources Conservation Service. 2010. *2007 Natural Resources Inventory: Soil Erosion on Cropland.* Retrieved January 11, 2011, from http://www.nrcs.usda.gov

Natural Resources Defense Council (NRDC). 2003a. "Danger in the Air: Thousands of Early Deaths Could Be Averted With Cleaner Air Standards." Retrieved August 22, 2003, from http://www.nrdc.org/air/pollution/nbreath.asp

Natural Resources Defense Council. 2003b. "Toxic Herbicide Atrazine Contaminating Water Supplies, While EPA Cuts Special Deal With Manufacturer." *Brief News: Toxic Chemicals & Health: Pesticides.* Retrieved October 24, 2003, from http://www.nrdc.org/health/pesticides/natrazine.asp

Nazarea-Sandoval, Virginia D. 1995. *Local Knowledge and Agricultural Decision Making in the Philippines: Class, Gender, and Resistance.* Ithaca, NY: Cornell University Press.

Nestle, Marion. 2002. *Food Politics: How the Food Industry Influences Nutrition and Health.* Berkeley: University of California Press.

New Delhi TV. 2010. "Bhopal Gas Tragedy Is a Closed Case Now: U.S." *Press Trust of India,* August 20. Retrieved February 9, 2011, from http://www.ndtv.com

Newport, Frank. 2010. "Americans' Global Warming Concerns Continue to Drop." Retrieved February 16, 2011, from http://www.gallop.poll

Nickell, David. 2007. "Between the Rivers: A Socio-Historical Account of Hegemony and Heritage." *Humanity & Society* 31(2–3):164–209.

Norberg-Hodge, Helena. 2009. *Ancient Futures: Lessons From Ladakh for a Globalizing World,* 2nd ed. San Francisco, CA: Sierra Club.

Northcott, Michael S. 1996. *The Environment and Christian Ethics.* Cambridge, UK, and New York: Cambridge University Press.

Notestein, Frank W. 1945. "Population: The Long View." Pp. 36–57 in *Food for the World,* edited by Theodore W. Schultz. Chicago, IL: University of Chicago Press.

Novek, Joel. 1995. "Environmental Impact Assessment and Sustainable Development Case Studies of Environmental Conflict." *Society and Natural Resources* 8:145–159.

Nowak, Peter, Bowen, Sarah, & Cabot, Perry E. 2006. "Disproportionality as a Framework for Linking Social and Biophysical Systems." *Society and Natural Resources* 19:153–173.

"Numerology: Asians Hit the Road." 2001. *Asiaweek.* Retrieved January 21, 2004, from http://www.asiaweek.com/asiaweek/magazine/nations/0,8782,165900,00.html

O'Brien, Timothy L. 2006. "What Happened to the Fortune Michael Jackson Made?" *New York Times,* May 14. Retrieved January 11, 2011, from http://www.nytimes.com

O'Connor, James. 1973. *The Fiscal Crisis of the State.* New York: St Martin's Press.

O'Connor, James. 1988. "Capitalism, Nature, Socialism: A Theoretical Introduction." *Capitalism, Nature, Socialism* 1(1):11–38.

Ogden, Cynthia L., & Carroll, Margaret D. 2010a. "Prevalence of Obesity Among Children and Adolescents: United States, Trends 1963–1965 Through 2007–2008." *National Center for Health Statistics.* Retrieved January 11, 2011, from http://www.cdc.gov

Ogden, Cynthia L., & Carroll, Margaret D. 201b. "Prevalence of Overweight, Obesity, and Extreme Obesity Among Adults: United States, Trends 1976–1980 Through 2007–2008." *National Center for Health Statistics.* Retrieved January 11, 2011, from http://www.cdc.gov

Olsen, Marvin E., Lodwick, Dora G., & Dunlap, Riley E. 1992. *Viewing the World Ecologically.* Boulder, CO: Westview Press.

"On Ethanol, Castro Is Right, Says the *Economist.*" 2007. *Economist,* April 6. Retrieved April 23, 2007, from http://www.soyatech.com/news_story.php?id=2077

Onions, C. T. [1933] 1955. *The Oxford Universal Dictionary on Historical Principles,* 3rd ed. Oxford, UK: Oxford at the Clarendon Press.

Oosterveer, Peter. 2006. "Environmental Governance of Global Food Flows: The Case of Labeling Strategies." Pp. 267–301 in *Governing Environmental Flows: Global Challenges to Social Theory,* edited by Gert Spaargaren, Arthur P. J. Mol, & Frederick H. Buttel. Cambridge, MA, and London: MIT Press.

Organic Monitor. 2010. "The Global Market for Organic Food and Drink: Business Opportunities and Future Outlook." Retrieved February 11, 2011, from http://www.organicmonitor.com

Organic Trade Association. 2010. "U.S. Organic Industry Overview." Retrieved February 9, 2011, from http://www.ota.com

Ostrom, Elinor. 1990. *Governing the Commons: The Evolution of Institutions for Collective Action.* Cambridge, UK, and New York: Cambridge University Press.

Oulan, Tarmo. (2003). "Pikaluistelun Historia: From the Dutch Canals to the Frozen Gulf of Bothnia." Retrieved August 20, 2003, from http://www.ouluntarmo.fi/lupy/pikaluistelu/historia/H/historia.htm

Pala, Christopher. 2006. "Once a Terminal Case, the North Aral Sea Shows New Signs of Life." *Science* 312:183.

Parfit, Derek. 1986. *Reasons and Persons.* Oxford, UK: Oxford University Press.

Parker, Barry R. 1996. *Chaos in the Cosmos: The Stunning Complexity of the Universe.* New York: Plenum Press.

Parkins, John R., & Mitchell, Ross E. 2005. "Public Participation as Public Debate: A Deliberative Turn in Natural Resource Management." *Society and Natural Resources* 18:529–540.

Parsons, Talcott. 1951. *The Social System.* New York: Free Press.

Pastor, Manuel, Jr., Morello-Frosch, Rachel, & Sadd, James L. 2005. "The Air Is Always Cleaner on the Other Side: Race, Space, and Ambient Air Toxics Exposures in California." *Journal of Urban Affairs* 27(2):127–148.

Pastor, Manuel, Jr., Sadd, James L., & Hipp, John. 2001. "Which Came First? Toxic Facilities, Minority Move-In, and Environmental Justice." *Journal of Urban Affairs* 23(1):1–21.

Pastor, Manuel, Jr., Sadd, James L., & Morello-Frosch, Rachel. 2002. "Who's Minding the Kids? Pollution, Public Schools, and Environmental Justice in Los Angeles." *Social Science Quarterly* 83(1):263–280.

Paul, Jim. 2003. "Warming, Smog Tested on Crops." Retrieved August 22, 2003, from http://stacks.msnbc.com:80/news/938545.asp?cp1=1

Pearce, Fred. 1995. "Acid Fallout Hits Europe's Sensitive Spots." *New Scientist* 147, July 8, p. 6.

Peerenboom, R. E. 1991. "Beyond Naturalism: A Reconstruction of Daoist Environmental Ethics." *Environmental Ethics* 13(1):3–22.

Pellow, David N., Weinberg, Adam, & Schnaiberg, Alan. 2001. "The Environmental Justice Movement: Equitable Allocation of the Costs and Benefits of Environmental Management Outcomes." *Social Justice Research* 14(4):423–439.

Peluso, Nancy Lee. 1996. "Reserving Value: Conservation Ideology and State Protection of Resources." Pp. 135–165 in *Creating the Countryside: The Politics of Rural and Environmental Discourse*, edited by Melanie Dupuis & Peter Vandergeest. Philadelphia, PA: Temple University Press.

Perrow, Charles.1999. *Normal Accidents*, Rev. ed. Princeton, NJ: Princeton University Press.

Petrzelka, Peggy, & Bell, Michael M. 2000. "Rationality and Solidarity: The Social Organization of Common Property Resources in the Imdrhas Valley of Morocco." *Human Organization* 59(3):343–352.

Pew Research Center. 2006. "Are We Happy Yet?" *Pew Research Center Publications*. Retrieved January 15, 2011, from http://www.pewresearch.org

Pew Research Center. 2007. *Global Unease With Major World Powers: Rising Environmental Concern in 47-Nation Survey*. Washington, DC: Author.

Pew Research Center. 2009a. "Global Warming Seen as a Major Problem Around the World: Less Concern in the U.S., China and Russia." Retrieved January 26, 2011, from http://pewresearch.org

Pew Research Center. 2009b. "Independents Take Center Stage in Obama Era." *Pew Research Center for the People and the Press*, May 21. Retrieved January 27, 2011, from http://people-press.org

Pew Research Center. 2010. "Wide Partisan Divide Over Global Warming." Retrieved January 26, 2011, from http://pewresearch.org

Pianin, Eric. 2003. "Study Finds Net Gain From Pollution Rules: OMB Overturns Past Findings on Benefits." *The Washington Post*, September 27, p. A1.

Picou, Steven J., & Gill, Duane A. 2000. "The *Exxon Valdez* Disaster as Localized Environmental Catastrophe: Dissimilarities to Risk Society Theory." Pp. 143–170 in *Risk in the Modern Age: Social Theory, Science, and Environmental Decision-Making*, edited by Maurie J. Cohen. New York: St. Martin's Press.

Pietila, Hillcka, & Vickers, Jeanne. 1994. *Making Women Matter: The Role of the United Nations*. London and Atlantic Highlands, NJ: Zed Books.

Pimentel, David. 2006. "Soil Erosion: A Food and Environmental Threat." *Environment, Development, and Sustainability* 8:119–137.

Pine, John C., Marx, Brian D., & Lakshmanan, Aruna. 2002. "An Examination of Accidental Release Scenarios From Chemical-Processing Sites: The Relation of Race to Distance." *Social Science Quarterly* 83(1):317–331.

Plant, Judith, ed. 1989. *Healing the Wounds: The Promise of Ecofeminism*. Philadelphia, PA: New Society.

Plato. [c. 360 BCE] 1965. *Timaeus and Critias*. London: Penguin.

Plato. [c. 399 BCE.] 1952. *Plato's Gorgias*, translated by W. C. Helmbold. New York: Liberal Arts Press.

Plato. [c. 399 BCE] 1985. *The Republic*, translated by Richard W. Sterling & William C Scott. New York: Norton.

Plato. [c. 399 BCE] 1997. *Defence of Socrates, Euthyphro, Crito*, edited and translated by David Gallup. Oxford, UK: Oxford University Press.

Platt, Harold L. 2010 "Exploding Cities: Housing the Masses in Paris, Chicago, and Mexico City, 1850–2000." *Journal of Urban History* 36(5):575–593.

Pliny the Elder. 1855. *The Natural History of Pliny*, translated by John Bostock & H. T. Riley. London: H. G. Bohn.

Plumwood, Val. 1994a. "The Ecopolitics Debate and the Politics of Nature." Pp. 64–87 in *Ecological Feminism*, edited by Karen Warren. London and New York: Routledge.

Plumwood, Val. 1994b. *Feminism and the Mastery of Nature*. London and New York: Routledge.

Plunket Research. N.d. "Automobile Industry Introduction." Retrieved January 21, 2011, from http://www.plunkett research.com

PolitiFact. 2011. "Michael Moore Says 400 Americans Have More Wealth Than Half of All Americans Combined." *Journal Sentinel PolitiFact Wisconsin*, March 10. Retrieved April 1, 2011, from http://www.politifact.com

Polletta, Francesca. 2002. *Freedom Is an Endless Meeting: Democracy in American Social Movements*. Chicago, IL: University of Chicago.

President's Cancer Panel. 2010. *Reducing Environmental Cancer Risk: What We Can Do Now*. Retrieved January 11, 2011, from http://deainfo.nci.nih.gov

Pretty, Jules. 2002. *Agri-Culture: Reconnecting People, Land and Nature*. London: Earthscan.

Pretty, J. N., Noble, A. D., Bossio, D., Dixon, J., Hine, R. E., Penning de Vries, F. W. T., et al. 2006. "Resource-Conserving Agriculture Increases Yields in Developing Countries." *Environmental Science and Technology* 40(4):114–119.

"Price of Rhino Horn Upstages Gold." 2009. *The Zimbabwe Mail.* Retrieved January 14, 2011, from http://newzim situation.com

Prince, S. H. 1920. *Catastrophe and Social Change.* New York: Columbia University Press.

Pring, George W., & Canan, Penelope. 1996. *SLAPPs: Getting Sued for Speaking Out.* Philadelphia, PA: Temple University Press.

Pritchard, Henderson W. 2010. "Race, Class and Environmental Equity: A Study of Disparate Exposure to Toxic Chemicals in the Commonwealth of Massachusetts." *Dissertation Abstracts International, A: The Humanities and Social Sciences,* 70(7):2738.

Public Agenda. 2003. "Environment: A Nation Divided?" *Environment.* Retrieved January 28, 2004, from http://www .publicagenda.org/issues/nation_divided_detail.cfm?issue_type=environment&list=1

Pucher, John, & Bijkstra, Lewis. 2003. "Promoting Safe Walking and Cycling to Improve Public Health: Lessons From the Netherlands and Germany." *American Journal of Public Health* 93(9):1509–1516.

Purcell, Kristen, Clark, Lee, & Renzulli, Linda. 2000. "Menus of Choice: The Social Embeddedness of Decisions." Pp. 62–79 in *Risk in the Modern Age: Social Theory, Science, and Environmental Decision Making,* edited by Maurie J. Cohen. New York: St. Martin's Press.

Putnam, Robert. 2000. *Bowling Alone: The Collapse and Revival of American Community.* New York: Simon & Schuster.

Quandt, Sara A., Chen, Haiying, Grzywacz, Joseph G., & Vallejos, Quirina M. 2010. "Cholinesterase Depression and Its Association With Pesticide Exposure Across the Agricultural Season Among Latino Farmworkers in North Carolina." *Environmental Health Perspectives* 118(5):635–640.

Quinnipiac Poll. 2008. "State Voters Back NYC Traffic Fee 2–1, If Funds Go to Transit, Quinnipiac University Poll Finds; Voters Back Millionaire's Tax 4–1." Retrieved February 8, 2011, from http://www.quinnipiac.edu

Rabelais, Francis. [1532–1552] 1931. *The Works of Francis Rabelais,* edited by Albert J. Nock & Catherine Rose Wilson. New York: Harcourt, Brace.

Rawls, John. 1971. *A Theory of Justice.* Cambridge, MA: Belknap Press of Harvard University Press.

Rawls, John. 1995. *Political Liberalism.* New York: Columbia University Press.

Reijnen, Rien, Foppen, Rudd, & Braak, Cajo-Ter. 1995. "The Effects of Car Traffic on Breeding Bird Populations in Woodland. III: Reduction of Density in Relation to the Proximity of Main Roads." *Journal of Applied Ecology* 32:187–202.

Rengasamy, Pichu. 2006. "World Salinization With Emphasis on Australia." *Journal of Experimental Botany* 57(5):1017–1023.

Renn, Ortwin. 1997. "The Demise of the Risk Society." Paper presented at the Annual Meeting of the American Sociological Association, Toronto, Canada.

Revkin, Andrew. 2006. "After 3,000 Years, Arctic Ice Shelf Broke Off Canadian Island, Scientists Find." *New York Times,* December 30. Retrieved June 25, 2007, from http://select.nytimes.com

Riddle, John M. 1997. *Eve's Herbs: A History of Contraception and Abortion in the West.* Cambridge, MA: Harvard University Press.

Rifkin, Jeremy. 1995. *The End of Work: The Decline of the Global Labor Force and the Dawn of the Post-Market Era.* New York: Putnam.

Roberts, Rebecca S., & Emel, Jacque. 1992. "Uneven Development and the Tragedy of the Commons: Competing Images for Nature-Society Analysis." *Economic Geography* 68:249–271.

Robinson, Emily. 2007. "Exxon Exposed." *Catalyst* 6(1). Retrieved January 15, 2011, from http://www.ucsusa.org

Rocheleau, Dianne, Thomas-Slayter, Barbara, & Wangari, Esther. 1996. *Feminist Political Ecology: Global Issues and Local Experiences.* New York: Routledge.

Rolston, Holmes, III. 1979. "Can and Ought We to Follow Nature?" *Environmental Ethics* 1(1):7–30.

Rommen, Heinrich A. [1936] 1947. *The Natural Law,* translated by Thomas R. Hanley. St. Louis, MO: B. Herder.

Rosa, Eugene A. 2000. "Modern Theories of Society and the Environment: The Risk Society." Pp. 73–101 in *Environment and Global Modernity,* edited by Gert Spaargaren, Arthur P. J. Mol, & Frederick Buttel. London: Sage.

Rubin, Lillian. 1994. *Families on the Fault Line: America's Working Class Speaks About the Family, the Economy, Race, and Ethnicity.* New York: HarperCollins.

Sachs, Aaron. 1996. "Dying for Oil." *Worldwatch* 9(3):10–21.

Sachs, Wolfgang, ed. 1992. *The Development Dictionary: A Guide to Knowledge as Power.* London and Atlantic Highlands, NJ: Zed Books.

Saha, Robin Kumar. 2002. "A Longitudinal and Historical Context Analysis of Racial and Socioeconomic Inequities in the Distribution of Hazardous Waste Facilities in Michigan." *Dissertation Abstracts International, A: The Humanities and Social Sciences* 63(2):765-A.

Sahlins, Marshall. 1972. "The Original Affluent Society." Pp. 1–39 in *Stone Age Economics,* by M. Sahlins. New York: Aldine.

Samet, Jonathan M., Dominici, Francesca, Curriero, Frank C., Coursac, Ivan, & Zeger, Scott L. 2000. "Fine Particulate Air Pollution and Mortality in 20 U.S. Cities, 1987–1994." *New England Journal of Medicine* 343(24):1742–1749.

Sarre, Philip, & Blunden, John. 1995. *An Overcrowded World? Population, Resources, and the Environment.* Oxford, UK, and New York: Oxford University Press and the Open University.

Saunders, Peter. 2000. "Use and Abuse of the Precautionary Principle." *Third World Network.* Retrieved January 31, 2004, from http://www.twnside.org.sg/titie/saunders.htm

Schaller, Thomas. 2010. "Jonah Goldberg, Anti-Maldistributionist." Retrieved January 11, 2011, from http://www.fivethirtyeight.com

Schiller, Ferdinand Canning Scott. [1903] 1912. *Humanism: Philosophical Essays,* 2nd ed. London: Macmillan.

Schmidt, Alfred. 1971. *The Concept of Nature in Marx,* translated by Ben Foukes. London: New Left Books.

Schnaiberg, Alan. 1980. *The Environment, from Surplus to Scarcity.* New York and Oxford, UK: Oxford University Press.

Schnaiberg, Alan, & Gould, Kenneth Alan. 1994. *Environment and Society: The Enduring Conflict.* New York: St. Martin's Press.

Schoenfish-Keita, Jennifer, & Johnson, Glenn S. 2010. "Environmental Justice and Health: An Analysis of Persons of Color Injured at the Work Place." *Race, Gender & Class* 17(1–2):270–304.

Schor, Juliet B. 1992. *The Overworked American: The Unexpected Decline of Leisure.* New York: Basic Books.

Schor, Juliet. 2010. "Beyond Business as Usual." *The Nation,* May 24. Retrieved January 15, 2011, from http://www.thenation.com

Schor, Juliet B. and Margaret Willis. 2008. "Conscious Consumption: Results From a Survey of New Dream Members." *Center for a New American Dream.* Retrieved March 30, 2011, from http://www.newdream.org/consumption/survey.pdf

Schrank, David, Lomax, Tim, & Turner, Shawn. 2010. *Urban Mobility Report 2010.* Texas Transportation Institute. Texas A&M University.

Schudson, Michael. 1984. *Advertising, the Uneasy Persuasion: Its Dubious Impact on American Society.* New York: Basic Books.

Schultz, T. Paul. 1981. *Economics of Population.* Reading, MA: Addison-Wesley.

Schutz, Alfred. 1962. *The Problem of Social Reality.* The Hague, the Netherlands: Martinus Nijhoff.

Schweikart, David. 1998. "Market Socialism: A Defense." Pp. 7–22 in *Market Socialism: The Debate Among Socialists,* edited by David Schweikart, James Lawler, Hillel Ticktin, & Bertell Ollman. New York and London: Routledge.

Schweikart, David. 2002. *After Capitalism.* Lanham, MD: Rowman & Littlefield.

Science and Environmental Health Network. 2004. "Precautionary Principle." Retrieved January 31, 2004, from http://www.sehn.org/precaution.html

Scientific Assessment Panel of the Montreal Protocol. 2010. *Scientific Assessment of Ozone Depletion, 2010: Executive Summary."* Retrieved January 10, 2011, from http://www.unep.org

Scott, James C. 1976. *The Moral Economy of the Peasant: Rebellion and Subsistence in Southeast Asia.* New Haven, CT: Yale University Press.

Scott, James C. 1986. *Weapons of the Weak: Everyday Forms of Peasant Resistance.* New Haven, CT: Yale University Press.

Scott, James C. 1990. *Domination and the Arts of Resistance: Hidden Transcripts.* New Haven, CT: Yale University Press.

Seager, Joni. 1993. *Earth Follies: Feminism, Politics, and the Environment.* London: Earthscan.

Seidman, Steven. 1994. *Contested Knowledge: Social Theory in the Post-Modern Era.* Oxford, UK: Blackwell.

Sen, Amartya. 1981. *Poverty and Famines: An Essay on Entitlement and Deprivation.* New York and Oxford, UK: Oxford University Press.

Sen, Amartya. 1992. *Inequality Reexamined.* Cambridge, MA: Harvard University Press.

Sen, Amartya. 1999. *Development as Freedom.* New York: Anchor Books.

Shiva, Vandana. 1988. *Staying Alive: Women, Ecology, and Development.* London: Zed Books.

Shprentz, Deborah Sheiman. 1996. *Breath-taking: Premature Mortality Due to Particulate Air Pollution in 239 American Cities.* Washington, DC: Natural Resources Defense Council.

Simmel, Georg. [1900] 1990. *The Philosophy of Money*, 2nd enl. ed., edited by David Frisby, translated by Tom Bottomore & David Frisby. London, and New York: Routledge.

Simon, Julian. 1981. *The Ultimate Resource*. Princeton, NJ: Princeton University Press.

Simon, Julian, ed. 1995. *The State of Humanity*. Oxford, UK, and Cambridge, MA: Blackwell.

Simon, Julian, & Kahn, Herman. 1984. *The Resourceful Earth: A Response to Global 2000*. Oxford, UK, and New York: Blackwell.

Sixbear, Jaipi. 2008. "Michael Jackson Sells Neverland Ranch: Really?" *Associated Press*. Accessed January 11, 2011, at http://www.associatedcontent.com

Slicer, Deborah. 1994. "Wrongs of Passage: Three Challenges to the Maturing of Ecofeminism." Pp. 29–41 in *Ecological Feminism*, edited by Karen Warren. London and New York: Routledge.

Slovic, Paul. 1987. "Perception of Risk." *Science* 236:280–285.

Slow Food International. (N.d.). "Our Mission." Retrieved March 29, 2011, from http://www.slowfood.com/international/2/our-philosophy

Smith, Lewis. 2006. "Experts Warn North Pole Will Be 'Ice Free' by 2040." *Times Online*. Retrieved December 12, 2006, from http://www.timesonline.co.uk/article/03–2499663,00.html

Smith, Tom W. 2007. "Trends in National Spending Priorities, 1973–2006." Chicago, IL: National Opinion Research Center, University of Chicago.

Snow, David A., & Benford, Robert D. 1988. "Ideology, Frame Resonance, and Participant Mobilization." *International Social Movement Research*, 1:197–217.

Socioeconomic Data and Applications Center (SEDAC). 2009. "Treaty Locator." *Environmental Treaties and Resource Indicators*. Retrieved January 27, 2011, from http://sedac.ciesin.org/entri/treatyTexts.jsp

Soper, Kate. 1995. *What Is Nature? Culture, Politics and the Non-Human*. Oxford, UK, and Cambridge, MA: Blackwell.

Spaargaren, Gert, & Cohen, Maurie J. 2009. "Greening Lifecycles and Lifestyles: Sociotechnical Innovations in Consumption and Production as Core Concerns of Ecological Modernization Theory." Pp. 257–274 in *The Ecological Modernization Reader: Environmental Reform in Theory and Practice*, edited by Arthur P. J. Mol, David A. Sonnenfeld, & Gert Spaargaren. London and New York: Routledge.

Spaargaren, Gert, Mol, Arthur P. J., & Buttel, Frederick H., eds. 2006. *Governing Environmental Flows: Global Challenges to Social Theory*. Cambridge, MA, and London: MIT Press.

Speth, James Gustave. 2009. *The Bridge at the Edge of the World: Capitalism, the Environment, and Crossing From Crisis to Sustainability*. New Haven, CT: Yale University Press.

Stallones, Lorann, & Beseler, Cheryl. 2001. "Pesticide Poisoning and Depressive Symptoms Among Farm Residents." *Annals of Epidemiology* 12(6):389–394.

Standing, Guy. 1989. "Global Feminization Through Flexible Labour." *World Development* 17(7):1077–1095.

Starr, Chauncey. 1969. "Social Benefit Versus Technological Risk: What Is Our Society Willing to Pay for Safety?" *Science* 165:1232–1238.

Stein, Dorothy. 1995. *People Who Count: Population and Politics, Women and Children*. London: Earthscan.

Steingraber, Sandra. 1997. *Living Downstream: An Ecologist Looks at Cancer and the Environment*. New York: Addison-Wesley.

Steingraber, Sandra. 2001. *Having Faith: An Ecologist's Journey to Motherhood*. New York: Berkley Books.

Steingraber, Sandra. 2010. *Living Downstream: An Ecologist Looks at Cancer and the Environment*, 2nd ed. Cambridge, MA: Da Capo Press.

Steinmetz, Charles. 2002. "Horace's Farmhouse Found Beneath Horace's Villa Site." *Backdirt*, Spring/Summer. Retrieved December 9, 2007, from http://www.ioa.ucla.edu/backdirt/spr02/steinmetz.html

Stevens, Wallace K. 1995. "Study of Cloud Patterns Points to Many Areas Exposed to Big Rises in Ultraviolet Radiation." *New York Times*, November 21, p. C4.

Stevens, Wallace K. 1997. "A Greener Green Belt Bears Witness to a Warming Trend." *New York Times*, April 22, p. B10.

Stevenson, Glenn G. 1991. *Common Property Economics: A General Theory and Land Use Applications*. Cambridge, UK, and New York: Cambridge University Press.

Stevis, Dimitris, & Bruyninckx, Hans. 2006. "Looking Through the State at Environmental Flows and Governance." Pp. 107–136 in *Governing Environmental Flows: Global Challenges to Social Theory*, edited by Gert Spaargaren, Arthur P. J. Mol, & Frederick H. Buttel. Cambridge, MA, and London: MIT Press.

Stewart, F. 1982. "Poverty and Famines: Book Review." *Disasters* 6(2):n.p.

Stiefel, Matthias. 1994. *A Voice for the Excluded: Popular Participation in Development: Utopia or Necessity?* London and Atlantic Highlands, NJ: Zed Books.

Stiglitz, Joseph E. 2002. *Globalization and Its Discontents.* New York: Norton.

Stiglitz, Joseph E., & Bilmes, Linda J. 2010. "The True Cost of the Iraq War: $3 Trillion and and Beyond." *Washington Post,* September 5, 2010. Retrieved May 16, 2010, from www.washingtonpost.com

Stiles, Kaelyn, Altiok, Ozlem, & Bell, Michael M. (2010). "The Ghosts of Taste: Food and the Cultural Politics of Authenticity." *Agriculture and Human Values.*

Stoecker, Randy. 2007. "Honey, Vinegar, Community Organizing, and the Political Opportunity Structure." Paper Presented to the Department of Rural Sociology, November 9, University of Wisconsin–Madison.

Stone, Richard. 1999. Aral Sea: Coming to Grips With the Aral Sea's Grim Legacy." *Science* 284(5411):30–33.

Streeter, Michael. 1996. "Record Haul of Rhino Horn Is Seized." *Independent,* September 4, p. 1.

Stretesky, Paul, & Hogan, Michael J. 1998. "Environmental Justice: An Analysis of Superfund Sites in Florida." *Social Problems* 45(2):268–287.

Stretesky, Paul, & Lynch, Michael J. 1999. "Environmental Justice and the Predictions of Distance to Accidental Chemical Releases in Hillsborough County, Florida." *Social Science Quarterly* 80(4):830–846.

Suplee, Curt. 1995. "Dirty Air Can Shorten Your Life, Study Says: Death Rate Higher in Worst Cities." *The Washington Post,* March 10, p. A1.

Swan, Shanna H., Brazil, Charlene, Drobnis, Erma Z., Liu, Fan, Kruse, Robin L., Hatch, Maureen, et al. 2003. "Geographic Differences in Semen Quality of Fertile U.S. Males." *Environmental Health Perspectives* 111(4):414.

Tagliabue, John. 2003. "Death Toll in Europe's Heat Wave Is Continuing to Climb." *New York Times,* August 14, 2003. Retrieved August 21, 2003, from http://www.nytimes.com/2003/08/14/international/europe/14CND-EURO.html

Tans, Pieter. N.d. "Trends in Atmospheric Carbon Dioxide: Mauna Loa." Retrieved December 18, 2007, from http://www.esrl.noaa.gov/gmd/ccgg/trends/

Taquino, Michael, Parisi, Domenico, & Gill, Duane A. 2002. "Units of Analysis and the Environmental Justice Hypothesis: The Case of Industrial Hog Farms." *Social Science Quarterly* 83(1):298–316.

Tarrow, S. (1994). *Power in Movement: Social Movements, Collective Action and Politics.* Cambridge, UK: Cambridge University Press.

Taylor, Dorceta E. 1989. "Blacks and the Environment: Toward an Explanation of the Concern and Action Gap Between Blacks and Whites." *Environment and Behavior* 21(2):175–205.

Taylor, Dorceta E. 2000. "The Rise of the Environmental Justice Paradigm: Injustice Framing and the Social Construction of Environmental Discourses." *American Behavioral Scientist* 43(4):508–580.

Taylor, Humphrey. 2003. "The Religious and Other Beliefs of Americans 2003." *Harris Poll Library.* Retrieved January 30, 2003, from http://www.harrisinteractive.com/harrispoll/index.asp

Taylor, Paul W. 1986. *Respect for Nature: A Theory of Environmental Ethics.* Princeton, NJ: Princeton University Press.

Taylor, Peter J. 2005. *Unruly Complexity: Ecology, Interpretation, Engagement.* Chicago and London: University of Chicago Press.

Taylor, Peter J., & Buttel, Frederick H. 1992. "How Do We Know We Have Global Environmental Problems? Science and the Globalization of Environmental Discourse." *Geoforum* 23:405–416.

Teitelbaum, Michael S. 1987. "Relevance of Demographic Transition Theory for Developing Countries." Pp. 29–36 in *Perspectives on Population: An Introduction to Concepts and Issues,* edited by Scott W. Menard & Elizabeth W. Moen. New York and Oxford, UK: Oxford University Press.

Thomas, Keith. 1983. *Man and the Natural World: Changing Attitudes in England, 1500–1800.* London: Allen Lane.

Thompson, Gary D., & Wilson, Paul N. 1994. "Common Property as an Institutional Response to Environmental Variability." *Contemporary Economic Policy* 12:12–21.

Thoreau, Henry David. [1854] 1962. *The Variorum Walden.* New York: Washington Square Press.

Thoreau, Henry David. [1862] 1975. "Walking." In *Excursions.* Gloucester, MA: Peter Smith.

Tocqueville, Alexis de [1835–1840] 1988. *Democracy in America.* Vols. I and II. New York: Harper & Row.

Tomlin, Clive. 2006. *The Pesticide Manual,* 14th ed. Newbury, UK: CPL Press.

Tomorrow Project. 2007. "The Story So Far." *Employment: Will People Have More Leisure?* Retrieved December 4, 2007, from http://www.tomorrowproject.net/pub/1-GLIM PSES/Employment/-285.html

Toro, Paul A., Tompsett, Carolyn J., Lombardo, Sylvie, et al. 2007. "Homelessness in Europe and the United States: A Comparison of Prevalence and Public Opinion." *Journal of Social Issues* 63(3):505–524.

"Traffic: Not Bothered." 1996. *Economist,* September 7, pp. 25–26.

Tuan, Yi-Fu. 1968. "Discrepancies Between Environmental Attitude and Behavior: Examples From Europe and China." *The Canadian Geographer* 12:176–191.

Tuan, Yi-Fu. 1974. *Topophilia*. Englewood Cliffs, NJ: Prentice Hall.

Twain, Mark. [1876] 1991. *The Adventures of Tom Sawyer*. Philadelphia, PA, and London: Running Press.

"Union Carbide's Factory in India: Still a Potential Killer." 2001. *Bhopal.net*. Retrieved October 12, 2003, from http://www.bhopal.net/oldsite/contamination.html

United Nations Development Programme (UNDP). 1992. *Human Development Report 1992*. New York: Oxford University Press.

United Nations Development Programme. 1994. *Human Development Report 1994*. New York: Oxford University Press.

United Nations Development Programme. 1999. *Human Development Report 1999*. New York: Oxford University Press.

United Nations Development Programme. 2000. *Human Development Report 2000*. New York: Oxford University Press.

United Nations Development Programme. 2003. *Human Development Report 2003*. New York: Oxford University Press.

United Nations Development Programme. 2006. *Human Development Report 2006: Beyond Scarcity—Power, Poverty, and the Global Water Crisis*. New York: Author.

United Nations Development Programme. 2010. *Human Development Report 2010*. New York: Author.

United Nations Educational, Scientific and Cultural Organization (UNESCO). 2004. "Threats: Disturbed Balance of the Lagoon." *Venice, Safeguarding Campaign*. Retrieved January 14, 2004, from http://www.unesco.org/culture/her1tage/tangible/venice/html_eng/menacelag.shtml

United Nations Educational, Scientific and Cultural Organization, United Nations Environment Programme (UNESCO-UNEP). 1978. "The Tbilisi Declaration: Final Report, Intergovernmental Conference on Environmental Education." Organized by UNESCO in cooperation with UNEP, Tbilisi, USSR, October 14–26, 1977, Paris, France. (UNESCO ED/MD/49).

United Nations Environment Programme. 2003. *Groundwater and Its Susceptibility to Degradation: A Global Assessment of the Problem and Options for Management*. Nairobi, Kenya: Author.

United Nations Human Settlements Program. 2003. *The Challenge of Slums: Global Report on Human Settlements 2003*. London and Sterling, VA: Earthscan.

United Nations Population Division. 2009. *World Population Prospects: The 2008 Revision*. New York: United Nations.

United Nations Programme on HIV/AIDS (UNAIDS). 2007. *AIDS Epidemic Update*. Geneva, Switzerland: UNAIDS and WHO.

Urry, John. 2000. *Sociology Beyond Societies: Mobilities for the Twenty-First Century*. New York: Routledge.

Urry, John 2003. *Global Complexity*. London: Polity.

U.S. Bureau of Labor Statistics. 1981. "Employee Benefits in Industry, 1980." Bulletin 2107. Retrieved February 11, 2008, from http://www.bls.gov/ncs/home.htm

U.S. Bureau of Labor Statistics. 2007. "National Compensation Survey: Employee Benefits in Private Industry in the United States, March 2007." Summary 07–05. Retrieved February 11, 2008, from http://www.bls.gov/ncs/home.htm

U.S. Bureau of Labor Statistics. 2010. "Employee Benefits in the United States, March 2010." News Release USDL-10-1044. Retrieved January 15, 2010, from http://www.bls.gov

U.S. Department of Transportation. 2003. "Average Annual Miles per Driver by Age Group." Retrieved April 2, 2011, from http://www.fhwa.dot.gov/ohim/onh00/bar8.htm

U.S. Department of Transportation. 2010. "Our Nation's Highways: 2010." *Federal Highway Administration*. Retrieved April 2, 2011, from http://www.fhwa.dot.gov

U.S. Census Bureau. 2007a. *Current Population Survey, Annual Social and Economic Supplements*. Retrieved November 10, 2007, from http://www.census.gov/hhes/www/income/histinc/p40.html

U.S. Census Bureau. 2007b. "50 Fastest-Growing Metro Areas Concentrated in West and South." *U.S. Census Bureau News*. Retrieved January 25, 2008, from http://www.census.gov/Press-Release/www/releases/archives/population/009865.html

U.S. Census Bureau. 2010. "Total Midyear Population for the World: 1950–2050." Retrieved January 22, 2011, from http://www.census.gov

U.S. Chemical Safety and Hazard Investigation Board (CSB). 2002. "French Chemical Plant That Exploded in Southern France Not to Reopen." *CSB Incident News Reports*. Retrieved October 22, 2003, from http://www.chemsafety.gov/circ/post.cfm?incident_id=5247

U.S. Department of Agriculture. 2007a. "The Emergency Food Assistance Program (TEFAP)." *Food and Nutrition Service Newsroom*. Retrieved December 19, 2007, from http://www.fns.usda.gov/cga/FactSheets/TEFAP_Quick_Facts.htm

U.S. Department of Agriculture. 2007b. "Rice Yearbook 2006." Retrieved December 7, 2007, from http://usda.mannlib.cornell.edu

U.S. Department of Agriculture. 2008. "U.S. Corn Crop a Record Breaker, USDA Reports: Cotton, Rice Yields Hit All-Time Highs." *National Agricultural Statistics Service Newsroom.* Retrieved February 7, 2008, from http://www.nass.usda.gov/Newsroom/ 2008/01_11_2008.asp

U.S. Department of Agriculture. 2011. "National School Lunch Monthly Data." Retrieved January 11, 2011 from http://www.fns.usda.gov/pd/36slmonthly.htm

U.S. Department of Energy. 2010. "Fact #617: April 5, 2010: Changes in Vehicles per Capita around the World." Retrieved January 21, 2011 from www1.eere.energy.gov

U.S. Department of Agriculture Economic Research Service (USDA ERS). 2010. "Real Historical Gross Domestic Product (GDP) Per Capita and Growth Rates of GDP Per Capita." Retrieved January 22, 2011, from http://www.ers.usda.gov/data/

Van der Heijden, H. A. 2006. "Globalization, Environmental Movements, and International Political Opportunity Structures." *Organization and Environment* 19(1):28–45.

Van Dyke, Fred. 1996. *Redeeming Creation: The Biblical Basis for Environmental Stewardship.* Downers Grove, IL: InterVarsity Press.

Veblen, Thorstein. [1899] 1967. *The Theory of the Leisure Class.* New York: Funk & Wagnalls.

Vermont Land Trust. 2007. "Vermont Land Trust Facts." Retrieved January 14, 2008, from http://www.vlt.org/fact.html

Vidal, John. 1995. "Black Gold Claims a High Price." *Guardian Weekly,* January 15, p. 7.

Vidal, John. 2008. "Scientists Take on Brown Over Nuclear Plans: Academics Say Safety Concerns of New Generation of Plants Not Yet Addressed." *Guardian* (Manchester), January 4. Retrieved January 14, 2007, from http://www.guardian.co.uk/environment/2008/jan/04/nuclearpower.greenpolitics

Von Weizsäcker, E., Lovins, Amory B., & Lovins, L. Hunter. 1998. *Factor Four.* London: Earthscan.

Voyles, Traci Brynne. 2010. "Decolonizing Cartographies: Sovereignty, Territoriality, and Maps of Meaning in the Uranium Landscape." *Dissertation Abstracts International, A: The Humanities and Social Sciences* 71(6): 2230.

Wachtel, Paul. 1983. *The Poverty of Affluence: A Psychological Portrait of the American Way of Life.* New York: Free Press.

Walby, Sylvia. 1996. "The 'Declining Significance' or the 'Changing Forms' of Patriarchy?" Pp. 19–33 in *Patriarchy and Development: Women's Positions at the End of the Twentieth Century,* edited by Valentine M. Moghadam. Oxford, UK: Clarendon Press.

Walton, Bryan K., & Bailey, Conner. 2005. "Framing Wilderness: Populism and Cultural Heritage as Organizing Principles." *Society and Natural Resources* 18(2):119–134.

Walton, John. 1994. *Free Markets and Food Riots: The Politics of Global Adjustment.* Oxford, UK, and Cambridge, MA: Blackwell.

Wang, Yanhui, Solberg, Svein, Yu, Pengtao, Myking, Tor, Vogt, Rolf D., & Du, Shicai. 2007. "Assessments of Tree Crown Condition of Two Masson Pine Forests in the Acid Rain Region in South China." *Forest Ecology and Management* 242(2–3):530–540.

Wardell, Mark. 1992. "Changing Organizational Forms: From the Bottom Up." Pp. 144–164 in *Rethinking Organization: New Directions in Organization Theory and Analysis,* edited by Michael Reed & Michael Hughes. London: Sage.

Wargo, John. 1998. *Our Children's Toxic Legacy: How Science and Law Fail to Protect Us From Pesticides,* 2nd ed. New Haven, CT: Yale University Press.

Warner, W. Keith, & England, J. Lynn. 1995. "A Technological Science Perspective for Sociology." *Rural Sociology* 60:607–622.

Warr, P., & Payne, R. 1982. "Experience of Strain and Pleasure Among British Adults." *Social Science and Medicine* 16:1691–1697.

Warren, Karen, ed. 1994. *Ecological Feminism.* London and New York: Routledge.

Warren, Karen. 1996. "Ecological Feminist Philosophies: An Overview of the Issues." Pp. ix–xxvi in *Ecological Feminist Philosophies,* edited by Karen Warren. Bloomington and Indianapolis: Indiana University Press.

Weber, Edward P. 2000. "A New Vanguard for the Environment: Grass-Roots Ecosystem Management as a New Environmental Movement." *Society and Natural Resources* 13:237–259.

Weber, Max. [1904–1905] 1958. *The Protestant Ethic and the Spirit of Capitalism.* New York: Charles Scribner.

Weber, Max. [1909] 1988. *The Agrarian Sociology of Ancient Civilizations.* London: Verso.

Weber, Max. [1922] 1967. *Economy and Society,* Vol. 1, edited by Guenther Roth & Claus Wittich. Berkeley: University of California Press.

Weinberg, Adam S. 1998. "The Environmental Justice Debate: A Commentary on Methodological Issues and Practical Concerns." *Sociological Forum* 13(1):25–32.

White, Lynn. 1967. "The Historical Roots of Our Ecological Crises." *Science* 155:1203–1207.

Wichterich, Christa. 1988. "From the Struggle Against 'Overpopulation' to the Industrialization of Human Production." *Reproductive and Genetic Engineering* 1(1):21–30.

Wilford, John Noble. 2000. "Open Water at Pole Not Surprising, Experts Say." *New York Times,* August 29. Retrieved February 2, 2004, from http://www.climateark.org/articles/2000/3rd/opwapole.htm

Williams, Jim. N.d. "Kentucky Woodlands Wildlife Refuge." *Kentucky National Wild Turkey Federation.* Retrieved December 10, 2007, from http://www.kentuckynwtf.com/history.html

Williams, Marilyn Marie. 2010. "Linking Health Hazards and Environmental Justice: A Case Study in Houston, Texas." *Dissertation Abstracts International, A: The Humanities and Social Sciences* 70(10): 4064.

Williams, Raymond, ed. [1972] 1980. "Ideas of Nature." Pp. 67–85 in *Problems in Materialism and Culture.* London: Verso.

Wilson, Harold Fisher. [1936] 1967. *The Hill Country of Northern New England.* New York: AMS Press.

Wilson, Randall, & Yaro, Robert D. 1988. *Dealing With Change in the Connecticut River Valley: A Design Manual for Conservation and Development.* Amherst: Center for Rural Massachusetts, University of Massachusetts.

Winchester, Paul D., Huskins, Jordan, & Ying, Jun. 2009. "Agrichemicals in Surface Water And Birth Defects in the United States." *Acta paediatrica* 98(4):664–669.

Wing, Steve, Richardson, David, Armstrong, Donna, & Crawford-Brown, Douglas. 1997. "A Reevaluation of Cancer Incidence Near the Three Mile Island Nuclear Plant: The Collision of Evidence and Assumptions." *Environmental Health Perspectives* 105(1):52–57.

Winner, Langdon. 1986. *The Whale and the Reactor: A Search for Limits in an Age of High Technology.* Chicago and London: University of Chicago Press.

Wolfers, Justin. 2008. "The Economics of Happiness, Part 1: Reassessing the Easterlin Paradox." *New York Times,* April 16. Retrieved January 13, 2011, from http://www.nytimes.com

Wolff, Edward N. 1995. *Top Heavy: A Study of the Increasing Inequality of Wealth in America.* New York: Twentieth-Century Fund Press.

Woodgate, Graham, & Redclift, Michael. 1998. "From a 'Sociology of Nature' to Environmental Sociology: Beyond Social Construction." *Environmental Values* 7:3–24.

World Bank. 2003. *World Development Report 2003: Sustainable Development in a Dynamic World.* New York: Oxford University Press.

World Bank. 2007a. "Facts and Figures from World Development Indicators 2007." Retrieved June 14, 2007, from http://siteresources.worldbank.org/DATASTATISTICS/Resources/reg_wdi.pdf

World Bank. 2007b. "Quick Reference Tables: GNI per capita 2005" (Atlas Method and PPP). Retrieved June 12, 2007, from http://go.worldbank.org/B5PYF93QF0

World Bank. 2007c. *World Development Report 2007: Development and the Next Generation.* Washington, DC: Author.

World Bank. 2009. "Protecting Progress: The Challenge Facing Low-Income Countries in the Global Recession." Retrieved January 11, 2011, from http://web.worldbank.org

World Bank. 2010a. Water and Development: An Evaluation of World Bank Support, 1997–2007. Washington, DC: World Bank.

World Bank. 2010b. *World Development Indicators: 2010.* Washington, DC: World Bank.

World Conservation Union. 2006. "West African Black Rhino Feared Extinct." Retrieved June 6, 2007, from http://www.iucn.org/en/news/archive/2006/07/7_pr_rhino.htm

World Health Organization (WHO). 2003a. "Obesity and Overweight: Global Strategy on Diet, Physical Activity, and Health Fact Sheet." Retrieved February 2, 2004, from http://www.who.int/hpr/gs.fs.obesity.shtml

World Health Organization. 2003b. "Road Traffic Injuries Fact Sheet." Retrieved September 12, 2003, from http://www.who.int/world-health-day/2004/en/

World Health Organization. 2003c. "20 Questions on Genetically Modified (GM) Food." Retrieved February 26, 2008, from http://www.who.int/foodsafety/publications/biotech/20questions/en/

World Health Organization. 2004. *World Report on Road Traffic Injury Prevention.* Geneva, Switzerland: Author.

World Health Organization. 2006a. "Obesity and Overweight Fact Sheet No. 311." Rome, Italy: Author. Retrieved December 19, 2007, from http://www.who.int/mediacentre/factsheets/fs311/en/print.html

World Health Organization. 2006b. *The WHO Recommended Classification of Pesticides by Hazard and Guidelines to Classification: 2004.* Corrigendum of June 28, 2006. Geneva, Switzerland: Author.

World Health Organization. 2007a. "Estimated Deaths and DALYs Attributable to Selected Environmental Risk Factors, by WHO Member State, 2002." Retrieved December 1, 2007, from http://www.who.int/quantifying_ehimpacts/countryprofilesebd.xls

World Health Organization. 2007b. *Indoor Air Pollution: National Burden of Disease Estimates.* Geneva, Switzerland: Author.

World Health Organization. 2009. *Protecting Health From Climate Change: Connecting Science, Policy and People.* Geneva, Switzerland: Author.

World Health Organization. 2010. *Trends in Maternal Mortality: 1990 to 2008.* Geneva, Switzerland: Author.

World Meteorological Organization. 2003a. "The Global Climate in 2002." *World Climate News* 23:4–5.

World Meteorological Organization. 2003b. "WMO Statement on the Status of the Global Climate in 2002" (WMO-No. 949). Geneva, Switzerland: Author.

World Meteorological Organization. 2003c. "WMO Statement on the Status of the Global Climate in 2003" (WMO-No. 702). Geneva, Switzerland: Author.

World Public Opinion. 2009. "Publics Want More Government Action on Climate Change: Global Poll." Retrieved January 25, 2011, from http://www.worldpublicopinion.org

World Values Survey. 2006. "Online Data Analysis." Retrieved November 11, 2007, from http://www.worldvaluessurvey.org

World Wind Energy Association. 2010. *World Wind Energy Report 2009.* Bonn, Germany: Author.

Worldwatch Institute. 2002. *Vital Signs 2002.* New York and London: Norton.

Worldwatch Institute. 2003. *Vital Signs 2003.* New York and London: Norton.

Worldwatch Institute. 2006. *Vital Signs, 2006–2007: The Trends That Are Shaping Our Future.* New York: Norton.

Worldwatch Institute. 2007. *Vital Signs, 2007–2008: The Trends That Are Shaping Our Future.* New York: Norton.

Worldwatch Institute. 2009. *Vital Signs 2009,* edited by Linda Starke. Washington, DC: Author.

Worthy, Trevor H., & Holdaway, Richard N. 2002. *The Lost World of the Moa: Prehistoric Life of New Zealand.* Bloomington: Indiana University Press.

Wright, Angus Lindsay. 1990. *The Death of Ramon Gonzalez: The Modern Agricultural Dilemma.* Austin: University of Texas Press.

WWF Australia. 2003. "Murray Darling Basin." Retrieved August 28, 2003, from http://www.wwf.org.au/default.asp

Yago, Glenn. 1984. *The Decline of Transit: Urban Transportation in German and U.S. Cities, 1900–1970.* Cambridge, MA, and London: Cambridge University Press.

Yearley, Steven. 1991. *The Green Case: A Sociology of Environmental Issues, Arguments, and Politics.* London: HarperCollins.

Yearley, Steven. 1996. *Sociology, Environmentalism, Globalization.* London: Sage.

Yearley, Steven. 2006. "Bridging the Science/Policy Divide in Urban Air-Quality Management: Evaluating Ways to Make Models More Robust Through Public Engagement." *Environment and Planning C: Government and Policy* 24:701–714.

Yearley, Steven, Cinderby, Steve, Forrester, John, Bailey, Peter, & Rosen, Paul. 2003. "Participatory Modeling and the Local Governance of the Politics of Air Pollution: A Three-City Case Study." *Environmental Values* 12:247–262.

Zavestoski, Stephen, Brown, Phil, & McCormick, Sabrina. 2004. "Gender, Embodiment, and Disease: Environmental Breast Cancer Activists' Challenges to Science, the Biomedical Model, and Policy." *Science as Culture* 13:563–586.

Zheng, T., Holford, T. R., Mayne, S. T., Ward, B., Carter, D., Owens, P. H., et al. 1999. "DDE and DDT in Breast Adipose Tissue and Risk of Female Breast Cancer." *American Journal of Epidemiology* 150:453–458.

Ziegler, Laura. 2006. "1986 Farm Protests." *KCUR News.* Retrieved December 14, 2007, from http://www.publicbroadcasting.net/kcur/news.newsmain?action=article&ARTICLE_ID=969223

Ziska, Lewis H., & Caulfield, Frances. 2000. "The Potential Influence of Rising Atmospheric Carbon Dioxide (CO_2) on Public Health: Pollen Production of Common Ragweed as a Test Case." *World Resource Review* 12(3):449.

Zwart, Ivan. 2003. "A Greener Alternative? Deliberative Democracy Meets Local Government." *Environmental Politics* 12(2):23–48.

Notes

Chapter 1

1. For an introduction to the literature on the realist–constructionist debate, see, on the realist side, Benton (1994, 2001a, 2001b); Dickens (1996); Dunlap (2010a, 2010b); Dunlap and Catton (1994); Martell (1994); Murphy (1994a, 1994b); and, on the constructionist side, Burningham and Cooper (1999); Cronon (1995b); Dupuis and Vandergeest (1996); Hajer (1995); Hannigan (1995); Mjoset (2001); and Yearley (1991, 1996). Within the United States, there was considerable debate over whether the position of Buttel, a leading figure in the field, was constructionist (for example, Buttel, 1992; P. Taylor and Buttel, 1992; see the discussion and critique in Dunlap and Catton, 1994). However, there have been a large number of attempts to reconcile the two sides, for example, M. Bell (1999); Buttel (1996); Carolan (2005); Lidskog (2001); Metzner (1997); Murphy (2004); Woodgate and Redclift (1998). I believe Burningham and Cooper can also be read as such an endeavor, but Dunlap (2010b) argues that it is not. In the past few years, there has also been much excitement among many environmental sociologists for the work of the French theorist Bruno Latour (1987, 1993, 1999, 2004, 2007) and his "actor-network" approach as a way to resolve the realist–constructionist debate. But some realists (for example, Dunlap) find his work too constructionist. See Chapters 3 and 8 for discussions of Latour.

2. The conceptual language of dialogue—or, more properly, of *dialogics*—is not common in sociology, although it is now common in the humanities. For an introduction to the concept of dialogics for the human sciences, including sociology, see Michael Bell and Gardiner (1998), which introduces the dialogic perspective of the Russian theorist Mikhail Bakhtin (1965/1984, 1981, 1986). For a discussion of the relevance of dialogism to creativity and possibility, see M. Bell (2011). The approach to environmental sociology I develop in the present book is heavily based on the work of Bakhtin, as well as that of the Brazilian dialogic theorist, Paulo Freire (1970/1993).

3. I give here a summary of a "holon" view of ecology, as sketched out in Bland and Bell (2007). But we don't need yet another term at this point in the book.

4. For an example, see Seidman (1994).

5. I mean *responsible* here in the sense of what I have elsewhere termed *response ability*—a word that encourages responses from others by giving those responses careful consideration (M. Bell 2004, 2011).

6. "Death Toll Rises" (2010).

7. Balmforth (2011).

8. This is the number of nuclear power plants in operation around the world as of January 10, 2011 (European Nuclear Society [2011]).

9. Federal Ministry for the Environment, Nature Conservation, and Nuclear Safety (2010), pp. 2–3.

10. World Wind Energy Association (2010).

11. AAE (Spanish Wind Energy Association) (n.d.).

12. Federal Ministry for the Environment, Nature Conservation, and Nuclear Safety (2010), p. 48.

13. Newport (2010). But the percentage has been falling in recent years, from 65 percent in 2008 to 53 percent in 2010.

14. Hansen et al. (2010).

15. Goddard Institute for Space Studies (2011).

16. Worldwatch Institute (2002), p. 51; Worldwatch Institute (2007).

17. Goddard Institute for Space Studies (2011).

18. Hansen et al. (2010).

19. Oulan (2003).

20. M. Bell (1985), 168.

21. Intergovernmental Panel on Climate Change (IPCC 2007b), Figure TS.5.

22. IPCC (2007a), Table SPM.3.

23. Worldwatch Institute (2003), 84.

24. IPCC (2007a), Figure SPM.7.

25. IPCC (2007a), Figure SPM.7.

26. IPCC (2007a), Figure SPM.7.

27. IPCC (2007a), Table SPM.2.

28. IPCC (2007a), Table SPM.2.

29. IPCC (2007a), Table SPM.2.

30. IPCC (2007a), Table SPM.2.

31. IPCC (2007a), Table SPM.2.

32. Charles Bullard (1997).

33. Long and Ort (2002); Paul (2003).

34. For example, the same study that found a 17 percent boost in soybean yields found a 20 percent drop due to ozone pollution, as discussed further on.

35. Buie (1995); Funk (1995); Klinenberg (2002), 9; MacKensie (1995).

36. World Meteorological Organization (2003b).

37. Bhattacharya (2003); "Death Toll" (2003); "France Ups Heat Toll" (2003); "French Minister Predicts" (2003); Tagliabue (2003).

38. Flaccus (2007).

39. Kim and Levitov (2010).

40. Worldwatch Institute (2009), 62, Figure 1.

41. Holland and Webster (2007).

42. Worldwatch Institute (2003), 93.

43. Worldwatch Institute (2003), 93; Worldwatch Institute (2007), 45; Worldwatch Institute (2009), 63.

44. Worldwatch Institute (2007), 45.

45. National Snow and Ice Data Center (2002); Revkin (2006).

46. National Snow and Ice Data Center (2010).

47. Kwok and Rothrock (2009).

48. Wilford (2000).

49. L. Smith (2006).

50. Hileman (1995).

51. Ziska and Caulfield (2000).

52. World Health Organization (WHO 2009), 10–11.

53. IPCC (2007a), 14.

54. Earth System Research Laboratory (2011).

55. NASA (2003).

56. Tans (n.d.).

57. IPCC (2007a), Table SPM.2.

58. IPCC (2007a).

59. Imbrie and Imbrie (1979), 178.

60. Unless, of course, your local utility uses nuclear or hydroelectric power. But these have problems of their own.

61. National Weather Service (2003).

62. C. S. Long et al. (2005), and the National Oceanic and Atmospheric Administration (NOAA 2005).

63. On cancer rates, see Intergovernmental Panel on Climate Change (2002b, 12.7.1). My children experienced the sun health classes when they attended school in New Zealand in 2001.

64. Worldwatch Institute (2002), 54.

65. On increases in ultraviolet radiation, see Stevens (1995).

66. Scientific Assessment Panel of the Montreal Protocol (2010).

67. Lohbeck (2004).

68. Greenpeace (2003).

69. Environmental Protection Agency (EPA 2007); Worldwatch Institute (2002), 54.

70. Mauzerall and Tong (2006).

71. M. Bell (2004).

72. Hudson et al. (2006).

73. Long et al. (2005); Paul (2003).

74. Karmel, FitzGibbon, and Cave (2002), citing an EPA study.

75. Mauzerall and Tong (2006).

76. Suplee (1995). See also Samet et al. (2000) in the *New England Journal of Medicine,* which found a similar result.

77. Compare G. Morgan et al. (1998) with Samet et al. (2000).

78. Gauderman et al. (2007).

79. Dominici et al. (2006).

80. Pearce (1995) for percentage of forests damaged, and European Commission (2000) for figures on defoliation.

81. Wang et al. (2007).

82. Environment Canada (2003).

83. National Atmospheric Deposition Program (n.d.)

84. General Accounting Office (2000).

85. General Accounting Office (2000).

86. European Commission (2000).

87. "Acid Rain Problem" (2000).

88. Freemantle (1995).

89. Environment Canada (2003).

90. Based on the figures in Pimentel (2006); but see the debate of Pimentel and Skidmore with Trimble in the November 19, 1999, issue of *Science.*

91. This is my calculation based on figures from the Natural Resource Conservation Service (2010), which reports that average soil erosion by wind and water in the United States was 4.8 tons per acre in 2007, or 9,600 pounds. Average corn yields in the United States are now trending up to about 150 bushels an acre, or 8,400 pounds (U.S. Department of Agriculture [USDA] 2008), as a bushel of corn is 56 pounds. So, actually it takes a bit more than a bushel of erosion to grow a bushel of corn most years.

92. Natural Resources Conservation Service (2010).

93. G. T. Miller (2005), 279.

94. World Bank (2003), 2.

95. All figures from United Nations Environment Programme (UNEP 2003), Table 28.

96. Rengasamy (2006).

97. World Bank (2010a) v.

98. United Nations Development Programme (UNDP 2006), Table 6.3.

99. I. Anderson (1995) gives a figure of one quarter; WWF (Worldwide Fund for Nature) Australia (2003) gives a figure of 21 percent.

100. N. Jones (2003); G. T. Miller (1994), 347; Stone (1999).

101. European Space Agency (2009).

102. European Space Agency (2009).

103. Pala (2006).

104. L. Brown, Lenssen, and Kane (1995), 123.

105. Columbia Water Center (2010).

106. G. T. Miller (1994), 351; UNEP (2003); Worldwatch Institute (2006), 104.

107. L. Brown, Renner, and Halweil (2000), 122.

108. UNEP (2003), 19.

109. L. Brown et al. (1995), 123; Worldwatch Institute (2006), 104.

110. United Nations Educational, Scientific and Cultural Organization (UNESCO; 2004).

111. UNESCO (2004), 122.

112. UNESCO (2004), 123.

113. UNEP (2003), 89.

114. Meadows, Meadows, and Randers (1992), 56.

115. Meadows et al. (1992), 54–56.

116. L. Brown et al. (1995), 41; Fresco (2003); G. T. Miller (1994), 362.

117. Kiely, Donaldson, and Grube (2004); data through 2001. This is a U.S. EPA study that the agency at one time repeated every 2 years but now appears to have been cancelled.

118. American Farmland Trust (2002). I am, however, no longer as confident about these numbers as I was in earlier editions of this book, as American Farmland Trust now reports substantially different figures than it did in its 1994 publication, my earlier source.

119. Worldwatch Institute (2006), 23.

120. Even Fidel Castro and the *Economist* agree on this one, as "On Ethanol, Castro Is Right, Says the *Economist*" (2007), shows.

121. Essential Action and Global Exchange (2000); Vidal (1995).

122. A. Sachs (1996).

123. A. Sachs (1996).

124. A. Sachs (1996).

125. Johansen (2003).

126. These events all took place in 2010, as reported in online issues of the *Ogoni Star*.

127. Shell Oil, belatedly, also agrees and has now modified its statement of business principles to include specific reference to human rights (Beavis and Brown 1996). However, it has only a spotty record of following through on its human rights commitment, such as the way it handled the April 29, 2010, oil blowout, taking days to stop it and months to even begin to clean it up, and its handling of the 2007 oil fires.

128. Here I am following, in modified form, the distinction Beck (1995) makes between "goods" and "bads." Beck's emphasis, however, is actually on the increasing equality he sees in the distribution of "bads," which is a central feature of what he terms the "risk society" of late modern societies. I review critiques of this conclusion in Chapter 9.

129. B. Goldman and Fitton (1994).

130. Mohai and Bryant (1992).

131. Boerner and Lambert (1995).

132. Heiman (1996); B. Goldman (1996).

133. Davidson and Anderton (2000). An earlier and very controversial study by this research team found no association with race at all and only a slight association with social class, as measured by blue-collar employment (Anderton et al. 1994).

134. cf. Downey (1998); Weinberg (1998).

135. Animashaun (2006); Cassidy, Judge, and Summers (2000); Daniels and Friedman (1999); Pastor, Morello-Frosch, and Sadd (2005); Stretesky and Hogan (1998); see also Hines (2001); Pastor, Sadd, and Hipp (2001); Pastor, Sadd, and Morello-Frosch (2002); Pine, Marx, and Lakshmanan (2002). Note that the last four did not study the question of class explicitly.

136. Denq, Constance, and Joung (2000); Derezinski, Lacy, and Stretesky (2003); Krieg (2005); Stretesky and Lynch (1999); Taquino, Parisi, and Gill (2002); see also Evans and Kantrowitz (2002) and Mennis (2002). Note that the last two did not study race explicitly.

137. Bolin, Grineski, and Collins (2005); Caron-Sheppard and Johnson (2003); Davidson and Anderton (2000); Downey (1998); Edwards and Ladd (2000); Faber and Krieg (2005); Fletcher (2004); Krieg (1998); Mitchell, Thomas, and Cutter (1999); Morello-Frosch, Pastor, and Sadd (2001); Saha (2002). Note that Davidson and Anderton found that class was far more significant, and the association they found with African Americans living in nonmetropolitan areas was not statistically significant at the 95 percent confidence level statisticians typically require to feel sure about their findings.

138. Grant et al. (2010); Hipp and Lakon (2010); Konisky and Schario (2010); Legot et al. (2010); Malin and Petrzelka (2010); Platt (2010); Pritchard (2010); Schoenfish-Keita and Johnson (2010); Voyles (2010); Williams (2010).

139. Pastor et al. (2002).

140. Pine et al. (2002).

141. Downey (1998).

142. Taquino et al. (2002).

143. Faber and Krieg (2005).

144. Evans and Kantrowitz (2002). It is sometimes asked which came first: Do the disadvantaged move into communities that are already contaminated, or are contaminants disproportionately sited in disadvantaged communities? There is, as yet, little research on this question, aside from Pastor et al. (2001), which found that disproportionate siting matters more than disproportionate move-in.

145. Schoenfish-Keita and Johnson (2010).

146. Hilz (1992).

147. World Bank (2010b), Table 1.1.

148. World Bank (2010b), Table 1.1.

149. UNDP (1994), Figure 2.6.

150. UNDP (1999), 3.

151. My figure is based on 2005 data from the World Bank (2007a), calculated by my research assistant Christine Vatovec for the third edition of this book. Given the Great Recession, the situation is almost certainly considerably worse today. Note that the UNDP (1999) gives a figure of 74 to 1 for 1997, the last year for which they have reported this item, but our methodology may have differed in a few respects. Christine derived the figure by first ranking countries by gross national income (GNI) per capita and then adding up their populations from the top and bottom until she arrived at one fifth of the world's population in 2005, and then adding together the GNI of each fifth. (This required taking in only a portion of the population of India, as it lies precisely at the cusp of the lowest 20 percent, and taking in its entire population would well exceed 20 percent of the world's population. Here, she added in the same proportion of the Indian GNI as the proportion of the population included.) I suspect that the UNDP figures reported higher inequality because they were based on gross national product (GNP, which includes overseas earnings) rather than GNI (which subtracts out indirect business taxes such as sales taxes), as GNI tends to lower the income of wealthy nations and to increase the income of poor nations, relative to GNP. In the second edition of this book, I gave a figure of 68 to 1, but that was based on GDP (gross domestic product), not GNI, which parses the data differently yet again.

152. UNDP (2003), Table 13, gives the figures for 107 countries around the world, and 48 of those round to 6 to 1 or less, and 60 to 7 to 1 or less.

153. UNDP (2003), Table 13.

154. Schaller (2010).

155. Wolff (1995), 21. The figures Wolff gives for the 1920s are for inequality in wealth, rather than income, presenting some problems of comparability with the figures I cite for the 1990s. In general, wealth inequalities are greater than income inequalities. However, the societies that are high in one are almost always high in the other, allowing the historical comparison to be made.

156. UNDP (2003), Table 13.

157. UNDP (2003), Table 13. For example, consider the ratio of income between the top and bottom 20 percent in the following: 5.9 to 1 in Jordan, 4.8 to 1 in Kazakhstan, 5.2 to 1 in Indonesia, and 4.8 to 1 in Pakistan. Turkey is a bit higher at 7.7 to 1, but still below the United States. Attention to the needs of the poor is a centuries-old Islamic tradition and is enshrined in the custom of *Zakat*, which means "giving alms," one of the Five Pillars of Islam. However, a number of Muslim countries in Africa have very high levels of inequality.

158. Korten (1995), 108; UNDP (1992). This is an extremely difficult number to gauge with any certainty.

159. World Bank (2010), 14; on the likely increase as a result of the Great Recession, see World Bank (2009).

160. World Bank (2010), 14.

161. World Bank (2010), 15.

162. World Bank (2007c), 6.

163. PolitiFact (2011).

164. Davies at el., (2008).

165. Miller and Kroll (2010).

166. World Bank (2007a), Table 1.

167. Comparing the $35 billion Slim was worth in 2009, according to the annual listing in *Forbes* magazine.

168. World Bank (2007a), Table 1.

169. Durning (1992), 50, Table 4-1.

170. To use the economic term, the demand for some of these items is *inelastic*—that is, there are limits to how much it can fluctuate, and it probably does not increase at the same rate as wealth.

171. UNDP (2003), 10.

172. Food and Agriculture Organization (FAO 2010a). Recent quarterly reports have, at this writing, always found at least 29 countries in need of external food assistance.

173. FAO (2010b), 9 and Table 1.

174. World Bank (2007c), Table 3.

175. World Bank (2007c), Table 3.

176. FAO (2006b), 18.

177. FAO (2005, 2006b).

178. World Bank (2010) 20.

179. United Nations Human Settlements Program (2003), xxv.

180. World Bank (2010) 21, figure for 2006.

181. World Bank (2010) 20, figure for 2006.

182. National Health Service (2010) 13.

183. National Health Service (2006, 2010).

184. Ogden and Carroll (2010a), Table 1.

185. Ogden and Carroll (2010a), Table 2, figures for those age 20 to age 74; Ogden and Carroll (2010b) Table 1, figures for children.

186. WHO (2006a).

187. Worldwatch Institute (2006), 120.

188. Worldwatch Institute (2006), 120.

189. WHO (2003a).

190. WHO (2003a).

191. UNDP (2010), Table 1, comparing the "least developed countries" with the "OECD (Organisation for Economic Co-operation and Development)" countries.

192. UNDP (2010), Table 1.

193. UNDP (2010), Table 14.

194. National Coalition for the Homeless (2009).

195. Toro et al. (2007).

196. Edgar and Meert (2006), data matrixes 1 and 2.

197. USDA (2011).

198. Anthanasiou (1996).

199. President's Cancer Panel (2010), ii.

200. Houlihan (2003).

201. Cone (2010).

202. President's Cancer Panel (2010), Letter to the President and ii.

203. Cone (2010).

204. Leopold ([1949] 1966), 262.

205. The first edition termed it "the beauty of nature," and the second and third editions termed it "the rights and beauty of habitat." It seems to me that "rights" is already implied with the notion of community that the word "ecology" embodies, so I have dropped it.

206. International Union for the Conservation of Nature (IUCN; 2010).

207. Worldwatch Institute (2003), 82.

208. Worthy and Holdaway (2002).

209. IUCN (2010).

210. IUCN (2010).

211. Leakey and Lewin (1996).

212. Worldwatch Institute (2006), 102, citing World Resources Institute estimates.

213. FAO (2010c).

214. FAO (2010c).

215. M. Bell and Laine (1985).

216. In the first two editions, I called this notion the "social organization of daily life." But on reflection, that may come across as one-sidedly materialistic. So in the third edition, I switched to the word *constitution* to represent the practices that come equally out of our material organization and the construction of our ideas.

Chapter 2

1. This number is disputed; some sources give a more modest figure of 350 million record sales.

2. O'Brien (2006).

3. O'Brien (2006).

4. Lewis (2005) and various flabbergasted sources on the web.

5. Julien's Auctions (2009).

6. On Jackson selling Neverland to himself, see Sixbear (2008).

7. Cited in Guha (1995).

8. Marx ([1844] 1972), 58; emphasis in the original.

9. Marx ([1844] 1972), 58.

10. In his more considered moments, Karl Marx recognized this interplay as well, offering the term *dialectics* to describe it. Drawing on Hegel, Marx described dialectics as an endless cycle of movement from thesis to antithesis to synthesis, with any synthesis becoming the thesis to which the next antithesis responds. Such an account, however, over-polarizes the explanation of social change. The concept of dialogue, which I take from the writings of the Russian social theorist Mikhail Bakhtin (1981, [1965] 1984, 1986), is an improvement because it emphasizes the mutual conditioning of social factors—a process that is not necessarily oppositional and polarized. For a more detailed critique of the concept of dialectics and a fuller explanation of the analytical advantages of dialogue, see Gardiner (1992). See also the closely related views of Weber ([1904–1905] 1958, p. 277, n. 84).

11. For a review, see Parker (1996).

12. Koestler (1967), 210.

13. In M. Bell et al. (2011), I call this invigorating delight in the creativity of the unexpected *diaesthesia*.

14. Maslow ([1943] 1970).

15. Maslow ([1943] 1970), 38.

16. Maslow ([1943] 1970), 100.

17. This point is also made by Inglehart (1977, 1990). For a detailed critique of Inglehart's particular application of this point, however, see Chapter 6 of this volume.

18. Sahlins (1972).

19. Sahlins (1972), 11.

20. Sahlins (1972), 4.

21. Sahlins (1972), 27.

22. Gusinde ([1937] 1961), 86–87, cited in Sahlins (1972), 13. The Yahgan called themselves the "Yamana" but have now virtually disappeared from the world. As of 1990, there remained only four native speakers of their language, four women living on Navarino Island in Chile's section of the Cape Horn region. For an excellent video on the subject, see *Homage to the Yahgans* (1990).

23. Sahlins (1972), 11.

24. Sahlins (1972), 37.

25. For more on the social creation of the concept of being poor, see Norberg-Hodge (2009).

26. Veblen ([1899] 1967).

27. For more on Veblen as an environmental sociologist, see R. Mitchell (2001).

28. Veblen ([1899] 1967), 83.

29. Hirsch (1977), 20.

30. Twain ([1876] 1991), 20.

31. See Katz (1994); R. Wilson and Yaro (1988).

32. Handwerk (2006). Also see the Greenpeace "seafood red list" of species, which are commonly sold in supermarkets and restaurants but have a high risk of being unsustainably harvested.

33. Fickling (2003).

34. Milliken, Emslie, and Talukdar (2009).

35. "Price of Rhino Horn Upstages Gold" (2009).

36. Milliken et al. (2009).

37. On the illegal trade in rhino horns, including a spectacular 1996 bust of a garage in Kensington, England, with 107 horns, see Streeter (1996).

38. Milliken et al. (2009) report that the population of white rhinos has tripled since 1991 to about 17,500 and the population of black rhinos has nearly doubled to about 4,200, by the end of 2007. But there are only a few hundred Sumatran rhinos and less than 50 lesser one-horned rhinos. On the extinction of the western black rhino, see World Conservation Union (2006).

39. Riddle (1997), 44–46. Riddle reports that, although silphium is now extinct, laboratory experimental tests on rats with other species of fennel have found them to be as high as 100 percent effective in preventing pregnancy if administered within 3 days of coitus.

40. Pliny the Elder (1855), XIX.15; Riddle (1997), 44–46.

41. For a discussion of the relationship among the ideas of Durkheim, Weber, and Tönnies on these points, see M. Bell (1998a).

42. Csikzentmihalyi and Rochberg-Halton (1981).

43. I take this and the following three paragraphs, in modified form, from M. Bell (1998a).

44. M. Bell (1998a).

45. I draw much of the section that follows from M. Bell (1997).

46. I take this quotation (which also appears in M. Bell, 1997) from Mauss ([1950] 1990), 11. I have also translated into English a few words left in Maori in the original and put back into Maori one word originally left in English.

47. Mauss ([1950] 1990), 20.

48. Douglas and Isherwood ([1979] 1996), 39.

49. There is a huge debate on just how effective ads are and how they are interpreted by the public. The classic work that refutes a Pavlovian response to ads is Schudson (1984).

50. Schudson (1984).

51. I quote this ad from an issue of the *International Herald Tribune* from sometime in the fall of 1996.

52. As advertised on www.gaiam.com on January 14, 2011.

53. For example, see Anthanasiou (1996).

54. Robinson (2007).

55. See the *Multinational Monitor*'s annual list of the "Ten Worst Corporations," which usually includes ExxonMobil.

56. ExxonMobile (n.d.).

57. For the observation about the similarity of BP's logo to that of the Canadian Green Party, see http://greenwashing .webs.com.

58. Landman (2010).

59. Wachtel (1983), 62, 64.

60. Schor (1992), 45.

61. Schor (1992); on other industrialized countries, see "Traffic" (1996); Huberman and Minns (2005), 26.

62. Schor (2010). The situation has been getting worse. In 1987, it was 163 hours more than in 1967, according to Schor (1992), 29.

63. Schor (1992) 32, 82, Table 3.2; Huberman and Minns (2005), 27.

64. Figures from the U.S. Bureau of Labor Statistics (1981, 2010).

65. For the figures on these, see Durning (1992), 43, 132, citing surveys by others, and Putnam (2000).

66. K. Fisher and Layte (2004).

67. On work time, see "Traffic" (1996); for the rest, see Knight and Stokes (1996).

68. Tomorrow Project (2007).

69. Schor (1992), 2.

70. For example, Douglas and Isherwood ([1979] 1996).

71. Douglas and Isherwood ([1979] 1996), 39.

72. For an account of class-bounded patterns of fellowship in Britain, see M. Bell (1994a); for the United States, see Rubin (1994).

73. Warr and Payne (1982), cited in Argyle (1987), 92.

74. Argyle (1987), 93, summarizing several studies.

75. Easterlin (1973) and Easterlin et al. (2010).

76. However, Gallup itself does not interpret the results this way, seeing their data as disconfirming the "Easterlin paradox" (Deaton 2008). Wolfers and Stevenson (2008) make the same point, and the *New York Times* and other papers in 2008 published this view widely. The *New York Times* even published a highly misleading graph of the Gallup data on a logarithmic scale that de-emphasized difference in results at higher incomes. Easterlin et al. (2010) present other data disputing Wolfers and Stevenson.

77. Schor (1992), 115, summarizing many surveys.

78. Pew Research Center (2006).

79. I take the widely used metaphor of a treadmill from Brickman and Campbell (1971); Durning (1992), p. 39; Schor (1992), p. 125; and Wachtel (1983); the phrase the "treadmill of consumption," however, originates with me, as far as I know.

80. Gallup Poll (1989), cited in Schor (1992), 126.

81. Bultena, Hoiberg, and Bell (1995).

82. Erikson (1976).

83. Erikson (1976).

Chapter 3

1. Hill (1988), cited in Irwin (1995), 2.

2. Schudson (1984) is the best account of our resistance to advertising, although he overstates our ability to resist because he concentrates on resistance to individual ads as opposed to the cumulative impact of being surrounded by ads constantly.

3. Korten (1995), 37.

4. Jacobs (1991).

5. M. Bell (2009).

6. Merton ([1968] 1973).

7. Douthwaite (1992), 18.

8. Closely related ideas can be found in the work of Schnaiberg (1980), Schnaiberg and Gould (1994), Galbraith (1958), Cochrane (1958), O'Connor (1973), and others.

9. See Chapter 1 for a discussion of growing inequality. During the late 1990s, income for most workers in the United States, at least, did indeed pick up for a time—only to be reversed in the early 2000s. More recently, the Great Recession has increased inequality even more.

10. On the rate of corporate profit, see Morin and Berry (1996).

11. Logan and Molotch (1987).

12. Logan and Molotch (1987).

13. Novek (1995).

14. Jacobs (1991), 25.

15. E. Becker (2002).

16. O'Connor (1988, 1991, 1998).

17. Foster (1999) and Foster, Clark, and York (2009).

18. NOAA (2010).

19. Lack, Short, and Willock (2003).

20. Food and Agriculture Organization (FAO 2007), 29.

21. Jevons ([1865] 2001).

22. Jevons ([1865] 2001), 100–103.

23. Fisher and Freudenburg (2004); Freudenburg (2005); Freudenburg and Nowak (2000); Nowak, Bowen, and Cabot (2006).

24. Freudenburg (2005).

25. The original source of the term is Freudenburg and Nowak (2000).

26. Freudenburg (2005).

27. Foster (2005).

28. Fischer and Dornbusch (1983), 14. For an overview of critiques of this idea, see Massey, Magaly Sanchez, and Behrman (2006).

29. I ask this question along with my colleague Philip Lowe in M. Bell and Lowe (2000). The argument in this section is mainly drawn from that paper.

30. Berlin [1958] 1969.

31. Including Berlin himself.

32. M. Bell and Lowe (2000).

33. I got these figures directly from the Office of the Federal Registrar. It is no longer possible to update these figures, though, as the Federal Registrar has reorganized the Federal Code into 50 titles and no longer keeps track of how many pages there are in all. I cannot help having a suspicion that someone took note of the political implications of the Code's length increasing despite supposed deregulation.

34. M. Bell (1995).

35. For example, see Cochrane (1958) and Rifkin (1995).

36. Warner and England (1995).

37. Latour (1987, 2007).

38. This phrasing of the "pragmatic maxim" of Charles Sanders Peirce is actually typically attributed to Alfred North Whitehead, who was not strictly speaking a pragmatist. But it probably predates Whitehead. Writing before Whitehead was well-known, and with no mention of Whitehead, Schiller ([1903] 1912) attributes the phrase to Aristotle.

39. It is worth noting, however, that ANT is often accused of insufficiently taking into account the political side of things.

40. Plunkett Research (n.d.).

41. Bureau of Transportation Statistics (2003).

42. On the size of the U.S. fleet in 2008, Bureau of Transportation Statistics (n.d.); on the number of licensed drivers in the United States in 2006, Federal Highway Administration (n.d.); on number of cars in the United States per 1,000 people, as of 2008, U.S. Department of Energy (2010).

43. Durning (1992), 80–81; French Road Federation (2006).

44. U.S. Department of Energy (2010).

45. Brown, Flavin, and Kane (1996), 84; Brown, Renner, and Flavin (1997), 74; Jianhua (2003).

46. Fontanella-Khan (2011) and "Numerology" (2001).

47. Gott (2008).

48. The precise figure for 2009 is 33,808 (National Highway Traffic Safety Administration [NHTSA] n.d.). But there has been fantastic improvement in this annual total, which was running around 40,000 or more as recently as 2007. Annual AIDS deaths are around 18,000.

49. Centers for Disease Control (2007d); National Center for Injury Prevention and Control (2007).

50. NHTSA (2010). Again, the good news is there has recently been substantial improvement in these figures, although they remain horrific.

51. Brown, Kane, and Roodman (1994), 132–133; International Road Traffic and Accident Database (IRTAD, 2003); National Safety Council figures.

52. World Health Organization (WHO 2004).

53. WHO (2003b, 2004).

54. "Roads Claim" (1996); Mead (1997). Mead estimates the road deaths of British birds to be between 3 and 60 million.

55. Royal Commission on Environmental Pollution study, cited in "Traffic" (1996).

56. Cited in Fischlowitz-Roberts (2002).

57. Shprentz (1996).

58. Total air pollution deaths are 2.6 million per year, but the majority of them are from indoor pollution due to the use of indoor open cooking fires in much of the developing world (WHO 2007b).

59. Energy Information Administration (2007).

60. Mead (1997); Reijnen, Foppen, and Braak (1995).

61. There was a spate of these studies in the 1990s. I cite the figures from perhaps the most comprehensive, albeit not the one with the biggest numbers: MacKensie, Dower, and Chen (1992).

62. MacKensie et al. (1992) here and for the subsidy numbers in the following paragraphs.

63. Trips figure from Bureau of Transportation Statistics (2003); parking spaces figure from MacKensie et al. (1992).

64. For reviews, see Alvord (2000); Duany, Plater-Zyberk, and Speck (2001); and Kay (1998). On the price of the Iraq War, see Stiglitz and Bilmes (2010).

65. NHTSA (2010).

66. I calculated the per car figure by dividing the $300 billion by the 190-million size of the U.S. car fleet at the time of the MacKensie et al. (1992) study, yielding $1,578.95 per car. I calculated the per gallon figure by dividing the 2 trillion miles Americans annually drove at that time by the fleet size, yielding 10,526.3 miles a year. Dividing this figure by a generous 25 (the mileage per gallon of the U.S. car fleet was actually around 21 at the time, and is now around the 22–23 mark) yields an average annual fuel use of 421.1 gallons. Dividing 421.1 into $1,578.95 yields $3.75.

67. Hart and Spivak (1993).

68. Yago (1984), 56–69.

69. Yago (1984), 60.

70. Merton (1948).

71. Winner (1986), 10.

72. Schutz (1962), 17.

73. On recipe knowledge, see Schutz (1962), xxix and 13–14.

74. Feder and Revkin (2000).
75. Mumford (1934), 365.
76. Mumford (1934), 6.

Chapter 4

1. Malthus ([1798] 1993).
2. Engels ([1844] 1987), 104.
3. For example, Wichterich (1988), cited in Duden (1992), 157. See discussion following.
4. G. Becker (1986). Similar views have been expressed by Simon (1981, 1995); Boserup (1965) represents a more moderate position. See later discussion in this chapter.
5. See, for example, Brown and Kane (1995); Catton (1982, 2008); McKibben (1998); and Meadows, Meadows, and Randers (2005).
6. Catton (2008), 8.
7. U.S. Census Bureau (2010).
8. U.S. Census Bureau (2010).
9. U.S. Census Bureau (2010).
10. Independent Commission on Population and Quality of Life (1996), frontispiece.
11. Brown and Kane (1995), 51. Because of annual fluctuations, a more precise figure is not possible.
12. Central Intelligence Agency (CIA n.d.), estimates for 2010.
13. CIA (n.d.), estimates for 2010.
14. CIA (n.d.), estimates for 2010.
15. I thank my son Sam for these calculations.
16. I thank my daughter, Eleanor, for these calculations.
17. United Nations Population Division (2009), Table II.1.
18. U.S. Census Bureau (2010).
19. U.S. Census Bureau (2010).
20. Catton (1982).
21. See Chapter 3.
22. Lappé and Schurman (1988).
23. Independent Commission on Population and Quality of Life (1996), 15. See that report for a more extensive discussion on this issue.
24. Stein (1995), 133.
25. Earth Policy Institute (n.d.).
26. Barraclough (1982), 102; McEvedy (1995).
27. For a more cynical, and I believe less accurate, interpretation, see Sachs (1992).
28. For example, see Lipset (1959, [1960] 1981).
29. Quoted in Esteva (1992), 6.
30. Brown, Lenssen, and Kane (1995), 72.
31. Based on UNDP (2010), Table 15. The precise figure is 16.4 percent, based on the average for the 39 "low human development" countries for which the table lists this figure.
32. For details, see Chapter 1.
33. UNDP (2010), Table 16.
34. UNDP (2010), Table 16.
35. Based on data from USDA Economic Research Service (2010), in constant 2005 dollars.
36. As of 1993, the World Bank had cleared $14 billion in profit for its investors (George and Sabelli 1994: 14). I have not been able to locate a more up-to-date figure, however.
37. World Bank (2010b), Tables 6.10 and 6.11.
38. World Bank (2010b), Table 6.11.
39. World Bank (2010b), Table 6.11.
40. UNDP (2010), Table 16.
41. For an introduction, see Hopkins and Wallerstein (1982) and Frank (1969). For a recent summary, see Hall (1996).

42. J. Cohen and Centano (2006), Table 9.

43. Brown et al. (1995), 72.

44. J. Walton (1994).

45. Herbert (1997).

46. Stiglitz (2002). See also Bruno and Squire (1996), an op-ed article by two high-ranking members of the World Bank's staff, arguing that inequality slows economic growth.

47. Schweikart (1998, 2002).

48. J. Cohen and Centano (2006).

49. Lappé and Shurman (1988), 12, Table 1.

50. Lappé, writing in 1977, used the figure of 2 pounds of grain or 3,000 calories a day, almost exactly what I came up with using the grain production figures for 2009 (FAO 2010a) and the U.S. Census Bureau (2010) figures for midyear world population. I use the ratio of grain to calories reported in Worldwatch Institute (2007), 20–21.

51. Brown et al. (1994), 66–67.

52. Brown et al. (1994), Table 4-1.

53. Lappé and Schurman (1988), 14.

54. Lappé and Schurman (1988), 21.

55. Sen (1981).

56. Sen engages in a little sleight of hand here, in my judgment. Because the third rice harvest, the *aman* harvest, lasts from November to January, it is difficult to decide how to divide its production between the 2 years. This is an important issue because Sen bases his food-availability figures on annual grain production. Unaccountably, however, Sen includes the *aman* crop entirely in the year in which the January part of the harvest falls, despite the fact that two-thirds of the harvest period is in the previous year (see Sen 1981, 137.) This allows him to effectively extend the size of the 1974 harvest by including the entire 1973–1974 *aman* crop in 1974 and displacing the entire 1974–1975 *aman* crop, which was badly damaged by the floods, into 1975. As a result, his figures for food availability in 1974 are suspiciously high.

57. Kinealy (1996).

58. Sen (1981), 129.

59. For example, see Devereux (1993), Sarre and Blunden (1995), and even such well-known neo-Malthusians as Meadows et al. (1992).

60. Brown and Kane (1995), 56; 1993 figures.

61. See Devereux (1993), 76–82, for an overview.

62. Stewart (1982), cited in Devereux (1993), 79.

63. Ehrlich (1968).

64. See Sarre and Blunden (1995), pp. 70–71, for a full account.

65. Simon (1981). See also Simon and Kahn (1984) and Simon (1995).

66. See Simon (1995).

67. See Chapter 1 for details.

68. As James Scott (1976) has argued, the poor tend to be averse to risk because their margin is already so tight to begin with.

69. Brown and Kane (1995), 22.

70. Brown and Kane (1995), 38.

71. IRRI (International Rice Research Institute) (2010). I used the USDA figures provided by IRRI, and compared it with the midyear world population figures of the U.S. Census Bureau, using the years 1986 and 2006 as the midpoints of the 1984–1988 and 2004–2008 periods.

72. Worldwatch Institute (2007).

73. Brown et al. (1996), 24.

74. For evidence that sustainable methods can boost production in developing countries by 79 percent, see Pretty et al. (2006).

75. Boserup (1981).

76. Boserup (1965).

77. Boserup (1981), 3.

78. See Scott (1986) for a study of the displacement of labor and subsequent impoverishment through agricultural intensification in Malaysia.

79. Boserup (1965), 118.

80. Devereux (1993) makes this point very well.

81. Barclay (2007).

82. Lash (2005).

83. Barclay (2007) reports that in 1994, the WHO ozone standard was exceeded on 340 days, as opposed to the 209 in 2006.

84. U.S. Census Bureau (2007b).

85. Jarvie (2007).

86. Notestein (1945).

87. Schultz (1981).

88. Stein (1995), 10.

89. Teitelbaum (1987), 31.

90. For a more detailed account of all of these, see Teitelbaum (1987).

91. Sachs (1992), 3–4.

92. Norberg-Hodge (2009).

93. Brown (2011); Brown and Kane (1995); Pretty (2002).

94. For a similar conclusion, see Teitelbaum (1987) and Sarre and Blunden (1995).

95. Pietila and Vickers (1994), 14.

96. Pietila and Vickers (1994), 15.

97. Boserup ([1970] 1989), 35.

98. UNDP (2003), Table 26.

99. Lappé and Schurman (1988), 25–26.

100. Stein (1995).

101. Walby (1996).

102. Moghadam (1996), 1.

103. Joekes (1987) and Standing (1989), cited in Moghadam (1996), 1.

104. Stein (1995); Moghadam (1996); Pietila and Vickers (1994).

105. WHO (2010).

106. I take this history from Stein (1995), 129–146, who should be consulted for a more detailed account.

107. Wichterich (1988), cited in Duden (1992), 157.

108. Ehrlich and Ehrlich (1990), 17. A similar emphasis on birth control can be found in Ehrlich (1968).

109. For examples of this critique, see Mamdani (1972) and Hartmann (1987), cited in Duden (1992), 156.

110. I take this account from Stein (1995), 20.

111. Cottrel (1955), 2; I have paraphrased Cottrel to eliminate his gender-specific usage.

Chapter 5

1. I give this account of the Bhopal tragedy based on Chauhan (1996); Delhi Science Forum (1984); Morehouse and Subramaniam (1986); and Weir (1986).

2. Timings from Morehouse and Subramaniam (1986).

3. Associated Press (2009).

4. Delhi Science Forum (1984) said a "conservative" figure was 5,000. Morehouse and Subramaniam (1986) agree, but report that some estimates put the figure at 10,000. Kurzman (1987) says 8,000. Websites say if those who died years afterward from the effects are included, the figure rises to 20,000 to 30,000.

5. One standard figure reported is 200,000, and some sources, like Kurzman (1987), put the figure at 300,000. But a government affidavit filed in a court case in 2006 put the total number of injuries at 558,125 (Dubey 2010).

6. I take the following quotations from Eklavya (1984).

7. "Union Carbide's Factory in India" (2001).

8. In February 2001, Union Carbide merged with Dow Chemical Company. Dow and Union Carbide now both claim that because of the merger, Dow has no liability in regard to what happened in Bhopal on December 3, 1984.

9. New Delhi TV (2010).

10. Klinenberg (2002), p. 11, goes on to note, "We can collectively unmake them, too, but only once we recognize and scrutinize the cracks in our social foundations that we customarily take for granted and put out of sight."

11. For more on this topic, see the discussion in Chapter 6, of Bakhtin's concept of the "classical body."

12. M. Bell and van Koppen (1998).

13. Stiles, Altiok, and Bell (2010).

14. Leopold ([1949] 1961).

15. Urry (2000, 2003).

16. Castells (1996).

17. Spaargaren, Mol, and Buttel (2006).

18. This is especially true in Europe, but increasingly so in North American scholarship as well.

19. Although mobilities theorists have not always given stabilities as much attention as they ought to, argues M. Bell (2011).

20. Steingraber (2010).

21. Berry (2003), 135.

22. Quoted in Erikson (1994), 38.

23. Erikson (1994).

24. Erikson (1994), 39.

25. For recent studies that show a clear negative association between who gets the bads and who gets the goods, see Edwards and Ladd (2000); Evans and Kantrowitz (2002); Faber and Krieg (2005); Mennis (2002); Morello-Frosch et al. (2001); Pastor et al. (2001); Pastor et al. (2002); Pellow, Weinberg, and Schnaiberg (2001); and Pine et al. (2002). More ambiguous results are found in Cassidy et al. (2000), who find effects of race and not of class; Denq et al. (2000), who find effects of class but not of race; and Taquino et al. (2002), who find that the unit of analysis has a strong impact on the results. For more detail, see the discussion of environmental justice in Chapter 1.

26. Erikson (1994), 34.

27. Erikson (1994).

28. Erikson (1994), 36.

29. Erikson (1994), 56.

30. U.S. Chemical Safety and Hazard Investigation Board (CSB 2002) and United Nations Environment Programme (2001) report far lower numbers—4,000 residences damaged and 500 left uninhabitable. The higher numbers come from a Toulouse government website (Mairie de Toulouse, 2003).

31. CSB (2002).

32. Arens and Thull (2001).

33. CSB (2002).

34. Environmental Protection Agency (EPA, n.d.), figure for 2000. The EPA has stopped measuring amounts and more recent figures are not available, as far as I can tell.

35. Tomlin (2006).

36. WHO (2006b). On a more optimistic note, WHO's 1998 assessment found that 74 percent of pesticides were hazardous; the 2006 assessment's finding of 55 percent hazardous represents a great improvement, albeit almost entirely through a huge drop in the "slightly hazardous" category.

37. I take these examples of pesticides and their links to birth defects from Steingraber (2001).

38. I also take this from Steingraber (2001).

39. Winchester, Huskins, and Ying (2009).

40. Alejandro, Spira, and Multigner (2001).

41. Swan et al. (2003).

42. Abell, Ernst, and Bonde (1994).

43. Greenlee, Arbuckle, and Chyou (2003).

44. Dick et al. (2007).

45. Bosma et al. (2000).

46. Stallones and Beseler (2001).
47. Charlier et al. (2003).
48. Alavanja et al. (2003).
49. Natural Resources Defense Council (2003b).
50. Quandt et al. (2010).
51. Harari et al. (2010).
52. Dennis et al. (2010).
53. D-H. Lee et al. (2010).
54. Zheng et al. (1999).
55. Brasher (2003).
56. Organic Trade Association (2010).
57. Organic Monitor (2010).
58. Harris Interactive (2011).
59. Wargo (1998).
60. EPA (2003b).
61. Steingraber (2001), 252.
62. Steingraber (2001), 251.
63. Cited in Steingraber (2001), 251–252.
64. Steingraber (2001), 253.
65. Bentham ([1779] 1996).
66. The origin of this phrase is also often attributed to Bentham, and he did use a similar version of it, but apparently borrowed it from elsewhere.
67. Parfit (1986).
68. I thank Jamie Mayerfeld for this observation.
69. Strictly speaking, Rawls did not himself see his theory as appealing to selfishness, however. His argument is that we should leave the whole question aside and start from the "original position" and see what principles of justice we would come up with then. His argument is, if we do so, we will be obliged to plan as if we were the worst-off person. But in my view, this nevertheless amounts to an appeal to our selfish concern that we might wind up among the disadvantaged. I again thank Jamie Mayerfeld for helping me think through this issue.
70. Rawls (1971), 60. Like many readers, I am sticking to the Rawls of 1971 rather than Rawls (1995), which subtly reformulated these principles.
71. Rawls (1971), 62.
72. Rawls (1971), 62.
73. I depart from calling Rawls's theory "liberalism," as is conventionally done, because the term *liberalism* is so easily misconstrued as a political commitment by those unfamiliar with the complexities of the term, as opposed to being a theory of justice—arenas that are closely connected but distinct.
74. "Act utilitarianism" is an example of a more egalitarian form of utilitarianism.
75. Sen (1992, 1999).
76. UNDP (2006).
77. Dworkin (2000).
78. For example, Rawls (1995).
79. The definition of *social power* has been debated long and contentiously within sociology. The definition I offer here is drawn from pragmatism and from Weber, with a hint of Foucault.
80. De Tocqueville ([1835–1840] 1988).
81. For more on isodemocracy, see M. Bell (2001b), from which I derive much of the argument here.
82. Abram (1996), 80.
83. Slow Food International (n.d.).
84. Dupuis (2000).

Chapter 6

1. Guha (1999).
2. Weber (1958), 60.
3. Weber ([1904–1905] 1958), 181.
4. Weber ([1904–1905] 1958), 181.
5. White (1967), 1207.
6. White (1967), 1204.
7. White (1967), 1205.
8. White (1967).
9. For examples, see Northcott (1996) and Van Dyke (1996).
10. I thank David Bylsma, a former student, for alerting me to green Christianity's alternative reading of "dominion."
11. Fung ([1948] 1966).
12. T. Merton (1965), 65, emphasis added.
13. T. Merton (1965), 136.
14. Tuan (1968).
15. Bellah (1969).
16. The following account is largely drawn from M. Bell (1994b) and Gardiner (1993).
17. Bakhtin ([1965] 1984).
18. Rabelais ([1532–1552] 1931).
19. Bakhtin ([1965] 1984).
20. Bakhtin ([1965] 1984).
21. Bakhtin ([1965] 1984), 28, 38.
22. On this point, see both M. Bell (1994b) and Cresswell (1996), 128.
23. Bakhtin ([1965] 1984), 474.
24. For these quotations, see Plumwood (1994b), p. 19, citing Morgan (1989).
25. Quoted in Menard (2001), 86–87.
26. For a review, see Warren (1996).
27. Plumwood (1994b), 13.
28. Quoted in Seager (1993), 221.
29. Handwerk (2011).
30. Kofman (2010).
31. Koch (2010).
32. Quoted in Seager (1993), 221.
33. See, for example, Cheney (1994).
34. Commission of the European Communities (2007), Table 1; U.S. Census Bureau (2007a), Table P-40.
35. Hochschild (1989).
36. Plumwood (1994b), 9.
37. Plumwood (1994b), 43.
38. Plumwood (1994a), 74.
39. Courtenay (2006), 141.
40. M. Bell (1994b), 210–224.
41. M. Bell (1994b), 218.
42. Bell (1994b), 219–220.
43. Seager (1993), 9.
44. Shiva (1988). In recent writings, Shiva remains unrepentant for her much-criticized idea of the "feminine principle."

45. For examples, see Plant (1989) and Diamond and Orenstein (1990).

46. See Seager (1993); Plumwood (1994a); Biehl (1991); and Slicer (1994). Biehl has been a strong critic from outside ecofeminism. The other three are from within ecofeminism. Buege (1994) is a critique of Biehl for tarring all ecofeminist writers with the same essentialist brush.

47. Slicer (1994).

48. Hierarchical categories also dialogically promote a hierarchical frame of mind. See M. Bell (1998a).

49. Goldman and Schurman (2000).

50. I have had a hand in this one, I confess. See Banerjee and Bell (2007).

51. There was an influential early use of this term in Rocheleau, Thomas-Slayter, and Wangari (1996).

52. Agarwal (2001); H. Campbell, Bell, and Finney (2006); Fortmann (1996); Harris (2006); Kalof, Dietz, and Guagnano (2002); Zavestoski, Brown, and McCormick (2004).

53. D. Bell (2006); Little (2007).

54. M. Bell (2011).

Chapter 7

1. Carson (1962), 261–262.

2. Carson (1962), 97.

3. See, for example, Guha (1999). Guha also notes that *Silent Spring* stayed on the *New York Times* best-seller list for 31 weeks, sold a half-million copies in hardback alone, and was soon published in a dozen countries.

4. World Values Survey (2006).

5. Gallup International (n.d.).

6. Inglehart (1995). See also the similar findings in Mertig and Dunlap (2001).

7. My reading of the Bible, however, suggests that environmental concern did not much cross the minds of the ancient Hebrews.

8. Horace ([c. 20 BCE] 1983), 215–216, Epistle I, 10.

9. Clare (1993), 48.

10. In making this argument, I draw heavily upon the analysis contained in M. Bell (1994a) and M. Bell (1999).

11. See M. Bell (1994a) for more detail.

12. Steinmetz (2002).

13. See M. Bell (1994a) and almost anything Michel Foucault ever wrote.

14. Collingwood ([1945] 1960), 45.

15. Lovejoy and Boas (1935), 104.

16. Finley (1963), 55; Burenhult (1994), p. 152, gives a figure of only 75,000 for the fifth century BCE, and 250,000 for Athens and its hinterlands, which is still a remarkable total.

17. Plato ([c. 399 BCE] 1952), 483.

18. Plato ([c. 399 BCE] 1985), 338e, 343b.

19. Rommen ([1936] 1947).

20. See Lovejoy and Boas (1935).

21. Plato ([c. 399 BCE] 1952), 483.

22. Plato ([c. 360 BCE] 1965), *Timaeus* sec. 4, 29.

23. See Aristotle (1987), *Parts of Animals*, 1.1, 639b, p. 20, and *Politics* 1.1, 1252a, pp. 0–5.

24. Bartlett ([1855] 1980), 3.

25. Bartlett ([1855] 1980), 3.

26. Proverbs 22:16, Revised Standard Version.

27. Lao Tzu (1963), v. 23.

28. Lao Tzu (1963), v. 156.

29. Lao Tzu (1963), v. 185a.

30. Lao Tzu (1963), vv. 43a, 81.

31. Lieh Tzu (1960), 135–136.

32. Fung ([1948] 1966), 18–19.

33. T. Merton (1965), 115.

34. Chuang Tzu (1968), 32–33.

35. Lao Tzu (1963), ch. 54, 126.

36. Lao Tzu (1963), ch. 56, 117.

37. Lao Tzu (1963), ch. 81, 143.

38. Fung ([1948] 1966), 177, 284; for a contrasting view, see Peerenboom (1991).

39. Fung ([1948] 1966), 101.

40. Thoreau ([1862] 1975), 164.

41. Thoreau ([1862] 1975), 176.

42. Thoreau ([1862] 1975), 170–171.

43. Thoreau ([1862] 1975), 185.

44. See M. Bell (1994a).

45. See *Defense of Socrates:* 21a–23c, Plato ([c. 399 BCE] 1997). It should be noted that ever since his own time, there has been a vigorous debate as to whether it is appropriate to consider Socrates a Sophist or not.

46. Thoreau ([1862] 1975), 203–204.

47. Thoreau ([1862] 1975), 175.

48. Thoreau ([1862] 1975), 200–201.

49. M. Bell (1994a), 147.

50. Inglehart (1995).

51. Downs (1972).

52. Pew Research Center (2009a).

53. Pew Research Center (2007).

54. Guadalupe (2007).

55. Downs (1972).

56. T. Smith (2007).

57. T. Smith (2007).

58. Global Strategy Group (2007).

59. J. Jones (2010).

60. Cotgrove (1982).

61. Jones and Dunlap (1992).

62. Mertig and Dunlap (2001).

63. Public Agenda (2003) reporting on an April 2000 ABC News/*Washington Post* poll.

64. Kalof, Dietz, and Guagnano (2002).

65. Kahan et al. (2007).

66. Kahan et al. (2007).

67. Douglas and Wildavsky (1982).

68. Kahan et al. (2007).

69. Dunlap (2010a).

70. Pew Research Center (2009b).

71. Dunlap (2008). See also McCright and Dunlap (2010).

72. Guha (1999).

73. For examples, see R. Bullard (1993, 1994a, 1994b); Hofrichter (1993); and D. Taylor (1989).

74. Cited in Guha (1999).

75. Abramson and Inglehart (1995), 1.

76. Inglehart (1977), 10.

77. Inglehart modifies Maslow, however, arguing that there is no particular order to the salience of the needs above the physical; see Inglehart (1990).

78. For simplicity's sake, I have reproduced here only the reduced version of Inglehart's scale, which contains three such sets of four choices. Postmaterialism researchers use the full scale whenever possible. But the single set of four choices I have included is the most widely used.

79. Abramson and Inglehart (1995), 19, Table 2-2.

80. Inglehart (1995), Figure 5.

81. Inglehart (1995), 68.

82. For a summary, see Inglehart (1990, 1995).

83. Brechin and Kempton (1994), Guha (1999), and the debate over the application of postmaterialism to global environmentalism in the spring 1997 issue of *Social Science Quarterly*.

84. Inglehart (1995).

85. Freudenburg (2008), 454.

86. Catton and Dunlap (1980); Cotgrove (1982); Dunlap (1980); Dunlap and Van Liere (1978, 1984); Milbrath (1984). See Dunlap et al. (2000) for a revised version of the NEP scale, plus a review of studies using the older version of the scale.

87. Not to mention the "transindustrial paradigm," the "metaindustrial paradigm," and the "ecological paradigm." See Olsen, Lodwick, and Dunlap (1992) for a complete review.

88. Olsen et al. (1992).

89. Olsen et al. (1992), 137.

90. For an example of research in this general tradition that uses a far more complex array of categories, in this case in relation to attitudes toward animals, see Kellert and Berry (1982).

91. Dunlap et al. (2000). Paradigm shift researchers have sometimes broken survey responses into different dimensions, although without any great consistency of results, as Dunlap et al. describe. Dunlap et al. argue that this lack of consistency is a sign that "multidimensionality" is not significant, however, and that overall the NEP hangs together in the public mind as a single, relatively unified construct. They also suggest that some of the phrasing of the original scale may have obscured the unity of the NEP. I confess to remaining dubious that the revised scale improves the validity of the single-construct view, although it may well increase the reliability of it.

92. Beus and Dunlap (1991); Dunlap and Van Liere (1978); Dunlap et al. (2000).

93. Dunlap et al. (2000) report the differences between 1976 and 1990 surveys of the state of Washington. The other example is Arcury and Christianson (1990).

94. Mol (2001), 59.

95. Mol (1996), 303.

96. Mol and Spaargaren (2000), 27.

97. Mol and Jänicke (2009), 24.

98. Pianin (2003).

99. Mol (2001).

100. International Organization for Standardization (2010).

101. For example, see Blowers (1997); Blühdorn (2000); Hannigan (1995); and Leroy and van Tatenhove (2000).

102. Mol (1996), 309.

103. Hannigan (1995), 184.

104. Spaargaren and Cohen (2009), 257.

105. Hajer (1995).

106. On "weak" versus "strong" ecological modernization, see Christoff (1996); "thin" and "thick" are my own suggestions.

107. Buttel (2003).

108. Socioeconomic Data and Applications Center (SEDAC, 2009).

109. See Mol, Sonnenfeld, and Spaargaren (2009) and Mol and Spaargaren (2000, 2005).

110. Thomas (1983), 16. Throughout this section, I draw heavily on Thomas's argument, but I depart in arguing for the importance not only of material comforts to the new sensibility of nature but also new democratic feelings toward human social relations.

111. Guha (1999); Mol (1995).

112. I depart here from what may appear to be the superficially similar argument of Nash (1989). My argument is not for a history of ethical extension through various social boundaries and thence into the natural world, but rather of the interactive development of both. Indeed, in many cases extension of rights into nature preceded extension of rights into society. For example, the Royal Society for the Prevention of Cruelty to Animals was founded some 100 years before the foundation of the Royal Society for the Prevention of Cruelty to Children. It is, yet again, a matter of dialogic causality.

113. Engels ([1845] 1973), 89–92.

114. Marsh ([1864] 1965), 3, 465.

115. Inglehart (1990).

116. Bridges and Bridges (1996).

117. Guha (1999).

118. Guha (1999).

Chapter 8

1. Rolston (1979), 9.
2. M. Bell (1994a, 1997). The distinction I am drawing has some parallels in what Rolston (1979) describes as "following nature in an absolute sense" versus "following nature in an artifactual sense," but differs in that Rolston does not consider issues of power and interest.
3. Erikson (1966).
4. The solution Rolston (1979), p. 12, offers is what he calls the "relative" sense of following nature, in which "we may conduct ourselves more or less continuously or receptively with nature as it is proceeding upon our entrance." I have elsewhere (M. Bell 1994a) described this as the "pastoral" solution, which places human ways on a gradient between the natural and the unnatural.
5. Aristotle (1987), *Physics,* 185b, 15–25.
6. In Aristotle's (1987) words, "A thing is due to nature, if it arrives, by a continuous process of change, starting from some principle in itself, at some end" (*Physics,* II.8, 199b, 15–20).
7. Kuo Hsiang's writings are intermingled with those of Chuang Tzu in the *Chuang Tzu,* which Kuo Hsiang edited and augmented with his own commentaries around 300 BCE. I take this version from Fung ([1934] 1953), pp. 216–217. Note that in the original in Fung, he gives "non-activity" where I have given the Chinese term, *wu wei.*
8. I thank a colleague from Hong Kong whose name I do not recall for this observation, made over beer at an Oxford pub.
9. See, for example, Beck (1996b); M. Bell (1996); Cantrill and Oravec (1996); Cronon (1995b); Dupuis and Vandergeest (1996); Eder (1996); Ellen and Fuhui (1996); Evernden (1992); Greider and Garcovich (1994); Hannigan (1995); Haraway (1991); Latour (1993); Soper (1995); and Yearley (1991). There is also a large older, largely nonsociological, literature that discusses the social construction of nature, for example, Collingwood ([1945] 1960); Hubbard (1982); Lovejoy (1936); Lovejoy and Boas (1935); Mill ([1874] 1961); and R. Williams ([1972] 1980), pp. 70–71. This older literature is often ignored by more recent writers, but I have found it enormously clarifying.
10. R. Williams ([1972] 1980), 70–71.
11. Meek (1971), 195.
12. Darwin (1958), 42–43, cited in Hubbard (1982), 24.
13. Hubbard (1982), 24.
14. Cited in Schmidt (1971), 47; see also Engels ([1898] 1940), 208.
15. Herrnstein and Murray (1994).
16. For example, Fraser (1995).
17. Cited in Gould ([1981] 1996).
18. Gould ([1981] 1996), Table 2.1.
19. Gould ([1981] 1996), 97.
20. Gould ([1981] 1996).
21. Gould ([1981] 1996), 93.
22. Huntington (1915).
23. Huntington (1915).
24. Scott (1990).
25. Dunlap and Catton (1994), 20.
26. Hannigan (1995), 3.
27. Burningham and Cooper (1999), 311.
28. Burningham and Cooper (1999), 312.
29. Cf. McCright and Dunlap (2000, 2003, 2010).
30. Latour (2004), 227.
31. Carolan (2005); Murphy (2004).
32. Dunlap (2010b), 28, as well as personal communications, many times over the years.
33. Gutro (2006).
34. Guha (1989, 1997).
35. Guha (1997), 15.
36. Peluso (1996).
37. Peluso (1996).

38. I draw this account from Banerjee (2006) and Nickell (2007).

39. J. Williams (n.d.). I also thank David Nickell (2006, personal communication) for clarifying the confusing history of the Coalins.

40. I thank David Nickell (2006, personal communication) for this observation.

41. Cronon (1995a), 81.

42. Urry (1990, 1995).

43. Greider and Garcovich (1994), 1.

44. M. Bell (1997).

45. For extensive critiques of heritage tourism, see Lowenthal (1985, 1997).

46. Hinrichs (1996), 259.

47. Barrett (1994), 256.

48. The letter also lists Linus Pauling, the two-time Nobel Prize winner, and Stewart Udall, former U. S. Secretary of the Interior, as members of the National Advisory Board of Population–Environment Balance.

49. Hardin (1977).

50. World Public Opinion (2009). This poll included most of the world's largest nations and represented over half of the world's population.

51. Pew Research Center (2009a).

52. Global Scan (2009).

53. Pew Research Center (2009a).

54. IPSOS (2010).

55. Pew Research Center (2010).

56. Dunlap (2008). See also McCright and Dunlap (2010).

57. Pew Research Center (2010). Among independents, 32 percent agreed.

58. Dunlap and McCright (2008).

59. Lichter (2008). Some 85 percent agreed that the danger of this increase is great or moderate, and 84 agreed that human activity is a significant factor.

60. Bray and Storch (2008) and Anderegg et al. (2010).

61. Dunlap and McCright (2008) and IPSOS (2010).

62. McCright and Dunlap (2010).

63. On the use of climate change to sort and polarize voters, see Dunlap and McCright (2008).

64. Latour (1993).

65. Freudenburg, Frickel, and Gramling (1995).

66. This story is also sometimes told in a way that is more flattering to Canute, namely, that he ordered the tide to stop as a deliberate lesson in the limits of the power of kings compared to the power of God.

67. See M. Bell (1994a), ch. 1.

Chapter 9

1. I base the following on the verbatim transcript of the April 16, 1996, *Oprah Winfrey Show* (retrieved on March 30, 2011, from http://www.madcowboy.com/02_OP_Transcript.000.html).

2. The nineteenth-century German philosopher, Georg Wilhelm Friedrich Hegel, believed that periods in history could be identified by a certain "spirit" that represents their age, a particular theme that captures the essence of an era. Eugene A. Rosa (2000) in particular has referred to "risk" as representing the spirit of our age.

3. Beck (1992, 1996a).

4. Centers for Disease Control (2007a).

5. National Creutzfeldt-Jakob Disease Surveillance Unit (NCJDSU, 2007).

6. Centers for Disease Control (2007b).

7. The UK Department of Health monthly reports list 3 deaths from nvCJD for 1995 and 10 for 1996 in the United Kingdom (retrieved January 31, 2004, from http://www.cjd.ed.ac.uk/figures.htm). I was unable to find yearly information for nvCJD deaths in other countries, but given that there have been only a few nvCJD deaths ever outside of the United Kingdom, the world figure at the time of Oprah's show is unlikely to have been more than 15.

8. In 1998, Oprah again stated publicly her intention to swear off hamburgers.

9. Although he is acknowledged on the title page and in the preface for his contributions to this chapter, I want to specifically thank Michael Carolan here for his suggestion of the distinction between *risk* and *risky.*

10. See Freudenburg (2000) for further discussion. Also see Perrow (1999) on "normal accidents"; Luhmann (1994) on "structural differentiation"; Giddens (1990) on "system complexity"; and Parsons (1951) on "disembedding of social systems."

11. This is an example of a "normal accident" (Perrow 1999), which is explained in more detail later in the chapter.

12. Maynard Haskins was my father's occasional pen name. AutoBAN is as well a fictitious group.

13. Haskins is, of course, best known for his other fictitious book, *The Frangibility of Change* (London: Soncino Press, 1972).

14. Centers for Disease Control (2008).

15. On U.S. traffic fatalities and injuries, see National Highway Traffic Safety Administration (2010). On worldwide traffic fatalities, see the World Health Organization (WHO 2004). On U.S. AIDS deaths, see Centers for Disease Control (2008). On worldwide AIDS deaths, see United Nations Programme on HIV/AIDS (UNAIDS 2007). For the figure on U.S. deaths from automotive air pollution, I divided the Natural Resources Defense Council's (NRDC) figure for all U.S. air pollution deaths in half (Shprentz 1996), based on the results of a study of air pollution in France, Switzerland, and Austria published in *The Lancet,* which found that half of air pollution mortality was due to motor vehicle exhaust (Künzli at al. 2002). (I am reluctant to use the NRDC figure, as it does not come from the peer-reviewed or governmental literature. However, there has been no peer-reviewed study for the United States, and peer-reviewed studies for Europe, such as Künzli et al., 2002, show a similar scale of deaths from air pollution.) For the global figure on deaths from automotive air pollution, I similarly divided the WHO figure of 865,000 annual outdoor air pollution deaths roughly in half (WHO 2007a; for a peer-reviewed version of the WHO methodology, see A. Cohen et al. 2005). There is, of course, considerable uncertainty in these figures, and they change from study to study over the years, necessitating considerable revision in this endnote from the previous editions of this book.

16. Elster (1989b), 37.

17. See, for example, Coleman (1990) and R. Hardin (1991, 1993).

18. Rosa (2000).

19. Kleim and Ludin (1997).

20. For example, see Elster (1989a).

21. Carolan (2002).

22. For example, see Elster (1989a).

23. See, for example, Douglas and Wildavsky (1982); Erikson (1994); Kleinhesselink and Rosa (1994); Kunreuther (1992); Perrow (1999); and Slovic (1987).

24. French philosopher Michel Foucault argues that the social relations of power and knowledge are inseparably constituted and thus refers to them as "power/knowledge." This chapter builds upon Foucault's (1969) notion of power/knowledge to include both "identity" and "trust." See M. Bell (2004); Carolan (2002); and Carolan and Bell (2003, 2004) for a more detailed discussion.

25. Erikson (1966).

26. Erikson (1966), 13.

27. Starr (1969) is credited with being the first to conceptually develop this distinction between "voluntary" and "involuntary" risk.

28. Purcell, Lee, and Renzulli (2000).

29. Freudenburg (2000); Purcell et al. (2000).

30. Onions ([1933] 1955) dates the word *risk* from 1661, when it was borrowed from the French *risque*—which was also at one time a common way *risk* was spelled in English. I draw much of this section, from pages 218 to 220, as well as some passages later on, from M. Bell and Mayerfeld (1998).

31. Simmel ([1900] 1990), 444–445.

32. J. Jones (2010). I get this figure by adding together the 15 percent who say that that "prayer is only effective for those holding certain beliefs" with the 80 percent who say that "prayer can be effective no matter what a person believes in." However, 61 percent believe that the frequency of prayer doesn't matter. Some 83 percent say there is a god who answers prayers—which suggests that some respondents think prayer can be effective without the intervention of a god. However, most probably link the effectiveness of prayer to a god.

33. H. Taylor (2003).

34. The continuing appeal of traditional institutions for dealing with uncertainty is largely the effect of the failure of rationalism to answer the question of *who* will get cancer and *when*. For a detailed critique of the limits of rationalism, see the next section.

35. See, for example, Barton (1969).

36. Couch, Kroll-Smith, and Kindler (2000); Edelstein (1988); Erikson (1994).

37. Freudenburg et al. (2007), 19.

38. Erikson (1976).

39. Erikson (1976), 30.

40. Prince (1920).

41. Freudenburg and Jones (1991).

42. Erikson (1994), 19.

43. Erikson (1994). Human-induced disasters are not exclusively new, however. During the early decades of the nineteenth century, the northeastern section of the United States experienced an unprecedented series of "natural" disasters, mainly in the form of flash floods, which wiped out mills and towns in numerous river valleys. Only later were these floods linked to deforestation and agriculture settlement (Lowenthal, 2000).

44. Indeed, recent studies have indicated that even the fourth generation of people (or the great-grandchildren of people) exposed to radiation could inherit unstable genomes, which was cited as a possible explanation for the leukemia cluster around Britain's Sellafield nuclear plant; see Dubrova et al. (2000). Childhood leukemia is around 10 times as common in Seascale in Cumbria, where many Sellafield workers live, as in Britain overall, although epidemiological studies have not conclusively linked this to the parents' exposure. See Muir (2002).

45. Beck (1996a).

46. Picou and Gill (2000).

47. Perrow (1999).

48. Perrow (1984), 90.

49. Erikson (1994), 139–140.

50. See Wing et al. (1997), and the controversy that ensued over this finding, still hotly disputed by many. Wing's work, however, remains highly regarded; in 2009, he was awarded the International Society for Environmental Epidemiology Research Integrity Award.

51. Taylor (2005).

52. On expecting the unexpected and bringing it into our analysis, also see Bell (2011).

53. Klinenberg (2002) makes this argument very well for the Chicago heat wave of 1995.

54. The imagery of the Frankenstein monster is best captured in what Beck (1992), p. 37, calls the "boomerang effect": "The formerly 'latent side effects' strike back even at the center of their production. The agents of modernization themselves are emphatically caught in the maelstrom of hazards that they unleash and profit from."

55. See, for example, Beck (1992, 1994, 1997, 1999).

56. Beck (1996a), 6. The distinction between "environmental goods" and "environmental bads," which has been used throughout the book, is drawn from Beck.

57. Beck (1992), 49.

58. Beck (2009), 8.

59. Beck (2009).

60. Beck (1992), 53; in the original, the second phrase is in italics.

61. Beck (1995), 16.

62. Beck (1994), 5.

63. Beck (1992), 1. This is also made in direct reference to Karl Marx. Thus, whereas Marx stands Hegel "on his head," here Beck does the same with Marx.

64. Alexander (1996); Buttel (1997); Freudenburg (2000); Renn (1997).

65. Beck (1992), 36.

66. As early as Beck (1992), pp. 35–36, he noted that risk is unequally distributed. Beck (2009) takes up the implications of what he calls the "inequality dynamics of risk" more vigorously.

67. Buttel (1997).

68. For more on such a conception of science, see M. Bell (2004). Some of the lines on science I include here are closely based on what appears there.

69. Irwin (1995).

70. P. Brown (1990).

71. For a study of participatory research in agriculture, see M. Bell (2004). For a study of participatory research in air pollution issues, see Yearley et al. (2003).

72. Reported in Nestle (2002), 163.

73. LaMotte and Davis (1998).

74. Nestle (2002), 164.

75. Pring and Canan (1996).

76. Beeman (2002). I checked the Agricola database on February 6, 2011, and Zahn had yet to have these results published. However, Zahn and his colleagues did publish in 2008 a related study (Angenent et al. 2008) that found that anaerobic manure digesters (a common solution swine producers seek for the problem of manure concentration) can increase the level of antibiotic-resistant bacteria in the waste.

77. Beeman (2002).

78. Beeman (2002).

79. Beeman (2002).

80. Scottish researchers fed the same potatoes to aphids and then fed the aphids to ladybugs. They found that the female ladybugs had their life spans cut in half and also suffered reproduction problems (Kirby, 1999).

81. Pusztai's article in *The Lancet* is Ewen and Pusztai (1999).

82. Gaskell et al. (2010).

83. Gallup Poll (2007).

84. Science and Environmental Health Network (2004).

85. Saunders (2000).

86. "How Safe Is GM Food?" (2002).

87. WHO (2003c). WHO has been listing this view on its website since at least 2003. See the answer for "Q8: Are GM Foods Safe?"

88. GMO-Free Europe (2010).

89. Cooper (1996), 9.

90. Lukaszewski (1992).

91. See Carolan (2002) and Carolan and Bell (2003, 2004).

Chapter 10

1. Gellius ([c. 150 CE] 1927), xi.

2. I base my telling of the story on many sources, including Black (1991); Bartlett ([1855] 1980); the original in Gellius, as translated in Gellius (([c. 150 CE] 1927); and my own demented imagination. I've also thrown in a bit from a similar Aesop's fable, "The Lion and the Mouse."

3. Caligula was often in this kind of trouble during his reign as emperor.

4. Okay, Okay, technically speaking, adult lions do not mew—only lion cubs do.

5. Adult lions do in fact purr, however—although there is some debate about this among wildlife biologists.

6. I take this line straight from Gellius, *Attic Nights*, ([c. 150 CE] 1927) 5.14. An online version, slightly different from what I use here, can be found in Gellius ([c. 150 CE] 2006).

7. This moral is actually from the telling of "The Lion and the Mouse" in Bartlett ([1855] 1980), p. 66, the Aesop fable that may have served as the original inspiration for "Androcles and the Lion." Black (1991) offers this conclusion instead: "And so this story teaches us that a good deed never goes unrewarded" (p. 24), which is a little too instrumental for my tastes.

8. But the main thing is togetherness, as I describe in my discussion of the etymology behind my choice of terms at the conclusion of the chapter.

9. UNESCO-UNEP (1978).

10. IPCC (2007a).

11. Gray and Leach (2010).

12. M. Bell (2004).

13. I take this example, largely verbatim, from M. Bell (2004), 221.

14. The end of the largely verbatim section from M. Bell (2004), 221.

15. M. Bell (2004), 157.

16. M. Bell (2004), 157.

17. M. Bell (2004), 153.

18. M. Bell (2004), 153.

19. M. Bell (2004), 155.

20. M. Bell (2004), 153.

21. Snow and Benford (1988) and many subsequent publications.

22. Aristotle (1987), *Rhetoric*, II.1.

23. Grey (2004).

24. D. Taylor (2000).

25. A. Jones (2007).

26. B. Walton and Bailey (2005).

27. Freire ([1970] 1993).

28. The above account of Glen is drawn from M. Bell (2004), 194–195.

29. Hardin (1968).

30. Shrank, Lomax, and Turner (2010), Exhibit 2.

31. To be precise, my search of Sociological Abstracts on February 7, 2011, using the search string "tragedy of the commons" in the "anywhere" option, returned 39 hits for 2010 and 346 for all of time.

32. For example, see Argyle (1991); Bromley (1992); Feeny, Hanna, and McEvoy (1996); Hinde and Groebel (1992); Ostrom (1990); Petrzelka and Bell (2000); R. Roberts and Emel (1992); Stevenson (1991); and Thompson and Wilson (1994). More recently, see Moritz et al. (2010).

33. On grazing land, see Thompson and Wilson (1994), Mearns (1996), and Moritz et al. (2010); on fisheries in India, see Gadgil and Guha (1995), 81–84; on fisheries in Brazil, see Begossi (1995).

34. See Andelson (2004); Argyle (1991); Ostrom (1990).

35. Petrzelka and Bell (2000).

36. For empirical support of this argument in actual commons, see Begossi (1995) on the role of kinship in the management of Brazilian fisheries, and Petrzelka and Bell (2000) on the importance of collective events, like daily communal dancing for herders in the Atlas Mountains of Morocco, discussed in detail following.

37. The following account is drawn from Petrzelka and Bell (2000).

38. Murphy (2004), 254.

39. Alinsky (1971), 21.

40. Eichler (1998).

41. Stoecker (2007).

42. M. Bell (2007).

43. Eisinger (1973), 25. Eisinger's original phrase was the "structure of political opportunities."

44. On the anti–nuclear power movement, see Kitschelt (1986); on the anti–global warming movement in the United States, see McCright and Dunlap (2003).

45. See especially Kitschelt (1986), on "input" versus "output" structures; Tarrow (1994), on the dynamism of political opportunity; Kriesi et al. (1992), on centralization of opportunity structures; and van der Heijden (2006), on the internationalization of opportunity structures.

46. On face in double politics, see M. Bell (2007).

47. Huntley and Creighton (1986); Manning (n.d.).

48. Manning (n.d.).

49. Ziegler (2006).

50. Manning (n.d.).

51. Berry (1977); Hightower (1973).

52. Busch and Lacy (1983, 1986); Dahlberg (1986).

53. I base the following account of PFI's double politics of engagement with Iowa State University on M. Bell (2004), ch. 7.

54. M. Bell (2007).

55. M. Bell (2004), 163, and M. Bell (2011).

56. On this, see M. Bell (2011) and my case for what I like to call "strangency" or "strange agency." But this chapter has had enough new terms already.

Chapter 11

1. U.S. Department of Transportation (2010) and NHTSA (n.d.). Men drive about 60 percent more than women, 16,550 miles versus 10,142 miles per year (U.S. Department of Transportation, 2003).

2. My electric utility informed me on February 8, 2011, that average electricity usage in Madison is 560 kilowatt hours per month for all households. Given that the average household size in Madison is 2.19 persons (as of the 2000 Census), that translates to 256 kilowatt hours per month per person for the whole city.

3. Brown (2006).

4. Food and Agriculture Organization (FAO 2006a).

5. The site was http://www.earthday.net/footprint/index.asp, which is no longer working as I write these lines in January 2008. (I took the survey on October 6, 2003.) But there are several others on the web.

6. In this paragraph, I am passing at great speed over an enormous amount of scholarly work. The "attitude–behavior relation," or "A–B split," as I am terming it here for reasons that the following should make clear, is assuredly one of the most researched topics in social science. For a review, see Kraus (1995). The work of Ajzen and Fishbein, who have popularized the phrase "attitude–behavior relation" in the literature, is particularly salient and widely cited. The classic work is Fishbein and Ajzen (1975), in which they propose their "theory of reasoned action," which basically accounts for the split by saying that social norms intervene between attitudes and behaviors; for a recent update, see Ajzen and Fishbein (2005). I take a more organizational approach, less oriented to predicting market behavior. Ajzen's later "theory of planned behavior" (Ajzen 1985), which adds in the importance of "perceived behavioral control," comes closer to what I have in mind but is still more psychological in orientation than is the sociologist's eye.

7. This is a point that most of the scholarly literature on the relationship between attitudes and behaviors misses. Social psychological theories like Festinger's ([1957] 1962) "cognitive dissonance" and Fishbein and Ajzen's (1975) "reasoned action" tend to individualize the question, to see it as a "micro" issue rather than placing it within a wider social context. For an environmental account of the attitude–behavior split as an effect of social structure, see Ungar (1994); my argument parallels his in several regards.

8. Cottrel (1955), p. 2, whom I also cited in this regard in Chapter 4. Once again, I have paraphrased Cottrel to eliminate his gender-specific usage.

9. There was recently, as I write in February of 2011, a huge controversy over Wisconsin governor Scott Walker's decision to return $870 million in stimulus dollars back to the U.S. federal government that had been intended to bring a train to Madison.

10. I take the following section, virtually verbatim in places, from M. Bell (2001a).

11. Hochschild (1997).

12. Pucher and Dijkstra (2003), Figure 1.

13. Pucher and Dijkstra (2003).

14. See, for example, Brecher (1994); Johnson (1992); Whitley (1992).

15. Lerner (2005).

16. National Oceanic and Atmospheric Administration (2010).

17. Hickman (2010).

18. Scott (1990).

19. Habermas (1975).

20. On "dialogic democracy," see Giddens (1994); on the process of "taking into account," see M. Bell (1998b).

21. Pew Research Center (2009a).

22. For example, Bryan (2004); Hurley and Walker (2004); Lane (2001); C. Lee (2007); Parkins and Mitchell (2005); E. Weber (2000); Zwart (2003).

23. For a sample of the extensive literature on participatory development, Brohman (1996); Ghai and Vivian (1992); Hobley (1996); Stiefel (1994). For an entry into the debate about whether participatory approaches have often been insufficiently sensitive to issues of power in development projects, see Cooke and Kothari (2001) and the response in Hickey and Mohan (2004).

24. See Bentley (1994, 2006); Bentley and Andrews (1991); Bentley and Melara (1991); Bentley, Rodriguez, and Gonzalez (1994).

25. For a compelling history of agricultural poisoning in Latin America, see Wright (1990).

26. There is a huge, and growing, literature on local knowledge. For useful introductions, see Brush and Stabinsky (1996); M. Campbell and Manicom (1995); Geertz (1983); Hassanein and Kloppenburg (1995); Nazarea-Sandoval (1995); and the recent debate between Cooke and Kothari (2001) and Hickey and Mohan (2004).

27. Bentley and Melara (1991), 43.

28. Yearley et al. (2003).

29. Yearley (2006), 711.

30. See C. Lee (2007) for a review.

31. Hurley and Walker (2004).

32. See C. Lee (2007) for a review.

33. C. Lee (2007).

34. Polletta (2002), 153.
35. On this, see what Lyon et al. (2010) call "maculate conceptions."
36. See Dorfman (2008) and Vidal (2008).
37. Cited in C. Lee (2007).
38. On participation fatigue, see C. Lee (2007); on the tyranny of participation, see Cooke and Kothari (2001); on participation-itis, speak to my good colleague at the University of Wisconsin-Madison, Ken Genskow (2007, personal communication).
39. Giddens (1984).
40. See Barnes (2006) and Speth (2009).
41. R. Williams ([1972] 1980), 84.
42. See Katz (1994); Kunstler (1993, 1996); and Langdon (1994).
43. Quinnipiac Poll (2008).
44. For an introduction, see Ayres and Ayres (1996); Graedel and Allenby (1995); or the *Journal of Industrial Ecology*.
45. Von Weizsäcker, Lovins, and Lovins (1998).
46. Jänicke (2006).
47. Fair Trade Labeling Organizations International (2007).
48. Lyson, Stevenson, and Welsh (2008). Their actual term is "values-based supply chains." But I find "values chains" less cumbersome and a nice contrast with the narrower term "value chain" in the singular.
49. H. Campbell (2005).
50. Busch and Bain (2004); Hatanaka, Bain, and Busch (2005).
51. Busch and Bain (2004), 342.
52. Barnes (2006), 158.
53. Land Trust Alliance (2007).
54. As of December 2007, the Vermont Land Trust had protected 470,000 acres in the state, which is 4.28 percent of Vermont's land area (Vermont Land Trust, 2007).
55. Anderson, Hartley, and Robinson (2007).
56. Deller et al. (2009).
57. Blackwood, Wing, and Pollock (2008).
58. Bureau of Economic Analysis (2009).
59. Daly (1991); Douthewaite (1992, 2000); Fotopoulos (2007); Latouche (2003, 2004, 2006).
60. Latouche (2003).
61. Schor (2010).
62. For example, Buttel (2006); Jänicke (2006); Stevis and Bruyninckx (2006).
63. Here I am giving the ideas of Oosterveer (2006) a bit of a twist.
64. On flickering, see Bland and Bell (2007).
65. Schor (2010) and Schor and Willis (2008).
66. The others weren't sure, I suspect in part because the poll phrased this survey item quite complexly. See Harris Interactive (2008).
67. Harris Interactive (2011), although this poll noted some slight declines from the year before in responses to the same questions.
68. Leopold ([1949] 1961), 239.
69. This point was famously made by Benedict Anderson ([1983] 1991).
70. P. Taylor (1986).

Index

About the Author

Michael Mayerfeld Bell is professor of community and environmental sociology at the University of Wisconsin-Madison. For his day job, Mike is principally an environmental sociologist and a social theorist, focusing on dialogics, the sociology of nature, and social justice. These concerns for the world have led him to studies of agroecology, the body, community, consumption, culture, development, food, democracy, economic sociology, gender, inequality, participation, place, politics, rurality, the sociology of music, and more.

Mike likes books and is the author or an editor of eight, three of which have won national awards: *Farming for Us All: Practical Agriculture and the Cultivation of Sustainability* (2004; winner of a 2005 Outstanding Academic Title award from the American Library Association*); Childerley: Nature and Morality in a Country Village* (1994; co-winner of the 1995 Outstanding Book Award of the Sociology of Culture Section of the American Sociological Association); and *The Face of Connecticut: People, Geology, and the Land* (1985; also a winner of an American Library Association award).

Mike has a second life as a composer and performer of contemporary classical and grassroots music. His classical compositions include pieces for the violin family, the mandolin family, solo piano, symphony orchestra, and various chamber ensembles. His recent compositions develop a dialogue between grassroots and classical traditions that he likes to call "class-grass" music. As a performer, Mike mainly favors the mandolin, but also plays guitar and occasionally imitates singing. He is an emeritus member of the Iowa-based Barn Owl Band, and he appeared with them on the long-running National Public Radio show *A Prairie Home Companion*. He currently performs with, and composes for, the class-grass group Graminy, as well as a variety of other ensembles, including a duo with his daughter.

You can learn more about Mike's work and passions at www.michaelbell.us.

4-51